An Introduction to Applied Probability

Ian F. Blake
University of Waterloo

JOHN WILEY & SONS

NEW YORK
CHICHESTER
BRISBANE
TORONTO

Library of Congress Cataloging in Publication Data:

Blake, Ian F
    An introduction to applied probability.

    Bibliography: p.
    Includes index.
    1. Probabilities. I. Title.
QA273.B586   519.2   78-11360
ISBN 0-471-03210-7
Printed in the United States of America
10 9 8 7 6 5 4 3 2 1

For Gwen Elizabeth

Preface

The purpose of this book is to provide an introduction to probability theory and its applications, and it is intended primarily for students of engineering, science, and management. The only prerequisite is a one-year course in calculus. Although some aspects of statistics are covered, such as hypothesis testing, confidence intervals, and regression analysis, they are included more as applications of probability likely to be of use to the intended audience rather than as an attempt to introduce the reader to statistical methods.

The book was designed for a one-semester course although it contains more material than could be covered in that time. This allows the instructor some flexibility in choosing the particular applications of probability to consider. The first seven chapters represent the core of probability theory. The eighth chapter considers sums of random variables and derives the sampling distributions to be used later as well as discussing the approximation of distributions. The last seven chapters deal with applications of probability, which fall into two categories: the statistical (Chapters 9, 10, 13, and 14) and the nonstatistical (Chapters 11, 12, and 15).

The notion of a random variable and its mean, variance, and cumulative distribution function are introduced early in the book, in Chapter 2. Chapters 3 and 4 are concerned with discrete random variables. Most of the discrete distributions are introduced in Chapter 3 by way of combinatorial situations such as games of chance and cell occupancy problems. Although many students find this material relatively difficult it tends to give a good appreciation of probability. Continuous random variables are considered in Chapter 5 with many of the distributions required for later use being introduced there. Chapter 6 considers bivariate random variables and Chapter 7 discusses transformations of random variables. The statistical applications of probability that are covered are estimation (Chapter 9), including a treatment of confidence intervals, hypothesis testing (Chapter 10), quality control and acceptance sampling (Chapter 13) and linear models for data (Chapter 14), which includes linear regression analysis. The nonstatistical applications of probability chosen for inclusion are queueing theory (Chapter 11), reliability theory (Chapter 12), and communication theory (Chapter 15). The last seven chapters are completely independent of each other with the following two exceptions: the last section of Chapter 12 uses the notions of confidence intervals and estimation on lifetime and reliability data; Chapter 14 uses the results of Chapters 9 and 10 to test hypotheses on, and obtain confidence intervals for, the regression coefficients and equation. Apart from these two exceptions, the chapters can be treated as independent. If Chapters 9 or 10 are to be covered, then at least the first section of Chapter 8 should be included. Typically, a course would cover the material in the first seven chapters and two or more of the last eight chapters depending on the time available and class interests.

There are many excellent books on probability and statistics and a few of these are listed in the references. This book is designed for a curriculum that contains only the single one-semester course on probability, such as found in many engineering and science programs. The question of what to cover in such a limited time for maximum benefit to the student is one of some importance. On the one hand, the course should develop the theory to an extent where the student obtains a thorough understanding of its methods. On the other, some serious applications of the theory should be considered so that its usefulness can be appreciated. To cover these applications without sufficient background in the theory can lead to a superficial computational type of course. It is hoped that the approach and coverage of this volume will make it a worthwhile addition to the literature.

Numerous worked examples throughout the text illustrate the material and each chapter concludes with a number of problems. These of course form an integral part of the text since they test the students' mastery of the subject and in some cases expand on the results of the chapter. The answers to those odd-numbered problems that have a numerical answer, or a mathematical expression for an answer, are given at the back of the book. A solutions manual for instructors is available from the publisher.

Finally, it is my pleasure to acknowledge the help and assistance of many. First and foremost among these is my family who have had to endure my preoccupation with this work for so long. I am very grateful for their patience, encouragement, and understanding. For typing and retyping the manuscript from difficult notes I greatly appreciate the efforts of Janet Schell and Lorna Westerveld. I am indebted to the editor of Biometrika, the Harvard University Press, and D. van Nostrand Co. for their permission to include the tables of the appendices and to the MIT Press and Dr. G. Raisbeck for permission to include Example 15.5. I am also grateful to the Literary Executor of the late Sir Ronald Fisher, F.R.S., and to Dr. Frank Yates, F.R.S., and to Longman Group Ltd., London, for permission to reprint a portion of Table III from their book *Statistical Tables for Biological, Agricultural and Medical Research* (6th edition, 1974). Two anonymous reviewers read various drafts of the book and I am in their debt. In many instances they saved me from error and found many deficiencies in style and content. Their comments improved the book considerably and I am grateful for their efforts. I assume the responsibility for the flaws and inadequacies that remain. Any comments or suggestions for the improvement of the book will be most welcome. And lastly, I would like to acknowledge the care, cooperation, and attention devoted to the preparation of the book by the staff of John Wiley and Sons, Inc. It has been very much appreciated.

IAN F. BLAKE

Contents

CHAPTER 1

A Notion of Probability

The study of probability stems from the analysis of certain games of chance popular in the sixteenth and seventeenth centuries. It has since found application in most branches of science and engineering and this breadth and depth of application makes it an interesting and important subject. The immediate purpose of this chapter is to lay a foundation for the study of probability and to introduce its notation and axioms. One of the basic tools required for this purpose is set theory. Although it is assumed that the reader has some acquaintance with sets, our actual requirements from set theory are minimal. A very brief introduction to sets, with an emphasis on those topics required here, is given in Appendix A.

1.1   *Random Experiments*

The idea of an experiment will be used extensively in the first few chapters of the book. Briefly, an experiment is any situation in which we observe an outcome. If a coin is tossed, we observe whether a head or tail is obtained and might describe this as the experiment of tossing a coin. There are essentially two types of experiments: deterministic and random. In a deterministic experiment, because of the physical situation, the observed result is not subject to chance. In other words, if we repeat a deterministic experiment under exactly the same conditions, we expect the same result. For example, if we have a length of straight wire and a ruler, measured in millimeters, an experiment might consist of asking an individual to measure the length of the wire. If the experiment is repeated under identical conditions, we expect the same result since experimental error should be negligible and the experiment is essentially deterministic.

In a random experiment the outcome is always subject to chance. If the experiment is repeated, the outcome may be different as there is some random phenomena or chance mechanism at work affecting the outcome. Classical examples of such experiments occur in gambling casinos where games involving dice rolling, cards, roulette, coin tossing, and so forth are to be found. A characteristic of these games is that each time they are repeated the outcome has the opportunity of being different. Indeed the participants wager money often on the basis of their intuitive notions as to how likely the various outcomes are. We will use several of these games as examples in later chapters as they provide excellent opportunities to gain insight into probabilistic notions.

To be a little more precise as to what we mean by a random experiment, we ask that they have the following features:

 (i) The experiment be repeatable under identical conditions
 (ii) The outcome on any particular trial is variable, dependent on some chance or random mechanism
(iii) If the experiment is repeated a large number of times, then some statistical regularity becomes apparent in the outcomes obtained.

The meaning of the first two conditions is fairly obvious. The term statistical regularity is an intuitive one, at this stage, which is explored further in Section 1.4. Its intent however is that, as an experiment is repeated a large number of times, the fraction of trials that a particular event A is observed tends to some limit.

We do not wish to dwell unduly on the notions of random and deterministic experiments. However, we should perhaps mention that at times an experiment may be viewed as random or deterministic, depending on the choice of the investigator and the aims he expects to achieve in performing the experiment. For example, in our experiment of measuring a length of straight wire, if we changed the ruler from one marked in millimeters to one marked more finely and asked a variety of individuals to measure the same wire, then the individual readings might exhibit a random character in the least significant digits. This randomness would be caused by a variety of effects such as parallax in the reading, different methods of holding the wire, the imperfections in the end of the wire, and so forth. There is, in fact, a theory concerning the distribution of errors in measurement. At a macroscopic level the experiment can be deterministic, while at a microscopic level it can be random. By and large, the less variation in the outcome of an experiment, the less one is tempted to think of the experiment as random.

To help readers familiarize themselves with the notion of random experiments and the scope of their application, a few examples of them are listed:

Medicine: The effect a drug has on a patient is usually random to some extent—it can be anywhere from ineffective to fatal. The success or failure of a difficult surgical procedure, the sex of a baby, and so forth are also considered random, in some sense.

Agriculture: The effect that a particular dosage of a fertilizer has on the yield of a crop can have important economic consequences. Similarly, the effect of various herbicides on weeds and plant diseases, the effect of certain animal feeding strategies, and so forth all fall into the class of random experiments.

Transportation systems: In designing a new transportation system, many random parameters involving customer arrival patterns, location, and cost are considered. The randomness of service is usually part of another question.

Computer systems: The performance of a computer system depends on many things, including its structure, the operating system, the amount and types of jobs being submitted, and so forth.

Probability is used in most areas of the sciences and engineering since many situations encountered have a sizable random component in them. The few examples indicated above could be extended considerably.

1.2   *The Sample Space*

Let E denote a random experiment. If we are going to analyze such experiments, we should first be aware of the various outcomes that it is possible to observe. The following definition introduces some convenient terminology.

Definition   For any random experiment E we define the sample space S to be the set of all possible outcomes of the experiment.

We should immediately point out that the set S depends entirely on what you wish to observe about the experiment E. Very often an experiment can have several sample spaces and the choice of which one to use depends on what aspect of the experiment you wish to analyze. This point is illustrated with some examples:

EXAMPLE 1.1

Consider the experiment E of tossing a coin three times. Describe two possible sample spaces for this experiment.

First, if we wish to observe the exact sequence of heads and tails obtained from one repetition of the experiment, then the corresponding sample space S_1 would contain the following eight points:

$$S_1 = \{TTT, TTH, THT, HTT, HHT, HTH, THH, HHH\}$$

where THH, for example, indicates a tail on the first throw and a head on each of the second and third throws. In passing, notice that if $S = \{T, H\}$ is the sample space corresponding to the experiment of tossing a coin once, then S_1 can be viewed as the Cartesian product $S \times S \times S$ (see the Appendix A for the definition of Cartesian product).

If the individual outcomes are not of importance to us, but we just wish to observe the number of heads in the three tosses, then a second sample space S_2 would be

$$S_2 = \{0, 1, 2, 3\}$$

where, for example, the outcome 2 indicates that two heads were obtained in the three tosses.

In this manner any particular experiment E can often have many different sample spaces, depending on the observations of interest. In this example the sample spaces S_1 and S_2 have a very natural correspondence, the point 0 of S_2 corresponds to TTT, 1 corresponds to the three points $\{TTH, THT, HTT\}$, 2 to the three points $\{HHT, HTH, THH\}$, and 3 to the point HHH.

EXAMPLE 1.2

An experiment E consists of throwing a dart onto a circular dart board marked into three concentric rings, the inner one being worth 3 points, the middle one 2 points, and the outer one 1 point. Describe a sample space for this experiment.

If the outcome of interest is the precise location of the dart (we assume the board is never missed), then the sample space S_1 would be the entire board. If the board has radius R, then the possible outcomes of the experiment are

$$S_1 = \{(r, \theta) | 0 \le \theta \le 2\pi, \ 0 \le r \le R\}$$

in polar coordinates, or

$$S_1' = \{(x, y) | x^2 + y^2 \le R\}$$

in rectangular coordinates, where the vertical line is to be read "given" or "such that." These are two different descriptions of the same sample space. As with the first example, if the outcome of interest was not the location of the dart but the number of points, then the appropriate sample space would be

$$S_2 = \{1, 2, 3\}$$

EXAMPLE 1.3

The experiment E consists of rolling a die until a 6 is obtained. Discuss possible sample spaces.

If we are interested in all possibilities, then the sample space would be

$$S_1 = \{ \quad 6,$$
$$16, 26, 36, 46, 56,$$
$$116, 126, 136, 146, 156,$$
$$216, 226, 236, 246, 256, \ldots \}$$

where the first line indicates the one way in which the experiment can be terminated in one throw, the second line indicates the five ways the experiment can be terminated in two throws, and the next few lines indicate the twenty-five ways in which the experiment can be terminated in three throws, and so forth.

A variation of this sample space would be to denote obtaining either 1, 2, 3, 4, or 5 (not a 6) by N. Since we are only interested in the first 6, an appropriate sample space would be

$$S_2 = \{6, N6, NN6, NNN6, \ldots\}$$

Again, there is a simple correspondence between S_1 and S_2. The sample space S_2 in effect just indicates the number of throws needed to get a 6 and thus is a version of the sample space

$$S_2' = \{1, 2, 3, 4, \ldots\}$$

There are three distinct kinds of sample spaces illustrated by the above examples. If we classify sample spaces according to the number of points that they contain, then the three possibilities are:

(i) finite
(ii) countable or denumerable
(iii) uncountable, or nondenumerable

A set is called countable if its elements can be placed in a one-to-one correspondence with a subset of the positive integers. An infinite set is uncountable if it is not countable. Sample spaces with either a finite or countable number of points will be called discrete sample spaces that will be considered extensively in Chapters 3 and 4. Sample spaces S_1 and S_2 of Example 1.1 and S_2 of Example 1.2 are finite. The sample spaces of Example 1.3 are all countable while the sample spaces S_1 and S_1' of Example 1.2 are uncountable. The next two examples illustrate slightly different types of finite sample spaces.

EXAMPLE 1.4

Two distinguishable balls are placed at random into 4 cells. Describe a sample space for this experiment.

To describe the possible outcomes consider a set of 2-tuples (x, y), where x represents the number of the cell in which the first ball is placed (numbers 1, 2, 3, 4) and y the number of the cell in which the second ball is placed. The sample space is

$$S = \{(1, 1)\ (1, 2)\ (1, 3)\ (1, 4)\ (2, 1)\ (2, 2)\ (2, 3)\ (2, 4)$$
$$(3, 1)\ (3, 2)\ (3, 3)\ (3, 4)\ (4, 1)\ (4, 2)\ (4, 3)\ (4, 4)\}$$

where, for example, $(4, 1)$ means the first ball is placed in cell 4 and the second in cell 1.

EXAMPLE 1.5

Four letters are placed at random into four addressed envelopes. Describe the sample space for this experiment.

Since any of the four letters, which we assume are numbered 1, 2, 3, and 4, can end up in any of the four envelopes, also numbered 1, 2, 3, and 4, consider the set of four tuples

$$S = \{(1, 2, 3, 4)\ (1, 2, 4, 3)\ (1, 3, 2, 4)\ (1, 3, 4, 2)\ (1, 4, 2, 3)\ (1, 4, 3, 2)$$
$$(2, 1, 3, 4)\ (2, 1, 4, 3)\ (2, 3, 1, 4)\ (2, 3, 4, 1)\ (2, 4, 1, 3)\ (2, 4, 3, 1)$$
$$(3, 1, 2, 4)\ (3, 1, 4, 2)\ (3, 2, 1, 4)\ (3, 2, 4, 1)\ (3, 4, 1, 2)\ (3, 4, 2, 1)$$
$$(4, 1, 2, 3)\ (4, 1, 3, 2)\ (4, 2, 1, 3)\ (4, 2, 3, 1)\ (4, 3, 1, 2)\ (4, 3, 2, 1)\}$$

where, for example, $(3, 2, 1, 4)$ indicates that letter 3 is placed in envelope 1, letter 2 in envelope 2, letter 1 in envelope 3, and letter 4 in envelope 4. All possible outcomes are represented by the 24 points and S is a suitable sample space.

1.3 Events

In most cases when we perform an experiment we will be interested in certain sets of outcomes, that is, subsets of S. For example, in playing poker we may be interested in the event that a certain player is dealt a full house.

Definition An elementary event of S is a single element of S corresponding to a particular outcome of the experiment.

Definition An event of the sample space S is any subset of S.

At times, elementary events will be referred to as points of S. In many instances an event is described by a word statement that translates into a subset of S. We will make no distinction between the subset and the statement, referring to them both as the event. If A is a subset of S, on any particular repetition of the experiment the outcome, which is an elementary event, may or may not be in A. We say that A occurred if the outcome is in A.

EXAMPLE 1.6

Two distinct dice are rolled and the numbers on their faces recorded. Describe the sample space. Define and describe the events

A = {the sum on the two dice is 6}

B = {both dice show the same number}

C = {at least one of the faces is divisible by 2}

Describe the events $A \cup B$, $A \cap B$, and $B \cap C$.

The sample space consists of the 36 points:

$$
\begin{array}{cccccc}
(1,1) & (1,2) & (1,3) & (1,4) & (1,5) & (1,6) \\
(2,1) & (2,2) & (2,3) & (2,4) & (2,5) & (2,6) \\
(3,1) & (3,2) & (3,3) & (3,4) & (3,5) & (3,6) \\
(4,1) & (4,2) & (4,3) & (4,4) & (4,5) & (4,6) \\
(5,1) & (5,2) & (5,3) & (5,4) & (5,5) & (5,6) \\
(6,1) & (6,2) & (6,3) & (6,4) & (6,5) & (6,6)
\end{array}
$$

The events A, B, and C are as follows:

$A = \{(5,1), (4,2), (3,3), (2,4), (1,5)\}$

$B = \{(1,1), (2,2), (3,3), (4,4), (5,5), (6,6)\}$

and

$C = \{(1,2), (1,4), (1,6), (2,1), (2,2), (2,3), (2,4), (2,5), (2,6),$
$\quad (3,2), (3,4), (3,6), (4,1), (4,2), (4,3), (4,4), (4,5), (4,6),$
$\quad (5,2), (5,4), (5,6), (6,1), (6,2), (6,3), (6,4), (6,5), (6,6)\}$

The translation from the word statement describing the event to the subsets was, in this case, a simple matter. It is possible to perform set operations on these sets but in some cases this leads to rather clumsy word descriptions. For example, $\bar{A}$ is the set containing all those points whose sum is not 6 and contains the 30 points or elements not in A. The set $A \cup B$ is the event that *either* the sum of the two faces is 6 or the two faces are the same, or both,

$A \cup B = \{(5,1), (4,2), (3,3), (2,4), (1,5),$
$\quad (1,1), (2,2), (4,4), (5,5), (6,6)\}$

The event $A \cap B$ can be described as the event that the sum of the two dice is 6 and the two faces are equal and these two conditions describe the single point, that is, an elementary event:

$A \cap B = \{(3,3)\}$

The event $B \cap C$ is the event that the two dice are the same and at least one of them is divisible by 2:

$B \cap C = \{(2,2), (4,4), (6,6)\}$

EXAMPLE 1.7

If A, B, and C are three events of the sample space S, express the following statements in terms of these sets:

(i) Exactly one of the events occurs.
(ii) Exactly two of the events occur.
(iii) All three events occur.
(iv) Either A or B occur, but not C.
(v) C occurs and either A or B, but not both.

The answer to part (iii) is simple, $A \cap B \cap C$. To answer part (i) consider that the set $A \cap (\overline{B \cup C})$ contains all points in A *and* in neither B nor C. Extending this idea gives the set

$$(A \cap (\overline{B \cup C})) \cup (B \cap (\overline{A \cup C})) \cup (C \cap (\overline{A \cup B}))$$

as the one containing outcomes lying in exactly one of the sets A, B, or C. Similarly, for the set containing outcomes lying in exactly two of the sets, we have

$$((A \cap B) \cap \bar{C}) \cup ((A \cap C) \cap \bar{B}) \cup ((B \cap C) \cap \bar{A})$$

for the solution to part (ii). Part (iv) is simply $(A \cup B) \cap \bar{C}$, while part (v) has the solution $C \cap ((A \cup B) \cap (\overline{A \cap B}))$.

In our definition of event we took *every* subset of S to be an event (including S and $\varnothing$, the empty set). When dealing with discrete sample spaces, this causes no problems. When dealing with nondiscrete sample spaces, a problem of a mathematical nature arises in that not every subset will be an event. We will ignore this problem since it will be of little significance for our treatment and, when we come to such sample spaces we will have developed alternative methods in which the problem does not arise.

It should be stressed again that, on any particular repetition of an experiment, the result is a single element of the sample space S. This result may or may not lie in A, an event of particular interest. From our insistence that, though the outcome of an experiment is random, there should be some statistical regularity to the experiment, we will see in the next section that it will be possible to assign a number to every event of the sample space. This number will indicate how likely it is that the event occurs on any repetition of the experiment, that is, we will be defining the probability that the event occurs.

In general, if A and B are events in S, then:

$\bar{A}$ = the event that the outcome is not in A (A did not occur);
$A \cup B$ = the event that either A or B or both occurred;
$A \cap B$ = the event that both A and B occurred.

Similarly, if $A_1, A_2, \ldots, A_n$ is a sequence of sets, then:

$$\bigcup_{i=1}^{n} A_i = \text{the event that at least one of the } A_i \text{ occurred;}$$

$$\bigcap_{i=1}^{n} A_i = \text{the event that all of the } A_i \text{ occurred.}$$

The following example generalizes the ideas encountered in Example 1.7 and although the expressions obtained are cumbersome, the underlying concepts are not difficult.

EXAMPLE 1.8

In a medical study 100 people are examined for obesity. If A_i is the event that the ith person examined is overweight, describe the following events in terms of the A_i, $i = 1, \ldots, 100$: (i) nobody is overweight; (ii) at least one person is overweight; (iii) at least two people are overweight; (iv) not more than two people are overweight; (v) exactly two people are overweight.

The natural sample space for such an "experiment" is the set of all 100-tuples where each position contains a U (for underweight) or O (for overweight) and, in this sample space, A_i is the set of all points with an O in the ith position. The event $\bigcup_{i=1}^{100} A_i$ is the set of points with at least one O in them, the answer to part (ii). The complement of this set $(\overline{\bigcup_{i=1}^{100} A_i})$ is the set of points with no O's, that is, the single point with all U's, the answer to part (i). An event of the form $A_i \cap A_j$ is the set of points with O's in the ith and jth places (i.e., the ith and jth persons were overweight) and if we take the union of such sets over all distinct pairs of integers i and j, the answer to part (iii) is obtained,

$$\bigcup_{\substack{i,j=1 \\ i \neq j}}^{100} (A_i \cap A_j)$$

The event

$$A_i \cap A_j \cap \overline{\left(\bigcup_{\substack{k=1 \\ k \neq i,j}}^{100} A_k \right)}$$

is the single point with O's in the ith and jth position and U's elsewhere, and the answer to part (v) is

$$\bigcup_{\substack{i,j=1 \\ i \neq j}}^{100} \left\{ A_i \cap A_j \cap \overline{\left(\bigcup_{\substack{k=1 \\ k \neq i,j}}^{100} A_k \right)} \right\}$$

From the answer to part (iii), the set

$$A = \bigcup_{i,j,k=1}^{100} (A_i \cap A_j \cap A_k)$$

is the event that at least three people are overweight. Its complement, $\bar{A}$, is the event that not more than two people are overweight.

1.4 The Notion and Axioms of Probability

Recall that in our definition of a random experiment we required that our experiment be repeatable and that the outcomes exhibit statistical regularity. In this section we are going to use these axioms in an intuitive way, to establish what we mean by the probability of an event.

Consider an experiment E with a sample space S and let A be a particular event in S. The experiment is repeated n times and the number of the experiments in which the outcome lies in A (i.e., A occurs) is noted. Denote this number by $n(A)$. We would like to conclude that, as n becomes large, the ratio $n(A)/n$ tends to some limit. In a physical situation one would intuitively expect this to be the case if the experiments are repeated independently and under identical conditions. We will say that an experiment exhibits statistical regularity if, for any event A, in its sample space we are able to conclude that for any sequence of n repetitions of the experiment the ratio $n(A)/n$ converges to the same limit as n becomes large. The ratio $n(A)/n$ is called the relative frequency of the event A. Thus, in a die rolling experiment you might expect that out of a million tosses, about one sixth of them will come up any particular number and that, if the actual ratio is very different from $1/6$, then the die, or the experiment, is suspect. It is tempting to define the probability of an event A, $P(A)$ using relative frequency,

$$P(A) = \lim_{n \to \infty} n(A)/n \tag{1.1}$$

In many physical problems, this would be an acceptable way of proceeding. Unfortunately, there are many situations in which the concept of repeatability is simply not valid. For example, one might want to analyze the stock market in a probabilistic sense, but how is repeatability to be achieved with such an experiment? Another vivid example of this was noted by Feller (1957), who calculated that there are about 10^{30} possible bridge hands. If we were to enlist the assistance of ten billion people who were capable of dealing a bridge hand every second, night and day, we would wait over 10^{12} years to see 10^{30} hands. Thus, even when experiments can be repeated the notion of repeatability may not be useful from a practical point of view.

There are many mathematical and philosophical pitfalls in using Eq. (1.1) to define the probability of the event A. Instead, we proceed by using the concept of relative frequency to derive those properties that we would expect any measure of probability to have, and then we will simply postulate these properties as axioms.

It is clear that, for any event A, the relative frequency of A will have the following properties:

(i) $0 \le n(A)/n \le 1$, where $n(A)/n = 0$ if A occurs in none of the trials and $n(A)/n = 1$ if A occurs in all of the n trials.

(ii) If A and B are disjoint events, then

$$n(A \cup B)/n = n(A)/n + n(B)/n$$

since A and B can never occur on the same trial, then the sum of the occurrences of $A \cup B$ is the sum of the occurrences of A and B.

Based on these two properties of relative frequencies, we will insist that the probabilities imposed on any sample space have the following properties.

AXIOMS OF PROBABILITY

The probability $P(A)$ assigned to any set A of a sample space S must satisfy the following axioms:

(i) $0 \le P(A) \le 1$, $P(S) = 1$.

(ii) If A and B are disjoint events, then

$$P(A \cup B) = P(A) + P(B)$$

(iii) If $A_1, A_2, A_3, \ldots$ is a finite or infinite sequence of disjoint events, then

$$P\left(\bigcup_{i=1}^{\infty} A_i \right) = \sum_{i=1}^{\infty} P(A_i) \qquad (1.2)$$

Some comments should be made on these important axioms. Since S is the set of all possible outcomes, it is often called the sure event. It occurs in every repetition of the experiment; thus $(n(S)/n) = 1$ and it is reasonable to insist that $P(S) = 1$. In the next section it will be shown that this implies that $P(\varnothing) = 0$. It is not valid, however, to assume that because $P(A) = 0$, then A never happens. For example, if we were asked to pick a point on the real line between 0 and 1 and if the choice were truly random, each point has the same probability of being chosen. If this probability is greater than 0, say p, then because there is an infinite number of points the total probability will exceed unity and we conclude that p must be 0. On the other hand, each point in the interval is possible and, after each experiment, one results (although there is a problem with what degree of accuracy the number is to be specified).

It follows from Axiom ii that if $A_1, A_2, \ldots, A_n$ is a *finite* sequence of disjoint sets of S, then

$$P\left(\bigcup_{i=1}^{n} A_i \right) = \sum_{i=1}^{n} P(A_i)$$

For reasons that we will not expand upon, this property does not follow for an infinite sequence of disjoint sets and, since it is required for a more rigorous treatment of probability, we include it as Axiom iii.

These axioms satisfy our intuitive notions of what we mean by a probability measure. These were determined largely by extrapolating from the notion of relative frequency. In many cases we will *assign* probabilities (rather than compute them) based on intuitive or physical reasoning. For example, when tossing a die we will assign each face a probability $1/6$ of showing up, this being based on our meaning of a fair die. No matter how probabilities are assigned to the events of a sample space, they must satisfy these three axioms.

For finite sample spaces, the situation is quite simple. Label the points of the space (the elementary events) $e_1, e_2, \ldots, e_n$. These are the possible individual outcomes of the experiment E. Suppose we assigned (either arbitrarily or based on intuitive argument) the probability p_i to e_i in such a way that

(i) $p_i \geq 0$, $i = 1, \ldots, n$
(ii) $p_1 + p_2 + \cdots + p_n = 1$.

It is a simple matter to check that these two conditions ensure that all of the axioms are satisfied, and hence is a valid assignment.

EXAMPLE 1.9

Let E be the experiment of throwing a single die and recording the outcomes. The sample space is

$$S = \{1, 2, 3, 4, 5, 6\}$$

and, if the die is fair and thrown at random, we would assign the probabilities $P(i) = 1/6$, $i = 1, \ldots, 6$. We might, however, wish to examine the effect that a biased die has on the probability a player wins a certain game using this die and, in this case, we might *assign* probabilities

$$P(1) = P(2) = P(3) = P(4) = P(5) = \tfrac{1}{8}, \quad P(6) = \tfrac{3}{8}$$

This is a perfectly valid way of assigning probabilities, since the assignment satisfies the axioms. It corresponds to the case where, on the average, a 6 comes up three times as often as any other number, and this fact may have been observed or postulated.

To recount, for any experiment E, there may be more than one sample space of interest. For any experiment E and sample space S, there may be more than one way of assigning probabilities to the events of S. The experiment, sample space, and probability assignment, as a collection, is often referred to as a probability space or probability system. For the remainder of the book we reserve the use of $P(\cdot)$ to mean the probability of the statement or event inside the brackets, regardless of how it is specified.

1.5 *Elementary Properties of Probability*

For any sample space S of an experiment E we will assume that, to each event A of S, we have assigned a number $P(A)$, the probability that the outcome of any particular experiment is in A, in a manner consistent with the axioms of probability. In this section we use the axioms to develop some other basic properties of any probability measure.

1. $P(\bar{A}) = 1 - P(A)$

Proof We can express the sample space S as the union of two disjoint events A and $\bar{A}$, $S = A \cup \bar{A}$. By Axiom ii of Eq. (1.2), $P(S) = 1 = P(A) + P(\bar{A})$ from which the result follows.

Observe also that since $S = S \cup \varnothing$, then $P(S) = 1 = P(S) + P(\varnothing) = 1 + P(\varnothing)$, implying that $P(\varnothing) = 0$.

2. If $A \subseteq B$, then $P(A) \leq P(B)$.

Proof Again we prove this by decomposing the set B into $A \cup (B \cap \bar{A})$, noting that A and $B \cap \bar{A}$ are disjoint. Using Axiom ii of Eq. (1.2) gives

$$P(B) = P(A) + P(B \cap \bar{A})$$

From Axiom i, $P(B \cap \bar{A}) \geq 0$, and so $P(B) \geq P(A)$ as required.

The concept of a partition of a sample space S will be useful in many instances, particularly in Chapter 4 where it will be used to derive Bayes theorem, which is an important result in probability.

Definition The sets $B_1, B_2, \ldots, B_n$ form a partition of the sample space S if

(i) $\displaystyle\bigcup_{i=1}^{n} B_i = S$

(ii) $B_i \cap B_j = \varnothing$, $i \neq j$

In words, the sets $B_1, B_2, \ldots, B_n$ form a partition of S if they are disjoint and their union is S. For example, we can partition the eight points of the sample space of the experiment of tossing a coin three times (sample space S_1 of Example 1.1) into the sets

$B_1 = \{TTT\}, \quad B_2 = \{TTH, THT, HTT\}, \quad B_3 = \{THH, HTH, HHT\},$
$B_4 = \{HHH\}$

where, in this case, the partitioning was done according to the number of heads in the outcome. In any partition of a sample space S, each point of S is in exactly one of the sets of the partition.

3. If $B_1, B_2, \ldots, B_n$ form a partition of the sample space, then, for any event $A \subseteq S$,

$$P(A) = \sum_{i=1}^{n} P(A \cap B_i)$$

Proof   The set A can be expressed as

$$A = \bigcup_{i=1}^{n} (A \cap B_i)$$

and since $B_i \cap B_j = \emptyset$, $i \neq j$, then $(A \cap B_i) \cap (A \cap B_j) = \emptyset$. Applying Axiom iii of Eq. (1.2) then establishes the result.

4. If S is a discrete sample space (finite or countable) with elementary events e_i, $i = 1, 2, 3, \ldots$, where e_i has probability $P(e_i)$, then, for any event $A \subseteq S$,

$$P(A) = \sum_{e_i \in A} P(e_i)$$

Proof   This is just a particular case of Property 3 where the sets of the partition are taken to be the elementary events themselves, that is, the e_i form a partition of S and

$$A \cap e_i = \begin{cases} e_i & \text{if } e_i \in A \\ \emptyset & \text{if } e_i \notin A \end{cases}$$

It readily follows that

$$P(A) = \sum_{e_i \in S} P(A \cap e_i) = \sum_{e_i \in A} P(e_i)$$

5. For any two sets A and B of S,

$$P(A \cup B) = P(A) + P(B) - P(A \cap B)$$

Proof   This is perhaps the first important nontrivial property. To prove this property, decompose the set $A \cup B$ into the disjoint union of A and $B \cap (\overline{A \cap B})$ and, using Axiom ii of Eq. (1.2),

$$P(A \cup B) = P(A \cup (B \cap (\overline{A \cap B})))$$

$$= P(A) + P(B \cap (\overline{A \cap B})) \tag{1.3}$$

In the same way the set B can be decomposed as the disjoint union of $(A \cap B)$ and $(B \cap (\overline{A \cap B}))$, or

$$P(B) = P(A \cap B) + P(B \cap (\overline{A \cap B}))$$

and, rewriting this last equation gives

$$P(B \cap (\overline{A \cap B})) = P(B) - P(A \cap B) \tag{1.4}$$

Substituting Eq. (1.4) into Eq. (1.3) results in

$$P(A \cup B) = P(A) + P(B) - P(A \cap B) \tag{1.5}$$

This property is a very useful relationship, particularly with discrete sample spaces. We can extend this relationship to any finite number of sets, but it is rather tedious work and requires the use of the induction principle, an important tool of mathematical proof. A brief outline of the proof is given since it will be of some importance.

Rather than doing the general case immediately, we consider first the case of the union of three arbitrary sets A_1, A_2, and A_3 and find an expression for $P(A_1 \cup A_2 \cup A_3)$. This will serve as an example for the proof of the general case. Let $B = A_1 \cup A_2$ and, applying Eq. (1.5) to $B \cup A_3$ results in

$$P(B \cup A_3) = P(B) + P(A_3) - P(B \cap A_3) \tag{1.6}$$

The set $B \cap A_3 = (A_1 \cup A_2) \cap A_3$ can also be expressed as $(A_1 \cap A_3) \cup (A_2 \cap A_3)$, and, applying Eq. (1.5) to this set also gives

$$P(B \cap A_3) = P(A_1 \cap A_3) + P(A_2 \cap A_3) - P((A_1 \cap A_3) \cap (A_2 \cap A_3))$$
$$= P(A_1 \cap A_3) + P(A_2 \cap A_3) - P(A_1 \cap A_2 \cap A_3) \tag{1.7}$$

Applying Eq. (1.5) to the set $B = A_1 \cup A_2$, we have

$$P(A_1 \cup A_2) = P(A_1) + P(A_2) - P(A_1 \cap A_2) \tag{1.8}$$

Substituting Eqs. (1.8) and (1.7) into (1.6), we have

$$P(A_1 \cup A_2 \cup A_3) = P(A_1) + P(A_2) + P(A_3) - P(A_1 \cap A_2) - P(A_1 \cap A_3)$$
$$- P(A_2 \cap A_3) + P(A_1 \cap A_2 \cap A_3) \tag{1.9}$$

This equation has a simple interpretation. The first three terms sum the probabilities of all elementary events in A_1, A_2, and A_3. An elementary event in, say, $A_1 \cap A_2$ will have its probability summed twice while an elementary event in $A_1 \cap A_2 \cap A_3$ will have its probability summed three times. We subtract the probabilities of points in at least two of the sets by subtracting $P(A_1 \cap A_2)$, $P(A_1 \cap A_3)$, and $P(A_2 \cap A_3)$. Now, however, an elementary event in all three sets has not had its probability counted at all (its probability has been added three times and subtracted three times), and so we add the term $P(A_1 \cap A_2 \cap A_3)$ to arrive at Eq. (1.9). This line of reasoning is an application of another important mathematical concept, the principle of inclusion-exclusion. It is just a method of counting to make sure each item in the set of interest is counted exactly once. The following property is a generalization of the present one.

6. If $A_1, A_2, \ldots, A_n$ are n arbitrary sets of S, then

$$P\left(\bigcup_{i=1}^{n} A_i\right) = \sum_{i=1}^{n} P(A_i) - \sum P(A_i \cap A_j) + \sum P(A_i \cap A_j \cap A_k)$$
$$- \cdots (-1)^{n-1} P(A_1 \cap \cdots \cap A_n) \tag{1.10}$$

where the sum of the second term is over all distinct pairs of sets, that of the third term over all distinct triples of sets, and so forth.

Proof We have already proven this for $k = 2$ and $k = 3$. We will prove this property using the principle of induction. We assume that Eq. (1.10) is true for any collection of j sets for any integer j, $1 \le j \le k$. We use this to prove the relationship for $(k + 1)$ sets. If this can be demonstrated, then the relationship will be true for any integer n. Only a sketch of the proof is given and the reader is urged to provide the details. Consider a collection of $(k + 1)$ sets $A_1, A_2, \ldots, A_{k+1}$ and assume that Eq. (1.10) is true for all collections of k or fewer sets. Let $B = \bigcup_{i=1}^{k} A_i$ and, using Property 5,

$$P\left(\bigcup_{i=1}^{k+1} A_i\right) = P(B \cup A_{k+1})$$

$$= P(B) + P(A_{k+1}) - P(A_{k+1} \cap B) \tag{1.11}$$

From the assumption that Eq. (1.10) is true for all collections of k or fewer sets,

$$P(B) = P\left(\bigcup_{i=1}^{k} A_i\right)$$

$$= \sum P(A_i) - \sum P(A_i \cap A_j) + \sum P(A_i \cup A_j \cap A_k)$$

$$- \cdots (-1)^{k-1} P(A_1 \cap A_2 \cdots \cap A_k) \tag{1.12}$$

The only term in Eq. (1.11) left to consider is $P(A_{k+1} \cap B)$. From set computations

$$P\left(A_{k+1} \cap \left(\bigcup_{i=1}^{k} A_i\right)\right) = P\left(\bigcup_{i=1}^{k} (A_i \cap A_{k+1})\right)$$

and again we can apply Eq. (1.10) to give

$$P\left(\bigcup_{i=1}^{k} (A_i \cap A_{k+1})\right) = \sum P(A_i \cap A_{k+1}) - \sum P(A_i \cap A_j \cap A_{k+1})$$

$$+ \cdots (-1)^{k-1} P(A_1 \cap A_2 \cap \cdots \cap A_{k+1}) \tag{1.13}$$

Substituting Eqs. (1.12) and (1.13) into (1.11) then gives the desired result.

A few examples will illustrate some of the properties developed in this section.

EXAMPLE 1.10

Given that $P(A) = p$, $P(B) = q$, and $P(A \cap B) = r$. Find (i) $P(A \cap \bar{B})$, (ii) $P(\bar{A} \cap \bar{B})$, and (iii) $P(A \triangle B)$.

(i) $A = (A \cap \bar{B}) \cup (A \cap B)$ and, since $(A \cap \bar{B}) \cap (A \cap B) = \varnothing$, we can write

$$P(A) = P(A \cap \bar{B}) + P(A \cap B)$$

or in terms of the given information

$$P = P(A \cap \bar{B}) + r \quad \text{or} \quad P(A \cap \bar{B}) = p - r$$

(ii) $\bar{A} \cap \bar{B} = \overline{A \cup B}$ and, since

$$P(A \cup B) = P(A) + P(B) - P(A \cap B) = p + q - r$$

it follows that $P(\overline{A \cup B}) = 1 - (p + q - r)$.

(iii) $A \triangle B = (A \cap \bar{B}) \cup (B \cap \bar{A})$ and, since these sets are disjoint,

$$P(A \triangle B) = P(A \cap \bar{B}) + P(B \cap \bar{A})$$

The first term on the right-hand side in this equation was found in part (i). The second term can be found, in a similar manner, to be

$$P(B \cap \bar{A}) = q - r$$

and so

$$P(A \triangle B) = p - r + q - r = p + q - 2r$$

EXAMPLE 1.11

Show that $P(A \cap B) \geq 1 - P(\bar{A}) - P(\bar{B})$.
From Eq. (1.5)

$$P(A \cup B) = P(A) + P(B) - P(A \cap B) \qquad (1.14)$$

and, for any sets A and B,

$$P(\bar{A}) = 1 - P(A), \qquad P(\bar{B}) = 1 - P(B)$$

so that, rearranging Eq. (1.14) gives

$$P(A \cap B) = P(A) + P(B) - P(A \cup B)$$

$$= 1 - P(\bar{A}) + 1 - P(\bar{B}) - P(A \cup B)$$

For any two sets A and B it is true that $P(A \cup B) \leq 1$ and using this fact in the last equation yields

$$P(A \cap B) \geq 1 - P(\bar{A}) - P(\bar{B})$$

EXAMPLE 1.12

In a sample space S, three events A, B, and C have the probabilities $P(A) = 1/3 = P(B)$, $P(C) = 1/4$, $P(A \cap B) = 1/6$, $P(A \cap C) = 1/8$, and $P(B \cap C) = 0$. Find the probability $P(A \cup B \cup C)$.

Using Eq. (1.10) we have

$$P(A \cup B \cup C) = P(A) + P(B) + P(C) - P(A \cap B) - P(A \cap C)$$

$$- P(B \cap C) + P(A \cap B \cap C)$$

and the only unknown term is $P(A \cap B \cap C)$. However, $A \cap B \cap C \subseteq B \cap C$ and, by Property 2, $P(A \cap B \cap C) \leq P(B \cap C) = 0$. Therefore,

$$P(A \cup B \cup C) = \tfrac{1}{3} + \tfrac{1}{3} + \tfrac{1}{4} - \tfrac{1}{6} - \tfrac{1}{8} = \tfrac{5}{8}$$

1.6   *Equally Likely Outcomes*

An important situation that will be considered extensively in Chapters 3 and 4 occurs when the sample space has only a finite number of points, say n. In this case, if we denote the elementary events e_i, $i = 1, \ldots, n$, then

(i) $0 \leq P(e_i) \leq 1$, $i = 1, \ldots, n$

(ii) $P(e_1) + P(e_2) + \cdots + P(e_n) = 1$

(iii) If $A = \bigcup_{i \in I} e_i$, where I is a collection of subscripts, then

$$P(A) = \sum_{e_i \in A} P(e_i)$$

A particularly simple and important case arises when each elementary event has the same probability. This often occurs in practice when we have no reason, physical or otherwise, to assume that any one elementary event is any more likely to occur than any other. For example, if we toss a coin, which we assume to be fair, then by definition each of the possible outcomes has the same probability. If we choose a ball at random from an urn containing n balls, then each ball has a probability of $1/n$ of being drawn. A consumer choosing one from n brands of soap may do so with equal probability. Discrete sample spaces in which each elementary event has the same probability are called "equally likely," and they occur frequently in games of chance, an area that provides considerable insight to probability theory.

Consider an equally likely sample space S containing n elements, each element having a probability of $1/n$. If A is any event of S containing k elements, then

$$P(A) = k/n$$

The simplicity of the situation makes the use of such spaces attractive.

In connection with equally likely sample spaces the phrase "choosing at random" is frequently used in the definition of the experiment. This simply means that among the various alternatives no one choice is more likely than another. The phrase in this context implies that each elementary event is equally likely. These simple but important points are illustrated in the following examples.

EXAMPLE 1.13

A bag contains 5 tags marked with the integers 1 through 5. Two tags are drawn at random, the first tag being replaced before the second is drawn. Find the probability of the events $A = \{$both tags drawn have the same number$\}$ and $B = \{$the second number drawn is strictly greater than the first number drawn$\}$.

The sample space of the experiment contains the twenty-five ordered pairs (i, j), $1 \leq i \leq 5$, $1 \leq j \leq 5$, where the first number of the pair indicates the first number drawn. Each such elementary event has the probability $1/25$. The event A contains the five elements (i, i), $i = 1, 2, \ldots, 5$, and so $P(A) = 5/25$.

The event B contains the points (i, j) for which $i < j$,

$$B = \{(1, 2), (1, 3), (1, 4), (1, 5), (2, 3), (2, 4), (2, 5), (3, 4), (3, 5), (4, 5)\}$$

and $P(B) = 10/25$.

EXAMPLE 1.14

The experiment of Example 1.13 is repeated, but without replacing the first tag. Find the probability of A and B in this case.

In this instance the sample space contains the 20 points (i, j), $i \neq j$, $1 \leq i \leq 5$, $1 \leq j \leq 5$ since we "choose at random without replacement." The event A is now the empty set $\varnothing$ since it is impossible to draw the same tag on the second draw as on the first and $P(\dot A) = 0$. The event B is unchanged and, since each elementary event now has probability $1/20$, $P(B) = 10/20$.

In the first experiment the sequential nature of it was implicit since we could not draw the second tag until the first had been replaced. The phrase "choosing at random" implied each of the resulting twenty-five points were equally likely. In the second experiment the phrase "choosing at random" implied that each of the possible twenty points were equally likely. This different meaning of the same phrase applied to similar (but essentially different) experiments should be carefully noted.

EXAMPLE 1.15

A secretary types 5 letters and 5 envelopes but puts the letters into the envelopes at random. What is the probability that at least one letter is put into its correct envelope.

To set up the sample space for this experiment we label each letter with a number 1, 2, 3, 4, and 5 and the corresponding envelope with the same number. We can then view an outcome of the experiment as a particular ordering of these five numbers. For example, the ordering 4 1 2 3 5 means letter 4 was placed in envelope 1, letter 1 in envelope 2, letter 2 in envelope 3, and so forth. The sample space then contains $5! = 5 \cdot 4 \cdot 3 \cdot 2 \cdot 1 = 120$ outcomes or elementary events and, from the fact that the letters were placed into the envelopes at random, each such outcome has the same probability, namely, $1/120$.

To find the probability that at least one letter is put into its correct envelope, let A_1 be the event that letter 1 was placed into envelope 1 and define $A_2, A_3, A_4,$ and A_5 in a similar manner. The event of interest, that at least one letter is put into its correct envelope, is then $A_1 \cup A_2 \cup A_3 \cup A_4 \cup A_5$ and using Eq. (1.10)

$$P\left(\bigcup_{i=1}^{5} A_i\right) = \sum_{i=1}^{5} P(A_i) - \sum_{\substack{i,j=1 \\ i \neq j}}^{5} P(A_i \cup A_j) + \sum_{\substack{i,j,k=1 \\ \text{distinct}}}^{5} P(A_i \cap A_j \cap A_k)$$

$$- \sum_{\substack{i,j,k,l=1 \\ \text{distinct}}}^{5} P(A_i \cap A_j \cap A_k \cap A_l)$$

$$+ P(A_1 \cap A_2 \cap A_3 \cap A_4 \cap A_5) \qquad (1.15)$$

Since each point in the sample space has the same probability, the remainder of the problem is just a counting argument. Consider first the number of elementary events in A_1. We require only that 1 appear in the first position and the remaining four elements appear in any of the 4! ways possible. As a result $|A_1| = 4! = 24$ and $P(A_1) = 24/120 = 1/5$. Precisely the same reasoning can be applied to each of the other sets and $P(A_i) = 1/5$, $i = 1, 2, 3, 4, 5$.

Consider now the set $A_1 \cap A_2$ specified as those elementary events containing a 1 in the first position and a 2 in the second position. Since the other three numbers can be ordered in any of 3! ways, $|A_1 \cap A_2| = 3!$ and $P(A_1 \cap A_2) = 3!/5! = 1/20$ and by an argument similar to that used in the previous case, $P(A_1 \cap A_2) = P(A_i \cap A_j)$, $i \neq j$. Proceeding in this fashion, $P(A_i \cap A_j \cap A_k) = 2!/5! = 1/60$, i, j, and k distinct, and $P(A_i \cap A_j \cap A_k \cap A_l) = 1/120 = P(A_1 \cap A_2 \cap A_3 \cap A_4 \cap A_5)$ since, if four of the letters are in their correct envelope, so is the fifth. Simple counting arguments show that there are 10 ways of choosing 2 distinct sets from 5, 10 ways of choosing 3 sets, and 5 ways of choosing 4 sets. Putting these results together into Eq. (1.15) gives

$$P\left(\bigcup_{i=1}^{5} A_i\right) = 5 \cdot (\tfrac{1}{5}) - 10 \cdot (\tfrac{1}{20}) + 10 \cdot (\tfrac{1}{60}) - 5 \cdot (\tfrac{1}{120}) + 1 \cdot (\tfrac{1}{120})$$

$$= \tfrac{19}{30}$$

This result could have been obtained by searching through the 120 points of the sample space and finding those with at least one of the numbers in its correct position, but this would have been tedious. Equation (1.10) can thus be effectively used to reduce the computation.

EXAMPLE 1.16

The thirteen diamonds from a deck of playing cards are separated and from these, two cards are drawn. Letting the jack represent 11, the queen 12, and the king 13, what is the probability of the two numbers being consecutive integers if: (i) the cards are drawn with replacement? (ii) the cards are drawn without replacement?

Considering (i): As the sample space for the experiment we take the $(13)^2$ ordered pairs of integers (i, j), $1 \leq i \leq 13$, $1 \leq j \leq 13$ since the drawing is with replacement. Pairs of the form $(i, i+1)$ or $(i+1, i)$ form consecutive integers, and it is readily verified that there are 24 such pairs and the probability of choosing one of these is $24/(13)^2$.

Considering (ii): When choosing without replacement, we can choose the first card in 13 ways and the second in 12 ways resulting in 13×12 ordered pairs (i, j), $1 \leq i \leq 13$, $1 \leq j \leq 13$, $i \neq j$. There are still 24 pairs corresponding to consecutive integers and the probability of choosing one of these is $24/13 \times 12 = 2/13$.

EXAMPLE 1.17

In an electrical network shown in Figure 1.1 the four relays are closed at a given time. If it is equally likely that a relay will work or not, what is the probability that a closed path will exist from left to right?

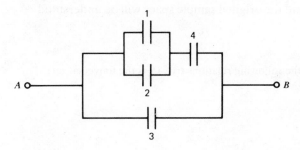

FIGURE 1.1

For this problem we can set up a sample space of which a typical point is (O, C, C, O) indicating that relay 1 is open, relays 2 and 3 closed and relay 4 is open. The sample space contains $2^4 = 16$ points and by assumption they are equally likely. If $A_i, i = 1, 2, 3, 4$ is the event that the ith relay is closed, then the set $A_1 \cap A_4$ contains all elementary events with a C in the first and fourth places, and corresponds to a closed path from A to B. The only other way a closed path can occur corresponds to an elementary event in one of the sets $A_2 \cap A_4$ or A_3. The probability of a closed path is then $P((A_1 \cap A_4) \cup (A_2 \cap A_4) \cup A_3)$. Applying Eq. (1.10) to this union of three sets gives

$$P((A_1 \cap A_4) \cup (A_2 \cap A_4) \cup A_3) = P(A_1 \cap A_4) + P(A_2 \cap A_4) + P(A_3)$$
$$-P((A_1 \cap A_4) \cap (A_2 \cap A_4))$$
$$-P((A_1 \cap A_4) \cap A_3) - P((A_2 \cap A_4) \cap A_3)$$
$$+P((A_1 \cap A_4) \cap (A_2 \cap A_4) \cap A_3)$$
$$= P(A_1 \cap A_4) + P(A_2 \cap A_4) + P(A_3)$$
$$-P(A_1 \cap A_2 \cap A_4) - P(A_1 \cap A_3 \cap A_4)$$
$$-P(A_2 \cap A_3 \cap A_4) + P(A_1 \cap A_2 \cap A_3 \cap A_4)$$

The number of elementary events in an intersection of j of the sets $j = 1, 2, 3$, or 4 is 2^{4-j} and, since each of these elementary events are equally likely,

$$P(\text{closed path}) = \tfrac{4}{16} + \tfrac{4}{16} + \tfrac{8}{16} - \tfrac{2}{16} - \tfrac{2}{16} - \tfrac{2}{16} + \tfrac{1}{16} = \tfrac{11}{16}$$

The notion of a sample space and a probability measure on it is central to all of probability theory. It will be used extensively, either explicitly or implicitly, in Chapters 3 and 4. With experience the reader will be able to tackle most problems without actually writing down the sample space, but there should always be one in mind. In the next chapter the notion of a sample space will be unified in that we will define functions, called random variables, that map our given sample space into a subset of the real line. With this approach we will be able to define useful functions on the real line that describe the probabilistic behavior of the experiment. After Chapter 4 it will be these functions that will be studied and the original sample space will be understood.

Problems

1. Under what conditions are the following relations true (S is the universal set)?
   (i) $A \cap B = S$
   (ii) $A \cup B = S$
   (iii) $A \cap B = \bar{A}$
   (iv) $A \cup B = \emptyset$
   (v) $A \cup B = A \cap B$

2. Two dice are tossed. Describe a suitable sample space and describe in detail the following events:

   $A = \{$the sum of two faces is a 7$\}$
   $B = \{$at least one of the faces is an even number$\}$
   $C = \{$one of the numbers divides the other$\}$

   Describe the events $A \cap B$ and $A \cap C$.

3. A student is chosen at random from a class. Let A be the event that the student is over 2 meters tall, B the event that the student has red hair, and C the event that the student weighs less than 70 kg.
   (i) Describe the event $A \cap B \cap C$.
   (ii) Under what conditions will $A \cap B \cap C = A$ hold?
   (iii) When will it be true that $\bar{C} \subseteq B$?
   (iv) When will it be true that $\bar{A} = B$?

4. A machine makes items of four different weights: 5, 10, 20, and 50 kg. If A_1, A_2, A_3, and A_4 are the events representing those items of weight not greater than 5, 10, 20, and 50 kgms, respectively, describe the following events in words.
   (i) $\bigcup_{i=1}^{4} A_i$
   (ii) $A_1 \cap A_3$
   (iii) $A_2 \cap \bar{A}_3$

5. Show that:
   (i) $A \cap (B \cup C) = (A \cap B) \cup (A \cap C)$
   (ii) $A \cup (B \cap C) = (A \cup B) \cap (A \cup C)$

6. Four items are tested and three events defined as follows:

   $A = \{$at least one of four items checked is defective$\}$
   $B = \{$all four items are good$\}$
   $C = \{$exactly two of the items are defective$\}$

   Interpret the following events:
   (i) $A \cup B$
   (ii) $A \cap B$
   (iii) $A \cup C$
   (iv) $A \cap C$

7. A number is selected at random. Let A be the event that the number is divisible by 2, B be the event that it ends in a 5, and C the event that it ends in a 0. Describe the following events:

(i) $A \cap C$
(ii) $B \cup C$
(iii) $A \cap B$

8. Four shoppers in a supermarket choose one of three brands of canned peas at random. Determine a sample space for such an experiment and describe the events:

   $A = \{$each brand was chosen by at least one shopper$\}$
   $B = \{$at least one brand was not chosen by any shopper$\}$
   $C = \{$all shoppers chose the same brand$\}$

   Describe the events $B \cap C$ and $\bar{A} \cap B$.

9. A large lot of items contains some that are defective. Items are chosen until the first defective one is obtained. Describe a sample space for this experiment. If items are chosen until the second defective one is obtained, describe a suitable sample space and the event $A = \{$fewer than 7 items were chosen$\}$. If in this second sample space $B = \{$no defective items were among the first 4 items chosen$\}$, describe $A \cap B$.

10. Prove the following set theory relations:
    (i) $(A \cap B) \cup (A \cap \bar{B}) = A$
    (ii) $A \cup B = B \cup (A \cap \bar{B})$

11. If A, B, and C are arbitrary events of a sample space S, express $A \cup B \cup C$ as the union of three disjoint sets.

12. A man deals himself a poker hand (5 cards from a 52 card deck). Describe a sample space for this situation. Describe the following events by giving typical points in the event and the number of points in the event:

    $A = \{$the hand contains exactly two distinct pairs$\}$
    $B = \{$the hand is a flush—five cards of the same suit, not necessarily in sequence$\}$
    $C = \{$the hand is a full house—three of a kind and a pair$\}$
    $D = \{$the hand contains exactly one pair$\}$

    Describe the events $B \cap D$, $A \cap B$, and $A \cap C$.

13. An electronic circuit contains 1 diode and 2 transistors and is designed to work if the diode and at least one of the transistors work. Let D be the event that the diode operates and T_i $(i = 1, 2)$ be the event that the ith transistor operates. If W is the event that the circuit works, express W and $\bar{W}$ in terms of D, T_1, and T_2.

14. Screws are inspected for three kinds of defects: the absence of a slot (type 1), defective thread (type 2), and thread length (type 3). Two hundred and fifty screws are inspected with the following results: 110 have a type 1 defect, 168 have a type 2 defect, and 103 have a type 3 defect; 53 have both type 1 and 2 defects, 59 have both type 2 and 3 defects, and 54 have both type 1 and 3 defects; 42 have all three types of defects. Show that there must have been a miscount.

15. Shoppers are questioned as to the acceptability of three brands, A, B, and C, of soap. Of those questioned 67 found brand A acceptable, 43 found brand B acceptable, and 38 found brand C acceptable. In addition 21 found both A and B acceptable, 14 found both B and C acceptable, and 14 found both A and C acceptable. If 108 shoppers were questioned, how many found all three brands acceptable?

16. A die is thrown until two sixes in a row are thrown. Describe a suitable sample space for this experiment.

17. A set of four cards marked A, B, C, and D are placed at random in a row. A second set with the same markings are then placed on top of them. What is the probability that there are no matchups?

18. A box of light bulbs contains 25, 60, and 120 watt bulbs. If it is assumed that there are many of each wattage in the box and an experiment consists of withdrawing three items sequentially, describe the following events:
   (i) $A = \{$at least one 60 watt bulb is drawn among the three$\}$
   (ii) $B = \{$the first bulb has a wattage greater than the other two$\}$
   (iii) $C = \{$all three bulbs have the same wattage$\}$
   (iv) $A \cap B$
   (v) $A \cap \bar{C}$

19. Show that $P(A \cap B) \leq P(A) \leq P(A \cup B) \leq P(A) + P(B)$ and discuss conditions under which equality is achieved in each case.

20. A box contains n items of which k are defective. Describe a sample space for each of the following experiments:
   (i) Two items are withdrawn without replacement.
   (ii) Two items are withdrawn with replacement.
   (iii) Items are withdrawn without replacement until a defective item is obtained.
   (iv) Items are withdrawn with replacement until a defective item is found.

21. For the four experiments in Problem 20 find the probabilities of the following events (part (i) refers to sample space (i) in Problem 20):
   (i) $A_1 = \{$the second item drawn is defective$\}$
   (ii) $A_2 = \{$the second item drawn is good$\}$
   (iii) $A_3 = \{$the first defective item occurs on the ith draw$\}$
   (iv) $A_4 = \{$at least i draws are required to find a defective item$\}$

22. If $P(A \cap \bar{B}) = .2$ and $P(\bar{B}) = .7$, find $P(A \cup B)$.

23. If A and B are disjoint events and $P(A) = .3$ and $P(B) = .45$, calculate (i) $P(A \cup B)$, (ii) $P(\bar{A} \cup \bar{B})$, (iii) $P(\bar{A} \cap B)$.

24. If, for two sets A and B, $P(A) = .4$, $P(B) = .42$, $P(A \cap B) = .25$, find (i) $P(A \cup B)$, (ii) $P(A \cap \bar{B})$, (iii) $P(\bar{A} \cup \bar{B})$.

25. If $P(A) = a$, $P(B) = b$, and $P(A \cap B) = ab$, find $P(A \cap \bar{B})$ and $P(\bar{A} \cap \bar{B})$.

26. If A, B, and C are three events, explain carefully why the following assignment of probabilities is not possible:
   (i) $P(A) = .65$, $P(A \cap \bar{B}) = .60$, $P(A \cap B) = .10$.
   (ii) $P(A) = .5$, $P(B) = .5$, $P(C) = .5$, $P(A \cap B) = 0$, and $P(A \cap B \cap C) = .2$.
   (iii) $A \cap B \cap C = 0$ and $P(A) = P(B) = .4$, $P(C) = .5$, $P(A \cup B \cup C) = .55$, and $P(A \cap C) = P(A \cap B) = 1$.

27. Is it possible to have an assignment of probabilities such that $P(A) = 1/2$, $P(A \cap B) = 1/3$, and $P(B) = 1/4$?

28. It is known that there are two defective items in a group of ten. One item is selected and then a second, without replacing the first item. What is the probability that (i) both items are good, (ii) the second one is defective, and (iii) both are defective.

29. On a particular exam the student is provided with n questions and n answers that must be matched to the questions. If the student does the matching at random, find:
   (i) the probability the first question is matched correctly;
   (ii) the probability that the first two questions are answered correctly;
   (iii) the probability that at least one of the questions is answered correctly.

30. A person throws three dice. Which of the events A or B is more likely where $A = \{$each die shows an odd number$\}$ and $B = \{$the sum of the three numbers is divisible by 9$\}$.

31. If A, B, and C are events for some random experiment, show that the probability that exactly one of the events A, B, or C occurs is

$$P(A) + P(B) + P(C) - 2P(A \cap B) - 2P(A \cap C) - 2P(B \cap C) + 3P(A \cap B \cap C)$$

32. Show the following inequalities:
    (i) If $A_1 \cap A_2 \subset A$, then $P(A) \geq P(A_1) + P(A_2) - 1$.
    (ii) By induction show that if $\bigcap_{i=1}^{n} A_i \subset A$, then $P(A) \geq \sum_{1}^{n} P(A_i) - (n-1)$.

33. A card is drawn from a deck of cards, its suit observed, and it is then replaced. If this is done four times, what is the probability that at least one of the cards was a club.

34. Ten people enter an elevator on the ground floor and get off, at random, at one of eight floors above ground. Describe a sample space for this situation and, if each point is equally likely (which will be reasonable only if you have chosen the appropriate sample space), what is the probability that exactly one person gets off at the fifth floor?

35. A coin is tossed until a head is obtained. Describe the appropriate sample space. If each point in the sample space requiring n tosses has probability $1/2^n$, what is the probability that the first head is obtained on an even-numbered toss?

36. An experiment with two possible outcomes A and B is performed repeatedly until the results on two consecutive trials are the same. Describe a sample space for such an experiment. If the probability of a point in the sample space corresponding to n experiments being performed has probability $1/2^n$, what is the probability the experiment ends on the tenth experiment?

37. A fair coin is tossed five times and each outcome is equally likely. What is the probability of obtaining a sequence of at least three consecutive heads?

38. Four workers leaving the office choose their time card from among the twenty in the rack at random. Describe a suitable sample space for this situation and determine the probability that at least one of them uses the right card.

39. Which is the more likely event, obtaining a 7 with a roll of 2 dice or 4 heads with 5 tosses of a coin?

CHAPTER 2

Random Variables and Their Distributions

In the first chapter the foundation of probability theory was set forth. The ideas of an experiment, a sample space corresponding to the experiment, and events in a sample space were introduced and the axioms of a probability measure on these events were postulated. In this chapter, the concept of a random variable and its distribution function is introduced. The main effect of using random variables will be to unify the study of probabilistic situations. This is achieved by mapping the original sample space to the real line, for any experiment, which means that we will only have to study the one sample space, the real line. This procedure, which might appear peculiar at first, yields many benefits. In Chapters 3 and 4 we will still take the approach of setting up a sample space, but beyond those chapters we will deal exclusively with random variables and the various functions associated with them.

There are basically two distinct types of random variables: discrete and continuous. In this and the next two chapters the emphasis will be on discrete random variables while much of the remainder of the text deals with problems associated with continuous random variables.

2.1 *The Concept of a Random Variable*

Definition Let S be the sample space associated with some experiment E. A random variable X is a function that assigns a real number $X(s)$ to each element $s \in S$.

EXAMPLE 2.1

As with Example 1.1 let E be the experiment of tossing a coin three times. Two sample spaces S_1 and S_2 were associated with this experiment:

$$S_1 = \{TTT, TTH, THT, HTT, HHT, HTH, THH, HHH\}$$

and

$$S_2 = \{0, 1, 2, 3\}$$

where $i \in S_2$ indicates the number of heads in three tosses. A random variable X on S_1 can be defined such that for $s \in S_1$, $X(s)$ is the number of heads in the point $s \in S_1$. Of course X maps S_1 into S_2, but S_2 is a subset of the real line and hence X is a random variable. Pictorially the random variable can be viewed as a mapping

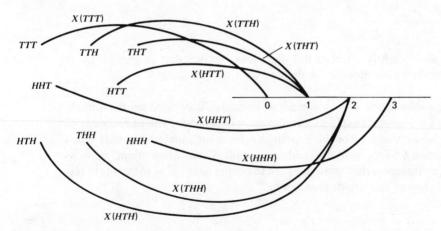

The first thing to notice about the definition of a random variable X is that X is not really a variable at all. It is a function that maps elements of S to the real line. It is also not random in the sense that once the outcome $s \in S$ is known, the real number $X(s)$ is completely determined. It is the outcome $s \in S$ that is random, not the mapping X. The term random variable however is now standard, and we often denote it simply by X, suppressing the dependence on the sample space.

Not every function from S to the real line will be allowed as a random variable. For example, a function that assigns more than one real number to any element of S (a one-to-many function) is unacceptable. Only functions that assign exactly one real number to each element of S are satisfactory. Of course, the same real number may be assigned to many elements of S (i.e., a many-to-one function). The domain of a random variable X is the sample space S and the range space R_X is the subset of the real line

$$R_X = \{r \in R \mid \text{there exists an } s \in S \text{ such that } X(s) = r\}$$

Just as there are often many sample spaces that can be of interest in an experiment, there can also be many different random variables of interest for the sample space. An example will illustrate the point.

EXAMPLE 2.2

Let E be the experiment consisting of rolling two dice and observing the outcomes, and let S be the sample space consisting of the 36 ordered pairs (i, j), $1 \le i \le 6$, $1 \le j \le 6$, where the first integer is the outcome of the first die and the second integer, the outcome of the second die. Let X be the random variable indicating the sum of the two faces, that is, for $(i, j) \in S$, $X(i, j) = i + j$. Let Y be the random variable indicating the outcome on the second die, that is, $Y(i, j) = j$. In this case the range spaces of the two random variables are

$$R_X = \{2, 3, 4, \ldots, 12\}$$

and

$$R_Y = \{1, 2, 3, 4, 5, 6\}$$

The example is slightly artificial but illustrates the fact that in general many different random variables can be defined on a sample space.

The range space R_X of a random variable X, defined on the sample space S, can be viewed simply as another sample space for the experiment E. We can argue as follows: For the experiment E we formulated a sample space S, then mapped S to R_X with the random variable X. Since our "final" result is in R_X, we can interpret the "outcomes" of the experiment E as elements in R_X. A pictorial view of the situation might be:

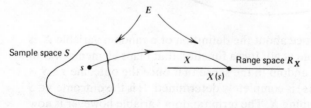

On each trial of the experiment we can view the situation in two ways. In one way we observe the outcome $s \in S$ and then map s to $X(s) \in R_X$. On the other hand, we might pretend there is another person present whose job it is to observe the outcome $s \in S$ and then relate to us, not s, but $X(s)$. As far as we are concerned, then, $X(s)$ is the outcome of the experiment and R_X the sample space.

Suppose that, because of the physical situation of the experiment, we know the probability of any event of S. Since we want to deal with the random variable X and its sample space R_X, how are we to assign probabilities in R_X?

Definition   The events $A \subset S$ and $B \subset R_X$ are called equivalent events if

$$A = \{s \in S \,|\, X(s) \in B\}$$

Equivalent events are to be thought of in the following way: Given a set $B \subset R_X$, the set of all those points of S that are mapped into B by the random variable X is the event $A \subset S$ equivalent to $B \subset R_X$. From our observer point of view he will indicate that an outcome is in B if and only if he sees it is in $A \subset S$. The elementary events of R_X are of the form $\{s \in S \,|\, X(s) = x \in R_X\}$.

The reason for concerning ourselves with equivalent events is that it provides a means for assigning a probability measure to R_X. It makes intuitive sense to assign the same probabilities to the sets $A \subset S$ and $B \subset R_X$ if they are equivalent events, since the outcomes in these events in their respective sample spaces are "coexistent."

Definition   If $A \subset S$ and $B \subset R_X$ are equivalent events, then we define the probability of the event B, $P(B)$, to be equal to $P(A)$.

Thus, in order to find the probability of an event B in R_X, we first find the equivalent event A in S, whose probability is assumed known, and use this as the probability of event B. Here we have *defined* $P(B)$ in this manner. Actually it is possible to *prove* that $P(B)$ must equal $P(A)$ and that such an assignment of probabilities in R_X satisfies all the axioms of probability on a sample space discussed in Section 1.4.

In defining the probabilities of equivalent events we are considering two different sample spaces S and R_X, using the known probability measure on S to induce one on R_X. It is therefore doubtful mathematical practice to use the single probability measure $P(\cdot)$ on both spaces. However, we have agreed that $P(\cdot)$ indicates the probability of the event inside the parentheses, regardless of which of the two spaces the event is in.

EXAMPLE 2.3

As with Example 2.1, let S_1 be the sample space containing the eight outcomes of tossing three coins. We assume that each of these eight points has the same probability, $1/8$. If X is the random variable giving the number of heads obtained, find $P(X = 2)$.

The elementary event $\{X = 2\}$ in R_X is equivalent to the event $A \subset S_1$, where

$$A = \{s \in S_1 | X(s) = 2\}$$
$$= \{HHT, HTH, THH\}$$

and since $P(A) = 3/8$, then $P(X = 2) = 3/8$.

It is conventional to use capital letters such as X, Y, S, T, ... to denote a random variable and the corresponding lower case letter, $x, y, s, t, \ldots$, to denote particular values in its range. For example, we might ask for $P(X \geq x)$, the probability that the random variable X is greater than or equal to $x \in R_X$. The X is to be thought of as random (a function of the random outcome of an experiment) while the x is to be thought of as nonrandom, an element in R_X.

The introduction of random variables unifies the study of probability since the probabilities are now always defined on subsets of the real line. The idea of the original sample space will not be used at all after Chapter 5. However, behind every random variable there is a sample space—in some cases the original sample space happens to be a subspace of the real line (e.g., throwing a die once). The advantages of this approach using random variables will become apparent in succeeding sections.

2.2   Cumulative Distribution Functions

For any random variable X we define the function $F_X(x)$ by the equation

$$F_X(x) = P(X \leq x) = P(\{s \in S | X(s) \leq x\})$$

where x is any real value. As mentioned at the end of the last section, we view $F_X(x)$ as simply an ordinary function of the real variable x. Since probabilities always lie between 0 and 1, $F_X(x)$ can also be viewed as a mapping of the whole line $(-\infty, \infty)$ to the interval $(0, 1)$. Notice that such a function is defined for any random variable and it will be the study of this and related functions, rather than the original sample space, that will tend to unify the study of probability. For any point x, $F_X(x)$ expresses a probability. If only the one random variable X is under consideration and there is no danger of confusion, we will write $F_X(x)$ simply as $F(x)$. The function $F_X(x)$ is called the cumulative distribution function (which will be abbreviated to cdf) of the random variable X.

There are several important properties of cdf's that are a consequence of the definition:

1. $0 \leq F(x) \leq 1$ for all $x \in (-\infty, \infty)$.   This is a consequence of the fact that for any x, $F(x)$ is a probability.

2. $\lim_{x\to\infty} F(x)=1$ and $\lim_{x\to-\infty} F(x)=0$. These properties of $F(x)$ can be proven using the axioms of probability. We will not prove them, but intuitively

$$\lim_{x\to\infty} F(x)= \lim_{x\to\infty} P(X\le x)\to P(X<\infty)=1$$

and

$$\lim_{x\to-\infty} F(x)= \lim_{x\to-\infty} P(X\le x)\to P(X<-\infty)=0$$

3. $F(x)$ is a nondecreasing function of x, that is, for two particular values of x, x_1 and x_2, $x_1\le x_2$, then $F(x_1)\le F(x_2)$.

To prove this property define two sets in R_X, $\{X\le x_1\}$ and $\{x_1<X\le x_2\}$. It is clear that

$$P(X\le x_2)= P(X\le x_1)+P(x_1<X\le x_2)$$

or

$$F(x_2)= F(x_1)+P(x_1<X\le x_2)$$

and since $P(x_1<X\le x_2)\ge 0$, then $F(x_2)\ge F(x_1)$.

4. $F(x)$ is continuous from the right. To explain this property, choose a particular value of x, say x_0. By the expression

$$\lim_{x\to x_0^+} F(x)=F(x_0)$$

we mean choose a value of x greater than x_0 and decrease it toward x_0 (x approaches x_0 from the positive side or from the right). Then for any x_0 the function $F(x)$ has the property

$$\lim_{x\to x_0^+} F(x)=F(x_0)$$

which we describe by saying that $F(x)$ is continuous from the right. It is not true that $F(x)$ is continuous from the left, that is,

$$\lim_{x\to x_0^-} F(x)\ne F(x_0)$$

for all points x_0. This is a result of the fact that $F(x)=P(X\le x)$. If we had defined $F(x)$ as $P(X<x)$ (strict inequality), this function would have been continuous on the left, but not on the right. A short example will illustrate the point.

EXAMPLE 2.4

Let X be a random variable that takes on the values 0, -1, and 1 with the probabilities $P(X=-1)=1/4$, $P(X=0)=1/2$, and $P(X=1)=1/4$. The cdf of X is shown in Figure 2.1, where $F(-1)=1/4$, $F(0)=3/4$, and $F(1)=1$.

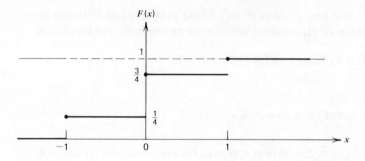

FIGURE 2.1

Since

$$\lim_{x \to 0^-} F(x) \neq F(0) = 3/4$$

and for any $-1 < x < 0$, $F(x) = 1/4$, the function is clearly not continuous from the left. The other properties of cdf's can also be readily verified.

The presence of jumps in the cdf, as shown for example in Figure 2.1, will be a distinguishing feature of random variables, that is, we will classify random variables shortly according to whether or not their cdf's contain such jumps. It can be shown that any cdf has at most a countable number of such jumps.

A cdf can be described as a nondecreasing function that is continuous from the right and with the property that $F(-\infty) = 0$, $F(\infty) = 1$. It turns out that the converse of this statement is also true, that is, any function satisfying these four properties is the cdf of some random variable. Also, any random variable has a unique cdf but it is possible that two different random variables have the same cdf.

Although it is beyond the level of this treatment, it can be shown that there are precisely three different types of distribution functions and any given distribution function $F(x)$ can be expressed as a sum of three such functions:

$$F(x) = F_1(x) + F_2(x) + F_3(x)$$

where

(i) $F_1(x)$ is a function that changes only in jumps (at most a countable number of them) and is constant between jumps;
(ii) $F_2(x)$ is a continuous function;
(iii) $F_3(x)$ is a singular function.

Singular distributions hardly ever arise in physical situations and we will ignore them.

Definition   If the cdf of the random variable X is of the type $F_1(x)$, that is, it changes values only in jumps, then X will be called a *discrete random variable*.

If the cdf of the random variable X is of the type $F_2(x)$, that is, a continuous function, we call X a *continuous random variable*. If the cdf of the random variable X is a sum of the two types, X is called a *mixed random variable*.

Discrete and continuous random variables will be studied separately in the next two sections. Mixed random variables occur only incidentally and are easily understood once the discrete and continuous cases are understood. Notice that it is the cdf which we used to classify the random variables. It is defined for all values of x on the real line and contains all the information of a probabilistic nature on X.

EXAMPLE 2.5

Let X be a continuous random variable with cdf:

$$F(x) = \begin{cases} 1 - e^{-x}, & x \geq 0 \\ 0, & x < 0 \end{cases}$$

What is the probability that X is greater than 10, in terms of $F(x)$?
From the definition of $F(x)$,

$$F(10) = P(X \leq 10)$$

and since

$$P(X \leq 10) + P(X > 10) = 1$$

we have

$$P(X > 10) = 1 - F(10) = 1 - (1 - e^{-10})$$
$$= e^{-10}$$

EXAMPLE 2.6

The random variable X has the cdf

$$F(x) = \begin{cases} \dfrac{x-a}{b-a}, & a \leq x \leq b \\ 0, & x \leq a \\ 1, & x \geq b \end{cases} \tag{2.1}$$

(i) Find the probability that $X \leq (2a+b)/3$. (ii) If $a = -3$ and $b = 4$, find the probability that $|X| \leq 1/2$.
    Considering (i): It is helpful in this case to sketch the function $F(x)$, as shown in Figure 2.2. Notice that $a \leq (2a+b)/3 \leq b$. Since $F(x) = P(X \leq x)$ for any real value of x, we have

$$P(X \leq (2a+b)/3) = F((2a+b)/3)$$

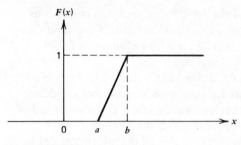

FIGURE 2.2

and to evaluate this expression we use the first line of Eq. (2.1) to give

$$P(X \le (2a+b)/3) = \frac{(2a+b)/3 - a}{b-a} = \frac{1}{3}$$

Considering (ii): The event $\{|X| \le 1/2\}$ can be translated to $\{-1/2 < X < 1/2\}$, and using the cdf we have

$$P(|X| \le 1/2) = P(X \le 1/2) - P(X \le -1/2) = F(1/2) - F(-1/2)$$

It is given that

$$F(x) = \frac{x+3}{7}, \quad -3 \le x \le 4$$

and since both $+1/2$ and $-1/2$ are in this interval,

$$P(|X| \le 1/2) = \frac{1/2+3}{7} - \frac{(-1/2)+3}{7} = \frac{1}{7}$$

2.3 Discrete Random Variables

A discrete random variable X has been defined as one whose cdf increases only in jumps and there are at most a countable number of such jumps. Another equally valid definition would be that X is a discrete random variable only if its range space R_X contains a finite or countably infinite number of points. Suppose the jumps in the cdf of a discrete random variable X occur at the points $x_1, x_2, x_3, \ldots$, where the sequence may be either finite or countably infinite, and we assume $x_i < x_j$ if $i < j$. If $F(x)$ is the cdf of this random variable, then

$$F(x_i) - F(x_{i-1}) = P(X \le x_i) - P(X \le x_{i-1})$$
$$= P(X = x_i)$$

since the x_i have been ordered. The sequence of probabilities $P(X = x_i)$, which will be written more conveniently as $p(x_i)$, $i = 1, 2, 3, \ldots$, is called the *prob-ability distribution* of the random variable X. The set of points $x_1, x_2, x_3, \ldots$ is the range space of X and are the only points on the real line that can have a

34 *Applied Probability*

positive probability of occurring. The jumps in the cdf also occur at these points. The notion of probability distributions of discrete random variables will be used extensively in the next three chapters.

The probability distribution $p(x_i)$, $i = 1, 2, 3, \ldots$, of any discrete random variable will satisfy two simple properties:

(i) $0 \le p(x_i) \le 1$, $i = 1, 2, 3, \ldots$
(ii) $\sum_i p(x_i) = 1$

where the summation is over all allowable values of i. Conversely, any sequence of real numbers $p(x_i)$, $i = 1, 2, 3, \ldots$, satisfying these two properties could qualify as the probability distribution of some discrete random variable.

EXAMPLE 2.7

In a large lot of items $100\theta\%$ are known to be defective. Items are drawn until a defective item is found, and we let X be the number of draws to find a defective item. We assume that X has the probability distribution function

$$P(X = i) = p(i) = (1 - \theta)^{i-1}\theta, \qquad i = 1, 2, 3, \ldots, \infty, \quad 0 < \theta < 1 \qquad (2.2)$$

(i) Show that this is a probability distribution function.
(ii) What is the probability that more than 10 draws will be needed to find a defective item?

Considering (i): Since both θ and $1 - \theta$ are positive and both less than 1, it is clear that $0 \le p(i) \le 1$, satisfying one of the properties we were required to show. It remains to show that

$$\sum_{i=1}^{\infty} p(i) = \sum_{i=1}^{\infty} (1 - \theta)^{i-1}\theta = 1$$

To prove this identity we recall some facts on geometric series. A series of the form $a, ar, ar^2, ar^3, \ldots$ is called a geometric series with first term a and the ratio of two successive terms r. The sum

$$a + ar + ar^2 + ar^3 + \cdots = \frac{a}{1-r} \quad \text{if } |r| < 1 \qquad (2.3)$$

as may easily be verified. To see this let

$$a + ar + ar^2 + \cdots = I$$

Since the sum is infinite, if we multiply by r and add a we have

$$a + r(a + ar + \cdots) = a + rI = a + ar + ar^2 + \cdots = I$$

and we immediately obtain

$$I = \frac{a}{1-r}$$

For $|r| > 1$ the series will not converge to a finite value. Also the finite sum $a + ar + \cdots + ar^n$ is

$$a + ar + \cdots + ar^n = \frac{a(1 - r^{n+1})}{1 - r}, \quad |r| < 1 \qquad (2.4)$$

Summing the distribution of Eq. (2.2) then yields

$$\theta + (1 - \theta)\theta + (1 - \theta)^2\theta + \cdots = \frac{\theta}{1 - (1 - \theta)} = \frac{\theta}{\theta} = 1$$

verifying that it is indeed a probability distribution function.

Considering (ii): The probability that more than 10 draws will be needed to find a defective item is

$$\sum_{i=11}^{\infty} P(X = i) = \sum_{i=11}^{\infty} (1 - \theta)^{i-1}\theta = \frac{\theta(1 - \theta)^{10}}{1 - (1 - \theta)} = (1 - \theta)^{10}$$

In this example we assumed a probability distribution rather than derived it by setting up a sample space and defining a random variable and equivalent events. This will be a common procedure and many important probability distributions will be either derived or postulated in this and subsequent chapters. The distribution of Example 2.7 is called the geometric distribution with parameter θ and will be encountered again.

The cdf $F(x)$ of X is given by

$$F(x) = P(X \le x) = \sum_{x_i \le x} p(x_i)$$

If A is an event of R_X, then the probability that on any repetition of the experiment the outcome $X(s)$ is in A, $P(X \in A)$, is

$$P(X \in A) = \sum_{x_i \in A} p(x_i)$$

If R_X is a finite sample space containing n elements $x_1, x_2, \ldots, x_n$ and each element has the same probability (i.e., an equally likely sample space), then $p(x_i) = 1/n$, $i = 1, 2, \ldots, n$, and if $|A| = k$, then $P(X \in A) = k/n$, as in Chapter 1.

A probability distribution is then a sequence of real numbers x_i, $i = 1, 2, \ldots$ either finite or countably infinite, together with a probability assigned to each of these points. A commonly used analogy is to think of a mass of material of unit weight distributed as point loads on a beam at the points x_i, $i = 1, 2, \ldots$ the fraction of material at x_i corresponding to $p(x_i)$.

In the next example an important distribution, the Poisson distribution, and some of its properties are introduced. This distribution is used frequently in traffic studies of various types, queueing theory, radioactive experiments, and so forth. Chapters 8 and 11 will make extensive use of this distribution.

EXAMPLE 2.8

The number of radioactive particles X observed in a 10 minute interval has a probability distribution function

$$P(X=j)=e^{-\lambda}\frac{\lambda^j}{j!}, \quad j=0, 1, 2, \ldots \tag{2.5}$$

where λ is a positive number that characterizes the distribution.

(i) Show that Eq. (2.5) defines a probability distribution.
(ii) What value or values of X are most likely to occur in a given 10 minute interval?
(iii) What is the probability that X is an even number (0 is even)?

Considering (i): To show that Eq. (2.5) is a probability distribution, we first note that each term is positive since $\lambda>0$. It remains to show that

$$\sum_{i=0}^{\infty} P(X=i)= \sum_{i=0}^{\infty} e^{-\lambda}\frac{\lambda^i}{i!}=1 \tag{2.6}$$

The series expansion about the origin of the exponential function, usually encountered in the first calculus course, is

$$e^x = \sum_{i=0}^{\infty} \frac{x^i}{i!} \tag{2.7}$$

and multiplying both sides of this equation by e^{-x} and replacing x by λ gives the required Eq. (2.6). Equation (2.7) should be remembered for future use.

Considering (ii): To find the positive integer j that maximizes $e^{-\lambda}\lambda^j/j!$, we consider the ratio

$$\frac{P(X=j+1)}{P(X=j)}=\frac{e^{-\lambda}\lambda^{j+1}}{(j+1)!} \cdot \frac{j!}{e^{-\lambda}\lambda^j}=\frac{\lambda}{j+1} \tag{2.8}$$

This ratio is greater than 1 for $\lambda>j+1$ and less than or equal to 1 otherwise. Let j_0 be the smallest value of j for which $\lambda/(j+1)$ is less than or equal to 1. For all smaller values of j the ratio is greater than 1. For values of j greater than or equal to j_0, the ratio of Eq. (2.8) is decreasing, with the possible exception that it could equal 1 for $j=j_0$. If the ratio does equal 1, then $\lambda=j_0+1$ and λ is an integer. If it does not equal 1, then the maximum occurs at j_0, which is also seen to be the largest integer less than λ. If λ is an integer, then $P(X=j_0)=P(X=j_0+1)$ and the distribution has two maxima. This argument should be carefully understood.

Considering (iii): The probability that X is even is given by the sum

$$\sum_{i=0}^{\infty} P(X=2i)= \sum_{i=0}^{\infty} e^{-\lambda}\frac{\lambda^{2i}}{(2i)!}$$

$$= e^{-\lambda}\left(\sum_{i=0}^{\infty} \frac{\lambda^{2i}}{(2i)!}\right) \tag{2.9}$$

To evaluate the summation in parentheses consider the following two expansions:

$$e^{\lambda} = 1 + \lambda + \frac{\lambda^2}{2!} + \frac{\lambda^3}{3!} + \frac{\lambda^4}{4!} + \cdots$$

$$e^{-\lambda} = 1 - \lambda + \frac{\lambda^2}{2} - \frac{\lambda^3}{3!} + \frac{\lambda^4}{4!} - \cdots$$

obtained by using Eq. (2.7), and, summing these gives

$$\sum_{i=0}^{\infty} \frac{\lambda^{2i}}{(2i)!} = \tfrac{1}{2}(e^{\lambda} + e^{-\lambda})$$

Substituting this into Eq. (2.9) gives

$$P(X = \text{even}) = e^{-\lambda} \tfrac{1}{2}(e^{\lambda} + e^{-\lambda}) = \tfrac{1}{2}(1 + e^{-2\lambda})$$

Many other important distributions, other than the geometric and Poisson distributions introduced here, will be encountered in the next few chapters. In particular, the binomial distribution, which is of central importance in probability, will be discussed in the next chapter. The Poisson process, important in many applications, will be introduced in Chapter 11. The distribution

$$P(X = j) = e^{-\lambda} \frac{\lambda^j}{j!}, \quad j = 0, 1, 2, \ldots$$

will be referred to as the Poisson distribution with parameter λ. Tables for its cumulative distribution function, for various values of λ, are contained in Appendix C and from these values, the actual probabilities can be obtained by subtraction.

2.4 *Continuous Random Variables*

Assume for the moment that X is a continuous random variable with cdf $F_X(x)$, which is not only continuous but also has a derivative

$$F'_X(x) = \frac{d}{dx} F_X(x) = f_X(x) \tag{2.10}$$

which is a continuous function. We first observe that the probability that X takes on a particular value, say x_0, must be zero since

$$P(X = x_0) \le P(x_0 \le X \le x_0 + \delta) = F_X(x_0 + \delta) - F_X(x_0)$$

for any $\delta \ge 0$ and, as $F_X(x)$ is continuous, the right-hand side can be made as small as desired by choosing δ small enough.

Since $f_X(x)$ and $F_X(x)$ have been assumed continuous and $F_X(-\infty) = 0$, it follows from elementary calculus that

$$F_X(x) = \int_{-\infty}^{x} f_X(y)\, dy \tag{2.11}$$

For any interval (a, b) we have

$$P(a < X < b) = \int_{-\infty}^{b} f_X(y)\, dy - \int_{-\infty}^{a} f_X(y)\, dy = \int_{a}^{b} f_X(y)\, dy$$

and since any single point has a probability of zero, this expression is the same as $P(a < X \le b)$, $P(a \le X < b)$, and $P(a \le X \le b)$.

In the previous paragraph it was assumed that $f_X(x)$ was continuous for convenience. It is not actually necessary that $f_X(x)$ be continuous everywhere. It is only necessary that the derivative $dF_X(x)/dx$ exist everywhere except at possibly a finite number of points and that this derivative is piecewise continuous. Such conditions will ensure that the function $f_X(x)$ have the following properties:

(i) $f_X(x) \ge 0$.
(ii) $\int_{-\infty}^{\infty} f_X(x)\, dx = 1$.
(iii) $f_X(x)$ is piecewise continuous.
(iv) For any interval (a, b),

$$P(a \le X \le b) = \int_{a}^{b} f_X(x)\, dx$$

The function $f_X(x)$ is called the *probability density function* (pdf) of the random variable X. As with cdf's, when the random variable X is understood, we will drop the subscript on the pdf. The value $f(x)$ of the function f at the point x is *not* a probability. It is only when the function is integrated over some interval that a probability is obtained, hence the name "density." However, the probability that X lies in some small interval $(x, x + \Delta x)$ can be approximated by

$$P(x \le X \le x + \Delta x) = \int_{x}^{x+\Delta x} f(y)\, dy \cong f(x)\, \Delta x$$

for Δx sufficiently small.

Notice that $f(x)$ need not be less than unity for all values of x. It need only be a positive, piecewise continuous function that has unit area and any such function is the pdf of some random variable. Also if x_0 is in the range of X, the fact that $P(X = x_0) = 0$ does not mean that x_0 cannot occur. Recall the example in the first chapter of choosing a point in the interval $(0, 1)$ at random. Each point has a probability of zero of being chosen on any particular trial, but on each trial some point is chosen.

As we did with discrete random variables, we mention the analogy of a pdf with a continuous loading of a beam. A mass of unit weight and uniform thickness is spread on a beam and the weight of mass in some interval is the probability that the random variable lies in that interval.

Before working some examples, we summarize the methods that have been adopted to describe the two different types of random variables. If X is a discrete random variable that can assume the values $x_1, x_2, \ldots$, then its probabilistic behavior is described by a probability distribution function (a term that will never be abbreviated in this book, to avoid confusion) $p(x_i)$, $i = 1, 2, \ldots$, where $p(x_i) = P(X = x_i)$. If X is a continuous random variable, then its probabilistic behavior is described by a probability density function (pdf). These probability distribution functions and pdf's are entirely equivalent to, but in some cases more convenient than, the corresponding cumulative distribution functions. Generally, the form of a pdf is assumed rather than derived from an experimental situation. Virtually all of the probability distribution functions can be derived from a description of some game or experimental situation. The following examples might clarify this point.

EXAMPLE 2.9

The cdf of the continuous random variable X is defined by

$$F(x) = \begin{cases} \dfrac{x}{1+x}, & x \geq 0 \\ 0, & x < 0 \end{cases}$$

Find the pdf of X and verify that it is a pdf. Find the number c such that $P(X \leq c) = 1/2$.

From Eq. (2.10),

$$f(x) = \frac{d}{dx} F(x) = \frac{d}{dx}\left(\frac{x}{1+x}\right) = \frac{1}{(1+x)^2}, \quad x \geq 0$$

$$= 0, \quad x \leq 0$$

and it is required to show that this function is a pdf. The functions $f(x)$ and $F(x)$ are sketched in Figure 2.3. The function $f(x)$ is certainly positive and piecewise

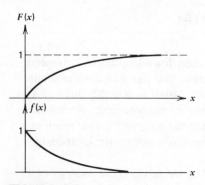

FIGURE 2.3

continuous (it has a single discontinuity at the origin). Since $F(-\infty)=0$ and $F(\infty)=1$, and since from Eq. (2.11)

$$F(\infty)=\int_{-\infty}^{\infty}f(x)\,dx$$

it is also clear that the area under $f(x)$ is unity and it is indeed a pdf. To find the constant c such that $P(X \le c)=1/2$, note that

$$P(X \le c)=F(c)=\frac{c}{1+c}=\frac{1}{2}$$

or

$$2c=1+c$$

which gives $c=1$ as the required value.

Notice in this example that there was no mention of a sample space or experimental situation from which the form of $F(x)$ or $f(x)$ could be derived. These functions were simply postulated and this will frequently be the case for continuous random variables.

EXAMPLE 2.10

The natural log of the height X of a certain kind of tree at maturity is assumed to be a random variable with pdf

$$f(x)=\begin{cases} \dfrac{1}{b-a}, & a \le x \le b \\ 0, & \text{elsewhere} \end{cases} \tag{2.12}$$

(i) Show that $f(x)$ is a pdf. (ii) Find the cdf of X. (iii) If $a=-b$, $b>0$, find the value of b such that $P(X<1)=3P(X>1)$.

Considering (i): The function is easily seen to be positive, piecewise continuous, and has area 1 and is a pdf.

Considering (ii): The cdf of X is

$$F(x)=\int_{-\infty}^{x}f(y)\,dy$$

$$=\begin{cases} 0 & \text{if } x<a \\ \dfrac{x-a}{b-a} & \text{if } a \le x \le b \\ 1 & \text{if } x>b \end{cases} \tag{2.13}$$

This is precisely the same cdf considered in Example 2.6 and shown in Figure 2.2. The pdf and cdf are shown in Figure 2.4.

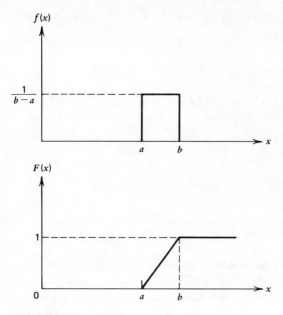

FIGURE 2.4

Considering (iii): The pdf is equal to $1/2b$ for $-b \le x \le b$ and 0 elsewhere. To determine the value of the variable b such that

$$P(X < 1) = 3P(X > 1)$$

we first note that b must be greater than 1, otherwise the equality is impossible. The left-hand side of this equation is simply $F(1) = (1+b)/2b$ while the right-hand side is $3(1 - F(1)) = 3(1 - (1+b)/2b)$. Setting these expressions equal gives

$$\frac{1+b}{2b} = 3\left(1 - \left(\frac{1+b}{2b}\right)\right).$$

or $b = 2$.

A random variable with the pdf of Eq. (2.12) (or, equivalently, the cdf of Eq. (2.13)) is called a *uniformly distributed random variable* since its pdf is uniformly distributed over the interval (a, b). It is the first important type of continuous random variable. It is often used where we have no prior knowledge of the actual pdf.

EXAMPLE 2.11

The lifetime of transistors obtained from a certain manufacturing process is a random variable X, measured in hours, which is assumed to have the pdf

$$f(x) = \begin{cases} \alpha e^{-\alpha x}, & x \geq 0 \\ 0, & x < 0 \end{cases} \qquad (2.14)$$

where α is a parameter that is some positive constant. Show that this function is a pdf and find its cdf.

It is only necessary to show that $f(x)$ has unit area since it clearly satisfies the other properties required of a pdf:

$$\int_{-\infty}^{\infty} f(x)\,dx = \int_{0}^{\infty} \alpha e^{-\alpha x}\,dx = -e^{-\alpha x}\Big|_{0}^{\infty} = 1$$

The cdf of this function is

$$F(x) = \int_{-\infty}^{x} f(y)\,dy = \begin{cases} 0, & x < 0 \\ \int_{0}^{x} \alpha e^{-\alpha y}\,dy, & x \geq 0 \end{cases}$$

or

$$F(x) = \begin{cases} 1 - e^{-\alpha x}, & x \geq 0 \\ 0, & x < 0 \end{cases}$$

The functions $f(x)$ and $F(x)$ are sketched in Figure 2.5. It should be stressed that the value of $F(x_0)$ is simply the area of $f(x)$ to the left of x_0 for any x_0 on the real line and Figure 2.5 attempts to illustrate this.

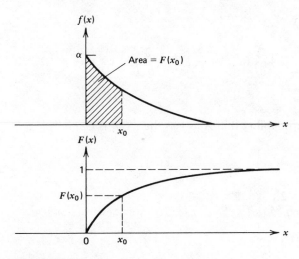

FIGURE 2.5

The random variable X of Example 2.11 is said to be exponentially distributed with parameter α, or is simply called an exponential random variable. It is the second important type of random variable encountered.

Another type of random variable has a two-sided exponential pdf:

$$f(x) = \tfrac{1}{2}\alpha\, e^{-\alpha|x|}, \quad -\infty < x < \infty \tag{2.15}$$

and it is a useful exercise to find its cdf. The term "exponentially" distributed random variable with parameter α will always be taken to mean one with a pdf given by Eq. (2.14).

EXAMPLE 2.12

The radiated energy from a particular atomic reaction is assumed to be a continuous random variable with pdf

$$f(x) = ax^2 e^{-bx}, \quad a, b > 0, \quad 0 \le x < \infty$$

(i) For a given value of b, what must a be? (ii) Find the cdf of X. (iii) What is the probability that X is less than $1/b$?

Considering (i): In order for $f(x)$ to be a pdf,

$$\int_0^\infty ax^2 e^{-bx}\, dx = a\left(\frac{2}{b^3}\right) = 1$$

which implies $a = b^3/2$.

Considering (ii): The cdf of X is given by

$$F(x) = \begin{cases} b^3 \displaystyle\int_0^x y^2 e^{-by}\, dy \\ \qquad = \dfrac{b^3}{2}\left(\dfrac{x^2}{b}\, e^{-bx} + \dfrac{2}{b^2}x\, e^{-bx} + \dfrac{2}{b^3}\, e^{-bx} + \dfrac{2}{b^3}\right), \quad x \ge 0 \\ 0, \quad x < 0 \end{cases}$$

Considering (iii): The probability that X is less than $1/b$ is given by

$$P\left(X < \frac{1}{b}\right) = F\left(\frac{1}{b}\right)$$

$$= \frac{b^3}{2}\left(\frac{1}{b^3}\, e^{-1} + \frac{2}{b^3}\, e^{-1} + \frac{2}{b^3}\, e^{-1} + \frac{2}{b^3}\right)$$

$$= \tfrac{1}{2}(5\, e^{-1} + 2)$$

Probably the most important type of continuous random variable is the normal random variable. Since the next two chapters are concerned more with discrete random variables than continuous, we defer discussion of this important type until Chapter 5.

2.5 Expectations, Means, and Variances

Historically, the term expectation derived from games of chance where gamblers were concerned with how much one might expect to win, in the long run, from a particular game. Since then it has become a very important notion central to much of probability theory and its applications. Typically a mean or average is calculated as the sum of the outcomes divided by the number of outcomes. In a gambling situation if a certain outcome, for which the winnings are $x_i \in R_X$, occurs a fraction $p(x_i)$ of the time, then the expected winnings of the gambler are

$$\sum_{i=1}^{n} p(x_i)x_i, \qquad R_X = \{x_1, x_2, \ldots, x_n\}$$

In this section the notion of the expectation of a random variable and of functions of a random variable are introduced and properties of the random variable analyzed in terms of these concepts. It is to be emphasized, however, that the probability distribution and density functions are complete characterizations of the probabilistic behavior of the random variable while means and variances are relatively weak, although nonetheless useful, characterizations. It will be seen that they give an indication of location and width or dispersion of the distributions.

Definition The expected value of the random variable X, which we denote by $E(X)$, is given by

$$E(X) = \sum_i x_i p(x_i) \tag{2.16}$$

if X is a discrete random variable and

$$E(X) = \int_{-\infty}^{\infty} xf(x)\,dx \tag{2.17}$$

if X is a continuous random variable, provided that the sum and integral, respectively, converge absolutely, that is,

$$\sum_i |x_i| p(x_i) < \infty \quad \text{and} \quad \int_{-\infty}^{\infty} |x| f(x)\,dx < \infty$$

If the sum, or integral, does not converge absolutely, then the expected value does not exist.

The expected value $E(X)$ of X is simply a constant that indicates what the statistical average of a large number of observations of X would be. It gives some indication as to the location of the probability distribution or density, although $E(X)$ need not even be in the range space of X. For example, if X is

Random Variables and Their Distributions 45

the outcome of rolling a dice, then

$$E(X) = \sum_{i=1}^{6} i \cdot \tfrac{1}{6} = 3.5$$

In continuing the analogy between distributions and densities with beam loadings, the expected value would be the first moment of inertia or the center of mass of the beam. However we interpret $E(X)$, it is a simple characterization of the distribution or density function, giving some information on X. It is somewhat standard to denote the expected value $E(X)$ by μ_x or simply μ if X is the only random variable under consideration. The expected value is also called the mean and we use these terms interchangeably.

As a single number the expected value of X, $E(X)$, cannot characterize a pdf very completely. For example, the two pdf's of Figure 2.6 would have the same mean μ but quite different appearances. Nonetheless, the expected value is a useful parameter to know for discussions of random variables.

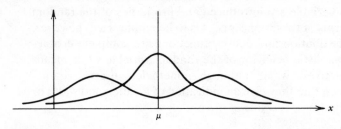

FIGURE 2.6

There are other simple characterizations that we will use less often. The *median* of a random variable X is the number c such that $P(X \geq c) = P(X \leq c)$. The *mode* of the random variable X is the number m for which the probability distribution or pdf achieves its maximum value. The following example is a mechanical exercise in finding the means of the distributions and densities encountered so far.

EXAMPLE 2.13

Find the expected value of the random variable X if:

(i) X is a discrete random variable and
    (a) has a geometric distribution with parameter θ (Eq. (2.2)), and
    (b) has a Poisson distribution with parameter λ (Eq. (2.5)).
(ii) X is a continuous random variable and
    (a) is uniformly distributed over the interval (a, b) (Eq. (2.12)), and
    (b) has an exponential pdf with parameter α (Eq. (2.14)).

Considering part (a) of (i): The random variable X has the probability distribution

$$P(X = j) = (1 - \theta)^{j-1}\theta, \quad j = 1, 2, 3, \ldots, \infty$$

and its expected value is given by

$$E(X) = \sum_{j=1}^{\infty} jP(X=j) = \theta \sum_{j=1}^{\infty} j(1-\theta)^{j-1} \qquad (2.18)$$

To evaluate this expression note that if

$$S = \sum_{j=1}^{\infty} (1-\theta)^j = \frac{1-\theta}{1-(1-\theta)}, \quad 0 < \theta < 1$$

then

$$\frac{dS}{d\eta} = \sum_{j=1}^{\infty} j(1-\theta)^{j-1} = \frac{d}{d\eta}\left(\frac{1-\theta}{1-(1-\theta)}\right) = \left(\frac{1}{1-(1-\theta)}\right)^2, \quad \eta = 1-\theta$$

and using this fact in Eq. (2.18) gives the result

$$E(X) = \theta \cdot \left(\frac{1}{1-(1-\theta)}\right)^2 = \frac{1}{\theta}$$

(b) If X is a Poisson random variable with parameter λ, then, by Eq. (2.5),

$$P(X=j) = e^{-\lambda}\frac{\lambda^j}{j!}, \quad j = 0, 1, 2, \ldots$$

and

$$E(X) = \sum_{j=0}^{\infty} j \cdot e^{-\lambda}\frac{\lambda^j}{j!} = e^{-\lambda} \sum_{j=1}^{\infty} \frac{\lambda^j}{(j-1)!}$$

since the $j=0$ term is zero. Changing the variable of summation from j to $k = j-1$ gives

$$E(X) = e^{-\lambda} \sum_{k=0}^{\infty} \frac{\lambda^{k+1}}{k!} = \lambda e^{-\lambda} \sum_{k=0}^{\infty} \frac{\lambda^k}{k!} = \lambda e^{-\lambda} e^{\lambda} = \lambda$$

where use has been made of Eq. (2.7).

Considering part (a) of (ii): The continuous random variable X has the pdf

$$f(x) = \begin{cases} \dfrac{1}{b-a}, & a \le x \le b \\ 0, & \text{elsewhere} \end{cases}$$

and

$$E(X) = \int_a^b x \cdot \frac{1}{b-a} \, dx$$

$$= \frac{1}{b-a}\left.\frac{x^2}{2}\right|_a^b$$

$$= \frac{1}{b-a} \cdot \frac{(b^2-a^2)}{2} = \frac{a+b}{2}$$

(b) The pdf of X is given by Eq. (2.14) as

$$f(x) = \begin{cases} \alpha e^{-\alpha x}, & x \geq 0 \\ 0, & x < 0 \end{cases}$$

and so

$$E(X) = \int_0^\infty x\alpha e^{-\alpha x}\, dx = -x e^{-\alpha x}\Big|_0^\infty + \int_0^\infty e^{-\alpha x}\, dx$$

$$= 1/\alpha$$

The techniques employed in this example will be applicable in many other situations and should be well understood.

It has been indicated that the expected value of a random variable, in some approximate sense, gives an idea of the location of the distribution or density function. It would be additionally helpful to have some indication of how widely spread out or dispersed the distribution is. The variance of a distribution or density is a number that does just this, but before we introduce it we will have to discuss functions of random variables.

The general situation we will be interested in is, given that the probability distribution or pdf of the random variable X is known, we would like to calculate the expected value $E(H(X))$ of some function of X, $H(X)$. An immediate question is how do such questions arise in practice? Consider a ball bearing manufacturer who is producing bearings from an exotic and expensive alloy. He may be interested in the amount of material being used per bearing. To obtain this he could measure the diameter of radius R of the bearing and set $V = \frac{4}{3}\pi R^3$. If he wanted to know $E(V)$, the average volume of material in the ball bearing, he could obtain this knowing the pdf of the radius R. A land surveyor measuring two sides of a large field might be interested in how errors in the side measurements affect the area enclosed. If the kinetic energy E of molecular particles is of interest, it may be experimentally expedient to measure velocity V and apply the transformation $E = \frac{1}{2}mV^2$, where m is the mass of a particle. Such transformations occur frequently in practice and will be the subject of Chapter 8.

If the probability distribution or pdf of $Y = H(X)$ were known rather than just the probability distribution or pdf of X, we could write

$$E(Y) = \sum y_i p_Y(y_i)$$

or

$$E(Y) = \int_{-\infty}^\infty y f_Y(y)\, dy$$

depending on whether Y is a discrete or continuous random variable. We claim, however, that it is not necessary to know the distribution or pdf of Y to find $E(Y)$; it suffices to know the distribution or pdf of X and that

$$E(Y) = E(H(X)) = \sum_i H(x_i) p_X(x_i) \tag{2.19}$$

if X is discrete or

$$E(Y) = E(H(X)) = \int_{-\infty}^{\infty} H(x) f_X(x)\, dx \qquad (2.20)$$

if X is continuous. The subscripts X and Y on the distributions and pdf's indicate which variable they refer to. We will only demonstrate the truth of Eq. (2.19) here. Equation (2.20) will follow for certain types of functions $H(X)$ from results in Chapter 7.

Assume that X is a discrete random variable with probability distribution $p(x_i)$, $i = 1, 2, \ldots$ and an expression for $E(H(X))$ is required. If $Y = H(X)$, suppose Y can take on the values $y_1, y_2, \ldots$ and let I_j be the set of subscripts such that if $k \in I_j$, then $H(x_k) = y_j$. A simple example of this is where X can take on the values $x_1 = -2$, $x_2 = -1$, $x_3 = 0$, $x_4 = 1$, $x_5 = 2$, and $Y = H(X) = X^2$. In this case Y can take on the values $y_1 = 0$, $y_2 = 1$, and $y_3 = 4$, and since $H(x_1) = H(x_5) = 4 = y_3$, $H(x_2) = H(x_4) = 1 = y_2$, and $H(x_3) = 0 = y_1$, these sets of subscripts are $I_1 = \{3\}$, $I_2 = \{2, 4\}$, and $I_3 = \{1, 5\}$. The situation is

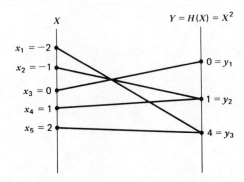

FIGURE 2.7

also shown in Figure 2.7. For this situation we can calculate $E(Y) = E(X^2)$ two ways. Using Eq. (2.19),

$$E(X^2) = \sum_i x_i^2 p_X(x_i)$$

$$= (-2)^2 \cdot p_X(-2) + (-1)^2 p_X(-1) + (1)^2 p_X(1) + (2)^2 p_X(1)$$

while calculating $E(Y)$ directly using the fact that $R_Y = (0, 1, 4)$ and $p_Y(0) = p_X(0)$, $p_Y(1) = p_X(-1) + p_X(1)$, $p_Y(4) = p_X(-2) + p_X(2)$,

$$E(Y) = 1 p_Y(1) + 4 p_Y(4)$$

$$= 1(p_X(-1) + p_X(1)) + 4(p_X(-2) + p_X(2))$$

$$= E(X^2)$$

Random Variables and Their Distributions 49

To show Eq. (2.19) in general, notice that the probability distribution of Y, $p_Y(y_j)$, $j = 1, 2, \ldots$ is determined as

$$p_Y(y_j) = \sum_{i \in I_j} p_X(x_i), \quad j = 1, 2, \ldots \tag{2.21}$$

The sum in Eq. (2.19) can be split according to the sets I_j as

$$\sum H(x_i)p_X(x_i) = \sum_j \left(\sum_{i \in I_j} H(x_i)p_X(x_i) \right)$$

But for each $i \in I_j$, $H(x_i) = y_j$, by definition of I_j, and thus

$$\sum_j y_j \left(\sum_{i \in I_j} p_X(x_i) \right) = \sum_j y_j p_Y(y_j) = E(Y)$$

where the first equality follows by using Eq. (2.21). The proof of Eq. (2.20) is more complicated and further discussion of it will be deferred to Chapter 7. We will however use both relationships.

Notice that if $H(X)$ and $G(X)$ are two functions of X, then

$$E(H(X) + G(X)) = \int (H(X) + G(X))f_X(x)\, dx$$

$$= \int H(x)f_X(x)\, dx + \int G(x)f_X(x)\, dx$$

$$= E(H(X)) + E(G(X))$$

and if C is a constant, then

$$E(CH(X)) = \int CH(x)f_X(x)\, dx$$

$$= C \int H(x)f_X(x)\, dx$$

$$= CE(H(X))$$

and these are two useful properties of the expectation operator to keep in mind.

The variance of a random variable or its probability distribution function or density function gives a simple measure of the spread or dispersion of the function.

Definition   The *variance* of the random variable X is

$$V(X) = E((X - E(X))^2)$$

If we let $H(X) = (X - E(X))^2 = (X - \mu)^2$, $\mu = E(X)$, then if X is discrete,

$$V(X) = \sum (x_i - \mu)^2 P_X(x_i)$$

while if X is continuous,

$$V(X) = \int_{-\infty}^{\infty} (x - \mu)^2 f_X(x)\, dx$$

The positive square root of $V(X)$ is defined to be the *standard deviation* of X and is denoted by σ_X or simply σ.

Notice that $V(X)$ is always a positive number. It is a weighted average of $(x - \mu)^2$ and this tends to weight large values of $(x - \mu)$ heavily; the more likely they are, the more they are weighted. In terms of our previous loaded beam analogy, it is the second central moment of inertia. In this sense, $V(X)$ gives a measure of dispersion of the distribution or density functions. To emphasize this point, two densities are sketched in Figure 2.8. Each has the same mean μ but $f_2(x)$ has the smaller variance. Of course each curve must have unit area.

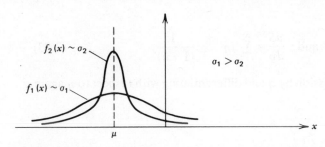

FIGURE 2.8

The expectation $E(X^k)$ of the random variable X is called the kth moment of X and $E((X - \mu)^k)$ the kth central moment, where $\mu = E(X)$. With this notation the variance is the second central moment of the distribution.

The next example is a continuation of Example 2.13 and determines the variance of the important distributions and densities encountered so far. It also is largely a mechanical exercise. Notice that since μ is a constant, if X is a continuous random variable, then

$$E((X - \mu)^2) = E(X^2 - 2\mu X + \mu^2)$$

$$= \int_{-\infty}^{\infty} (x^2 - 2\mu x + \mu^2) f(x)\, dx$$

$$= \int_{-\infty}^{\infty} x^2 f(x)\, dx - 2\mu \int_{-\infty}^{\infty} x f(x)\, dx + \mu^2 \int_{-\infty}^{\infty} f(x)\, dx$$

The first integral is $E(X^2)$, the second integral is $E(X)$, and the last one is unity, which gives

$$V(X) = E((X - \mu)^2)$$

$$= E(X^2) - 2\mu^2 + \mu^2$$

$$= E(X^2) - \mu^2$$

$$= E(X^2) - (E(X))^2 \tag{2.22}$$

which is a useful form for determining the variance.

EXAMPLE 2.14

Determine the variances for the distribution and density functions considered in Example 2.13.

Since $E(X)$ has already been determined for each of the distribution and density functions, it remains to find $E(X^2)$ for each of them.

(i) (a) $E(X^2) = \sum_{j=1}^{\infty} j^2 \eta^{j-1} \theta, \quad \eta = 1 - \theta$

To evaluate this expression we use a similar technique employed in finding the mean of X. Let

$$S = \sum_{j=1}^{\infty} \eta^j = \frac{\eta}{1-\eta} \quad \text{and} \quad \frac{dS}{d\eta} = \sum_{j=1}^{\infty} j\eta^{j-1} = \frac{1}{(1-\eta)^2} \qquad (2.23)$$

Multiplying this expression by η and differentiating with respect to η again gives

$$\frac{d}{d\eta}\left(\eta \frac{dS}{d\eta}\right) = \frac{d}{d\eta}\left(\sum_{j=1}^{\infty} j\eta^j\right)$$

$$= \sum_{j=1}^{\infty} j^2 \eta^{j-1}$$

$$= \frac{d}{d\eta}\left(\frac{\eta}{(1-\eta)^2}\right)$$

$$= \frac{1+\eta}{(1-\eta)^3}$$

Using this expression in $E(X^2)$,

$$E(X^2) = \left(\frac{1+\eta}{(1-\eta)^3}\right)\theta = \frac{1+\eta}{(1-\eta)^2}, \quad 1 - \eta = \theta$$

and finally

$$V(X) = E(X^2) - (E(X))^2$$

$$= \frac{1+\eta}{(1-\eta)^2} - \frac{1}{(1-\eta)^2}$$

$$= \frac{\eta}{(1-\eta)^2} = \frac{(1-\theta)}{\theta^2} \qquad (2.24)$$

(b) For X a Poisson random variable with parameter λ,

$$E(X^2) = \sum_{j=0}^{\infty} j^2 e^{-\lambda} \frac{\lambda^j}{j!} = e^{-\lambda} \sum_{j=1}^{\infty} j \frac{\lambda^j}{(j-1)!}$$

and, setting $k = j - 1$, as in Example 2.13,

$$E(X^2) = e^{-\lambda} \lambda \sum_{k=0}^{\infty} (k+1) \frac{\lambda^k}{k!}$$

$$= \lambda e^{-\lambda} \sum_{k=0}^{\infty} k \frac{\lambda^k}{k!} + \lambda e^{-\lambda} \sum_{k=0}^{\infty} \frac{\lambda^k}{k!}$$

The first summation on the right-hand side is λe^{λ} and the second summation is e^{λ}, both of which follow from Example 2.13. Thus

$$E(X^2) = \lambda^2 + \lambda$$

and

$$V(X) = E(X^2) - (E(X))^2 = \lambda^2 + \lambda - (\lambda)^2 = \lambda$$

and we have shown that the Poisson distribution has the remarkable property that its mean is equal to its variance.

(ii) (a) If X is uniformly distributed over (a, b), then

$$E(X^2) = \int_a^b x^2 \frac{1}{(b-a)} \, dx = \frac{1}{b-a} \left. \frac{x^3}{3} \right|_a^b = \frac{1}{(b-a)} \frac{1}{3}(b^3 - a^3)$$

and

$$V(X) = \frac{b^3 - a^3}{3(b-a)} - \left(\frac{a+b}{2} \right)^2$$

$$= \frac{4(b^3 - a^3) - 3(a+b)^2(b-a)}{12(b-a)}$$

$$= \frac{4(b^2 + ab + a^2) - 3(a+b)^2}{12} = \frac{(b-a)^2}{12}$$

(b) For X exponentially distributed with parameter α,

$$E(X^2) = \int_0^{\infty} x^2 \alpha e^{-\alpha x} \, dx$$

and, by successive differentiation by parts, this can be shown to be

$$E(X^2) = \frac{2}{\alpha^2}$$

and, consequently,

$$V(X) = \frac{2}{\alpha^2} - \left(\frac{1}{\alpha} \right)^2 = \frac{1}{\alpha^2}$$

There are several properties of means and variances that will be useful. We will prove them only for the case where X is a continuous random variable with pdf $f(x)$. The proofs are similar when X is a discrete random variable and the results are identical.

1. For any two constants a, b, $E(aX + b) = aE(X) + b$.

Proof From Eq. (2.19), letting $H(X) = aX + b$, it follows that

$$E(aX + b) = \int_{-\infty}^{\infty} (ax + b) f(x)\, dx$$

$$= a \int_{-\infty}^{\infty} x f(x)\, dx + b \int_{-\infty}^{\infty} f(x)\, dx$$

$$= aE(X) + b$$

providing $E(X)$ exists.

2. For any two constants a, b, $V(aX + b) = a^2 V(X)$.

Proof

$$V(aX + b) = E((aX + b - E(aX + b))^2)$$

$$= E((aX + b)^2) - (E(aX + b))^2$$

$$= a^2 E(X^2) + 2abE(X) + b^2 - a^2(E(X))^2 - 2ab(E(X))^2 - b^2$$

$$= a^2(E(X^2) - (E(X))^2) = a^2 V(X)$$

From Property 1, if $a = 1$, $E(X + b) = E(X) + b$, and the mean of a variable translated by b is its mean translated by b. It is also not difficult to see that if $f(x)$ is the pdf of X and $Y = X + b$, then the pdf of Y is $f(y - b)$. On the other hand, $V(X) = V(X + b)$, which is intuitively satisfying since translating a pdf does not change its shape, only its location, and hence its variance should be unchanged. We conclude this section with some relatively straightforward applications of this material.

EXAMPLE 2.15

The number of customers that enter a store in a one-hour interval is a Poisson random variable X for which it has been observed that $P(X = 0) = .00012$. Determine the mean and variance of X.

Since X is assumed to have a Poisson distribution function

$$P(X = 0) = e^{-\lambda} = .00012$$

or $\lambda = -\ln(.00012) \approx 9$ and it has already been noted that this parameter of the Poisson distribution is equal to both its mean and variance.

EXAMPLE 2.16

A small telephone exchange is capable of handling 10 calls. It has been observed that the switchboard gets jammed about 1% of the time. If it is assumed that the number of callers requesting a line at any one time, X, is a Poisson random variable, find its distribution function.

From the information given, the probability of the exchange receiving requests for more than 10 lines at any one instant is .01, and so

$$\sum_{i=11}^{\infty} e^{-\lambda} \frac{\lambda^i}{i!} = .01 \quad \text{or} \quad \sum_{i=0}^{10} e^{-\lambda} \frac{\lambda^i}{i!} = .99$$

From the table in Appendix C, this implies that $\lambda \approx 4.7$, using linear interpolation, and so the probability distribution function of X is

$$P(X = i) = e^{-4.7} \frac{(4.7)^i}{i!}, \quad i = 0, 1, 2, \ldots$$

EXAMPLE 2.17

The pdf of a continuous random variable X is

$$f(x) = \begin{cases} |x|, & |x| \leq 1 \\ 0, & \text{otherwise} \end{cases}$$

Find the mean, variance, and cdf of X.

The mean of X is

$$E(X) = \int_{-1}^{1} x|x| \, dx$$

which is immediately seen to be zero since x is an odd function about the origin, $|x|$ is an even function about the origin, and the product of an even and an odd function is an odd function. The integral of an odd function over a symmetric interval about the origin is always zero. The variance of X is

$$
\begin{aligned}
V(X) &= E(X^2) - E^2(X) \\
&= E(X^2) \\
&= \int_{-1}^{1} x^2|x| \, dx \\
&= \int_{0}^{1} x^3 \, dx + \int_{-1}^{0} x^2(-x) \, dx \\
&= \left. \frac{x^4}{4} \right|_0^1 - \left. \frac{x^4}{4} \right|_{-1}^0 \\
&= \left(\frac{1}{4} - 0 \right) - \left(0 - \frac{1}{4} \right) = \frac{1}{2}
\end{aligned}
$$

The cdf of X is

$$F(x) = \int_{-\infty}^{x} f(s)\, ds$$

$$= 0 \quad \text{if } x < -1$$

$$= \int_{-1}^{x} (-s)\, ds = -\frac{s^2}{2}\Big|_{-1}^{x} = \frac{1}{2} - \frac{x^2}{2}, \quad -1 < x < 0$$

$$= \frac{1}{2} + \int_{0}^{x} s\, ds = \frac{1}{2} + \frac{s^2}{2}\Big|_{0}^{x} = \frac{1}{2} + \frac{x^2}{2}, \quad 0 < x < 1$$

$$= 1, \quad x > 1$$

and is sketched in Figure 2.9.

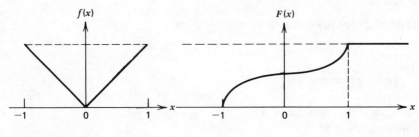

FIGURE 2.9

2.6 The Moment Generating Function

In the previous section some properties of random variables in terms of moments and central moments were examined. It happens that under certain circumstances, all such moments can be derived in a straightforward manner from the moment generating function.

Definition The moment generating function (mgf) of a random variable X, $m_X(t)$, is defined as the expected value of the function e^{tX}, where t is a real variable, that is,

$$m_X(t) = E(e^{tX}) = \begin{cases} \sum_{x_i \in R_x} p(x_i)\, e^{tx_i}, & X \text{ discrete} \\ \int_{R_x} f(x)\, e^{tx}\, dx, & X \text{ continuous} \end{cases} \tag{2.25}$$

If the moment generating function exists for all values of t in some small interval $(-\delta, \delta)$, $\delta > 0$ about the origin, then it can be shown that the probability distribution or density function can be obtained from it. If the mgf's of two random variables exist and are equal, then they must have the same probability distribution or density functions. In this sense the moment generating function is equivalent to the distribution or density function.

Suppose that the mgf of X exists (does not take on infinite values) in a region of t about the origin. In this region we expand the exponential function in the series

$$e^{tx} = 1 + tx + \frac{1}{2!}t^2x^2 + \frac{1}{3!}t^3x^3 + \cdots$$

and assuming X is continuous, the mgf can be written

$$m_X(t) = E(e^{tX}) = \int_{R_x} \left(\sum_{i=0}^{\infty} \frac{(tx)^i}{i!} \right) f(x)\, dx$$

and, assuming the order of summation and integration can be interchanged,

$$m_X(t) = \sum_{i=0}^{\infty} \frac{t^i}{i!} \int_{R_x} x^i f(x)\, dx$$

$$= \sum_{i=0}^{\infty} \frac{t^i}{i!} E(X^i)$$

whence the name moment generating function. The first derivative of $m_X(t)$ can be written

$$m_X'(t) = E(X) + tE(X^2) + \frac{t^2}{2!}E(X^3) + \cdots$$

and, setting t to zero in this expression yields

$$m_X'(t)|_{t=0} = m_X'(0) = E(X)$$

Similarly, the second derivative can be written

$$m_X^{(2)}(t) = E(X^2) + tE(X^3) + \cdots$$

and setting t to zero in this expression gives

$$m_X^{(2)}(t)|_{t=0} = m_X^{(2)}(0) = E(X^2)$$

In general,

$$m_X^{(k)}(t) = \sum_{i=k}^{\infty} \frac{t^{i-k}}{(i-k)!} E(X^i)$$

and

$$E(X^k) = m_X^{(k)}(0), \quad k = 1, 2, \ldots \tag{2.26}$$

The relationship (2.26) can be very useful in determining the moments of a random variable since, once its mgf is known, it is simply a matter of differentiating and, for most functions of interest, this is easier than the usual method of involving summation or integration.

A random variable may not always possess an mgf. For example, a random variable X with pdf

$$f(x) = \begin{cases} 1/x^2, & x \geq 1 \\ 0, & x < 1 \end{cases}$$

satisfies all the requirements of a pdf, but all of its moments $E(X^k)$, $k \geq 1$, are infinite and it does not have an mgf. In general, the mgf will exist only for those values of t for which the sum or integral converges absolutely. For most of the cases of importance to us, the mgf will exist and, although we will not make extensive use of it, it has some very interesting and useful properties. For example, suppose X is a random variable with mgf $m_X(t)$ and we define a new random variable $Y = aX + b$. The mgf of Y is then given by the expression

$$m_Y(t) = E(e^{tY}) = E(e^{t(aX+b)}) = e^{tb}E(e^{(at)X})$$
$$= e^{tb}m_X(at)$$

With this observation the kth central moment $E((X - \mu)^k)$ can be obtained by letting $a = 1$ and $b = -\mu$, that is, if $Y = X - \mu$, then

$$E((X - \mu)^k) = m_Y^{(k)}(0) = \frac{d^k}{dt^k}(e^{-\mu t}m_X(t))\bigg|_{t=0}$$

Further properties of the mgf will be mentioned in later chapters.

We should mention that a function closely related to the moment generating function is the *characteristic function* defined for a continuous random variable by

$$C_X(t) = E(e^{itX}) = \int_{-\infty}^{\infty} e^{itx}f(x)\,dx$$

where $i = \sqrt{-1}$ and t is a real variable. (The characteristic function is actually the Fourier transform of the pdf for those familiar with the term.) As

$$|C_X(t)| = \left|\int_{-\infty}^{\infty} e^{itx}f(x)\,dx\right|$$
$$\leq \int_{-\infty}^{\infty} |e^{itx}|f(x)\,dx = \int_{-\infty}^{\infty} f(x)\,dx = 1$$

and since $|e^{itx}| = 1$ for all values of t, it follows that $C_X(t)$ exists for all values of t. This fact makes it a more attractive function, mathematically, to use, and many books on probability theory use the characteristic function rather than the moment generating function. However, use of the characteristic function requires some knowledge of functions of a complex variable and we preferred not to assume such a prerequisite here.

EXAMPLE 2.18

Determine the moment generating functions for the random variables of Example 2.13.

(i) (a) If X has a geometric distribution with parameter θ, then

$$E(e^{tX}) = \sum_{j=1}^{\infty} e^{jt}\theta\eta^{j-1}$$

$$= \theta e^t \sum_{j=1}^{\infty} (\eta e^t)^{j-1}, \quad \eta = 1 - \theta$$

and using the geometric series summation of Eq. (2.3) gives

$$E(e^{tX}) = \frac{\theta e^t}{1 - \eta e^t} = \frac{\theta e^t}{1 - (1-\theta) e^t} \tag{2.29}$$

(b) Let X be a Poisson random variable with parameter λ:

$$E(e^{tX}) = \sum_{j=0}^{\infty} e^{jt} e^{-\lambda} \frac{\lambda^j}{j!}$$

$$= e^{-\lambda} \sum_{j=0}^{\infty} \frac{(\lambda e^t)^j}{j!}$$

$$= e^{-\lambda} e^{\lambda e^t} = e^{\lambda(e^t - 1)} \tag{2.30}$$

(ii) (a) The random variable X is uniformly distributed over (a, b) and

$$E(e^{tX}) = \int_a^b \left(\frac{1}{b-a}\right) e^{tx} dx$$

$$= \frac{1}{(b-a)} \frac{1}{t} e^{tx} \Big|_a^b$$

$$= \frac{e^{tb} - e^{ta}}{t(b-a)} \tag{2.31}$$

(b) The random variable X is exponentially distributed with parameter α and

$$E(e^{tX}) = \int_0^{\infty} \alpha e^{-\alpha x} \cdot e^{tx} dx$$

$$= \frac{\alpha}{(t-\alpha)} e^{(t-\alpha)x} \Big|_0^{\infty}$$

$$= \left(\frac{\alpha}{\alpha - t}\right)$$

An interesting exercise is to verify the means and variances of these distributions and densities obtained in Examples 2.13 and 2.14 by using the moment generating functions of Example 2.18 and Eq. (2.26).

Problems

1. Verify that the following functions are cdf's and find the corresponding pdf's:

(i)

$$F(x) = \begin{cases} 0, & x < 0 \\ x^2, & 0 \le x < \frac{1}{2} \\ 1 - \frac{3}{2}(1-x), & \frac{1}{2} \le x \le 1 \\ 1, & x > 1 \end{cases}$$

(ii)

$$F(x) = \begin{cases} 0, & x < 0 \\ \frac{1}{2}, & 0 \le x < \frac{1}{2} \\ x, & \frac{1}{2} \le x \le 1 \\ 1, & x > 1 \end{cases}$$

2. Show that the following functions are pdf's and find the corresponding cdf's:

(i)

$$f(x) = \begin{cases} 2(1-x), & 0 < x < 1 \\ 0, & \text{elsewhere} \end{cases}$$

(ii)

$$f(x) = \begin{cases} 10 e^{-10x}, & x \ge 0 \\ 0, & x < 0 \end{cases}$$

3. Determine appropriate constants so that the following are discrete probability distribution functions:

(i)

$$P(X = i) = \begin{cases} A \cdot i, & i = +1, +2, \ldots, +n \\ 0, & \text{elsewhere} \end{cases}$$

(ii)

$$P(X = i) = \begin{cases} A/2^i, & i = 1, 2, \ldots \\ 0, & \text{elsewhere} \end{cases}$$

(iii)

$$P(X = i) = \begin{cases} A/3^i, & i = 1, 3, 5, 7, \ldots \\ A/4^i, & i = 2, 4, 6, \ldots \\ 0, & \text{elsewhere} \end{cases}$$

4. Determine which of the following functions are pdf's for an appropriate choice of constant. For those that are, determine the constant, find the cdf, mean, and variance of the corresponding random variable:

(i)

$$f(x) = \begin{cases} A|x|, & 0 < |x| < 2 \\ 0, & \text{elsewhere} \end{cases}$$

(ii)

$$f(x) = \begin{cases} Ax e^{-x^2}, & x \ge 0 \\ 0, & x < 0 \end{cases}$$

(iii)
$$f(x)=\begin{cases} A\cos(x), & -\pi/2 \le x \le \pi/2 \\ 0, & \text{elsewhere} \end{cases}$$

5. The random variable X has a pdf

$$f(x)=\begin{cases} \frac{3}{2}x^2, & -1<x<1 \\ 0, & \text{elsewhere} \end{cases}$$

Determine:
 (i) $P(|X|<\frac{1}{2})$
 (ii) $P(-\frac{1}{4}<X<\frac{2}{3})$
 (iii) $P(X \in (-\frac{3}{4}, -\frac{1}{4}) \cap (-\frac{1}{2}, \frac{1}{2}))$

6. On a particular airline route the number of minutes that an airplane is late, X, is a random variable with cdf

$$F(x)=\begin{cases} 1-e^{-(.2)x}, & x \ge 0 \\ 0, & x < 0 \end{cases}$$

(i) Find the pdf of X.
(ii) What is the probability that an airplane is more than 10 minutes late?

7. Verify that the following functions are pdf's and find their cdf's.
(i)
$$f(x)=\begin{cases} 1/2\sqrt{x}, & 0<x<1 \\ 0, & \text{elsewhere} \end{cases}$$

(ii)
$$f(x)=\begin{cases} \frac{3}{2}(x+1)^2, & -1<x<0 \\ \frac{3}{2}(1-x)^2, & 0<x<1 \\ 0, & \text{elsewhere} \end{cases}$$

8. The time to failure of a component T is assumed to have a pdf

$$f_T(t)=\begin{cases} \alpha e^{-\alpha t}, & t \ge 0 \\ 0, & t < 0 \end{cases}$$

(i) What is the probability that the component fails between k and $k+1$, where k is an integer?
(ii) If the mean time to failure of the component is 100 hours, what is the probability that any particular component will last 200 hours?

9. In a large lot of items a certain fraction θ are defective. Given the number of draws required to obtain the first defective is a random variable X with probability distribution function

$$P(X=j)=A(.95)^{j-1}, \quad j=1, 2, \dots$$

(i) Find the constant A.
(ii) What is the fraction defective θ?
(iii) What is the probability that more than 20 items will have to be drawn before the first defective is obtained?

10. If X is a uniformly distributed random variable with pdf

$$f(x)=\begin{cases} 5/A, & -A/10 \le x \le A/10 \\ 0, & \text{elsewhere} \end{cases}$$

where A is a constant, find this constant if $P(|X|<2)=2P(|X|>2)$.

11. If X is a uniformly distributed random variable over the interval $(1, 4)$, find the probability that $Y < 0$, where $Y = X^2 - 4$.

12. If X is a continuous random variable with a pdf that is symmetric about the origin and $P(X < -2) = 0.1$, find $P((X + 1)(X - 1) > 3)$.

13. Let X be a continuous random variable with pdf

$$f(x) = \begin{cases} .1, & -5 \leq x \leq 5 \\ 0, & \text{elsewhere} \end{cases}$$

If $Y = X^2$, find the cdf and the pdf of Y.

14. Repeat Problem 13 if the pdf of X is

$$f(x) = \begin{cases} .1, & -1 \leq X \leq 9 \\ 0, & \text{elsewhere} \end{cases}$$

15. If X is a continuous random variable with pdf,

$$f(x) = \frac{1}{\sqrt{2\pi}} e^{-x^2/2}, \quad -\infty < x < \infty$$

find the pdf of $Y = 4X + 7$.

16. For the continuous random variable X, it is given that

$$P(X > t) = e^{-\mu t}(\mu t + 1), \quad \mu > 0, \ t \geq 0$$

Find:
(i) $F(x)$, the cdf of X.
(ii) $f(x)$, the pdf of X.
(iii) $P(X > 1/\mu)$.

17. If n customers choose among one of three brands of cola, let X be the number of brands chosen by no one. Show that

$$P(X = 0) = 1 - \frac{(2^n - 1)}{3^{n-1}}, \quad P(X = 1) = \frac{2^n - 2}{3^{n-1}}, \quad P(X = 2) = \frac{1}{3^{n-1}}$$

18. The number of cars crossing a bridge during a specified period of time is assumed to be a Poisson random variable. If the probability of no arrivals in this period is $1/4$, find an expression for the probability of at least two arrivals.

19. A newspaper boy buys his papers for 12 cents each and sells them for 20 cents each. Unsold papers are given back to the distributor and he receives 10 cents for each of these. Suppose the probability distribution of the demand D is

$$P(D = k) = e^{-10}\frac{10^k}{k!}, \quad k = 0, 1, 2, \ldots$$

Describe the random variable that represents the daily profit if he buys 10 papers each day.

20. The number of machine failures per day in a certain plant has a Poisson distribution with parameter $\lambda = 3$. Present maintenance facilities can repair 3 machines per day. Failures in excess of three are repaired by a contractor.
(i) On a given day what is the probability of having machines repaired by the contractor?
(ii) If the maintenance facilities could repair four machines per day, what would the probability in (i) be?
(iii) What is the most probable number of machine failures per day?

(iv) What is the expected number of machines repaired at the plant each day?

(v) What is the expected number of machines repaired by the contractor each day?

21. Let X be a Poisson random variable with parameter λ. Show that

$$P(X \text{ is even}) = \tfrac{1}{2}(1 + e^{-2\lambda})$$

22. Let X be a random variable with a Poisson distribution. If $P(X = 1) = P(X = 2) = .270$, find the mean and variance of X.

23. An experiment that measures the number of radioactive particles emitted during a fixed time period is repeated 100 times and the number of these experiments in which k particles were recorded is as follows:

k	0	1	2	3	4
Number	35	38	16	8	3

(i) Decide on the Poisson distribution that would best fit these data.

(ii) Devise a test to determine whether this fit is good.

24. A book has 500 pages with 100 known typographical errors. If the number of such errors per page has a Poisson distribution, what is the probability that a particular page contains no such error? One error? Two errors?

25. The demand for a certain type of film from a supplier averages 20 rolls a week and is assumed to be a Poisson random variable. How big a stock should he carry at the start of every week so that he has less than a 10% chance of running out?

26. For a given (fixed) value of b, for what values of a will the function

$$f(x) = \begin{cases} a e^{-bx}, & x > 1/b > 0 \\ 0, & \text{elsewhere} \end{cases}$$

be a pdf. Find the cdf, mean, and variance of this density function.

27. The probability density function of the yearly incomes of persons who must pay income tax is given by

$$f(x) = \begin{cases} \dfrac{11}{2} \dfrac{(5000)^{11/2}}{x^{6.5}}, & x \geq 5000 \\ 0, & x < 5000 \end{cases}$$

Find the yearly income that is exceeded by a randomly selected taxpayer with probability $1/2$.

28. Find the mean, variance, and cdf of the random variable x that has the pdf

$$f(x) = \begin{cases} 3/x^4, & x \geq 1 \\ 0, & \text{elsewhere} \end{cases}$$

29. The probability distribution of the discrete random variable X is given by

$$P(X = k) = C, \quad k = A + 1, \ldots, A + n$$

where A is a fixed positive integer. Find the constant C and the mean and variance of X.

30. The ratio of the deviations from the nominal length and width of silicon chips has the pdf

$$f(x) = \frac{a}{1 + x^2}, \quad -\infty < x < \infty$$

(i) For what value of a is this a pdf?

(ii) Find the cdf of x.

(iii) Find the probability that X will fall in the interval $(-1, 1)$.

(iv) Investigate the moments of X.

31. The demand for a new product of a company is assumed to be a random variable with a probability density function of the form

$$f(x) = \begin{cases} \dfrac{x}{a^2}\exp\left(-\dfrac{x^2}{2a^2}\right), & x \geq 0 \\ 0, & x < 0 \end{cases}$$

Find the mean and variance of this random variable and find the probability that it will exceed a.

32. The percentage impurity in a high grade alloy is a random variable with pdf

$$f(x) = \begin{cases} \sqrt{\dfrac{2}{\pi}}\,\dfrac{x^2}{a^3}\exp\left(-\dfrac{x^2}{2a^2}\right), & x \geq 0 \\ 0, & x < 0 \end{cases}$$

Find the mean and variance of this pdf.

33. Determine the mean and variance of the continuous random variable X with pdf

$$f(x) = \begin{cases} \frac{1}{6}, & 1 < |x| < 2 \\ \frac{1}{3}, & 0 < |x| < 1 \\ 0, & \text{elsewhere} \end{cases}$$

34. A point X is picked at random on a rod of length 1 meter. What is the probability that the area of the rectangle, with sides of lengths chosen as the lengths from X to the ends of the rod, exceeds $1/5$?

35. The random variable X is uniformly distributed over the interval (a, b). Find an expression for the nth moment $E(X^n)$ and nth central moment $E((X - \mu)^n)$, $\mu = E(X)$.

36. Let X be a random variable with mean μ and variance σ^2.

(i) Evaluate the expression $E((X - C)^2)$ for an arbitrary constant C.

(ii) For what value of C is $E((X - C)^2)$ minimized?

37. The random variable X has the pdf

$$f(x) = \begin{cases} 2x, & 0 < x < 1 \\ 0, & \text{elsewhere} \end{cases}$$

Find the moment generating function of X and, from the mgf, determine its mean and variance.

38. The time between arrivals T, of customers in a store, has an exponential pdf

$$f_T(t) = \begin{cases} 10\exp(-10t), & 0 \leq t \\ 0, & 0 > t \end{cases}$$

in hours. Determine the first four moments of T from its mgf.

39. The random variable X has the pdf

$$f(x) = \begin{cases} x\,e^{-x}, & x \geq 0 \\ 0, & x < 0 \end{cases}$$

Find the mgf of X.

40. The pdf of a random variable X is given by
$$f(x) = \begin{cases} \frac{1}{4}, & 2 \le x \le 4 \\ \frac{1}{2}, & 4 < x \le 5 \\ 0, & \text{elsewhere} \end{cases}$$

Find the mean and variance of X from its mgf.

CHAPTER 3

Elements of Combinatorial and Geometrical Probability

A combinatorial problem is simply one associated with the selection and/or the arrangement of objects according to a set of specified rules. As such, these problems often lead to interesting questions in probability. Using combinatorial arguments many problems associated with games and gambling can be answered. Historically, the subject of probability arose from such considerations. Our purpose in this chapter is to look at some challenging probabilistic situations to gain insight and experience with probabilistic methods. Most of the chapter deals only with discrete random variables. The final section will briefly describe geometric probability, an approach that is useful in some cases.

3.1 *Permutations, Combinations, and Enumeration*

Consider the following simple question: In how many ways can you order the integers 1, 2, 3, 4? To answer this question we think of placing the integers, one at a time into "pigeon holes":

working from left to right. In the first hole we can place any of the four integers. After placing one in the first hole we have three left, any one of which we can place in the second hole. There are thus $4 \times 3 = 12$ ways of filling the first two holes. For each of these ways there are two ways of filling the third hole, corresponding to the two integers left at this stage. There are $4 \times 3 \times 2 = 12 \times 2 = 24$ ways of filling the first three holes. Of course, once these are filled we are left with only one integer and there is only one way to fill the remaining hole. In total, there are $4 \times 3 \times 2 \times 1 = 24$ ways of ordering the four integers corresponding to:

1	2	3	4		2	1	3	4		3	1	2	4		4	1	2	3
1	2	4	3		2	1	4	3		3	1	4	2		4	1	3	2
1	3	2	4		2	3	1	4		3	2	1	4		4	2	1	3
1	3	4	2		2	3	4	1		3	2	4	1		4	2	3	1
1	4	2	3		2	4	1	3		3	4	1	2		4	3	1	2
1	4	3	2		2	4	3	1		3	4	2	1		4	3	2	1

This simple example illustrates two important points. The first is that it introduces the notion of a permutation. We call 2341 a permutation, or a particular ordering, of the elements 1, 2, 3, 4. If we had n distinct elements, say the integers $1, 2, \ldots, n$, then, using the same reasoning as before, there would be $n \times (n-1) \times (n-2) \times \cdots \times 2 \times 1$ ways of ordering them. We denote this quantity by $n!$ and in words by n factorial.

The second point raised in this example is fundamental to many counting or enumeration arguments. It says simply that if we are going to perform two procedures in succession and if the first procedure can be performed in m ways, and if, for each of these ways, the second procedure can be performed in n ways, then there are mn ways in which the two procedures can be performed successively. Of course, if we are performing k procedures successively, the ith procedure capable of being performed in m_i ways regardless of the ways in which the first $(i-1)$ procedures were performed, then the k procedures can be performed in $m_1 m_2 \cdots m_k$ ways. The following example will illustrate variations on this theme.

EXAMPLE 3.1

A series of five switches controlling a mechanism can each be set in one of six positions 1, 2, 3, 4, 5, 6. (i) How many different switch settings are there? (ii) How many different settings are there if we insist that adjacent switches not be in the same position?

Considering (i): Again, imagine that each switch corresponds to a pigeon hole. In the first hole we can put one of six numbers corresponding to the switch setting we select. Regardless of what number we place in the first hole, we can place six in the second hole (notice the difference from the problem of permutations). There are 6×6 ways of filling the first two holes and continuing in this manner we see there are $6 \times 6 \times 6 \times 6 \times 6 = 6^5$ ways of setting the five switches.

Considering (ii): With the restriction that adjacent switches not be in the same position, we proceed as follows. There are six ways of filling the first hole. When it comes to filling the second hole we must avoid the number we used in the first hole, that is, we can only place one of five numbers in the second hole, regardless of which number we used in the first hole. There are $6 \times 5 = 30$ ways of filling the first two holes. In filling the third pigeonhole we again only have to avoid one number, that used in the second hole, that is, we can fill the third hole in five ways. Continuing the argument we see that there are 6×5^4 ways of setting the switches so that no two adjacent switches are in the same position.

Let us continue the problem by asking the following question: If the five switches are set at random, what is the probability that no two adjacent switches will be in the same position? As our sample space we take the 6^5 possible settings for the switches that we determined in part (i). The meaning of the words "at random" is then simply that each of these 6^5 settings has the same probability, namely $1/6^5$. The event A that no two adjacent switches are in the same position, corresponds to the 6×5^4 settings found in part (ii). Since the sample space is equally likely, the probability of the event A is simply

$$P(A) = \frac{6 \times 5^4}{6^5} = \left(\frac{5}{6}\right)^4$$

Notice how the original combinatorial problem was translated into a probabilistic one by the words "at random." Essentially, this told us how to assign probabilities in the sample space.

As a modification of the problem of ordering n distinct elements, we ask the following question: In how many distinct ways can you write down ordered sets (permutations) of k distinct objects chosen from n distinct objects? Returning to our pigeon holes, we now only have k holes to fill. We can fill the first hole in n ways, the second in $(n-1)$ ways, the third in $(n-2)$ ways, and so forth. The last hole, the kth, we can fill in $(n-k+1)$ ways. Thus the total number of ways of ordering k elements, chosen from n, is

$$n(n-1)(n-2) \cdots (n-k+1)$$

a quantity that can also be expressed as $n!/(n-k)!$ as seen by writing out each factorial and canceling identical terms. This quantity is sometimes denoted by P_k^n, the number of permutations of n things taken k at a time.

EXAMPLE 3.2

How many computer passwords containing eight distinct letters from the 26 of the alphabet can be constructed?

The answer is simply $P_8^{26} = 26!/8!$, the number of ways of ordering 26 elements taken eight at a time.

So far we have been assuming that all elements under consideration are distinct. What happens if they are not? Consider the following example.

EXAMPLE 3.3

A ship has six signaling flags, three of which are blue, two yellow, and one red. In how many distinct ways could these six flags be ordered?

If the six flags were distinct, there would be 6! ways of ordering them. To analyze the situation we first make all the colors distinct by adding a subscript to the colors—thus we have colors B_1, B_2, B_3, Y_1, Y_2, and R. Consider one of the orderings of the flags, say, $B_1 Y_2 R B_2 Y_1 B_3$. If we permute the B's among themselves, we obtain five more permutations:

$$B_1 \ Y_2 \ R \ B_3 \ Y_1 \ B_2$$
$$B_3 \ Y_2 \ R \ B_1 \ Y_1 \ B_2$$
$$B_3 \ Y_2 \ R \ B_2 \ Y_1 \ B_1$$
$$B_2 \ Y_2 \ R \ B_1 \ Y_1 \ B_3$$
$$B_2 \ Y_2 \ R \ B_3 \ Y_1 \ B_1$$

If we removed the subscripts from the colors, these six permutations would cease to be distinct. Thus each of the 6! permutations falls naturally into a group of $3! = 6$ in this way. Removing the subscripts from the B's, but not the Y's, then results in only $6!/3!$ permutations. Repeating the argument for the Y's, for which there are only two subscripts that can be ordered in 2! ways, we conclude that the number of distinct orderings of the six colors is

$$\frac{6!}{3!2!}$$

The reader should convince himself, by extending the arguments used in the previous example, that the number of permutations of n objects, n_1 of which are of type 1, n_2 of which are of type 2, ..., n_s of which are of type s (different types are distinct), is

$$\frac{n!}{n_1!n_2!\cdots n_s!}, \qquad n_1+n_2+\cdots+n_s=n$$

Of course, if $n_1 = n_2 = \cdots = n_s = 1$ ($s = n$), then this number is simply $n!$.

Combinations are permutations in which order is not important. Specifically, we seek the number of ways we can choose k objects from n, where the order in which we choose them is irrelevant. We know that we can fill k holes with objects drawn from n distinct objects in $n!/(n-k)!$ ways and each of these orderings is distinct. If we are now only concerned with the k objects chosen and not with the way in which they are written down, then each ordering can be permuted $k!$ ways and each of these $k!$ orderings corresponds to the same set of elements. Consequently, it is possible to choose, without regard to order, k elements from n distinct elements in

$$\frac{n!}{k!(n-k)!}$$

ways. This is an important quantity in combinatorics and we will denote it by $\binom{n}{k}$. It is also called a binomial coefficient and is sometimes written as C_k^n to reflect the fact that it is the number of ways of choosing k things from n.

There are a large number of properties of binomial coefficients. For example, it is readily seen that

$$\binom{n}{k} = \binom{n}{n-k}$$

since the number of ways you can choose k items from n is the number of ways you can choose $n - k$ items to leave behind (i.e., not choose). A slightly more intricate identity claims that

$$\binom{n}{k} = \binom{n-1}{k-1} + \binom{n-1}{k} \tag{3.1}$$

To verify this, we distinguish a single element of the n elements in some way. The number of choices of k elements containing this distinguished element is $\binom{n-1}{k-1}$. The number of choices of k elements not containing this element is $\binom{n-1}{k}$. Since any particular selection either contains this distinguished element or does not, the result follows. These results may also be verified by calculation. The reader will find other identities involving the binomial coefficients, of varying degrees of difficulty to prove, in the problems.

EXAMPLE 3.4

In how many ways can a standard deck of 52 playing cards be dealt into four hands, 13 cards each?

We can look on the situation as follows: The first player chooses a hand first and this can be done in $\binom{52}{13}$ ways. The second player then chooses a hand in $\binom{39}{13}$ ways. The third player chooses a hand in $\binom{26}{13}$ ways, and the fourth player receives the 13 remaining cards. The total number of ways of dividing the deck is then

$$\binom{52}{13}\binom{39}{13}\binom{26}{13} \approx 5.3645 \times 10^{28}$$

The following theorem will be used extensively throughout the remainder of the book.

THEOREM 3.1 (The Binomial Theorem)

$$(a+b)^n = \sum_{k=0}^{n} \binom{n}{k} a^k b^{n-k}$$

Proof Two proofs will be given to illustrate the different approaches possible. Note that the binomial coefficients $\binom{m}{0}$ and $\binom{m}{m}$, for any integer m, are always taken to be equal to unity and $\binom{m}{j}$ for $j<0$ or $j>m$ is, by convention, also always taken to be equal to zero.

(i) We can regard the expansion of

$$(a+b)^n = (a+b)(a+b)\cdots(a+b) \quad (n \text{ factors})$$

as choosing a or b from each factor in the multiplication. The only way we can arrive at a term $a^k b^{n-k}$ is to choose the "a" term from k of the factors and the "b" term from the remaining $n-k$ factors. Since we can choose these k factors from the n in precisely $\binom{n}{k}$ ways, the coefficient of the term $a^k b^{n-k}$ in the expansion must be $\binom{n}{k}$, as it was required to show.

(ii) In this proof we use the principle of induction encountered in the first chapter. We first observe that, since

$$(a+b)^2 = a^2 + 2ab + b^2 = \binom{2}{2}a^2 + \binom{2}{1}ab + \binom{2}{0}b^2$$

the theorem is true for $n=2$. Suppose now that the theorem is true for $n-1$, where n is an arbitrary integer greater than three. We use this information to prove that it must be true for the next largest integer n. Since it is true for $n-1$, we have

$$(a+b)^{n-1} = \sum_{k=0}^{n-1} \binom{n-1}{k} a^k b^{n-1-k}$$

Multiplying both sides of this equation by $(a+b)$ yields

$$(a+b)(a+b)^{n-1} = (a+b)^n$$

$$= (a+b) \sum_{k=0}^{n-1} \binom{n-1}{k} a^k b^{n-1-k}$$

$$= \sum_{k=0}^{n-1} \binom{n-1}{k} a^{k+1} b^{n-1-k} + \sum_{k=0}^{n-1} \binom{n-1}{k} a^k b^{n-k}$$

In the first summation replace k by $j-1$, that is, $k=j-1$ or $j=k+1$, and in the second summation replace k by j. Rewriting the equation gives

$$(a+b)^n = \sum_{j=1}^{n} \binom{n-1}{j-1} a^j b^{n-j} + \sum_{j=0}^{n-1} \binom{n-1}{j} a^j b^{n-j} \qquad (3.2)$$

The coefficient of $a^j b^{n-j}$ is, for $1 \le j \le n-1$,

$$\binom{n-1}{j-1} + \binom{n-1}{j}$$

which, by Eq. (3.1), is just $\binom{n}{j}$. From Eq. (3.2) it follows that

$$(a+b)^n = \sum_{j=0}^{n} \binom{n}{j} a^i b^{n-i}$$

as required.

Both of these proofs require some thought to fully understand them if the reader is not familiar with these ideas. An extension of the binomial theorem states that

$$(x_1 + x_2 + \cdots + x_s)^n = \sum \binom{n}{k_1 k_2 \cdots k_s} x_1^{k_1} x_2^{k_2} \cdots x_s^{k_s} \qquad (3.3)$$

where $k_1 + k_2 + \cdots + k_s = n$, and

$$\binom{n}{k_1 k_2 \cdots k_s} = \frac{n!}{k_1! k_2! \cdots k_s!}$$

and the summation is over all possible sets of integers $(k_1, k_2, \ldots, k_s)$ such that (i) $0 \le k_i \le n$, $i = 1, 2, \ldots, s$ and (ii) $k_1 + k_2 + \cdots + k_s = n$. Equation (3.3) is called the multinomial theorem and can be proven using the same technique as in proof (i) of the previous theorem. For example, we expand the left-hand side of (3.3) as

$$(x_1 + x_2 + \cdots + x_s)^n$$

$$= (x_1 + x_2 + \cdots + x_s)$$

$$\times (x_1 + x_2 + \cdots + x_s) \cdots (x_1 + x_2 + \cdots + x_s) \quad (n \text{ factors})$$

and inquire as to how many distinct ways the term $x_1^{k_1} x_2^{k_2} \cdots x_s^{k_s}$ can arise (i.e., what is the coefficient of $x_1^{k_1} x_2^{k_2} \cdots x_s^{k_s}$ in the expansion). In expanding the product, a term is chosen from each bracket. To arrive at $x_1^{k_1}$ in the product, we must choose the term x_1 from k_1 of the brackets and this can be done in $\binom{n}{k_1}$ ways. To obtain $x_2^{k_2}$ we must choose k_2 of the remaining $(n - k_1)$ brackets from which to take the x_2 term. This process is continued and the coefficient of $x_1^{k_1} x_2^{k_2} \cdots x_s^{k_s}$ must be

$$\binom{n}{k_1}\binom{n-k_1}{k_2}\binom{n-k_1-k_2}{k_3} \cdots \binom{n-k_1-k_2-\cdots-k_{s-1}}{k_s}$$

$$= \frac{n!}{k_1!(n-k_1)!} \cdot \frac{(n-k_1)!}{k_2!(n-k_1-k_2)!} \cdots \frac{(n-k_1-\cdots-k_{s-1})!}{k_s! 0!}$$

$$= \frac{n!}{k_1! k_2! \cdots k_s!} = \binom{n}{k_1 k_2 \cdots k_s}$$

as claimed.

One consequence of the binomial theorem that we will have frequent use for is the following: If θ and η are two real numbers, $0 < \theta < 1$, $0 < \eta < 1$, $\theta + \eta = 1$, then

$$(\theta + \eta)^n = 1^n = 1 = \sum_{k=0}^{n} \binom{n}{k} \theta^k \eta^{n-k} \qquad (3.4)$$

As a final problem for this section, assume that we have n distinct items in some container. It is very popular to use urns as containers in probability, so we will assume the n items are in an urn. We will assume that the items are numbered $1, 2, \ldots, n$. If k items from the n are chosen at random, then the appropriate sample space contains $\binom{n}{k}$ equally likely points. Suppose further that items $1, 2, \ldots, r$ are red and the remaining $n - r = s$ items are blue. What is the probability that, in a draw of k items, we obtain exactly j red items. Let X be the number of red items obtained. There are $\binom{r}{j}$ ways of choosing j red items. For each of these ways there are

$$\binom{s}{k-j} = \binom{n-r}{k-j}$$

ways of choosing the remaining $k - j$ items from the $n - r = s$ blue items. Consequently, we have

$$P(X = j) = \frac{\binom{r}{j} \binom{s}{k-j}}{\binom{n}{k}}, \quad j = 0, 1, 2, \ldots, \min(k, r) \qquad (3.5)$$

since each of the $\binom{n}{k}$ ways of choosing the k items is equally likely. This is referred to as the hypergeometric distribution and it will be encountered again.

Notice that we could have phrased the question asked in the preceding paragraph as "what is the probability of choosing exactly j items of type 1 in k drawings from an urn containing r items of type 1 and $s = n - r$ items of type 2." In this formulation it is not stated that the items are distinct. Indeed, it is not necessary that they be distinct. It is clear, however, that the solution to the problem (Eq. (3.5)) does not depend on this assumption. The assumption simply assures a consistent solution.

To evaluate the mean of the hypergeometric distribution, we first observe that since it is a distribution (if our derivation was correct), then it must sum to unity and so

$$\binom{n}{k} = \sum_{j=0}^{r} \binom{r}{j} \binom{n-r}{k-j}$$

If $k < r$, then this sum actually only extends to k since then

$$\binom{n-r}{k-j} = 0 \quad \text{for } j > k$$

For convenience we will assume $k > r$ as the same result will be obtained for $k < r$. The above equation is valid for any integers k, r, n such that $0 < r < n$, $0 < k < n$, and this is important to realize. For example,

$$\binom{n-1}{k-1} = \sum_{j=0}^{r-1} \binom{r-1}{j} \binom{n-r}{k-1-j}$$

obtained by replacing each parameter by one less. The mean of the distribution is simply

$$E[X] = \sum_{j=0}^{r} j \frac{\binom{r}{j}\binom{n-r}{k-j}}{\binom{n}{k}} = \frac{1}{\binom{n}{k}} \sum_{j=1}^{r} \frac{j \cdot r!}{j!(r-j)!} \binom{n-r}{k-j}$$

$$= \frac{k!(n-k)!}{n!} \sum_{j=1}^{r} \frac{r(r-1)!}{(j-1)!(r-j)!} \binom{n-r}{k-j}$$

$$= \frac{rk}{n} \cdot \frac{1}{\binom{n-1}{k-1}} \sum_{j=1}^{r} \binom{r-1}{j-1} \binom{n-r}{k-j}$$

If we let $l = j - 1$ in this expression and sum over l from 0 to $r-1$ (write out a few terms to show that we have not changed anything), then

$$E[X] = \frac{rk}{n} \sum_{l=0}^{r-1} \frac{\binom{r-1}{l}\binom{n-r}{k-1-l}}{\binom{n-1}{k-1}} = \frac{rk}{n}$$

since we have already seen the summation is unity. The variance can be found in the same manner but the calculations are tedious. The result is

$$V(X) = \frac{kr}{n}\left(1 - \frac{r}{n}\right)\left(\frac{n-k}{n-1}\right)$$

Before proceeding further, it is noted that there are two ways of drawing from an urn—either with or without replacement. Drawing k items from n with replacement simply means that after an item is withdrawn its color, or another attribute, is noted, and it is then returned. Drawing without replacement implies that the chosen item remains out of the urn. In the previous discussion it was clear that items were withdrawn without replacement. It will usually be stated explicitly which of the two regimens is in effect.

EXAMPLE 3.5

In error, 30 undersize washers have been mixed in with 200 ready for shipment. What is the probability that if 20 washers are drawn, exactly k of them will be undersize?

We can regard the 30 undersize washers as being of type 1 and the 200 good washers as being of type 2. The question then is what is the probability that k of the 20 washers will be of type 1? This is precisely the reasoning behind the hypergeometric distribution in Eq. (3.5). If X is the number of undersize washers obtained, then

$$P(X=k) = \frac{\binom{30}{k}\binom{200}{20-k}}{\binom{230}{20}}, \quad k = 1, 2, \ldots, 20$$

3.2 *Combinatorial Probability*

With the combinatorial principles established in the previous section, many interesting problems can be considered. Basically our approach will be to use combinatorial methods to enumerate the number of points in the sample space. Probabilities will then be assigned, usually by intuitive reasoning. Examples will be used to establish important general results. We begin with a simple coin tossing experiment.

EXAMPLE 3.6

A coin is tossed n times. What is the probability of obtaining k heads?

To set up the sample space we record the outcome of a particular experiment as an ordered sequence of heads (H) and tails (T), for example, HTHHTT $\cdots$ HT. On the first toss there are two possible outcomes H and T. For each of these there are two possible outcomes on the second toss and a total of 2×2 outcomes for the first two tosses. For each of these four outcomes there are again two possible outcomes for the third toss, and hence $2 \times 2 \times 2 = 8$ possible outcomes for the first three tosses. Proceeding in this manner, we conclude that there are 2^n possible outcomes, that is, sequences of H's and T's. From an intuitive feeling, we have no reason to expect that any one of these outcomes is any more likely than any other, hence we assume that all of the 2^n outcomes are equally likely. Let X be the number of the n tosses that result in a head. Finding the probability that $X = k$, $P(X = k)$ reduces to finding the number of outcomes containing k H's (and hence $n - k$ T's). This number is just $\binom{n}{k}$ corresponding to the number of ways we can choose k positions out of n to place H's (and fill the remaining $n - k$ positions with T's). It follows easily that

$$P(X=k) = \binom{n}{k}/2^n$$

There is an important generalization to this example. Consider performing some experiment and observing only whether or not the event A occurs. If A occurs we call the experiment a success, and if it does not occur ($\bar{A}$ occurs), a failure. Suppose the probability that A occurs is $P(A) = \theta$ and hence $P(\bar{A}) = \eta = 1 - \theta$. We repeat this experiment n times (trials) under the following assumptions:

 (i) On each trial the event A has the same constant probability of occurring.
(ii) The n trials are independent, that is, the outcome of one experiment has no effect on the outcome of another. (The notion of independent events will be formalized in the next chapter.)

Such an independent repetition of the same experiment under identical conditions is called a Bernoulli trial and is typified by coin tossing, dice tossing experiments, and so forth.

A point in the sample space now is a sequence of n A's and $\bar{A}$'s. A point with k A's and $(n-k)$ $\bar{A}$'s will be assigned a probability of $\theta^k \eta^{n-k}$. We attempt to justify this as follows. Consider the case $n = 2$, where the sample space contains only the four points AA, $A\bar{A}$, $\bar{A}A$, and $\bar{A}\bar{A}$. On the first experiment the outcome A appears the fraction θ of the time. Of this fraction θ, the outcome A appears the fraction θ of the time on the second experiment—thus the outcome AA of the two experiments appears, on the average, the fraction $\theta \times \theta = \theta^2$ of the time. Similarly, the outcome $A\bar{A}$ will appear, on the average, the fraction $\theta \times \eta = \theta\eta$ of times. Extending this argument to n repetitions results in assigning the probability $\theta^k \eta^{n-k}$ to any point of the sample space containing k A's and $(n-k)$ $\bar{A}$'s. This result can be obtained in a more rigorous manner using concepts introduced in Chapter 4. For the moment we accept this assignation.

There is an easy way to calculate the mean and variance of a binomial distribution, which will be given in Chapter 8. We present a hard way here, which is instructive. We could, of course also find the moment generating function and differentiate it (see Problem 16). The mean of the distribution is

$$E[X] = \sum_{k=0}^{n} k \binom{n}{k} \theta^k \eta^{n-k}, \quad \theta + \eta = 1$$

$$= \sum_{k=0}^{n} k \cdot \frac{n!}{k!(n-k)!} \theta^k \eta^{n-k}$$

$$= \sum_{k=1}^{n} \frac{n \cdot (n-1)!}{(k-1)!(n-k)!} \theta^k \eta^{n-k}$$

where the change in the lower limit of the summation from 0 to 1 is permissible since the $k = 0$ term is zero. In the last expression change the summation variable k to $j = k - 1$ and sum over j from 0 to $n - 1$ to yield

$$E[X] = n \sum_{j=0}^{n-1} \frac{(n-1)!}{j!(n-1-j)!} \theta^{j+1} \eta^{n-1-j}$$

$$= n\theta \sum_{j=0}^{n-1} \binom{n-1}{j} \theta^j \eta^{n-1-j}$$

But this last summation must be one, and to see this, we consider Eq. (3.4) again. As noted, when summed over k from 0 to n, we always obtain unity and this result is true for *any* positive integer n, as long as θ and η are such that $0 < \theta < 1$, $0 < \eta < 1$, $\theta + \eta = 1$. In particular, it is true for the integer $n - 1$ and we conclude that

$$E[X] = n\theta \qquad (3.6)$$

The variance may be calculated in a similar manner and we omit the details. The result is that

$$V(X) = n\theta\eta = n\theta(1 - \theta) \qquad (3.7)$$

Thus the mean and variance of any binomial random variable with parameters n and θ are given by Eq. (3.6) and (3.7), respectively. The next example looks at various other questions that can be asked about a coin tossing experiment.

EXAMPLE 3.8

A fair coin is tossed until it comes up the same twice in succession. Find the probability of the following events:

(i) The experiment ends at the sixth toss.
(ii) An even number of tosses is required.
(iii) The experiment ends on the kth toss.

Find the expected number of tosses to end the game.

We first determine the sample space of this experiment. Every point in the space, which will be a sequence of H's and T's indicating the outcomes of the individual tosses, must end with HH or TT. The first few points of the sample space are

HH
TT
THH
HTT
THTT
HTHH
$\vdots$

The sample space is countably infinite and to a point terminating at the nth toss we associate the probability $1/2^n$. To answer part (i) of the question, there are two points in space corresponding to the event the experiment ends at the sixth toss, namely,

HTHTHH   and   THTHTT

and the probability of this event is $2 \times 1/2^6 = 1/2^5$. In fact, the probability the experiment ends on the kth toss is just $2 \times 1/2^k = 1/2^{k-1}$, which answers part (iii). Notice that if X is the number of tosses required to complete the experiment, then

$$P(X = j) = 1/2^{j-1}, \quad j = 2, 3, \ldots \tag{3.8}$$

is, in fact, a valid probability distribution since

$$\sum_{j=2}^{\infty} P(X = j) = \sum_{j=2}^{\infty} \frac{1}{2^{j-1}} = \frac{1/2}{1 - 1/2} = 1$$

The probability the experiment ends on an even-numbered toss is just

$$P(X = \text{even}) = \sum_{\substack{j=2 \\ j\,\text{even}}}^{\infty} \frac{1}{2^{j-1}} = \frac{1/2}{1 - 1/4} = \tfrac{2}{3}$$

The expected number of tosses to the end of the game, or the mean of the distribution (3.8) is

$$E[X] = \sum_{j=2}^{\infty} j \frac{1}{2^{j-1}}$$

To evaluate this expression we note that for $0 < x < 1$,

$$\sum_{j=2}^{\infty} x^i = \frac{x^2}{1-x}$$

using the summation formula for a geometric series. Differentiating this expression with respect to x gives

$$\frac{d}{dx}\left(\sum_{j=2}^{\infty} x^i \right) = \sum_{j=2}^{\infty} jx^{i-1} = \frac{d}{dx}\left(\frac{x^2}{1-x} \right) = \frac{x(2-x)}{(1-x)^2}$$

and, evaluating this last expression at $x = 1/2$ gives

$$E[X] = 3$$

There are several other distributions associated with Bernoulli trials, apart from the binomial. Suppose we repeat an experiment, independently, until the event A (success) occurs. What is the probability that the first success occurs on the jth try if, on any one try, $P(A) = \theta$? The sample space for such an experiment consists of the points

A
$\bar{A}A$
$\bar{A}\bar{A}A$
$\vdots$
$\vdots$

and it is clear that in order to obtain the first success on the jth try, the first $(j-1)$ tries must end in failure ($\bar{A}$). If X is the number of tries to obtain the first success, then

$$P(X=j)=\theta\eta^{j-1}, \quad j=1,2,\ldots, \theta+\eta=1 \tag{3.9}$$

Equation (3.9) is referred to as the geometric distribution first encountered in the previous chapter. Its mean and variance were determined in Chapter 2 as

$$E(x)=1/\theta, \qquad V(X)=\eta/\theta^2 \tag{3.10}$$

A slight modification of the previous situation will yield another interesting distribution. Rather than inquire about when the first success was obtained, we let X be the number of trials to obtain the kth success (event A). What, then, is the probability that X equals j? We could describe the sample space but in this case it is cumbersome. Consider a point that achieves its kth success on the jth trial. The probability of any such point is $\theta^k\eta^{j-k}$ since it contains k successes and $j-k$ failures. How many points in the sample space are there that terminate on the jth try? Since one of the successes occurs on the jth trial, there are $k-1$ distributed in some manner among the first $j-1$ positions and there are $\binom{j-1}{k-1}$ ways of distributing these successes, corresponding to the number of ways we can choose the $k-1$ positions from $j-1$ in which to put the A's or successes. We have shown that

$$P(X=j)=\binom{j-1}{k-1}\theta^k\eta^{j-k}, \quad j=k,k+1,\ldots,\infty$$

This is the Pascal distribution—since we have done the mathematics correctly it must be a distribution, and so

$$\sum_{j=k}^{\infty}\binom{j-1}{k-1}\theta^k\eta^{j-k}=1$$

Again, this equation is true for any integer $k>0$, for values of θ, η such that $0<\theta<1$ and $\theta+\eta=1$. In particular, it is true for the value $k+1$, that is,

$$\sum_{j=k+1}^{\infty}\binom{j-1}{k}\theta^{k+1}\eta^{j-k-1}=1$$

a fact that will be useful in calculating the mean of X, which we now do:

$$
\begin{aligned}
E[X] &= \sum_{j=k}^{\infty} j\cdot\binom{j-1}{k-1}\theta^k\eta^{j-k} \\
&= \sum_{j=k}^{\infty} \frac{j\cdot(j-1)!}{(k-1)!(j-k)!}\theta^k\eta^{j-k} \\
&= \frac{k}{\theta}\sum_{j=k}^{\infty} \frac{j!}{k\cdot(k-1)!(j-k)!}\theta^{k+1}\eta^{j-k} \\
&= \frac{k}{\theta}\sum_{j=k}^{\infty}\binom{j}{k}\theta^{k+1}\eta^{j-k}
\end{aligned}
$$

Letting $j' = j + 1$ and summing over j' from $k + 1$ to ∞ yields

$$E[X] = \frac{k}{\theta} \sum_{j'=k+1}^{\infty} \binom{j'-1}{k} \theta^{k+1} \eta^{j'-1-k} = \frac{k}{\theta}$$

since the summation has already been shown to equal unity. The variance can be calculated in the same manner as $k\eta/\theta^2$ and the computation is omitted.

Another distribution associated with the binomial is the multinomial. Suppose an experiment has s possible (disjoint) outcomes $A_1, A_2, \ldots, A_s$ rather than two as for the binomial. The experiment is repeated n times independently and on each repetition exactly one of the events A_i occurs $i = 1, \ldots, s$ (i.e., $\bigcup_1^s A_i = S$, the sample space, $A_i \cap A_j = \varnothing, i \neq j$) and $P(A_i) = \theta_i, i = 1, \ldots, s$ and $\theta_1 + \theta_2 + \cdots + \theta_s = 1$. If X_i is the number of times that event A_i occurred among the n repetitions, then what is the probability that $X_1 = k_1$, $X_2 = k_2, \ldots, X_s = k_s$, $k_1 + k_2 + \cdots + k_s = n$? Since any point in the sample space of the n repetitions is a sequence of A_i's, and since any sequence with k_1 A_1's, k_2 A_2's, $\ldots, k_s A_s$'s has probability $\theta_1^{k_1}\theta_2^{k_2} \cdots \theta_s^{k_s}$, this is just the same as counting the number of distinct points with k_1 A_1's, k_2 A_2's, and so forth. We can choose the k_1 places in which the A_1's occur in $\binom{n}{k_1}$ ways, the k_2 places in which the A_2's occur in the $\binom{n-k_1}{k_2}$ ways, and so on. Consequently, there are

$$\binom{n}{k_1}\binom{n-k_1}{k_2} \cdots \binom{n-k_1-k_2-\cdots-k_{s-1}}{k_s} = \frac{n!}{k_1!k_2!\cdots k_s!}$$

$$= \binom{n}{k_1k_2\cdots k_s}$$

such points and

$$P(X_1 = k_1, X_2 = k_2, \ldots, X_s = k_s) = \binom{n}{k_1k_2\cdots k_s}\theta_1^{k_1}\theta_2^{k_2}\cdots\theta_s^{k_s}$$

This is referred to as the multinomial distribution. Of course, it reduces to the binomial distribution in the case $s = 2$.

The following example illustrates the different situations that give rise to the binomial, hypergeometric, multinomial, and a fourth nonstandard distribution.

EXAMPLE 3.9

An urn contains six red, four white, and eight blue balls. Consider the following situations:

(i) Five balls are drawn with replacement. What is the probability of obtaining three red?

(ii) Five balls are drawn without replacement. What is the probability of obtaining three red?

(iii) Five balls are drawn with replacement. What is the probability of obtaining two red, two white, and one blue?

(iv) Five balls are drawn without replacement. What is the probability of obtaining two red, two white and one blue?

Drawing with replacement simply means that each time you draw the situation is the same. In part (i) we are only interested in whether or not you draw a red ball, that is, a success (red ball) versus failure (other color of ball). If X is the number of red balls drawn, then X is the binomial random variable with parameters $n = 5$ and $\theta = 6/18 = 1/3$ (the number of red balls divided by the total number of balls). The answer to part (i) is then

$$P(X = 3) = \binom{5}{3}(\tfrac{1}{3})(\tfrac{2}{3})^2$$

So far as part (ii) of the problem is concerned, there are only two types of balls present (red and not red), and we draw five without replacement. In this instance the hypergeometric distribution applies and

$$P(X = 3) = \frac{\binom{6}{3}\binom{12}{2}}{\binom{18}{5}}$$

In part (iii) of the problem we are repeating an identical experiment (drawing a ball) five times but we are interested in three possible outcomes (red, white, or blue), not merely two (red or not red), as in part (i). In this instance the multinomial distribution applies and we have

$$P(\text{two red, two white, one blue}) = \frac{5!}{2!\,2!\,1!}\left(\frac{6}{18}\right)^2\left(\frac{4}{18}\right)^2\left(\frac{8}{18}\right)^1$$

where, on each draw, 6/18 is the probability of drawing a red, 4/18 the probability of a white, and 8/18 the probability of a blue.

The last part of the question involves a generalization of the hypergeometric distribution. Since we are drawing without replacement and are concerned with three possible outcomes, using similar arguments to those used in deriving the hypergeometric, we have

$$P(\text{two red, two white, one blue}) = \frac{\binom{6}{2}\binom{4}{2}\binom{8}{1}}{\binom{18}{5}}$$

EXAMPLE 3.10

An engineer has determined that 80% of all motor failures are due to either leaking joints or bearing overheating and that these two occur with equal probability. The remaining failures are due to miscellaneous causes. Among ten failures, what is the probability that six of them were due to bearing overheating, two due to leaking joints, and two due to other causes.

From the data given the probability, on any given failure, that the failure is due to leaking joint is 0.4, due to bearing overheating is 0.4, and due to a miscellaneous cause is 0.2. The distribution is thus multinomial and

$$P(6 \text{ overheat, 2 leak, 2 misc.}) = \frac{10!}{6!2!2!}(0.4)^6 \ (0.4)^2 \ (0.2)^2$$

$$= .033$$

The next example for this section involves no particular standard distribution but is good preparation for the remainder of this section where some card and dice games will be considered.

EXAMPLE 3.11

A person throws two dice repeatedly and observes the sum of the two faces. What is the probability that they throw a four before a seven?

There are essentially only three outcomes of interest; a four, a seven, or neither a four nor a seven. We denote these by 4, 7, and N (for neither), respectively. On any given throw, $P(4) = 3/36$, $P(7) = 6/36$, and $P(N) = 27/36$. Since the experiment terminates when either a four or a seven is thrown, an appropriate sample space is

$$\{4, 7, N4, N7, NN4, NN7, \ldots\}$$

The outcomes favorable to the event of throwing a four before a seven are simply those ending with a four and the sum of their probabilities is

$$P(4 \text{ before a } 7) = \tfrac{3}{36} + \tfrac{27}{36} \cdot \tfrac{3}{36} + \left(\tfrac{27}{36}\right)^2 \tfrac{3}{36} + \cdots$$

$$= \sum_{j=0}^{\infty} \tfrac{3}{36}\left(\tfrac{27}{36}\right)^j$$

$$= \frac{3/36}{1 - 27/36} = \frac{1}{3}$$

As a review, the important discrete distributions encountered in this and the previous chapter along with their mean, variance, and generating functions are listed in Table 3.1. Much of the information contained in this table was not derived in the text, and it is a worthwhile exercise for the reader to derive those that were not.

As mentioned in the Introduction, probability theory has its roots in the analysis of games of chance. There are many books on the subject of gambling and the various strategies one can adopt for maximizing expected profit. Unfortunately, these strategies are often quite complicated functions of previous rounds of the game (for example) and a discussion of them is beyond our

TABLE 3.1 *Parameters of Some Important Discrete Distributions*

DISTRIBUTION		MEAN	VARIANCE	MOMENT GENERATING FUNCTION
Binomial	$P(X=j) = \binom{n}{j}\theta^j \eta^{n-j}$ $j=0,1,\ldots,n,\ \theta+\eta=1$	$n\theta$	$n\theta\eta$	$(\theta e^t + \eta)^n$
Geometric	$P(X=j) = \theta\eta^{j-1}$ $j=1,\ldots,\infty,\ \theta+\eta=1$	$\dfrac{1}{\theta}$	$\dfrac{\eta}{\theta^2}$	$\theta e^t(1-\eta e^t)$
Poisson	$P(X=j) = \dfrac{\lambda^j}{j!}e^{-\lambda}$ $j=0,1,\ldots,\infty,\ \lambda>0$	λ	λ	$e^{\lambda(e^t-1)}$
Hypergeometric	$P(X=j) = \dfrac{\binom{r}{j}\binom{n-r}{k-j}}{\binom{n}{k}}$ $j=0,1,\ldots,r,\ 0\le r\le n,$ $1\le k\le n$	$\dfrac{kr}{n}$	$\dfrac{kr}{n}\left(1-\dfrac{r}{n}\right)\left(\dfrac{n-k}{n-1}\right)$	
Pascal	$P(X=j) = \binom{j-1}{k-1}\theta^k\eta^{j-k}$ $j=k,k+1,\ldots,\infty,\ \theta+\eta=1$	$\dfrac{k}{\theta}$	$\dfrac{k\eta}{\theta^2}$	$\left(\dfrac{\theta}{1-\eta e^t}\right)^k$

scope. In the remainder of this section a few games of chance are analyzed combinatorially and this is done informally by examples. The book by Epstein (1967) contains further analyses on many of these problems.

EXAMPLE 3.12   (Odd man out)

Each of n players tosses a fair coin and if his outcome is different from the $(n-1)$ others (i.e., he has a head and all others have tails or vice versa), then he wins. They continue to throw until somebody wins. If X is the number of throws required to terminate the game, find the probability distribution of X and its expected value.

On any given trial, the probability that the game ends is just the probability of obtaining one head and $(n-1)$ tails

$$\left(\binom{n}{1} (\tfrac{1}{2})(\tfrac{1}{2})^{n-1} \right)$$

or one tail and $(n-1)$ heads

$$\left(\binom{n}{1} (\tfrac{1}{2})^{n-1}(\tfrac{1}{2}) \right)$$

Thus the probability the game terminates on a given trial is

$$\theta = 2 \cdot n \cdot (\tfrac{1}{2})^n = n \cdot 2^{-(n-1)}$$

and the probability it continues is $\eta = 1 - \theta$. The probability that the game terminates on the jth throw is then

$$P(X = j) = \theta \eta^{j-1}, \quad j = 1, 2, \ldots$$

which is just the geometric distribution encountered in the previous section. The mean of this distribution is just $1/\theta$ or $2^{n-1}/n$.

The game of poker has been the object of much combinatorial analysis. The probability of the various types of hands are calculated in the following example.

EXAMPLE 3.13

Calculate the probabilities of the following poker hands: (i) one pair; (ii) two pairs; (iii) three of a kind; (iv) a straight; (v) a flush; (vi) a full house; (vii) four of a kind; (viii) a straight flush; and (ix) a royal flush.

In describing the hand, for example, as containing a single pair, it is assumed that it does not contain a stronger hand (e.g., two pairs). In discussing cards we will call the face value of the card ace, 2, 3, . . . , 10, jack, queen, king as the designation. Thus there are thirteen designations and four suits. Since there are $\binom{52}{5}$ possible poker hands we need only enumerate the number of possibilities for obtaining the various hands to determine their probabilities.

(i) A single pair. There are $\binom{13}{1}$ ways of choosing the designation for the pair and, for each of these, $\binom{4}{2}$ ways of choosing the two suits. We have to fill the hand out with three other cards and these must have different designations. There are $\binom{12}{3}$ ways of choosing these since we cannot choose the designation of the pair. For each of these three designations we can choose one of the four suits and hence the number of ways of obtaining a poker hand containing exactly one pair is

$$\binom{13}{1}\binom{4}{2}\binom{12}{3}\binom{4}{1}^3$$

and

$$P(\text{exactly one pair}) = \frac{\binom{13}{1}\binom{4}{2}\binom{12}{3}\binom{4}{1}^3}{\binom{52}{5}} = .423$$

(ii) Two pairs. We can choose the two designations of the pairs in $\binom{13}{2}$ ways and, for each of the two designations we can choose the two suits in $\binom{4}{2}$ ways. The remaining card can be chosen in $\binom{44}{1} = \binom{11}{1}\binom{4}{1}$ ways and

$$P(\text{exactly two pairs}) = \frac{\binom{13}{2}\binom{4}{2}^2\binom{11}{1}\binom{4}{1}}{\binom{52}{5}} = .048$$

(iii) Three of a kind. The designation can be chosen in $\binom{13}{1}$ ways and the three suits in $\binom{4}{3}$ ways. The remaining two cards must not be a pair and not be of the same designation as the three and can be chosen in $\binom{12}{2}\binom{4}{1}^2$ ways:

$$P(\text{three of a kind}) = \frac{\binom{13}{1}\binom{4}{3}\binom{12}{2}\binom{4}{1}^2}{\binom{52}{5}} = .021$$

(iv) A straight (five cards in sequence in two or more suits). Since the designations must be in sequence and we assume that an ace can be counted as either high or low, there are ten possible beginning cards for the straight. For each of the five cards the suit can be chosen in $\binom{4}{1}$ ways. If all five cards have the same suit, it is called a straight flush and is a stronger hand. The number of straight flushes is $10\binom{4}{1}$, and so

$$P(\text{straight}) = \frac{10\binom{4}{1}^5 - 10\binom{4}{1}}{\binom{52}{5}} = .004$$

and

$$P(\text{straight flush}) = \frac{10\binom{4}{1}}{\binom{52}{5}} = 0.000015$$

(v) A flush (five cards of the same suit, not necessarily in sequence). The suit can be chosen in $\binom{4}{1}$ ways and, for this suit, the five cards can be chosen in $\binom{13}{5}$ ways:

$$P(\text{flush}) = \frac{\binom{4}{1}\binom{13}{5}}{\binom{52}{5}} = .002$$

(vi) Full house (three of one kind and two of another). The two designations can be chosen in $\binom{13}{2}$ ways and, for three of a kind the three suits can be chosen in $\binom{4}{3}$ ways while for two of a kind the two suits can be chosen in $\binom{4}{2}$ ways. For the $\binom{13}{2}$ choices for the designation we can associate the triple with either of the two choices and the pair with the remaining one:

$$P(\text{full house}) = \frac{2\binom{13}{2}\binom{4}{3}\binom{4}{2}}{\binom{52}{5}} = .0014$$

(viii) Four of a kind

$$P(\text{four of a kind}) = \frac{\binom{13}{1}\binom{4}{4}\binom{48}{1}}{\binom{52}{5}} = .00025$$

(ix) A royal flush (five cards in suit and sequence, ace high)

$$P(\text{royal flush}) = \frac{4}{\binom{52}{5}} = 1.5 \times 10^{-6}$$

In Example 3.10 we found the probability of throwing a four before a seven with two dice. This is the kind of calculation required in analyzing the game of craps described in the following example.

EXAMPLE 3.14

In the game of craps a player throws two dice and observes the sum. If, on this first throw, he throws a seven or an eleven, he wins outright. If he throws a two, three, or twelve, he loses outright. Otherwise the sum he obtains (4, 5, 6, 8, 9, 10) becomes his point and, in order to win the game now, he must continue to throw and obtain his point before he throws a seven. If he throws a seven before his point, he loses. What is the probability of winning at craps?

It should be mentioned that spectators to the game can also bet on each throw of the dice, on a variety of possible outcomes. We consider only the simple question of the thrower's probability of winning.

The probability of a win on the first toss is just the probability of either a seven or an eleven $(6/36 + 2/36)$. Suppose a four (the player's "point") is thrown on the first toss. The player can now only win by throwing a four before a seven, and, by Example 3.10 the probability of this is $1/3$. Therefore the probability of a player winning by throwing a four on the first throw is $(3/36) \cdot (1/3)$. If the player throws a five on the first throw, then the probability of his throwing a five before a seven on subsequent throws can be found in the same manner as in Example 3.10 as

$$P(\text{five before a seven}) = \frac{4}{36} + \frac{26}{36} \cdot \frac{4}{36} + \left(\frac{26}{36}\right)^2 \cdot \frac{4}{36} + \cdots$$

$$= \frac{4/36}{1 - 26/36}$$

$$= \frac{4}{10}$$

For the remaining possible points it can be shown that

$$P(\text{six before a seven}) = P(\text{eight before a seven}) = \frac{5}{11}$$

and that

$$P(\text{nine before a seven}) = P(\text{five before a seven}) = \frac{4}{10}$$

and

$$P(\text{ten before a seven}) = P(\text{four before a seven}) = \frac{1}{3}$$

Thus the probability of winning at craps is

$$P(\text{winning}) = P(\text{seven or eleven on first throw}) + P(\text{four on first throw})$$

$$\times P(\text{four before a seven on succeeding throws}) + \cdots$$

$$+ P(\text{ten on first throw})$$

$$\times P(\text{ten before a seven on succeeding throws})$$

$$= \frac{6}{36} + \frac{2}{36} + 2\left(\frac{3}{36} \cdot \frac{1}{3} + \frac{4}{36} \cdot \frac{4}{10} + \frac{5}{36} \cdot \frac{5}{11}\right)$$

$$= \frac{1952}{36 \times 110} \cong 0.493$$

3.3 *Cell Occupancy and Matching Problems*

Two problems are considered in this section. The first concerns distributing balls in cells at random. This is a classical problem in probability theory and versions of it occur in a wide variety of applications. Matching problems involve ordering a set of distinct objects in two ways, one of them being random, and observing the number of objects lying in the same position in both orderings (matches). Both problems rely heavily on combinatorial methods for solution.

We consider first the problem of randomly placing n balls in m cells, a problem often referred to as the classical cell occupancy problem. Many difficult questions can be asked about such a simple situation and even the relatively straightforward ones investigated here require some involved computations.

To describe the problem we imagine placing n balls, one at a time, into m cells that are somehow distinguished. We assume the cells are tagged with the numbers $1, 2, \ldots, m$. A sample space for the experiment will consist of n-tuples $(a_1, a_2, \ldots, a_n)$, where each a_i can take on the values $1, 2, \ldots, m$. If a_j is k, then the jth ball is placed into the kth cell. For example, if we place five balls in three cells, then the 5-tuple (2, 1, 2, 3, 1) indicates that the first ball is placed in cell 2, the second ball in cell 1, the third ball in cell 2, and so forth. In the general situation of placing n balls in m cells, since we can place m possible numbers (the cell numbers) in each of n-places (corresponding to the ball numbers), there are m^n possible n-tuples. A simple example of this will be useful.

EXAMPLE 3.15

Describe the sample space of placing 2 balls into 4 cells at random. What is the probability that (i) both balls are placed in the same cell, (ii) cell number 3 is empty, and (iii) cell number 1 contains exactly one ball?

The sample space consists of the $4 \times 4 = 16$ 2-tuples:

(1, 1) (2, 1) (3, 1) (4, 1)
(1, 2) (2, 2) (3, 2) (4, 2)
(1, 3) (2, 3) (3, 3) (4, 3)
(1, 4) (2, 4) (3, 4) (4, 4)

There are 16 points in this sample space and the term "at random" is interpreted as meaning that each of these points has the same probability. The event $A = \{$both balls in the same cell$\}$ is the set $\{(1, 1), (2, 2), (3, 3), (4, 4)\}$ and has probability $4/16 = 1/4$. The event $B = \{$cell number 3 is empty$\}$ consists of those 2-tuples not containing a 3 in either position. There are nine such points and $P(B) = 9/16$. The event $C = \{$cell number 1 contains exactly 1 ball$\}$ consists of those containing exactly one 1 and there are six such points giving $P(C) = 6/16$.

We should notice immediately that the assumption of m^n equally likely points in the sample space implies that we are placing n distinguishable balls into m distinguishable cells. In the previous example both points (1, 2) and (2, 1) mean cells 1 and 2 each contain one ball, but they are not the same point. The point (1, 2) means ball 1 is in cell 1 and ball 2 in cell 2 while (2, 1) has the reverse meaning. We will comment on other ways of placing balls in cells later.

The first problem of interest is to determine the probability distribution of the number of balls in a particular cell that we take as cell 1 since all cells are, from a probabilistic point of view, the same. Let X be the number of balls in cell 1. To determine $P(X = j)$ we have to know the number of points in the sample space containing exactly j 1's. We can choose the j places in which to place 1's in $\binom{n}{j}$ ways. In the remaining $(n - j)$ places, any cell number but 1 can appear and there are $(m - 1)^{n-j}$ ways of filling these $(n - j)$ places. Thus there is a total of $\binom{n}{j}(m - 1)^{n-j}$ points in the sample space containing exactly j 1's and

$$P(X = j) = \binom{n}{j} \frac{(m - 1)^{n-j}}{m^n}, \quad j = 0, 1, \ldots, n \tag{3.12}$$

Writing this equation in the form

$$P(X = j) = \binom{n}{j}\left(\frac{1}{m}\right)^i \left(1 - \frac{1}{m}\right)^{n-i}, \quad j = 0, 1, \ldots, n$$

it is clear that X is a binomial random variable with parameters n and $\theta = 1/m$. But this is clear since each ball has a probability of $1/m$ of being placed in cell 1 (success) and the experiment involves repeating this placement n times.

EXAMPLE 3.16

Forty customers at a bank are waiting for the doors to open, whereupon they will each choose, at random, one of five tellers (and after choosing they do not change lines). If no more customers, after the forty, arrive: (i) What is the most likely number of customers that teller three has to serve? (ii) What is the expected length of the queue of teller one? (iii) What is the probability that teller two has no customers?

(i) If X is the number of customers in the queue of teller three, then X has a binomial random variable with parameters $n = 40$ and $\theta = 1/5$. As shown in Problem 25 the most likely value of X is the smallest integer greater than $n\theta - (1 - \theta)$, in this case, 8. (ii) The expected length of any queue is $n\theta = 8$. (iii) The probability that teller three has no customers is just $(4/5)^{40}$.

A more complicated question is that of determining the distribution of the number of empty cells when placing n balls at random among m cells. Letting Y be the number of empty cells, we first determine $P(Y = 0)$. If A_i is the event that the ith cell is empty (the set of all points in the sample space containing no i's), then $B = \bigcup_{i=1}^{m} A_i$ is the event that at least one cell is empty. The event $\bar{B}$ is then the event that every cell has at least one ball in it, that is, no cell is empty. Thus we have

$$P(Y = 0) = P(\bar{B}) = 1 - P(B)$$

where, by Property 6 of Section 1.5,

$$P(B) = P\left(\bigcup_{i=1}^{m} A_i\right)$$

$$= \sum_{i=1}^{m} P(A_i) - \sum_{i \neq j} P(A_i \cap A_j) + \cdots + (-1)^{m-1} P(A_1 \cap \cdots \cap A_m)$$

$$(3.13)$$

and the reader should recall the inclusion-exclusion principle used there.

The probability that cell i is empty, $P(A_i)$, is just $(1 - 1/m)^n$. The probability that both cells i and j are empty (regardless of what is in the others) is

$$\left(\frac{m-2}{m}\right)^n = \left(1 - \frac{2}{m}\right)^n$$

and, the probability that a fixed set of j cells is empty is $(1-j/m)^n$. Equation (3.13) with these substitutions becomes

$$P(B) = \binom{m}{1}\left(1-\frac{1}{m}\right)^n - \binom{m}{2}\left(1-\frac{2}{m}\right)^n + \cdots + (-1)^m \binom{m}{m}\left(1-\frac{m}{m}\right)^n$$

$$= \sum_{i=1}^{m} (-1)^i \binom{m}{i}\left(1-\frac{i}{m}\right)^n$$

and, consequently,

$$P(Y=0) = 1 - P(B)$$

$$= 1 - \sum_{i=1}^{m} (-1)^i \binom{m}{i}\left(1-\frac{i}{m}\right)^n$$

$$= \sum_{i=0}^{m} (-1)^i \binom{m}{i}\left(1-\frac{i}{m}\right)^n \tag{3.14}$$

which gives the probability that no cells are empty.

To find the probability that exactly j cells are empty it is, first of all, clear that these j cells could be chosen in $\binom{m}{j}$ ways. For each of these ways the probability that these j cells are empty is $(1-j/m)^n$ and the probability that the remaining $(m-j)$ cells are nonempty is given by (3.14) with m replaced by $(m-j)$. Putting all these parts together gives

$$P(Y=j) = \binom{m}{j}\left(1-\frac{j}{m}\right)^n \sum_{i=0}^{m-j} (-1)^i \binom{m-j}{i}\left(1-\frac{i}{m-j}\right)^n$$

$$= \binom{m}{j} \sum_{i=0}^{m-j} (-1)^i \binom{m-j}{i}\left(1-\frac{(i+j)}{m}\right)^n \tag{3.15}$$

EXAMPLE 3.17

Twelve people enter an elevator on the ground floor. If people leave the elevator at random and there are seven floors above ground, what is the probability: (i) Exactly two people leave on the fourth floor? (ii) At least one person leaves on each floor? (iii) All people exit on the same floor?

The translation of this problem to a cell occupancy problem is clear: The cells are the floors and the balls, the people. (i) On any one floor, the number of people exiting is binomially distributed with parameters $n = 12$ and $\theta = 1/7$. The probability of exactly two people exiting on the four floor is

$$\binom{12}{2}\left(\frac{1}{7}\right)^2\left(1-\frac{1}{7}\right)^5$$

(ii) Translated, we want the probability that no cell is empty, which, by Eq. (3.14) is

$$\sum_{i=0}^{7} (-1)^i \binom{7}{i}\left(1-\frac{i}{7}\right)^{12}$$

(iii) Of the 7^{12} points in the equally likely sample space, exactly seven correspond to all people exiting at the same floor and the probability of this event is $7/7^{12} = 1/7^{11}$.

The cell occupancy problem can also be posed as a multinomial distribution problem. Essentially the problem of placing n balls in m cells is the same as placing one of m numbers in each of n places. If X_i is the number of balls in cell i, then $P(X_1 = k_1, X_2 = k_2, \ldots, X_m = k_m)$, $k_1 + k_2 + \cdots + k_m = n$ is simply $(1/m^n)$ times the number of ways of placing k_1 1's, k_2 2's, $\ldots$, k_m m's into n places. We can place the k_1 1's in the n places in $\binom{n}{k_1}$ ways, the k_2 2's in the remaining $(n - k_1)$ places in $\binom{n - k_1}{k_2}$ ways, and so on. This is precisely the argument used in developing the multinomial distribution and we have

$$P(X_1 = k_1, X_2 = k_2, \ldots, X_m = k_m) = \frac{n!}{k_1! k_2! \cdots k_m!} \left(\frac{1}{m}\right)^{k_1} \left(\frac{1}{m}\right)^{k_2} \cdots \left(\frac{1}{m}\right)^{k_m}$$

$$= \binom{n}{k_1 k_2 \cdots k_m} \left(\frac{1}{m}\right)^n$$

Unfortunately, this formulation is not useful for many of the questions of interest in cell occupancy problems.

When the number of balls is less than the number of cells ($n < m$), we can ask for the probability that no two balls end up in the same cell. To count the number of points in the sample space corresponding to this event we argue as follows: The first coordinate position can be filled in m ways, the second in $(m - 1)$ since there is to be no repetition, the third in $(m - 2)$ ways, and so forth. The probability of no two balls in the same cell is then

$$\frac{m(m-1)(m-2) \cdots (m-n+1)}{m^n} = \frac{m!}{(m-n)! \, m^n} \tag{3.16}$$

EXAMPLE 3.18

In a gathering of n people, $n \leq 365$, what is the probability that at least two people have the same birthday (month/day), assuming a 365 day year?

The probability that at least two people have the same birthday is one minus the probability that there are no multiple birthdays and, using Eq. (3.16),

P(at least 2 people have the same birthday)

$$= 1 - \frac{365 \cdot 364 \cdot 363 \cdot \cdots \cdot (365 - (n-1))}{365^n}$$

These probabilities are tabulated in Table 3.2 for n from 1 to 40 and note that with only 20 people in the room there is a better than 50-50 chance that two of them were born on the same day of the year.

TABLE 3.2

1	0
2	0.0027
3	0.0082
4	0.0164
5	0.0271
6	0.0405
7	0.0562
8	0.0743
9	0.0946
10	0.1169
11	0.1411
12	0.167
13	0.1944
14	0.2231
15	0.2529
16	0.2836
17	0.315
18	0.3469
19	0.3791
20	0.4114
21	0.4437
22	0.4757
23	0.5073
24	0.5383
25	0.5687
26	0.5982
27	0.6269
28	0.6545
29	0.681
30	0.7063
31	0.7305
32	0.7533
33	0.775
34	0.7953
35	0.8144
36	0.8322
37	0.8487
38	0.8641
39	0.8782
40	0.8912

The prototype of the matching problem is to place n cards, marked with the integers 1 through n, at random into n cells also marked with the integers 1 through n. If card i is placed into the ith cell, it is called a match. The problem is to determine the probability of exactly k matches $0 \leq k \leq n$. There are $n!$ ways of placing the n cards into the n cells, corresponding to the $n!$ permutations of the integers $1, \ldots, n$. These $n!$ permutations form the points of the sample space S that we assume to be equally likely.

Let X be the random variable giving the number of matches. Before finding the distribution of X we first determine an important piece of information, the number of the $n!$ permutations in which none of the n cards fall into their "natural" cell (i.e., the number of permutations corresponding to no matches). Let A_i be the event that i falls into the ith cell. The event

$$B = \bigcup_{i=1}^{n} A_i$$

is the event that at least one of the integers falls into its natural cell, that is, at least one match. The event $\bar{B}$ contains those points in which none of the integers fall in their natural cell. Again using Property 6 of Section 1.5,

$$P\left(\sum_{i=1}^{n} A_i\right) = \sum_{i=1}^{n} P(A_i) - \sum_{i \neq j} P(A_i \cap A_j)$$

$$+ \sum_{i \neq j \neq k} P(A_i \cap A_j \cap A_k) - \cdots (-1)^{n+1} P(A_1 \cap \cdots \cap A_n)$$

$$(3.17)$$

The number of points in A_i is the number of ways we can place i in cell i and the other $(n-1)$ integers, in any order, in the other $(n-1)$ cells—there may be more than one match. This can be done in $(n-1)!$ ways. For any i, $P(A_i) = (n-1)!/n! = 1/n$. The set $A_i \cap A_j$ contains those points for which i is in the ith cell and j in the jth cell. There are $(n-2)$ ways of permuting the remaining $(n-2)$ numbers and $P(A_i \cap A_j) = (n-2)!/n!$, $i \neq j$. Proceeding in the same manner we can determine that $P(A_i \cap \cdots \cap A_{i_k}) = (n-k)!/n!$ for any set of k distinct subscripts $(i_1, \ldots, i_k)$. Equation (3.17) then reduces to

$$P\left(\bigcup_{i=1}^{n} A_i\right) = n \cdot \frac{1}{n} - \binom{n}{2}\frac{(n-2)!}{n!} + \binom{n}{3}\frac{(n-3)!}{n!} - \cdots (-1)^{n+1}\binom{n}{n}\frac{(n-n)!}{n!}$$

$$= \sum_{i=1}^{n} (-1)^{i+1}\binom{n}{i}\frac{(n-i)!}{n!}$$

$$= \sum_{i=1}^{n} (-1)^{i+1}\frac{1}{i!}$$

It also follows that the number of points in $B = \bigcup_{i=1}^{n} A_i$ is

$$\sum_{i=1}^{n} (-1)^{i+1}\binom{n}{i}(n-i)!$$

and so the number of points in $\bar{B}$, corresponding to those containing no matches, is

$$M_n = n! - \sum_{i=1}^{n} (-1)^{i+1}\binom{n}{i}(n-i)!$$

$$= \sum_{i=0}^{n} (-1)^{i}\binom{n}{i}(n-i)! \qquad (3.18)$$

To evaluate $P(X = k)$, the probability of exactly k matches on any particular outcome of the experiment, we need only count the number of outcomes containing exactly k matches. To count these points we first choose k positions from the n, which can be done in $\binom{n}{k}$ ways, and count the number of points containing no matches in the remaining $(n - k)$ positions. But this is just the number of ways of permuting $(n - k)$ numbers and having no matches, which is given by Eq. (3.18) with n replaced by $(n - k)$, that is, M_{n-k}. The probability of exactly k matches is then

$$P(X = k) = \binom{n}{k} \frac{M_{n-k}}{n!}$$

$$= \frac{n!}{(n - k)!\, k!} \cdot \frac{1}{n!} \cdot \sum_{i=0}^{n-k} (-1)^i \binom{n - k}{i} (n - k - i)!$$

$$= \frac{1}{k!} \sum_{i=0}^{n-k} \frac{(-1)^i}{i!}, \quad k = 0, 1, \dots, n \qquad (3.19)$$

Recalling that a series expansion for the inverse of the exponential function is given by

$$e^{-1} = \sum_{i=0}^{\infty} \frac{(-1)^i}{i!}$$

then for n large, it is apparent that we can approximate the distribution of Eq. (3.19) by

$$P(X \cong k) = \frac{e^{-1}}{k!}, \quad k = 0, 1, \dots, n$$

which, if n were infinity, would just be a Poisson distribution with parameter $\lambda = 1$. The mean of the distribution of Eq. (3.19) is, by direct calculation,

$$E[X] = \sum_{k=0}^{n} k P(X = k)$$

$$= \sum_{k=0}^{n} k \cdot \frac{1}{k!} \left(\sum_{i=0}^{n-k} \frac{(-1)^i}{i!} \right)$$

$$= \sum_{k=1}^{n} \frac{1}{(k - 1)!} \left(\sum_{i=0}^{n-k} \frac{(-1)^i}{i!} \right)$$

By setting $l = k - 1$ and summing over values of l from 0 to $n - 1$, this equation reduces to

$$E[X] = \sum_{l=0}^{n-1} \frac{1}{l!} \left(\sum_{i=0}^{n-l-1} \frac{(-1)^i}{i!} \right)$$

$$= 1 \qquad (3.20)$$

where the last equality results from the fact that Eq. (3.19) with n replaced by $n-1$ is a distribution and thus its sum over all permissible values as in Eq. (3.20), is unity.

It might be noted in Eq. (3.19) that $P(X = n-1)$ is zero since it is impossible to have $(n-1)$ matches without having n matches. It is also interesting to compare cell occupancy and matching problems with urn drawing problems. Suppose we have an urn containing m tags marked 1 through m. If we draw out m of them, one at a time with replacement, then we can view an outcome of such an experiment as an n-tuple, each position of which can take one of m values. The equivalence to the cell occupancy problem is clear. If drawing is without replacement, and all m tags are withdrawn, then there are precisely $m!$ possible outcomes and the equivalence to the matching problem becomes clear. The next example is a simple generalization of the standard matching problem.

EXAMPLE 3.19

The n numbers $1, 2, \ldots, n$ are written down in order. Under these, the same set of numbers is written down at random and under these the same set is again written down at random. What is the probability that all three numbers in exactly k of the columns are identical?

Let Y be the number of double matches obtained. We first determine $P(Y = 0)$. Let A_i be the event that all three numbers in column i are i (a double match). The event $B = \bigcup_{i=1}^{n} A_i$ is the event that there is at least one double match. The sample space consists of $(n!)^2$ points consisting of the $n! \times n!$ possible double permutations. In a calculation similar to the single matching case,

$$P(A_i) = \left(\frac{(n-1)!}{n!}\right)^2 \quad \text{and} \quad P(A_i \cap A_j) = \left(\frac{(n-2)!}{n!}\right)^2, \text{etc.}$$

From Eq. (4.6)

$$P\left(\bigcup_{i=1}^{n} A_i\right) = \sum_{i=1}^{n} (-1)^{i+1} \binom{n}{i} \left(\frac{(n-i)!}{n!}\right)^2$$

and

$$P\left(\overline{\bigcup_{i=1}^{n} A_i}\right) = \sum_{i=0}^{n} (-1)^{i} \binom{n}{i} \left(\frac{(n-i)!}{n!}\right)^2$$

It also follows that the number of points in the sample space containing *no* double matches is

$$M'_n = \sum_{i=0}^{n} (-1)^{i} \binom{n}{i} ((n-i)!)^2 \tag{3.21}$$

To find the probability of exactly k double matches we first observe there are $\binom{n}{k}$ ways of choosing the k places for the double match to occur. The remaining $(n-k)$ positions must contain no double matches and this number is given by Eq. (3.21) with n replaced by $n-k$. The probability of exactly k double matches is then

$$P(Y=k)=\binom{n}{k}\sum_{i=0}^{n-k}(-1)^i\binom{n-k}{i}\left(\frac{(n-k-i)!}{n!}\right)^2$$

$$=\frac{1}{k!n!}\sum_{i=0}^{n-k}(-1)^i\frac{(n-k-i)!}{i!},\quad k=0,1,2,\ldots,n$$

EXAMPLE 3.20

A computer operator has misfiled the files of n users. He attempts to cover his mistake by attaching n passwords and n names to the n files at random. What is the probability that: (i) No file will contain its correct password-name combination? (ii) No file will contain even one of its correct attributes (either name or password)?

(i) The problem corresponds to a double matching problem where the first row corresponds to the n files laid out in order, the second row corresponds to the n names passed-out at random, and the third row corresponds to the n passwords also passed out at random. This first part asks for the probability of no double matches and, by Eq. (3.21) this is

$$P(Y=0)=\frac{1}{n!}\sum_{i=0}^{n}(-1)^i\frac{(n-i)!}{i!}$$

(ii) The number of ways that each name can be assigned a file not its own is, by Eq. (3.18)

$$\sum_{i=0}^{n}(-1)^i\binom{n}{i}(n-i)!$$

and for each of these ways the passwords can be assigned to the files in the same number of ways. Dividing by $(n!)^2$, the total number of ways of assigning these two quantities gives

P(no file will contain even one correct attribute)

$$=\left(\frac{1}{n!}\right)^2\left(\sum_{i=0}^{n}(-1)^i\binom{n}{i}(n-i)!\right)^2$$

$$=\left(\sum_{i=0}^{n}(-1)^i\frac{1}{i!}\right)^2$$

Another typical problem to which this theory relates is the psychic or ESP research problem where, for example, a set of n cards, marked with the integers $1, 2, \ldots, n$ is placed face down by some random order. It is desired to test a subjects' ESP by asking him to decide on the number on each card. There are various strategies the subject may follow, but if he chooses all of the n

numbers once during the course of each experiment (he does not have to—if he says each card contains a 1, he will be guaranteed of exactly one match), we can use this theory of matching to compare his performance with a "random guess" strategy.

3.4 *Geometric Probability*

The term geometric probability refers to a class of problems for which probability is assigned by geometric reasoning. There is no theory as such—for each problem, the statement must be carefully read to determine the most reasonable approach. The subject will be illustrated by a series of examples.

EXAMPLE 3.21

What is the probability that the sum of two numbers x and y randomly chosen on the interval $(0, 1)$ is greater than 1 while the sum of their squares is less than 1?

Let the two random numbers be X and Y. The probability that the pair (X, Y) falls in any section of the square shown in Figure 3.1 is proportional to the area of that section. In fact, since the total area of the square is unity, the probability that the pair (X, Y) falls into some region of the square of area A is just A. The curves $x + y = 1$ and $x^2 + y^2 = 1$ are shown and the region of interest is shaded. The area of this region is just $\pi/4 - \frac{1}{2}$, the probability of the event that $(x + y) > 1$ and $(x^2 + y^2) < 1$.

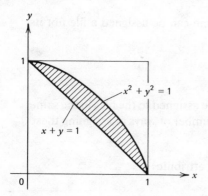

FIGURE 3.1

EXAMPLE 3.22

A game at a carnival consists of throwing dimes onto a tabletop marked in squares of side a. If the dime does not touch a line, the player wins. If r is the radius of a dime, what should the ratio r/a be if the game operator wants a player to have a chance of less than .25 of winning?

We can approximate the tabletop by an infinite grid of squares of side a, as shown in Figure 3.2a. For any particular square, as in Figure 3.2b, if the

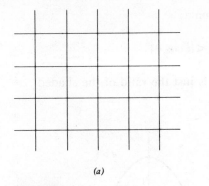
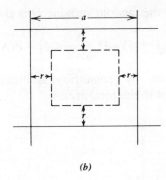

(a)   (b)

FIGURE 3.2

center of the dime lies within r of a side, then it will touch and so, in order not to touch it must lie within the square of side $(a - 2r)$. The probability of this event is just the ratio of the areas of the two squares:

$$P(\text{dime does not touch a line}) = \frac{(a - 2r)^2}{a^2} = \left(1 - \frac{2r}{a}\right)^2$$

For this quantity to be less than .25 requires that

$$\left(1 - 2\left(\frac{r}{a}\right)\right)^2 < \frac{1}{4}$$

which implies the ratio r/a should be greater than $1/4$.

EXAMPLE 3.23

A negative ion is attracted toward a positive plate. If the ion is injected at a distance X from the plate, where X is uniformly distributed between 0 and 1 cm, and the angle with which it proceeds to the plate is uniformly distributed between $-\pi/2$ and $\pi/2$, what is the cumulative distribution function of the actual distance the ion travels to the plate (see Figure 3.3)?

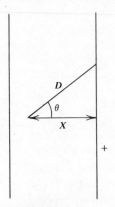

FIGURE 3.3

The cdf of the random variable D is given as

$$F_D(d) = P(D \le d) = P\left(\frac{X}{\cos \theta} < d\right) = P(X < d \cos \theta)$$

This last probability, by geometric reasoning, is just the ratio of the shaded areas in Figure 3.4 to the total area π.

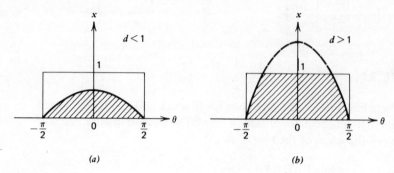

FIGURE 3.4

(i) $0 < d < 1$, $P(X < d \cos \theta) = \dfrac{2d}{\pi} \displaystyle\int_0^{\pi/2} \cos \theta \, d\theta = \dfrac{2d}{\pi}$

(ii) $d > 1$, $P(X < d \cos \theta) = \dfrac{2d}{\pi} \displaystyle\int_{\text{arc cos}(1/d)}^{\pi/2} \cos \theta \, d\theta + \dfrac{2}{\pi} \, \text{arc cos}\left(\dfrac{1}{d}\right)$

$$= \frac{2d}{\pi}\left(1 - \frac{\sqrt{d^2 - 1}}{d}\right) + \frac{2}{\pi} \, \text{arc cos}\left(\frac{1}{d}\right)$$

which gives the required cdf.

The next example is a classical problem (i.e., it's given a name) and is similar in nature to the previous example.

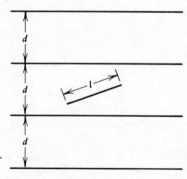

FIGURE 3.5

EXAMPLE 3.24 (Buffon's Needle Problem)

Consider an infinite set of infinite parallel lines on a smooth surface, the lines being distance d apart as shown in Figure 3.5. A needle of length l is dropped onto this surface at random. What is the probability that it will cross one of the lines if $l < d$?

To attack this problem it is necessary to make some assumptions as to what is meant by the phrase "at random" in the problem statement. We interpret the phrase to mean that the lowest part of the needle is at a position x that is uniformly distributed between the lines and the angle θ, as in Figure 3.6,

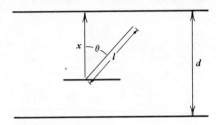

FIGURE 3.6

taken from the horizontal, is uniformly distributed over $(-\pi/2, \pi/2)$. It can be seen that these assumptions are reasonable and equivalent to any other reasonable interpretation of the statement. The probability that the needle does not cross a line is the probability that $l \cos \theta < x$. Since the assumptions indicate that θ and x are uniformly distributed over the rectangle $-\pi/2 < \theta < \pi/2, 0 < x < d$, the probability of the needle not crossing a line is the ratio of the shaded area in Figure 3.7 to the total area of the rectangle:

$$\frac{\int_{-\pi/2}^{\pi/2} l \cos \theta \, d\theta}{\pi d} = \frac{2l}{\pi d}$$

The case where $l > d$ is handled in a similar manner.

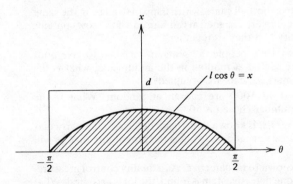

FIGURE 3.7

The final example of this section describes an example superficially different from the previous ones, but it quickly reduces to a similar situation.

EXAMPLE 3.25

What is the probability that the equation $x^2 + 2\sqrt{A}x - (B-1) = 0$ will have no real roots if A and B are jointly, uniformly distributed over the quarter circle $A^2 + B^2 < 1$, $A > 0$, $B > 0$?

The equation will have no real roots if and only if the discriminant $(2\sqrt{A})^2 + 4(B-1)$ is less than zero or, equivalently, $A + B < 1$. The probability that the equation has no real roots is the ratio of the two areas in Figure 3.8, which in this case is

$$\frac{1/2}{\pi/4} = \frac{2}{\pi}$$

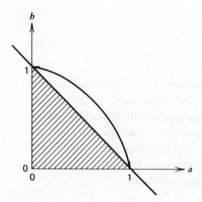

FIGURE 3.8

Problems

1. Four friends are given tickets to a football game, which happen to all be in the same row. If, within the row, the tickets were assigned at random and if the row contains 12 seats, what is the probability that they sit together?

2. In an auditorium that can seat 1000 people, 50 seats are marked to give door prizes. If $n < 1000$ people are seated at random in the auditorium, what is the probability that all 50 of the marked seats are occupied?

3. Four distinct numbers from 1 to 1000 are chosen at random. What is the probability that the smallest number chosen is 50?

4. On a day's production of 800 cars, it is known that 70 are defective. If a particular car dealer chooses 40 of these cars at random, what is the probability that he will get nine defective cars?

5. In a lot of 100 items, five are known to be defective. As a quality control measure, 20 items are tested. If at most one defective item is found, the lot is accepted. What is the probability of accepting the lot?

6. A baker gets his old stock of 80 loaves mixed up with his 160 freshly baked loaves. If each customer chooses their loaf at random, what is the probability that the one hundredth customer gets a stale loaf?

7. A student decides that half of the $2n$ questions on a true–false exam are true. If she decides to answer half the questions true, in how many ways could she fill out the exam? If half of the questions are in fact true, what is the probability that she gets the exam 100% correct? If the questions each have three possible answers and she chooses at random but decides that it is a bad strategy to answer adjacent questions the same, in how many ways could she answer the questions?

8. A collection of n tags are each numbered 1 through n. If k of the tags are drawn without replacement, what is the probability that they are consecutive integers? If the k tags are drawn with replacement, what is the probability that they are consecutive integers?

9. A bucket contains a large number of nails of various sizes. Half of the nails are $\frac{1}{2}$ cm in length, one quarter of length 1 cm, and the remaining quarter of length 2 cm. Fifteen nails are drawn at random from the bucket. What is the probability of:
   (i) obtaining four $\frac{1}{2}$-cm, eight 1-cm, and three 2-cm nails?
   (ii) having at least one nail of each kind?
   (iii) having only two of the three kinds of nails?

10. If n tags numbered 1 to n are in a bag and $m < n$ are drawn with replacement, what is the probability that no two of them are the same?

11. A lazy professor decides to mark only four questions, chosen at random from among the eight questions on an exam. A student, unaware of this marking strategy, answers only six questions, but knows that he answered them correctly. What is the probability distribution of his mark, assuming all questions were of equal value? Will he do better or worse on the average than if the professor marked all the questions?

12. A random number generator gives an integer between 0 and 9 every millisecond. How long is it necessary to wait in order for the probability of obtaining a 3 to be at least 0.95?

13. An urn contains n balls of which k_1 are of color 1, k_2 of color 2, ..., k_s of color s, $k_1 + k_2 + \cdots + k_s = n$. If m balls are chosen at random without replacement, what is the probability of obtaining j_1 of color 1, j_2 of color 2, ..., j_s of color s, $j_1 + \cdots + j_s = m$?

14. (i) What would the answer of Problem 13 be if the balls were chosen with replacement?
    (ii) What would the answer be if it was a large lot of which a fraction θ_i of the balls were of color i, $i = 1, \ldots, s$ and choice was with replacement? Without replacement?

15. If four dice are tossed, what is the probability that the sum of their faces is k, $4 \le k \le 24$.

16. Find the moment generating function of the binomial probability distribution function with parameters n and θ and from it show that the mean and variance of the distribution are $n\theta$ and $n\theta(1 - \theta)$, respectively.

17. A supermarket stocks n different items. How many ways could a shopper choose k not necessarily distinct items?

18. At a government tax office n people are given tags numbered 1 to n and asked to wait. The k inspectors, however, each choose a different person at random to help. What is the probability that the jth smallest tag number contains the number l, $1 \le l \le n$?

19. A coin is tossed five times and two heads are observed. Is it more likely that the coin is fair or that the probability of a head is .4?

20. How many numbers should be selected from a table of random numbers so that the probability of finding at least one even number among them is at least 0.9?

21. Ten electronic components are put on a life test. The time T to failure of a component is a random variable that is exponentially distributed with parameter α, that is, the pdf of T is $f_T(t) = \alpha e^{-\alpha t}$, $t > 0$. Find an expression that gives the probability that at least eight of these components will still be operative after 150 hours.

22. Using the mgf of a binomial distribution, show that

$$E((X - \mu)^3) = n\theta\eta(\eta - \theta)$$

where $\mu = E(X)$ and X is a binomial random variable with parameters n and θ, $0 < \theta < 1$, $\eta = 1 - \theta$.

23. If the probability of a male birth is .55 and that of a female birth is .45, what is the probability that a family with six children will have three girls and three boys?

24. A machine produces resistors with a nominal value of 1000 ohm but whose actual value is a random variable uniformly distributed from 950 to 1050 ohm. If ten resistors are taken at random from the output of this machine, what is the probability of obtaining at least two whose actual values are within $\pm 1\%$ of 1000 ohm?

25. If X is a binomial random variable with parameters n and θ, for what value of k will $P(X = k)$ be a maximum?

26. For a given set of parameters of the hypergeometric distribution, for what value of the random variable is the maximum achieved?

27. A fair die is tossed 12 times. What is the most probable number of 6's obtained? Two fair dice are thrown 17 times. What is the most probable number of times that their sum was either 7 or 11?

28. A machine produces items that, on the average, are 10% defective. Every hour, on the hour, an inspector chooses ten items from the output of that machine and if he finds two or more defectives he shuts the machine down. What is the expected length of time the machine runs between shutdowns?

29. If n points are chosen at random on a line of length unity, how many points must be chosen in order that there is a probability of .95 that at least one of them is greater than .9?

30. A machine produces relays. The probability that a machine breaks down on the ith closing is given by

$$P(X = i) = (1/10)(9/10)^{i-1}, \quad i = 1, 2, \ldots$$

If five relays are chosen at random from the machines output, what is the probability that exactly four of them will each operate 20 times or more?

31. The Pascal probability distribution is given by

$$P(X = j) = \binom{j-1}{k-1}\theta^k \eta^{j-k}, \quad 0 < \theta < 1, \quad \eta = 1 - \theta, \quad j = k, k+1, \ldots$$

Verify the mean, variance, and mgf for this distribution given in Table 3.1.

32. If X has the Pascal distribution of Problem 31, it has the interpretation of the number of trials to obtain the kth "success" when the probability of success on each trial is θ. What is the probability of obtaining the third 6 on the jth throw of a die? Show that when $k = 1$ the Pascal distribution becomes the geometric distribution function.

104 *Applied Probability*

33. Given m distinguishable balls distributed at random among n cells:
   (i) In how many distinct ways can this be accomplished?
   (ii) In how many ways can it be done if the balls are indistinguishable?

34. In how many ways can m indistinguishable balls be placed into n cells if no cell is to be empty?

35. Three women agree to meet in a downtown store. Unfortunately, this store has four branches. What is the probability that none of them meet? What is the probability that they all meet?

36. Cards labeled $1, 2, \ldots, n$ are placed down in a random order. What is the probability that no card occupies its "natural" position?

37. Twelve telephone booths at an airport are being used by seven people. If the seven people chose their booth at random, subject to the constraint of no more than one person per booth, what is the probability that three particular booths will be empty?

38. A flustered librarian reshelves N books on a rack in the N spaces allotted for them, but does so at random. What is the probability that no book is in its correct position?

39. Six limousines are waiting at an airport. The limousine company has a policy of not letting a limousine depart until it contains five passengers. If the cars are initially empty and one customer arrives every five minutes and chooses one of the limousines at random, what is the probability that a particular limousine is still waiting for passengers after 1 hour? (Assume that as a limousine leaves, another takes its place, and answer the question only for one of the original six limousines).

40. A bottle contains 100 cubic centimeters of milk with 30 bacteria dispersed at random through it. If 10 cubic centimeters are drawn off, what is the probability that this sample will contain ten of the bacteria?

41. A swarm of 100 fruit flies descends upon a tree containing 12 pieces of fruit. Each fly chooses a piece of fruit to land on. If exactly one piece of fruit has been poisoned by the farmer, what is the expected number of flies killed?

42. A bubblegum company includes pictures of football players on cards with each packet of bubblegum. If there are m different cards and the probability of getting any one of them with a purchase is the same, what is the probability that in $k < m$ purchases all the cards will be distinct? What is the probability that with $l > m$ purchases we will have at least one copy of each card?

43. Twenty people at a party compare birthdays. What is the probability that ten of the people have their birthday in January?

44. If n balls are placed at random into n cells, what is the probability that each cell contains one ball? What is the probability that there is exactly one empty cell?

45. Tags marked with the integers $1, \ldots, n, n > 4$, are placed in a random order. What is the probability that the numbers 1, 2, 3, and 4 appear next to each other and in their correct order.

46. In a mass of 100 kgm of molten steel, 30 cinders are distributed at random. From the steel, 100 1-kgm steel castings are made, and if a casting contains a cinder, it is considered defective. Find an expression for the probability that 30 castings will be defective.

47. From a random group of four people, what is the probability that two or more have been born on the same day of the week (assuming all seven days are equally likely to be a birthday)?

48. Ten hunters, all perfect shots, shoot at a flock of ten ducks flying overhead. If each hunter fires only the one shot and chooses a target at random, what is the probability distribution of the number of ducks shot?

49. Players A and B each toss a coin. If they each get heads or each get tails, they tie. Otherwise, the player who throws a head wins. Out of n throws what is the probability that A wins i, ties j, and loses $k = n - i - j$.

50. Two players A and B take turns to throw two dice. The first player to throw a 7 wins. What is the probability that A wins if A starts the game?

51. What is the probability of obtaining the third six on the sixteenth roll of a die?

52. Two players A and B play a coin game with a biased coin that has a probability θ of a head. Each player tosses until a head is obtained and then passes the coin to the other. If only n tosses are allowed and the player with the coin after the nth toss loses and A starts the game, what is the probability A wins?

53. What is the expected number of throws of a die to obtain three sixes?

54. It is known that for a particular unfair die it is three times as likely to obtain a 6 as any other number. What is the probability of obtaining exactly two 6's in four throws of the die? Compare this number with that expected for a fair die.

55. Two points are chosen at random on an interval of unit length. Find the mean, variance, and pdf of the distance between them.

56. Two points A and B are chosen at random on the unit interval $(0, 1)$. What is the probability that A is chosen closer to 0 than it is to B?

57. If in Example 3.22 the player wins if the dime touches either no line or one line, but not two, what is her probability of winning?

58. Two pieces of electrical machinery each operate once an hour for exactly five minutes each. Due to poor design, if they both operate at the same time, a fuse is blown. If each piece of machinery starts up at random, on the interval $(0, 55 \text{ min})$, what is the probability of not blowing a fuse?

59. Choose two points X and Y at random and independently on a line AB. Find the probability that the distance from Y to X is at least twice that from A to X.

60. Two numbers are chosen at random between 0 and 1. What is the probability that the reciprocal of their product is greater than 4.

61. Two men agree to meet on a given corner sometime between 12:00 and 1:00 p.m. but they each forget the exact time. If they each arrive at random and wait for only ten minutes for the other to show up, what is the probability that they meet?

62. The coefficients A and B in the quadratic equation $x^2 + Ax + B = 0$ are chosen at random in the intervals $(-1, 1)$ and $(-1, 1)$, respectively. What is the probability that the roots of this equation are real? positive?

CHAPTER 4

Conditional Probability, Bayes Theorem, and Independence

4.1   *Conditional Probability*
4.2   *Bayes Theorem*
4.3   *Independent Events*

The ideas of independence and conditional probability play a central role in probability theory. In fact, we have been using the notion of independence informally in the past few chapters and one of our tasks here will be to formalize it. Conditional probability is the study of how additional information can change our notion of how likely another event is to occur. The two notions of independence and conditional probability are closely related and this relationship will also be considered. Bayes theorem is a particular application of conditional probability, allowing problems that appear to have complicated statements to be solved in a straightforward manner.

4.1   *Conditional Probability*

A simple example will introduce the idea behind conditional probability.

EXAMPLE 4.1

An urn contains one white ball and two black balls. Two balls are drawn, without replacement. What is the probability of drawing a white on the second draw?

In order to get a white on the second draw, we must draw a black on the first draw. The probability of this is 2/3. If the first draw was a black, then the probability of drawing a white on a second draw is 1/2. Thus the probability of a white on the second draw is $(2/3) \cdot (1/2) = 1/3$. Suppose, however, that we have an observer telling us, after the first draw, what the color obtained was. This information is going to change our estimate of the probability of obtaining a white on the second draw. If he tells us the first draw is white, the probability of the second draw being white is zero. If he tells us the first draw is black, the probability of the second draw being white is $1/2$. In either case the additional information changed our probabilities. It is the manner in which the additional information changes the probabilities that is the study of conditional probability.

We consider a more general situation and let A and B be two events in a sample space S. Before we are told anything about what actually happened on a particular trial of the experiment, the probability that the outcome will be in A is $P(A)$. If we have an observer at the experiment who relays to us the information that the outcome is in the event B, how will this information affect our estimate of the probability that A also occurred? We will denote this probability by $P(A|B)$, read "the probability of A given B," also referred to as the conditional probability of A given B.

There are a few simple observations that can be made: If $B \subset A$, then the knowledge that B occurred implies that A must also have occurred and so $P(A|B) = 1$, regardless of what our prior estimate $P(A)$ of the probability A was. On the other hand, if A and B are disjoint events, $A \cap B = \varnothing$, then knowledge that B occurred implies that A could not have occurred and so $P(A|B) = 0$. It is what happens in the cases intermediate to these "extreme" cases that will be of interest to us.

To investigate the general situation where A and B intersect, we consider again the relative frequency approach. In defining $P(A|B)$ it is clear that we will only consider those outcomes in which the event B occurred. Those experiments in which the outcomes did not lie in B are completely ignored, as far as calculating $P(A|B)$ is concerned. Using the arguments of Section 1.5 it seems reasonable to define $P(A|B)$ as the fraction of the number of times that *both* A and B occurred relative to the number of times that B occurs. If we denote the three possible outcomes of an experiment as (i) $A \cap B$, (ii) B or, (iii) N depending on whether (i) both A and B occurred, (ii) B but not A occurred, or (iii) B did not occur, then a particular sequence of outcomes might be

$$N, N, A \cap B, N, A \cap B, B, B, A \cap B, N, N, B, B, N, B, A \cap B, \ldots$$

In a long string of outcomes it is reasonable to define $P(A|B)$, the probability that A occurred given that B occurred, as the ratio of the number of times that both A and B occurred to the number of times that B occurred (i.e., $A \cap B$ or B), $n(A \cap B)/n(B)$. This is our intuitive definition of conditional probability. If we now invoke our insistence that the experiment have some statistical regularity to it, then, for large n, we would expect that

$$\lim_{n \to \infty} \frac{n(A \cap B)}{n(B)} = \lim_{n \to \infty} \frac{n(A \cap B)/n}{n(B)/n} \cong \frac{P(A \cap B)}{P(B)}$$

For these reasons we define conditional probability in the following manner.

Definition   The conditional probability of A given B is

$$P(A|B) = \frac{P(A \cap B)}{P(B)} \qquad (4.1)$$

provided $P(B) \neq 0$. If $P(B) = 0$, then $P(A|B)$ is undefined.

In defining the probability of the occurrence of event A, given that event B occurred, we took the ratio of the number of joint occurrences of A and B to those of B. This is essentially the same as ignoring any outcome of the experiment that does not fall into B and hence we can view B as our new sample space, that is, knowing that the outcome is in B reduces our sample space from S to B. However, in order to normalize probabilities in this new or reduced sample space so that the probability of the whole new space is unity, we must divide all probabilities by its probability, $P(B)$. In this way we can view $P(A|B) = P(A \cap B)/P(B)$ as a probability on the reduced space B. The situation is shown in Figure 4.1. It is an easy matter to verify that it does satisfy

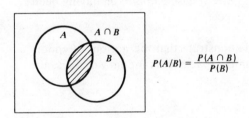

FIGURE 4.1

all the axioms required for it to be a probability measure (see Problem 4.15). To reinterpret the two extreme cases, if $A \cap B = \varnothing$, then $P(A|B) = P(A \cap B)/P(B) = 0$ and if $B \subset A$, then $P(A|B) = P(A \cap B)/P(B) = P(B)/P(B) = 1$ and both of these results agree with our earlier intuitive notions.

EXAMPLE 4.2

If the probability of hitting a target is $1/5$ and ten shots are fired, then: (i) What is the probability that the target will be hit at least twice? (ii) What is the conditional probability of the target being hit at least twice given that at least one hit was scored?

The purpose of the problem is to investigate the effect that the additional information—that at least one hit was scored—has on the probability that at least two hits were scored. If X is the number of hits, then it is a binomial random variable with parameters $n = 10$ and $\theta = 1/5$ and, consequently, the probability that at least two hits were scored, $P(X \geq 2) = 1 - P(X=0) - P(X=1)$ is

$$P(X \geq 2) = 1 - \binom{10}{0}\left(\frac{1}{5}\right)^0\left(\frac{4}{5}\right)^{10} - \binom{10}{1}\left(\frac{1}{5}\right)^1\left(\frac{4}{5}\right)^9 = .623$$

The conditional probability we seek is $P(X \geq 2 | X \geq 1)$, that is, the probability of the event $A = \{X \geq 2\}$ conditioned on the event $B = \{X \geq 1\}$. Since $A \cap B = \{X \geq 2\}$, the conditional probability is simply

$$P(X \geq 2 | X \geq 1) = \frac{P(X \geq 2)}{P(X \geq 1)}$$

$$= \frac{1 - \binom{10}{0}\left(\frac{1}{2}\right)^0\left(\frac{4}{5}\right)^{10} - \binom{10}{1}\left(\frac{1}{5}\right)^1\left(\frac{4}{5}\right)^9}{1 - \binom{10}{0}\left(\frac{1}{5}\right)^0\left(\frac{4}{5}\right)^{10}}$$

$$= \frac{.623}{.892} = .698$$

and our estimate of the probability that $\{X \geq 2\}$ has increased on having been told that $\{X \geq 1\}$, which is as we would expect.

The next example is simply to demonstrate that the notions of conditional probability are applicable to all sample spaces and random variables, not just discrete ones.

EXAMPLE 4.3

The lifetime T of a certain type of light bulb is a random variable with the exponential pdf.

$$f_T(t) = .001 \exp(-(.001)t), \quad t \geq 0$$

(i) Find the probability that a bulb will last 2000 hours. (ii) Find the probability that a bulb will last 2000 hours given that it has lasted 1000 hours.
   (i) From the definition of a pdf,

$$P(T \geq 2000) = \int_{2000}^{\infty} (.001) \exp(-(.001)t) \, dt$$

$$= \exp(-(.001) \times 2000) = e^{-2}$$

(ii) From the definition of conditional probability,

$$P(T \geq 2000 | T \geq 1000) = \frac{P((T \geq 2000) \cap (T \geq 1000))}{P(T \geq 1000)}$$

$$= \frac{P(T \geq 2000)}{P(T \geq 1000)}$$

and, from part (i) this is just

$$P(T \geq 2000 \mid T \geq 1000) = \frac{e^{-2}}{e^{-1}} = e^{-1}$$

Using the notion of conditional probability we can derive a very useful formula. Recall from Section 1.5 that the collection of sets $\{B_i\}$, $i = 1, \ldots, n$ forms a partition of the sample space S if (i) $B_i \cap B_j = \varnothing$, $i \neq j$ and (ii) $\bigcup_{i=1}^{n} B_i = S$. For any event $A \subset S$ we can use this partition to write

$$A = \bigcup_{i=1}^{n} (A \cap B_i)$$

which is another way of saying that every element of A is in exactly one of the sets B_i. Furthermore, since $(A \cap B_i) \cap (A \cap B_j) = \varnothing$, $i \neq j$, the sets $(A \cap B_i)$, $i = 1, \ldots, n$ are disjoint, which allows us to write

$$P(A) = \sum_{i=1}^{n} P(A \cap B_i) \tag{4.2}$$

From the equation defining conditional probability

$$P(A \cap B_i) = P(A \mid B_i)P(B_i) \tag{4.3}$$

for any two sets A, B_i, provided that $P(B_i) > 0$ for all i. Substituting Eq. (4.3) into Eq. (4.2) results in

$$P(A) = \sum_{i=1}^{n} P(A \mid B_i)P(B_i) \tag{4.4}$$

an equation that is often referred to as the theorem on total probability. It can be a very useful tool in the solution of some problems as it has the effect of decomposing a difficult question into several relatively simple ones. The following examples will illustrate this point.

EXAMPLE 4.4

Two names are to be drawn out of a hat containing n names, the drawing being sequential and without replacement. If $m < n$ of the names are those of women, what is the probability that the second name drawn will be that of a woman?

We first solve the problem without resort to the theorem on the total probability. It will be convenient in this instance to set up a sample space, which we will write as

$$S = \{(W, W), (W, M), (M, W), (M, M)\}$$

where, for example, (W, M) means that a woman's name was obtained on the first draw and a man's on the second. These points are not equally likely and the probability of the point (W, M) is calculated as follows: The probability of drawing a woman's name on the first draw is m/n. The probability of then drawing a man's name is $(n-m)/(n-1)$ and so the probability of the point (W, M) is

$$\frac{m}{n} \cdot \frac{n-m}{n-1}$$

The probabilities of the other points are calculated in a similar manner so that

$$P((W, W)) = \frac{m(m-1)}{n(n-1)}, \quad P((M, W)) = \frac{(n-m)m}{n(n-1)}, \quad \text{and} \quad P((M, M))$$

$$= \frac{(n-m)(n-m-1)}{n(n-1)}$$

We are interested in the probability of the event $A = \{$the second name drawn is that of a woman$\}$. Since $A = \{(W, W), (M, W)\}$, we have

$$P(A) = P((W, W)) + P((M, W))$$

$$= \frac{m(m-1)}{n(n-1)} + \frac{(n-m)m}{n(n-1)} = \frac{m}{n} \qquad (4.5)$$

This result was not difficult to achieve without using the theorem on total probability. It is perhaps conceptually easier to use it, as we now show. Choose the partition of the sample space S as $M_1 = \{(M, M), (M, W)\}$ and $W_1 = \{(W, M), (W, W)\}$ according to whether a man's or woman's name was obtained on the first draw. Using Eq. (4.4) we can write

$$P(A) = P(A|M_1)P(M_1) + P(A|W_1)P(W_1) \qquad (4.6)$$

and we claim that all of these terms are easy to calculate. Clearly $P(M_1)$ is just $(n-m)/n$ and $P(W_1)$ is m/n. The term $P(A|M_1)$ is the probability of obtaining a woman's name on the second draw, given that a man's was drawn on the first and this is $m/(n-1)$. Similarly, $P(A|W_1) = (m-1)/(n-1)$. Equation (4.6) then becomes

$$P(A) = \frac{m}{(n-1)} \cdot \frac{(n-m)}{n} + \frac{(m-1)}{(n-1)} \cdot \frac{m}{n} = \frac{m}{n}$$

as before. Equation (4.6) is, of course, just (4.5) with a different interpretation, but in problems of greater complexity this use of the theorem on total probability can simplify the situation greatly. It is an interesting exercise to go on and show that the probability of drawing a woman's name on any draw is m/n if no information on what was drawn on previous draws is available.

EXAMPLE 4.5

A pizza delivery boy finds that, each time he returns to the shop to pick up pizzas for delivery, the number of pizzas waiting is equally likely to be 1, 2, 3, 4, or 5. If it is also equally likely that a customer does or does not give him a tip, find the probability that he gets at least two tips on any trip.

It is possible to set up a sample space for this experiment (set up one that has $2+2^2+2^3+2^4+2^5$ points). Use of Eq. (4.4) makes this unnecessary. Let A be the event of obtaining at least two tips on any trip. Let B_i, $i = 1, \ldots, 5$ be the number of pizzas he began the trip with. Certainly these sets B_i will form a partition of any sample space and from Eq. (4.4)

$$P(A) = \sum_{i=1}^{5} P(A|B_i)P(B_i) \qquad (4.7)$$

From the information given, $P(B_i) = 1/5$. If the trip starts with i pizzas, the number of tips is then a binomial random variable with parameters $n = i$ and $p = 1/2$. The conditional probabilities are then

$$P(A|B_1) = 0$$

$$P(A|B_2) = \left(\frac{1}{2}\right)^2$$

$$P(A|B_3) = \left\{\binom{3}{2} + \binom{3}{3}\right\}\frac{1}{2^3}$$

$$P(A|B_4) = \left\{\binom{4}{2} + \binom{4}{3} + \binom{4}{4}\right\}\frac{1}{2^4}$$

$$P(A|B_5) = \left\{\binom{5}{2} + \binom{5}{3} + \binom{5}{4} + \binom{5}{5}\right\}\frac{1}{2^5}$$

Using Eq. (4.7) we obtain

$$P(A) = \frac{1}{5} \sum_{j=2}^{5} \frac{1}{2^j}\left\{\sum_{i=2}^{j} \binom{j}{i}\right\} = \frac{9}{20}$$

It is to be emphasized that this problem could have been solved by other means, but the use of the theorem of total probability reduced it to some simple computations.

The final example of this section applies the theorem of total probability to discrete distributions to derive a general result of some importance.

EXAMPLE 4.6

The number of particles emitted during a 10-min interval has a Poisson distribution with parameter λ. Due to the measuring apparatus, each particle has a probability p of being detected. What is the probability distribution of the number of particles actually detected during the 10-min interval?

Let X be the random variable giving the number of particles emitted during the 10-min period and Y the number of particles actually detected. The random variable X has a Poisson distribution

$$P(X = j) = \frac{\lambda^j e^{-\lambda}}{j!}, \quad j = 0, 1, \ldots, \infty$$

During any 10-min interval any number of particles, from zero to infinity, could be emitted. Suppose n particles are emitted. Since each particle has a probability θ of being detected, then the number of particles actually detected is binomially distributed with parameters n and θ, that is,

$$P(Y=j\,|\,X=n)=\binom{n}{j}\theta^j\eta^{n-j}, \quad j=0,1,\ldots,n$$

Using the theorem on total probability and identifying the partition $B_i = \{X=i\}$ and $A=\{Y=j\}$, then

$$P(Y=j)=\sum_{i=0}^{\infty} P(Y=j\,|\,X=i)P(X=i)$$

$$=\sum_{i=j}^{\infty} P(Y=j\,|\,X=i)P(X=i)$$

as $P(Y=j\,|\,X=i)=0$ for $i<j$. Substituting for the distribution gives

$$P(Y=j)=\sum_{i=j}^{\infty}\binom{i}{j}\theta^j(1-\theta)^{i-j}\frac{\lambda^i}{i!}e^{-\lambda}$$

$$=\sum_{i=j}^{\infty}\frac{i!}{j!\,(i-j)!}\theta^j(1-\theta)^{i-j}\frac{\lambda^i}{i!}e^{-\lambda}$$

$$=\frac{e^{-\lambda}}{j!}\left(\frac{\theta}{1-\theta}\right)^j\sum_{i=j}^{\infty}\frac{(\lambda(1-\theta))^i}{(i-j)!}$$

Changing the variable of summation from i to $k=i-j$ results in the equation

$$P(Y=j)=\frac{e^{-\lambda}}{j!}\left(\frac{\theta}{1-\theta}\right)^j\sum_{k=0}^{\infty}\frac{(\lambda(1-\theta))^{k+j}}{k!}$$

$$=\frac{e^{-\lambda}}{j!}\left(\frac{\theta}{1-\theta}\right)^j(\lambda(1-\theta))^j\sum_{k=0}^{\infty}\frac{(\lambda(1-\theta))^k}{k!}$$

$$=\frac{e^{-\lambda}}{j!}(\lambda\theta)^j e^{\lambda(1-\theta)}$$

$$=e^{-\lambda\theta}\frac{(\lambda\theta)^j}{j!}, \quad j=0,1,\ldots,\infty$$

where the equality

$$e^{\lambda(1-\theta)}=\sum_{k=0}^{\infty}\frac{(\lambda(1-\theta))^k}{k!}$$

was used. The distribution of the number of particles actually detected, Y, is Poisson with parameter $\lambda\theta$.

4.2 Bayes Theorem

The second important result on conditional probability is called Bayes Theorem and finds uses in situations where quantities of the form $P(A|B_i)$ and $P(B_i)$ are known and we wish to determine $P(B_i|A)$. This theorem can also be viewed as a device for updating the probabilities $P(B_i)$ given the information A. Several examples of this will be given later. Again let $\{B_i\}$, $i = 1, \ldots, n$ be a partition of the sample space S and A an event in S. From the definition of conditional probability

$$P(B_i|A) = P(A \cap B_i)/P(A) \tag{4.8}$$

and

$$P(A \cap B_i) = P(A|B_i)P(B_i) \tag{4.9}$$

while, from the theorem on total probability

$$P(A) = \sum_j P(A|B_j)P(B_j) \tag{4.10}$$

Substituting Eqs. (4.9) and (4.10) into (4.8) we find that

$$P(B_i|A) = \frac{P(A|B_i)P(B_i)}{\sum_j P(A|B_j)P(B_j)} \tag{4.11}$$

a result known as *Bayes theorem*. The terms on the right-hand side are all conditioned on the events B_i while that on the left is conditioned on A. This reversal of conditioning can be surprisingly useful as will be shown in the examples. The first example is typical of the application of Bayes theorem.

EXAMPLE 4.7

Of 100 patients in a hospital with a certain disease, ten are chosen to undergo a drug treatment that increases the percentage cured rate from 50% in the untreated case to 75%. If a doctor later encounters a cured patient, what is the probability that he received the drug treatment?

Let C be the event that a patient is cured. Let D be the event the patient received the drug treatment, and N the event he did not. Since the events D and N certainly partition the sample space, we can use Bayes theorem to write

$$P(D|C) = \frac{P(C|D)P(D)}{P(C|D)P(D) + P(C|N)P(N)}$$

From the information given $P(D) = 10/100$ and $P(N) = 90/100$. Given that a patient received the drug treatment, the probability of him being cured is .75, that is, $P(C|D) = .75$ and similarly $P(C|N) = .50$. Using these values

$$P(D|C) = \frac{(.75) \times (.1)}{(.75) \times (.1) + (.50) \times (.9)} = \frac{1}{7}$$

The important point in the previous example is that while it is not clear intuitively how to calculate $P(D|C)$ directly from the problem statement, Bayes theorem allowed the conditioning to be reversed so that all terms needed were relatively simple to calculate. The next example is often cited as one where the use of Bayes theorem seems to go against intuition.

EXAMPLE 4.8   (Bertrand's Box Paradox)

Three boxes each contain two coins. In one box, B_1, both coins are gold, in another, B_2, both are silver, and in the third, B_3, one is gold and the other is silver. A box is chosen at random and from it a coin is chosen at random. If this coin is gold, what is the probability that it came from the box containing two gold coins?

This problem is solved in almost exactly the same way as the previous problem. Let $P(B_i) = 1/3$ be the probability that B_i was chosen and G and S be the event that a gold and silver coin, respectively, was chosen. Again using Bayes theorem the desired probability is

$$P(B_1|G) = \frac{P(G|B_1)P(B_1)}{P(G|B_1)P(B_1) + P(G|B_2)P(B_2) + P(G|B_3)P(B_3)}$$

$$= \frac{1 \cdot 1/3}{1 \cdot 1/3 + 0 \cdot 1/3 + 1/2 \cdot 1/3} = \frac{2}{3}$$

This is the correct answer to the problem. However, someone not familiar with the subject might argue as follows. After drawing a gold coin there are two possible boxes it could have come from, B_1 and B_3. Since we are choosing boxes at random, with probability $1/2$ it came from B_1. The problem with this line of reasoning is that it is far more likely to draw a gold coin from B_1 than from B_2 and this argument fails to take this into account.

The next two examples illustrate other uses for Bayes theorem.

EXAMPLE 4.9

A committee of $4n$ people contains half men and half women. A woman is elected to choose, at random, from the remaining $4n - 1$ people, a committee of n people. If the n people chosen happen to be of the same sex, what is the probability they are all male?

Let M be the event that all n people chosen are male, S the event that all people chosen are of the same sex, and F the event that all n people chosen are female. By Bayes theorem

$$P(M|S) = \frac{P(S|M)P(M)}{P(S|M)P(M) + P(S|F)P(F)} \tag{4.12}$$

and in this expression $P(S|M) = P(S|F) = 1$ while

$$P(M) = \frac{\binom{2n}{n}}{\binom{4n-1}{n}} \quad \text{and} \quad P(F) = \frac{\binom{2n-1}{n}}{\binom{4n-1}{n}}$$

which when used in Eq. (4.12) yields

$$P(M|S) = \frac{\binom{2n}{n} / \binom{4n-1}{n}}{\left(\binom{2n}{n} / \binom{4n-1}{n}\right) + \left(\binom{2n-1}{n} / \binom{4n-1}{n}\right)}$$

$$= \frac{1}{1 + \binom{2n-1}{n} / \binom{2n}{n}} = \frac{1}{1 + 1/2} = \frac{2}{3}$$

giving the rather surprising result that the probability is independent of n.

EXAMPLE 4.10

Of three large lots of items, each lot of equal size, two contain 5% defective items and one 10% defective. Two of the lots chosen at random are mixed together and from this mixture five items are drawn out and one is found to be defective. What is the probability that one of the lots used in the mixing was the 10% defective one?

Either the two lots chosen for mixing were 5% defective or one was 5% and the other 10% defective. Denote the first event by L_1 and the second by L_2 and, since the lots were chosen at random (two items from three) the distribution of 5% lots among the two chosen has a hypergeometric distribution and $P(L_1) = 1/3$ and $P(L_2) = 2/3$. Denote the event of obtaining one defective item from a choice of five by D. If the five items were chosen from the combined lot of 5% lots (in which case 5% of items are defective), the probability of obtaining i defective items among five drawn is given by the binomial distribution with parameters $n = 5$ and $\theta = .05$. In particular, we have

$$P(D|L_1) = \binom{5}{1}(.05)(.95)^4$$

If the combined lot contains a 5% and a 10% defective lot, then it has 7.5% of its items defective since the original lots were assumed to be of equal size. It follows that

$$P(D|L_2) = \binom{5}{1}(.075)(.925)^4$$

and, from Bayes theorem,

$$P(L_2|D) = \frac{P(D|L_2)P(L_2)}{P(D|L_1)P(L_1) + P(D|L_2)P(L_2)}$$

$$= \frac{\binom{5}{1}(.075)(.925)^4 \cdot \frac{2}{3}}{\binom{5}{1}(.05)(.95)^4 \cdot \frac{1}{3} + \binom{5}{1}(.075)(.925)^4 \cdot \frac{2}{3}} \approx .7695$$

This quantity is to be compared with the "prior" probability $P(L_2) = 2/3$.

4.3 Independent Events

The previous two sections were primarily concerned with how some additional information about the outcome of a particular trial of an experiment affected the probability of some other outcome. In some situations, however, knowledge that the outcome is in the event B does not change the probability that A occurred, that is, $P(A|B) = P(A)$. Intuitively, this is what one would mean by two events being independent. Since the condition $P(A|B) = P(A)$ implies that $P(A \cap B) = P(A)P(B)$, and since we encounter some trouble with defining $P(A|B) = P(A \cap B)/P(B)$, if $P(B)$ is zero, we define independence as follows:

Definition The events A and B will be called independent if

$$P(A \cap B) = P(A)P(B) \tag{4.13}$$

It follows immediately that if A and B are independent events, then $P(A|B) = P(A)$ and $P(B|A) = P(B)$.

We have been making implicit use of the notion of independence in the past few chapters. To see how, consider an experiment that we repeat twice, each time only observing whether A occurred or not, $P(A) = \theta$. The sample space of such an experiment is $S = \{AA, A\bar{A}, \bar{A}A, \bar{A}\bar{A}\}$. Define two events on S: $A_1 = \{A$ occurred on the first trial$\}$ and $A_2 = \{A$ occurred on the second trial$\}$. In order to assign probabilities to the elements of S, we stated that the experiments were repeated independently, without formally defining what we meant by independently. We can now justify this informal and intuitive approach; if, by assumption, A_1 and A_2 are independent events, then

$$P(A_1 \cap A_2) = P(\text{event } A \text{ occurs on both first and second trials})$$

$$= P(A_1 \cap A_2) = P(A_1)P(A_2)$$

$$= P(\text{event } A \text{ occurs in first trial})$$

$$\cdot P(\text{event } A \text{ occurs on second trial})$$

$$= \theta^2$$

Thus using the assumption of independence between trials and knowledge of probabilities of outcomes on each trial, we can assign probabilities for outcomes on n repetitions of the experiment. For the above situation, $P(A_1 \cap \bar{A}_2) = \theta(1 - \theta)$, $P(\bar{A}_1 \cap A_2) = (1 - \theta)\theta$, and $P(\bar{A}_1 \cap \bar{A}_2) = (1 - \theta)^2$. Our previous use of the notion of independence is now formalized.

EXAMPLE 4.11

A lot of ten items contains four defective and six good items. (i) If two items are drawn sequentially without replacement and D_1 is the event of obtaining a defective item of the first draw and D_2 a defective item on the second draw, are D_1 and D_2 independent? (ii) If the drawings are with replacement, are D_1 and D_2 independent?

(i) If the drawings are without replacement and N_1 is the event of not getting a defective item on the first draw then

$$P(D_2) = P(D_2|N_1)P(N_1) + P(D_2|D_1)P(D_1)$$

$$= \frac{4}{9} \cdot \frac{6}{10} + \frac{3}{9} \cdot \frac{4}{10} = \frac{36}{90} = \frac{4}{10}$$

Notice that this is the same as $P(D_1)$. On the other hand, $P(D_2|D_1) = 3/9 \neq 4/10 = P(D_2)$ and D_1 and D_2 are not independent.

(ii) If the drawings are with replacement, then $P(D_1) = P(D_2) = 4/10$ as before, while $P(D_1 \cap D_2) = (4/10)^2$, and D_1 and D_2 are independent. This last calculation used the fact that D_1 and D_2 are independent in that in choosing the second item, it was assumed the outcome of the first draw had no effect. Thus, this is little more than a verification of the assumption of independence between draws.

There is a strong connection between the concepts of disjoint events and independent events. The natural inclination on first considering the connection might be to assume that disjoint events must be independent. Exactly the opposite is true and to see this let A and B be two events for which $P(A) > 0$ and $P(B) > 0$. If A and B are disjoint, $A \cap B = \varnothing$, then $P(A \cap B) = 0 \neq P(A) \cdot P(B) > 0$ and so disjoint events with positive probability cannot be independent. The intuitive reasoning here is also clear. If A and B are disjoint, then $P(A|B) = P(A \cap B)/P(B) = 0$ meaning that if B occurs, then A cannot have occurred and so the knowledge that B occurs very definitely affects the probability that A also occurred. It would seem that two events A and B can be independent only if they intersect each other in a certain way, namely, in such a manner that Eq. (4.13) is satisfied. Thus disjointedness of sets is a property of the sets themselves while independence is a multiplicative property of their probabilities. As we have seen in this and previous chapters, independence of certain events is usually an assumption made, based on our intuition on how the experiment or experiments are to be conducted, and this assumption of independence allows probabilities to be assigned.

EXAMPLE 4.12

Ten television sets are life tested under extreme conditions of temperature. The time to failure of each set is a random variable T with pdf (in hours)

$$f_T(t) = (10^{-4})\exp(-(10^{-4})t), \quad t \geq 0$$

If the times to failure are independent, find an expression for the probability that exactly two of the sets are still operating after 20,000 hours.

Suppose we set up a sample space consisting of 2^{10} points of 10-tuples containing F (for failed) or O (for operating) in the ith position indicating whether or not the ith television set had failed by 20,000 hours. From the information given, the sets $A_i = \{i$th television has failed by 20,000 hours$\}$ are then independent events. The probability that a set has failed by 20,000 hours is then

$$\theta = P(T \leq 20,000)$$

$$= \int_0^{20,000} (10^{-4})\exp(-(10^{-4})t)\, dt$$

$$= 1 - e^{-2}$$

and a little thought will reveal that the number of failed sets is a binomial random variable with parameters $n = 10$ and θ. We can think of testing the ten sets, one at a time, the outcome on each trial being independent of the others and the probability that the set failed before 20,000 hours (a "success") being constant for each set. Thus the probability that exactly two of the sets are still operating after 20,000 hours (i.e., exactly eight have failed) is

$$\binom{10}{8}(1-e^{-2})^8(e^{-2})^2$$

Notice that independence here was an assumption that allowed a simple solution. It may be of course that all sets contain the same faulty component in which case the assumption would not be justified. Often the assumption is justified with either experience of intuition or else is made because without it the problem cannot be tackled.

EXAMPLE 4.13

If A and B are two independent events in a sample space and the probability that both A and B occur is .16 while the probability that neither occur is .36, what can you say about $P(A)$ and $P(B)$?

From the problem statement $P(A \cap B) = P(A)P(B) = .16$ and $P(\overline{A \cup B}) = .36$, from which it follows that

$$P(A \cup B) = .64$$

$$= P(A) + P(B) - P(A \cap B)$$

$$= P(A) + P(B) - .16$$

These relationships give two equations with the unknowns $P(A)$ and $P(B)$:

(i) $P(A)+P(B)=.80$
(ii) $P(A)P(B)=.16$

Substituting $P(A)=.80-P(B)$ into (ii) yields the quadratic equation

$$P(B)(.80-P(B))=.16$$

or

$$P(B)^2-.80P(B)+.16=0$$

which has the solution $P(B)=.40$, which implies from (i) that $P(A)=.40$.

It is not difficult to show that if A and B are independent events, then so are A and $\bar{B}$, $\bar{A}$ and B, and $\bar{A}$ and $\bar{B}$. The concept of independence of a pair of events extends readily to independence of a sequence of events A_i, $i = 1, 2, \ldots, n$ although the number of conditions that have to be checked grows exponentially.

Definition The events A_i, $i = 1, 2, \ldots, n$ are independent if and only if for any collection of k distinct integers $(i_1, i_2, \ldots, i_k)$ chosen from the set $\{1, 2, \ldots, n\}$ it is true that

$$P(A_{i_1} \cap A_{i_2} \cap \cdots \cap A_{i_k}) = P(A_{i_1})P(A_{i_2}) \cdots P(A_{i_k})$$

for all integers k, $2 \le k \le n$.

For a fixed integer k there are $\binom{n}{k}$ conditions to be checked and since k can vary between 2 and n the total number of conditions to be checked is

$$\sum_{k=2}^{n} \binom{n}{k}$$

From the binomial theorem

$$(1+1)^n = 2^n = \sum_{k=0}^{n} \binom{n}{k}$$

it is seen that the total number of conditions that must be checked to ensure that n events $A_1, A_2, \ldots, A_n$ are independent is

$$\sum_{k=2}^{n} \binom{n}{k} = 2^n - n - 1 \tag{4.14}$$

EXAMPLE 4.14

If A, B, and C are independent events such that $P(A)=.2$, $P(B)=.1$, and $P(C)=.4$, find $P(A \cup B \cup C)$.

This is a simple exercise in definitions. From Section 1.5,

$$P(A \cup B \cup C) = P(A) + P(B) + P(C) - P(A \cap B) - P(A \cap C)$$
$$- P(B \cap C) + P(A \cap B \cap C)$$

which, since the events are independent, reduces to

$$\text{LHS} = P(A) + P(B) + P(C) - P(A)P(B) - P(A)P(C) - P(B)P(C)$$
$$+ P(A)P(B)P(C)$$

$$= .2 + .1 + .4 - .2 \times .1 - .2 \times .4 - .1 \times .4 + .2 \times .1 \times .4$$

$$= .568$$

It is interesting to note that the events A_i, $i = 1, \ldots, n$ can be pairwise independent (i.e., $P(A_i \cap A_j) = P(A_i)P(A_j)$, $i \neq j$) without being independent, as the following example illustrates.

EXAMPLE 4.15

Let A_1, A_2, and A_3 be three events that are pairwise independent, with the additional property that $A_1 \cap A_2 = A_1 \cap A_3 = A_2 \cap A_3 = A_1 \cap A_2 \cap A_3$. If $0 < P(A_i) < 1$, $i = 1, 2, 3$, then show that these sets cannot be independent.

Since we are given the information that the sets are pairwise independent, for complete independence it would only be necessary to check that

$$P(A_1 \cap A_2 \cap A_3) = P(A_1)P(A_2)P(A_3)$$

As we are also given that $A_1 \cap A_2 \cap A_3 = A_1 \cap A_2$, this equation can be expressed as

$$P(A_1 \cap A_2 \cap A_3) = P(A_1 \cap A_2) = P(A_1)P(A_2)P(A_3)$$

and, as $P(A_1 \cap A_2) = P(A_1)P(A_2)$ we can cancel these factors to give $P(A_3) = 1$, contrary to assumption.

As a more concrete example as to how this situation can arise, let S be the sample space consisting of the 36 points representing the possible outcomes of throwing two dice. Define the events:

$A_1 = \{$1st dice shows a 1$\}$

$\quad = \{(1, 1), (1, 2), (1, 3), (1, 4), (1, 5), (1, 6)\}$

$A_2 = \{$2nd dice shows a 1$\}$

$\quad = \{(1, 1), (2, 1), (3, 1), (4, 1), (5, 1), (6, 1)\}$

$A_3 = \{$the two dice show the same number$\}$

$\quad = \{(1, 1), (2, 2), (3, 3), (4, 4), (5, 5), (6, 6)\}$

As each set contains six points, $P(A_i) = 1/6$, $i = 1, 2, 3$, and any two of the sets intersect precisely in the point $(1, 1)$ and so $P(A_i \cap A_j) = 1/36$, $i \neq j$. We conclude immediately that A_i and A_j are independent sets for $i \neq j$. However, since $A_1 \cap A_2 \cap A_3 = A_1 \cap A_2 = A_1 \cap A_3 = A_2 \cap A_3$, we have seen that the three sets cannot be independent since

$$P(A_1 \cap A_2 \cap A_3) = P((1, 1))$$

$$= \frac{1}{36}$$

$$\neq P(A_1)P(A_2)P(A_3)$$

$$= \left(\frac{1}{6}\right)^3 = \frac{1}{216}$$

In most situations the assumption that a collection of sets are independent will be justified on either physical or intuitive grounds. The assumption invariably simplifies the computation of probabilities.

To distinguish once again between the disjointness and independence of a collection of sets we summarize:

(i) If A_i, $i = 1, \ldots, n$ is a sequence of disjoint events, then

$$P\left(\bigcup_{i=1}^{n} A_i\right) = \sum_{i=1}^{n} P(A_i)$$

(ii) If A_i, $i = 1, \ldots, n$ is a sequence of independent events, then

$$P\left(\bigcap_{i=1}^{n} A_i\right) = \prod_{i=1}^{n} P(A_i)$$

and a similar equality holds for any subcollection of the sets. The additive and multiplicative nature, respectively, of these two relationships should be noted.

Problems

1. Among 12 candidates, five are women and seven are men. One person is chosen to speak at a rally first and then a second is chosen. If the second person chosen was a woman, what is the probability that the first one was also?

2. Of five stores in a block, four have six employees each, four men and two women, while the fifth has five men and two women. One of the employees of these stores is involved in an accident and it happens to be a man. What is the probability that he came from the store with five male employees?

3. In a collection of 15 urns, ten contain four red balls and six yellow balls each and the other five contain five red and five yellow each. A ball is drawn and found to be yellow. What is the probability that it came from one of the ten urns containing six yellow?

4. From a lot containing three 1-cm, eight 2-cm, and five 3-cm nails, three are chosen at random. What is the probability that they are of the same length?

5. A bankroll contains ten one-dollar bills and five ten-dollar bills. Two of these are stolen and their denomination is unknown. If a bill is chosen at random from those left, what is the probability it will be a ten-dollar bill?

6. Two baseball teams are drawn from grade 6 and 7 students. Initially team 1 has six grade 6 and four grade 7 while team 2 has four grade 6 and four grade 7 players. A player is chosen at random from team 1 and transferred to team 2. If a player is then chosen at random from team 2 to toss the coin, and he happens to be in grade 7, what is the probability that the transferred player was in grade 7?

7. A political committee of size four is to be drawn from 20 people representing four different regions, five from region 1, seven from region 2, three from region 3, and five from region 4. If four are chosen at random from the 20, what is the probability that each region is represented?

8. Among a set of 12 coins there is exactly one that weighs either more or less, but not the same, as the other 11. Devise a weighing procedure that will find the odd coin in three weighings. Can you find a procedure that will find the coin in two weighings, with probability 3/4, if it is known that the coin is heavy?

9. If a fair coin is tossed n times, find the probability that any particular throw was a head, given that a total of k heads was obtained in the n throws.

10. A random variable X has the pdf

$$f(x) = \tfrac{1}{2} e^{-|x|}, \quad -\infty < x < \infty$$

   (i) Find the cdf of X.
   (ii) What is the probability that out of ten independent determinations of X, exactly four will have values greater than 1?
   (iii) Find an expression for $P(X > 2 \,|\, |X| > 1)$.

11. A newspaper vendor observes that the number of people passing his stand between the hours of 5 p.m. and 6 p.m. has a Poisson distribution with parameter 1000. If there is a probability of .1 of a person who passes the stand buying a paper, derive an expression for the distribution of the number of papers Y he sells in the given time interval and find the expected value of Y.

12. Three lots of screws contain 100, 150, and 200, respectively, and contain 1%, 2%, and 3% defectives, respectively. If a screw is chosen from one of the bins and it turns out to be defective, what is the probability that it came from the lot containing the 200?

13. A message is coded into the binary symbols 0 and 1 with probability of transmission $P(0) = .4$ and $P(1) = .6$. The communication channel introduces error such that a "0" is distorted into a "1" with probability .2 and a "1" into a "0" with probability .1. Find the probability that:
   (i) A 0 is received.
   (ii) A 1 is received.

14. Three urns U_i, $i = 1$, 2, and 3 contain red and yellow balls in the following proportions:

	U_1	U_2	U_3
Red	4	3	4
Yellow	2	3	7

If a ball is withdrawn from each urn, what is the probability that they are all of the same color?

15. Let A, B, and C be three events, each with probability greater than zero, in a sample space S. Show that the sample space S conditioned on the event A yields a valid probability measure by showing the following properties:

(i) $P(S|A) = 1$.

(ii) $P(B|A) \leq P(C|A)$ if $B \subset C$.

(iii) $P(B|A) + P(C|A) = P(B \cup C|A)$ if $B \cap C = \varnothing$.

Show also that $P(B|A) + P(\bar{B}|A) = 1$.

16. A grade school class contains ten 7 year olds and twelve 8 year olds. Four children are chosen at random, sequentially. Find the probability that the second child is 7 years old if it is known that three of the four children chosen are 7 years old and the children are chosen (i) with replacement and (ii) without replacement.

17. A coin is tossed n times. What is the probability of obtaining k heads if the first j tosses gave heads?

18. For n arbitrary sets in a sample space S, show that

$$P(A_1 \cap A_2 \cap \cdots \cap A_n)$$
$$= P(A_n|A_1 \cap \cdots \cap A_{n-1})P(A_{n-1}|A_1 \cap \cdots \cap A_{n-2}) \cdots P(A_2|A_1)P(A_1)$$

19. A coin is tossed n times and the result is all heads. If the probability that the coin is two headed is twice the probability that it is a fair coin, what is the probability that the coin used to get the n heads is two headed?

20. A lot contains ten items each of which may be either good or defective. The number of defective items in the lot is unknown, but it is assumed that all possible numbers of defective items is equally likely. An item is drawn at random and found to be defective. What is the most likely number of defective items in the original lot?

21. A handyman keeps nails and screws in two jars. One jar contains the same number of nails and screws while the other contains 75% nails. If he chooses a jar at random and chooses an item from that jar at random, what is the probability he chooses a nail?

22. The following data represent the results of inspection of bearings supplied by three different manufacturers:

	Company A	Company B	Company C
Satisfactory	3000	2500	1500
Oversized	880	100	120
Undersized	200	150	250

Find the probability that
(i) a defective part is oversized,
(ii) Company A supplied a defective part, and
(iii) a part is both defective and from Company C.

23. A baseball team owner decides to have a bat day to give a free bat to any boy under 12 attending the game. The probability distribution of the number of people attending the game is modeled by a Poisson distribution with parameter 50,000. It is assumed that each person arriving at the gate has a probability of 1/4 of being a boy under 12. What is the probability distribution of the number of boys under 12 attending the game? What is the mean and variance of this distribution?

24. Three stores have 8, 12, and 14 employees of whom 4, 7, and 10, respectively, are women.
(i) A store is chosen at random and from that store an employee is chosen at random. If this employee is a woman, what is the probability she came from the store with 12 employees?

(ii) If a second employee is chosen from the same store in (i), what is the probability that a woman will be chosen?

25. The number of defective items in a lot is a binomial random variable with parameters n and θ. If an item is drawn and found to be defective, what is the probability that the lot contains i defective items? Compare this with the unconditional probability that the lot contains i defective items.

26. Given that a person has scored at least three hits on a target in ten shots, find the probability they scored six hits if the probability of scoring a hit on any one shot is .1.

27. A student on a multiple choice test containing ten questions for which each question has five possible answers, has a probability θ of knowing the answer to any particular question. For a question to which he does not know the answer he guesses and has a probability $1/5$ of guessing correctly. If each question is worth ten marks and is assigned a mark of 0 or 10, what is his expected mark?

28. W_1 and W_2 are two weather stations 100 km apart. Let R_1 and R_2 denote the occurrence of rain at W_1 and W_2, respectively. From observation it is found that $P(R_1) = .4$ and $P(R_2) = .35$. If the events R_1 and R_2 are independent, what is the probability of rain at either W_1 or W_2 or both?

29. Show that if A and B are independent events, then A and $\bar{B}$ are independent events.

30. Show that if $A_1, A_2, \ldots, A_n$ are independent events, then

$$P(A_1 \cup A_2 \cup \cdots \cup A_n) = 1 - (1 - P(A_1))(1 - P(A_2)) \cdots (1 - P(A_n))$$

31. If A and B are independent events, is it necessarily true that $P(A \cap B | C) = P(A | C) P(B | C)$?

32. A certain mechanism has four components that fail independently with probabilities 0.1, 0.2, 0.25, and 0.3. What is the probability that exactly one of the components is working? What is the probability that three of the components are working?

33. Let $A_1, A_2, \ldots, A_n$ be events in the sample space S.
    (i) If $\sum_{i=1}^{n} P(A_i) = 1$, under what conditions will $\bigcup_{i=1}^{n} A_i = S$?
    (ii) If $A_1, \ldots, A_n$ are independent and $P(A_i) = \theta$, $i = 1, \ldots, n$, find an expression for $P(\bigcup_{i=1}^{n} A_i)$.

34. A man tosses a coin three times and records the result. He repeats this experiment a large number of times and decides that the probabilities of the various outcomes are:

Outcome	HHH	HHT	HTH	THH	HTT	THT	TTH	TTT
Probability	1/12	2/12	1/12	1/12	2/12	2/12	1/12	2/12

Based on these results, would you say that his tosses are independent?

35. The sample space S consists of the interval $[0, 1]$. If sets of equal length have the same probability, find necessary and sufficient conditions for two events to be independent.

CHAPTER 5

Continuous Random Variables

The emphasis of the last few chapters has been on discrete random variables, their properties, and applications. In this chapter and in much of the rest of the text the emphasis will be on continuous random variables. In many of the problems associated with discrete random variables, a probability distribution function can be derived from the assumptions and physical description of the experiment. The situation is usually different for continuous random variables in that very often the probability distribution function of the random variable under consideration is assumed to be of a certain form. The form of the density assumed takes cognizance of known relevant facts but is not usually derived from some physical model. For example, in testing the lifetime of light bulbs, a certain failure pdf may be assumed, taking into account that certain measurements, such as expected lifetime, have been made. It is often not the result of setting up a model as to how the light bulbs actually fail. In this chapter some of the more important continuous random variables and their probability density functions will be studied. The reason for the importance of some of them will be found in later chapters.

5.1   *Continuous Random Variables*

We first review the essential properties of continuous random variables discussed in Chapter 2. The random variable X is continuous if it has a cdf $F(x)$, which is continuous. In virtually all problems encountered in this text, we will assume additionally that the first derivative of $F(x)$ is also continuous so that the pdf

$$f(x) = \frac{d}{dx} F(x)$$

is continuous. The "inverse" relationship to this is $F(x) = \int_{-\infty}^{x} f(s) \, ds$. The mean of a continuous random variable X with pdf $f(x)$ is

$$E(X) = \int_{-\infty}^{\infty} x f(x) \, dx = \mu$$

The nth central moment of X is defined as

$$E((X-\mu)^n) = \int_{-\infty}^{\infty} (x-\mu)^n f(x) \, dx$$

and the second central moment is called the variance $V(X)$,

$$V(X) = E((X-\mu)^2) = \int_{-\infty}^{\infty} (x-\mu)^2 f(x) \, dx$$
$$= E(X^2) - \mu^2$$

a quantity that is sometimes denoted by σ_X^2.

The median of a random variable X is that value x_0 for which $P(X \le x_0) = F_X(x_0) = 1/2$. The mode of a continuous random variable is the value x_1 that maximizes the pdf, and we note immediately that it may not be unique. Along with the mean and variance of a random variable the mean and mode are simple parameters that give rough indications of the shape and location of the pdf.

If the random variable X has the cdf $F_X(x)$ and the pdf $f_X(x)$, it is often useful to derive expressions for the random variable $Y = aX + b$. From the definition of the cdf of Y,

$$F_Y(y) = P(Y \le y) = P(aX + b \le y) = P(aX \le y - b)$$
$$= P\left(X \le \frac{y-b}{a}\right) = F_X\left(\frac{y-b}{a}\right)$$

and so knowing the cdf of X gives the cdf of $Y = aX + b$. By differentiating we can also relate their pdf's as

$$f_Y(y) = f_X\left(\frac{y-b}{a}\right) \frac{1}{a}$$

assuming that a is positive. If a is negative, then

$$P(aX + b \le y) = P\left(X \ge \frac{y-b}{a}\right)$$

128 *Applied Probability*

and we would have

$$f_Y(y) = -f_X\left(\frac{y-b}{a}\right)\frac{1}{a}$$

These two equations can be combined into the single relationship

$$f_Y(y) = f_X\left(\frac{y-b}{a}\right)\frac{1}{|a|} \tag{5.1}$$

and this is a simple example of more involved computations that will be encountered in Chapter 7. If $m_X(t)$ is the moment generating function of X, then the mgf of $Y = aX + b$ is

$$M_Y(t) = E(e^{tY})$$
$$= E(e^{t(aX+b)})$$
$$= e^{tb}E(e^{atX}) = e^{tb}m_X(at)$$

If X has mean μ_x and variance σ_x^2, then the mean and variance of $Y = aX + b$ are computed as follows:

$$\mu_y = E(Y) = E(aX + b) = E(aX) + b = aE(X) + b = a\mu_x + b \tag{5.2}$$
$$\sigma_y^2 = V(Y) = V(aX + b) = V(aX) = a^2V(X) = a^2\sigma_x^2 \tag{5.3}$$

where the properties of expectation developed in Chapter 2 have been used. It is often useful, given the random variable X with mean μ_x and variance σ_x^2, to determine the linear transformation $Y = aX + b$, which gives Y with mean 0 and variance 1. From Eqs. (5.2) and (5.3) we determine

$$\mu_y = 0 = a\mu_x + b$$
$$\sigma_y^2 = 1 = a^2\sigma_x^2$$

and so

$$a = 1/\sigma_x \text{ (which is always positive)} \quad \text{and} \quad b = -a\mu_x = -\mu_x/\sigma_x \tag{5.4}$$

The required transformation then is $Y = (X - \mu_x)/\sigma_x$ and the mean and variance of Y are 0 and 1, respectively, a useful fact to keep in mind.

EXAMPLE 5.1

A professor, on grading an exam, finds that the marks have a mean of 62% and a standard deviation of 4%. Feeling the exam may have been unduly difficult, he would prefer to report a mean of 70% and a standard deviation of 5% and wants to achieve this by a linear transformation on the marks. What transformation should he use?

Using Eqs. (5.2) and (5.3) we can write $70 = a62 + b$ and $25 = a^2 16$ or $a = 5/4$ and $b = 15/2$ and the transformation should be $Y = 5/4x - 15/2$.

EXAMPLE 5.2

The demand for a soap product is known to be uniformly distributed over the range (9500, 10,500) in cases per month. In order to analyze the data in a more convenient form, a linear transformation is applied to it so the resulting data is standardized to have mean zero and variance 1. What is the pdf of the resulting data?

Let X represent the data uniformly distributed over (9500, 10,500). From Chapter 2, $\mu_x = 10,000$ and $\sigma_x^2 = (1000)^2/12$ and, from Eq. (5.4) the required transformation is

$$Y = \frac{X - 10,000}{(1000/\sqrt{12})}$$

where Y is the standardized data. Equation (5.1) gives the pdf of the random variable Y as

$$f_Y(y) = f_X\left(\frac{1000}{\sqrt{12}}y - 10,000\right) \cdot \frac{1000}{\sqrt{12}}, \quad 9500 \le \frac{1000}{\sqrt{12}}y - 10,000 \le 10,500$$

$$= \frac{1}{1000} \cdot \frac{1000}{\sqrt{12}}, \quad \frac{-500 \cdot \sqrt{12}}{1000} \le y \le \frac{500 \cdot \sqrt{12}}{1000}$$

$$= \frac{1}{\sqrt{12}}, \quad -\sqrt{3} \le y \le \sqrt{3}$$

and such a pdf clearly has a mean of 0 and variance of 1.

There are many interactions between discrete and continuous variables and the next example illustrates just one way such an interaction can arise.

EXAMPLE 5.3

The times between machine breakdowns is modeled as an exponential random variable, with pdf

$$f(x) = \alpha e^{-\alpha x}, \quad x \ge 0$$

in hours. (i) If it is desired to have at least 90% of the times between failure, longer than 100 hours, what should the expected time to failure be? (ii) If, rather than the exact time of failure only the hour in which it failed (measured from its previous failure) is recorded, what fraction of the time will it take j hours to fail? (iii) If a machine is still operating after 200 hours from its previous failure, find the pdf of its time to failure (measured from the 200 hour mark).

(i) The mean of the exponential pdf with parameter α was shown to be $1/\alpha$ in Chapter 2. The probability that the time to failure exceeds t is

$$P(T \ge t) = \int_t^\infty \alpha e^{-\alpha x}\, dx = e^{-\alpha t}$$

and if this expression is to be .90 for $t = 100$, the solution to the equation $e^{-100\alpha} = .90$ is $\alpha = 1.05 \times 10^{-3}$ and the expected time to failure is $(1/1.05) \times 10^3$ hours.

(ii) If Y is the fraction that fails during the jth hour, then

$$P(Y = j) = \int_{j}^{j+1} \alpha e^{-ax} \, dx$$

$$= -e^{-ax}\big|_{j}^{j+1}$$

$$= -e^{-\alpha(j+1)} + e^{-\alpha j}$$

$$= e^{-\alpha j}(1 - e^{-\alpha}) = \theta^j(1-\theta), \quad j = 0, 1, 2, \ldots$$

and Y is seen to have a geometric distribution with parameter $\theta = e^{-\alpha}$.

(iii) Given that a machine has not failed before 200 hours, we want the probability it fails in the interval $(200 + t, \ 200 + t + dt)$, a quantity we will express by the "conditional" function $g(t) \, dt$. Thus

$$g(t) \, dt = P(200 + t \le T \le 200 + t + dt | T \ge 200)$$

$$= \frac{P(200 + t \le T \le 200 + t + dt)}{P(T \ge 200)}$$

$$= \frac{\alpha e^{-\alpha(200+t)} \, dt}{e^{-\alpha 200}}$$

$$= \alpha e^{-\alpha t} \, dt$$

and we conclude that $g(t) = f(t)$. The implication of this is that a machine that has survived without a breakdown to 200 hours faces exactly the same future, probabilistically speaking, as one that has just been repaired. Another way of expressing this is by saying that the machine is as good as new until it fails. This type of behavior is examined in more detail in Chapter 12 where reliability theory is discussed.

5.2 Normal Random Variables

A random variable X is said to be normally distributed if it has the pdf

$$f_X(x) = \frac{1}{\sqrt{2\pi}\sigma} \exp\left(-\frac{(x-\mu)^2}{2\sigma^2}\right), \quad -\infty \le x \le \infty, \ \sigma > 0 \tag{5.5}$$

This is called the normal density function, but is also referred to as the Laplace–de Moivre law and the Gaussian density function. Notice that it has two parameters, μ and σ. It will be shown later in the section that $E(X) = \mu$ and $V(X) = \sigma^2$. For convenience we will often write $X \sim N(\mu, \sigma^2)$ as a shorthand notation to mean that X has the normal pdf given in Eq. (5.5).

Normal random variables play a predominant role in probability theory, being both practically and theoretically important. They arise frequently in practice and it will be seen in Chapter 8 that any random variable that is the sum of a large number of independent random variables will, under certain conditions, have a cdf "close" to that of a normal cdf. Normal random variables have some peculiar and distinctive properties. In some aspects they are mathematically easy to deal with, but in others, very difficult.

We first show that $f_X(x)$ given in Eq. (5.5) is indeed a pdf. It is positive everywhere and it is only necessary to show that it integrates to unity. Unfortunately, there is no simple function that is the indefinite integral of $f_X(x)$ and we use a circuitous route to show that it has unit area. First note that by changing the variable $y = (x - \mu)/\sigma$ the integral of $f_X(x)$ becomes

$$\int_{-\infty}^{\infty} \frac{1}{\sqrt{2\pi}\sigma} \exp\left(-\frac{(x-\mu)^2}{2\sigma^2}\right) dx = \int_{-\infty}^{\infty} \frac{1}{\sqrt{2\pi}} \exp\left(-\frac{y^2}{2}\right) dy$$

and it is sufficient to show that this last integral is unity. Letting

$$I = \int_{-\infty}^{\infty} \frac{1}{\sqrt{2\pi}} \exp\left(-\frac{y^2}{2}\right) dy \qquad (5.6)$$

and rather than showing that $I = 1$ we show that $I^2 = 1$. Since I is, by definition, a positive, real number, this implies that $I = 1$. We can write

$$I^2 = \left(\int_{-\infty}^{\infty} \frac{1}{\sqrt{2\pi}} \exp\left(-\frac{y^2}{2}\right) dy\right)^2$$

$$= \int_{-\infty}^{\infty} \frac{1}{\sqrt{2\pi}} \exp\left(-\frac{y^2}{2}\right) dy \cdot \int_{-\infty}^{\infty} \frac{1}{\sqrt{2\pi}} \exp\left(-\frac{z^2}{2}\right) dz$$

$$= \frac{1}{2\pi} \int_{-\infty}^{\infty} \int_{-\infty}^{\infty} \exp\left(-\frac{(y^2+z^2)}{2}\right) dy\, dz$$

The reason for proceeding in this fashion is that we can now evaluate I by changing to polar coordinates. The transformation from rectangular to polar coordinates is accomplished by letting

$$r = \sqrt{x^2 + y^2} \quad \text{and} \quad \theta = \arctan(x/y)$$

and, in terms of differentials, $r\, dr\, d\theta = dx\, dy$. The expression for I^2 becomes

$$I^2 = \frac{1}{2\pi} \int_0^{\infty} \int_0^{2\pi} \exp\left(-\frac{r^2}{2}\right) r\, dr\, d\theta$$

where r ranges over the interval $(0, \infty)$ and θ over $(0, 2\pi)$. Integrating over θ first gives

$$I^2 = \frac{1}{2\pi} \int_0^{\infty} r \exp\left(-\frac{r^2}{2}\right) dr \int_0^{2\pi} d\theta$$

$$= \int_0^{\infty} r \exp\left(-\frac{r^2}{2}\right) dr$$

and now integrating over r results in

$$I^2 = -\exp\left(-\frac{r^2}{2}\right)\Big|_0^\infty = 1$$

As I is a positive real number we conclude that $I = +1$.

If we assumed a knowledge of the gamma function, then there is a simpler way of showing that $I = 1$. Making the substitution $u = y^2/2$ in Eq. (5.6), $du = y\,dy = \sqrt{2u}\,dy$ gives

$$I = 2\int_0^\infty \frac{1}{\sqrt{2\pi}}e^{-u}\frac{du}{\sqrt{2u}}$$

$$= \frac{1}{\sqrt{\pi}}\int_0^\infty \frac{e^{-u}}{\sqrt{u}}\,du = \frac{1}{\sqrt{\pi}}\Gamma(\tfrac12) = 1$$

since $\Gamma\left(\dfrac{1}{2}\right)$ is $\sqrt{\pi}$, where $\Gamma(x)$ is the gamma function evaluated at x, as discussed in the next section.

It is very important to realize that a useful fact has been established, namely, that for *any* finite real number a and *any* finite positive real number b, we have shown that the function

$$f(x) = \frac{1}{\sqrt{2\pi b}}\exp\left(-\frac{(x-a)^2}{2b^2}\right), \quad -\infty < x < \infty$$

is a pdf and no other computation is required. The following simple example illustrates the point.

EXAMPLE 5.4

For what values of a is the function

$$f(x) = \frac{1}{\sqrt{\pi}}\exp(-x^2 + x - a), \quad -\infty < x < \infty$$

a pdf?

The function $f(x)$ can be written in the form

$$f(x) = \left(\frac{1}{\sqrt{\pi}}\exp(-(x-\tfrac12)^2)\right) \cdot \exp(-a + \tfrac14)$$

and since the function in the brackets is the pdf of a random variable with pdf $N(\tfrac12, \tfrac12)$, we have

$$\int_{-\infty}^\infty f(x)\,dx = \exp(-a + \tfrac14)\int_{-\infty}^\infty \frac{1}{\sqrt{\pi}}\exp(-(x-\tfrac12)^2)\,dx$$

$$= \exp(-a + \tfrac14)$$

This equation implies that $f(x)$ is a pdf only when $a = \tfrac14$.

Now suppose X is a normal random variable with pdf as given in Eq. (5.5) (i.e., $X \sim N(\mu, \sigma^2)$). We show that $E(X) = \mu$ and $V(X) = \sigma^2$. From the definition

$$E(X) = \int_{-\infty}^{\infty} x \frac{1}{\sqrt{2\pi}\sigma} \exp\left(-\frac{(x-\mu)^2}{2\sigma^2}\right) dx$$

and changing the variable of integration to $y = (x - \mu)/\sigma$,

$$E(X) = \int_{-\infty}^{\infty} (\sigma y + \mu) \frac{1}{\sqrt{2\pi}} \exp\left(-\frac{y^2}{2}\right) dy$$

$$= \frac{\sigma}{\sqrt{2\pi}} \int_{-\infty}^{\infty} y \exp\left(-\frac{y^2}{2}\right) dy + \mu \int_{-\infty}^{\infty} \frac{1}{\sqrt{2\pi}} \exp\left(-\frac{y^2}{2}\right) dy$$

The first integral is zero since its integrand is an odd function. The second integral is unity since its integrand is the pdf of a normal random variable $N(0, 1)$. This results in

$$E(X) = \mu$$

To find the variance we first find $E(X^2)$:

$$E(X^2) = \int_{-\infty}^{\infty} x^2 \frac{1}{\sqrt{2\pi}\sigma} \exp\left(-\frac{(x-\mu)^2}{2\sigma^2}\right) dx$$

and substituting $y = (x - \mu)/\sigma$ gives

$$E(X^2) = \int_{-\infty}^{\infty} (y\sigma + \mu)^2 \frac{1}{\sqrt{2\pi}} \exp\left(-\frac{y^2}{2}\right) dy$$

$$= \int_{-\infty}^{\infty} (\sigma^2 y^2 + 2\mu\sigma y + \mu^2) \frac{1}{\sqrt{2\pi}} \exp\left(-\frac{y^2}{2}\right) dy \qquad (5.7)$$

We evaluate the first integral, integrating by parts,

$$\frac{\sigma^2}{\sqrt{2\pi}} \int_{-\infty}^{\infty} y^2 \exp\left(-\frac{y^2}{2}\right) dy = -\frac{\sigma^2}{\sqrt{2y}} y \exp\left(-\frac{y^2}{2}\right)\Big|_{-\infty}^{\infty}$$

$$+ \sigma^2 \int_{-\infty}^{\infty} \frac{1}{\sqrt{2\pi}} \exp\left(-\frac{y^2}{2}\right) dy$$

$$= \sigma^2$$

Equation (5.7) can then be written as

$$E(X^2) = \sigma^2 \int_{-\infty}^{\infty} y^2 \frac{1}{\sqrt{2\pi}} \exp\left(-\frac{y^2}{2}\right) dy + 2\mu\sigma \int_{-\infty}^{\infty} y \frac{1}{\sqrt{2\pi}} \exp\left(-\frac{y^2}{2}\right) dy$$

$$+ \mu^2 \int_{-\infty}^{\infty} \frac{1}{\sqrt{2\pi}} \exp\left(-\frac{y^2}{2}\right) dy$$

$$= \sigma^2 + 0 + \mu^2 = \sigma^2 + \mu^2$$

and finally, the variance is

$$V(X) = E(X^2) - E^2(X)$$
$$= \sigma^2 + \mu^2 - \mu^2 = \sigma^2$$

Thus the mean and variance of a normally distributed random variable uniquely determine the pdf, as specified in Eq. (5.5). By saying that X is a normally distributed random variable with mean μ and variance σ^2, or $X \sim N(\mu, \sigma^2)$, means precisely that X has the pdf of Eq. (5.5).

Again we emphasize that some very useful facts have been established in that for any real finite number a and any real finite positive number b, we have the following three integrals:

(i) $\displaystyle\int_{-\infty}^{\infty} \frac{1}{\sqrt{2\pi}b} \exp\left(-\frac{(x-a)^2}{2b^2}\right) dx = 1$

(ii) $\displaystyle\int_{-\infty}^{\infty} \frac{x}{\sqrt{2\pi}b} \exp\left(-\frac{(x-a)^2}{2b^2}\right) dx = a$ (5.8)

(iii) $\displaystyle\int_{-\infty}^{\infty} \frac{x^2}{\sqrt{2\pi}b} \exp\left(-\frac{(x-a)^2}{2b^2}\right) dx = a^2 + b^2$

EXAMPLE 5.5

Find the mean and variance for the pdf of Example 5.4
For the value of a established in that example the pdf is

$$f(x) = \frac{1}{\sqrt{\pi}} \exp(-x^2 + x - \tfrac{1}{4}), \quad -\infty < x < \infty$$

$$= \frac{1}{\sqrt{2\pi} \cdot (1/\sqrt{2})} \exp\left(-\frac{(x-\tfrac{1}{2})^2}{2(\tfrac{1}{2})}\right)$$

and, using Eq. (5.8), identifying a with $1/2$ and b with $1/2$ we find directly that

$$E(X) = \frac{1}{2} = a$$

$$E(X^2) = \frac{1}{2} + \frac{1}{2} = a^2 + b^2 = 1$$

$$V(X) = b^2 = \frac{1}{4}$$

and no other computations are required.

To examine the shape of the normal density function of Eq. (5.5) further it is simple to verify that both the mean and the mode of this function occur at $x = \mu$. The maximum value of the function is $1/(\sqrt{2\pi} \cdot \sigma)$ at $x = \mu$. For values of

creasing from μ the derivative increases negatively to a point of inflection
and then decreases out to zero again at infinity. The points of inflection are
found by equating the second derivative to zero and are $\mu \pm \sigma$. Figure 5.1 gives
the general shape of the normal density function $N(\mu, \sigma^2)$, for two values of σ^2.

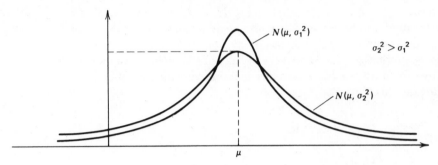

FIGURE 5.1

It is a relatively straightforward matter to find the moment generating
function of a random variable $X \sim N(\mu, \sigma^2)$. From the definition

$$m_X(t) = E(e^{tX})$$

$$= \int_{-\infty}^{\infty} \exp(tx) \frac{1}{\sqrt{2\pi}\sigma} \exp\left(-\frac{(x-\mu)^2}{2\sigma^2}\right) dx$$

and again making the substitution $y = (x - \mu)/\sigma$ gives

$$m_X(t) = \int_{-\infty}^{\infty} e^{t(y\sigma+\mu)} \frac{1}{\sqrt{2\pi}} \exp\left(-\frac{y^2}{2}\right) dy$$

$$= \exp(t\mu) \int_{-\infty}^{\infty} \frac{1}{\sqrt{2\pi}} \exp\left(-\frac{y^2}{2} + t\sigma y\right) dy$$

and completing the square in the exponent of the integrand,

$$m_X(t) = \exp\left(t\mu + \frac{t^2}{2}\sigma^2\right) \int_{-\infty}^{\infty} \frac{1}{\sqrt{2\pi}} \exp\left(-\frac{(y - t\sigma)^2}{2}\right) dy$$

Since the integrand is the pdf $N(t\sigma, 1)$ (viewing t as a constant so far as the
integral is concerned), it follows that

$$m_X(t) = \exp\left(t\mu + \frac{t^2\sigma^2}{2}\right) \tag{5.8}$$

With Eq. (5.8) it is easy to verify the mean and variance of $N(\mu, \sigma^2)$
since from Chapter 2 it was shown that

$$E(X^n) = m_X^{(n)}(0) = \frac{d^n}{dt^n} m_X(t)\Big|_{t=0}$$

and it simply remains to compute these functions for $n = 1$ and 2 for the moment generating function of Eq. (5.8):

$$E(X) = \frac{d}{dt} \exp\left(t\mu + \frac{t^2\sigma^2}{2}\right)\bigg|_{t=0}$$

$$= (\mu + t\sigma^2) \exp\left(t\mu + \frac{t^2\sigma^2}{2}\right)\bigg|_{t=0}$$

$$= \mu$$

and

$$E(X^2) = \frac{d^2}{dt^2} \exp\left(t\mu + \frac{t^2\sigma^2}{2}\right)\bigg|_{t=0}$$

$$= (\sigma^2 + (\mu + t\sigma^2)^2) \exp\left(t\mu + \frac{t^2\sigma^2}{2}\right)\bigg|_{t=0}$$

$$= \sigma^2 + \mu^2$$

which yields the results as found by integration.

The mgf is also quite convenient to use for establishing other results. For example, by successive differentiation of the mgf of Eq. (5.8) with $\mu = 0$ and $\sigma^2 = 1$, the nth moment of a normal random variable $N(0, 1)$ can be expressed as

$$E(X^n) = \begin{cases} \dfrac{(2n)!}{2^n n!}, & n \text{ even} \\[2mm] 0, & n \text{ odd} \end{cases}$$

It was observed that there is no simple function giving the indefinite integral of Eq. (5.5) and this fact causes a problem. In particular, if $X \sim N(\mu, \sigma^2)$, then the cdf of X is

$$F_X(x) = \int_{-\infty}^{x} \frac{1}{\sqrt{2\pi}\sigma} \exp\left(-\frac{(y-\mu)^2}{2\sigma^2}\right) dy \tag{5.9}$$

and there is no simple expression for this function. Consequently, this function must be tabulated but it is, of course, an impossible task for all values of μ and σ that are likely to be of interest since both of these functions have infinite ranges. The problem is overcome by making the change of variable $z = (y - \mu)/\sigma$ in Eq. (5.9) to give

$$F_X(x) = \frac{1}{\sqrt{2\pi}} \int_{-\infty}^{(x-\mu)/\sigma} \exp\left(-\frac{z^2}{2}\right) dz \tag{5.10}$$

It is now clear that we do not have to tabulate $F_X(x)$ in Eq. (5.9) for various values of μ and σ. We need only tabulate the *single* function

$$\Phi(x) = \frac{1}{\sqrt{2\pi}} \int_{-\infty}^{x} \exp\left(-\frac{z^2}{2}\right) dz \tag{5.11}$$

This function is the cdf of a normal random variable with mean zero and variance one ($N(0, 1)$). If $Z \sim N(0, 1)$, then Z is called a *standard normal random variable* and if $X \sim N(\mu, \sigma^2)$, $Y = (X - \mu)/\sigma$ is a standard normal random variable. By combining Eqs. (5.10) and (5.11),

$$F_X(x) = \Phi\left(\frac{x - \mu}{\sigma}\right) \tag{5.12}$$

and *so having a tabulation for* $\Phi(x)$ *yields values for the cdf of any normal random variable*, an important fact to bear in mind. The function $\Phi(x)$, as defined in Eq. (5.11), has been extensively tabulated to varying degrees of accuracy and a short tabulation of it is given in Appendix D. The use of Φ to denote the cdf of a standard normal random variable is relatively standard and it will be used only for this purpose in this text.

In the first section of this chapter we observed that if X has the pdf $f_X(x)$, then $Y = aX + b$ has the pdf

$$f_Y(y) = f_X\left(\frac{y - b}{a}\right) \frac{1}{|a|}$$

If X has the pdf $N(\mu, \sigma^2)$, then Y has the pdf

$$f_Y(y) = \frac{1}{\sqrt{2\pi}\sigma} \exp\left(-\frac{1}{2\sigma^2}\left(\left(\frac{y - b}{a}\right) - \mu\right)^2\right) \cdot \frac{1}{|a|}, \quad -\infty < y < \infty$$

$$= \frac{1}{\sqrt{2\pi}|a|\sigma} \cdot \exp\left(-\frac{(y - a\mu - b)^2}{2a^2\sigma^2}\right), \quad -\infty < y < \infty$$

which is the pdf $N(a\mu + b, a^2\sigma^2)$. Thus if X has the pdf $N(\mu, \sigma^2)$, then $Y = aX + b$ has the pdf $N(a\mu + b, a^2\sigma^2)$. We verified earlier that $E(Y) = a\mu + b$ and $V(Y) = a^2\sigma^2$ for any density function of X, and the point of this exercise is to note that a linear transformation of a normal random variable is again a normal random variable.

For the standard normal cdf $\Phi(x)$, it will be convenient to define the parameter z_α by the relation

$$\Phi(z_\alpha) = \alpha$$

that is, z_α is the ordinate such that the area under the standard normal pdf to the left of z_α is precisely α as shown in Figure 5.2. This parametric relationship will be convenient in many instances later.

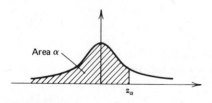

Area α

z_α

FIGURE 5.2

EXAMPLE 5.6

The line of sight distance by which a missile misses its target is normally distributed with mean -15 meters and variance 25 meters^2. What is the probability that the missile misses its target by more than 20 meters?

If X is the distance by which the missile misses its target, then $X \sim N(-15, 25)$ and we want to calculate

$$P(|X| > 20) = 1 - P(-20 < X < 20)$$

where

$$P(-20 < X < 20) = F_X(20) - F_X(-20)$$

From Eq. (5.12) this expression can be written as

$$P(-20 < X < 20) = \Phi\left(\frac{20 - 15}{5}\right) - \Phi\left(\frac{-20 - 15}{5}\right)$$

$$= \Phi(1) - \Phi(-7)$$

which, from the table of Appendix D is

$$P(-20 < X < 20) = .841 - .000 = .841$$

and so

$$P(|X| > 20) = 1 - .841 = .159$$

Certain rule of thumb estimates, when dealing with a normal random variable, are useful to keep in mind: The probability that a normal random variable will fall within

(i) $1\,\sigma$ of its mean is

$$\Phi\left(\frac{\mu + \sigma - \mu}{\sigma}\right) - \Phi\left(\frac{\mu - \sigma - \mu}{\sigma}\right) = \Phi(1) - \Phi(-1) = .6826$$

(ii) $2\,\sigma$ of its mean is

$$\Phi\left(\frac{\mu + 2\sigma - \mu}{\sigma}\right) - \Phi\left(\frac{\mu - 2\sigma - \mu}{\sigma}\right) = \Phi(2) - \Phi(-2) = .9544$$

(iii) $3\,\sigma$ of its mean is

$$\Phi\left(\frac{\mu + 3\sigma - \mu}{\sigma}\right) - \Phi\left(\frac{\mu - 3\sigma - \mu}{\sigma}\right) = \Phi(3) - \Phi(-3) = .9974$$

and these figures can be useful reference points to keep in mind. The next example is simply an illustration of the fact that the distribution of a normal random variable is completely determined by two parameters, its mean and variance.

EXAMPLE 5.7

The maximum temperature on June 1 in a certain locality has been recorded over the years. About 15% of the time it has exceeded 30°C and about 5% of the time it has been less than 20°C. If this temperature is modeled as a normally distributed random variable, what is its mean and variance?

Suppose the temperature T has the density $N(\mu, \sigma^2)$. We are told that, approximately,

$$P(T \le 20) = .05 \quad \text{and} \quad P(T \ge 30) = .15$$

or, equivalently,

$$P\left(\frac{T-\mu}{\sigma} \le \frac{20-\mu}{\sigma}\right) = .05 \quad \text{and} \quad P\left(\frac{T-\mu}{\sigma} \le \frac{30-\mu}{\sigma}\right) = .85$$

From the tables of $\Phi(x)$ in Appendix D, we must have

$$\Phi\left(\frac{20-\mu}{\sigma}\right) = 0.05 \quad \text{or} \quad \frac{20-\mu}{\sigma} = -1.645$$

and

$$\Phi\left(\frac{30-\mu}{\sigma}\right) = .85 \quad \text{or} \quad \frac{30-\mu}{\sigma} = 1.036$$

Solving these two equations gives $\mu = 26.12$ and $\sigma = 3.73$.

The following example is a variation on the same theme.

EXAMPLE 5.8

Suppose the professor of Example 5.1 with average marks of 62% and standard deviation 4% now decides that in his large class he would like to fail no more than 15% of his class while still maintaining an average of 70%, by using a linear transformation on the marks. For this purpose he assumes the marks are normally distributed. What should his transformation be?

If X represents a random variable $N(62, 16)$ (of which the marks are thought of as repeated observations) and $Y = aX + b$, the marks after transformation have the pdf $N(62a + b, 16a^2)$. The mean of the transformed marks is 70 and the variance must be adjusted so that

$$P(Y \le 50) = .15 \quad \text{or} \quad P\left(Y - \frac{62a-b}{4a} \le \frac{50-62a-b}{4a}\right) = .15$$

These two facts translate into the equations

$$62a + b = 70 \quad \text{and} \quad \frac{50-62a-b}{4a} = -1.037$$

using the tables of Appendix D. These are solved to give $a = 4.82$ and $b = -228.84$.

EXAMPLE 5.9

A particular metal oxide comes in two types, depending on the level of impurities present. The level of impurity is regarded as a normal random variable. If type A has an impurity level, measured in hundredths of a percent, with a pdf $N(75, 9)$, and type B has a pdf $N(78, 3)$ and in a particular application it is very important that the level of impurity not exceed .80%, which type should be used?

For a sample of the type A oxide, the probability that its impurity level does not exceed .80% is

$$P_A(X \le 80) = P_A\left(\frac{X-75}{3} \le \frac{80-75}{3}\right)$$

$$= \Phi(1.667) = .9522$$

while for type B this probability is

$$P_B(X \le 80) = P_B\left(\frac{X-78}{\sqrt{3}} \le \frac{80-78}{\sqrt{3}}\right)$$

$$= \Phi(1.155) = .8760$$

and the type A oxide should be employed.

Sometimes it is convenient to approximate the pdf of a positive random variable (one that can assume only positive values) with a normal pdf. Since the normal pdf is nonzero for all negative values, this might appear strange, but the point is that the area under the normal curve for negative values should be negligible if the approximation is to be of any use. The following two examples illustrate this point.

EXAMPLE 5.10

The lifetime T of a transistor is normally distributed with mean 200 hours and standard deviation 9. What is the probability that a particular transistor will last more than 220 hours? What is the probability that a transistor which lasts 210 hours will last 220 hours?

The actual pdf of T, $f_T(t)$, is zero for t less than zero. We approximate $f_T(t)$ by the pdf $f'_T(t)$

$$f'_T(t) = \frac{1}{\sqrt{2\pi}9} \exp\left(-\frac{(t-200)^2}{2.9^2}\right), \quad -\infty < t < \infty$$

and note that the integral

$$\int_{-\infty}^{0} f'_T(t)\, dt$$

is equal to $\Phi(-200/9)$, which for any practical purpose is zero. The probability that T exceeds 220 is

$$P(T>220)=1-P(T\leq 220)=1-\Phi\left(\frac{220-200}{9}\right)$$

$$=1-\Phi(2.22)\cong .0132$$

Given that a transistor has lasted 210 hours, the probability that it will still be operable another 10 hours is

$$P(T\geq 220|T\geq 210)=\frac{P(T\geq 220)}{P(T\geq 210)}$$

$$=\frac{1-\Phi\left(\dfrac{220-200}{9}\right)}{1-\Phi\left(\dfrac{210-200}{9}\right)}$$

$$=\frac{.0132}{.133}\cong .10$$

illustrating again the effect that conditional information has on the probabilities, that is the probability a transistor will operate for 220 hours is .0132 while given that it has operated for 210 hours the probability that it will operate for 220 hours is .10.

EXAMPLE 5.11

The diameter of ball bearings in a manufacturing process is a normal random variable with mean .20 cm and variance σ^2. What should σ^2 be (i.e., how accurate should the manufacturing process be) if the probability of the diameter deviating from the mean by .02 cm or more is to be less than .05.

If X is the diameter, then $X \sim N(0.20, \sigma^2)$ and we want to determine σ^2 such that

$$P(|X-.20|\leq .02)=.95$$

or

$$\Phi\left(\frac{.20+.02-.20}{\sigma}\right)-\Phi\left(\frac{.20-.02-.20}{\sigma}\right)=\Phi\left(\frac{.02}{\sigma}\right)-\Phi\left(\frac{-.02}{\sigma}\right)$$

From the tabulation of $\Phi(x)$ we have that $.02/\sigma=1.96$ or $\sigma\cong .01$.

5.3 Other Important Continuous Random Variables

The uniform, exponential, and normal random variables have been considered in Chapter 2 and the previous section and their means, variances, and cdf's discussed. Some of this information is tabulated in Table 5.1. In this section we

TABLE 5.1

	pdf	MEAN	VARIANCE	MOMENT GENERATING FUNCTION
Uniform	$f(x)=\dfrac{1}{b-a}$, $a\leq x\leq b$	$\dfrac{a+b}{2}$	$\dfrac{(b-a)^2}{12}$	$\dfrac{e^{tb}-e^{ta}}{t(b-a)}$
Exponential	$f(x)=\alpha\,e^{-\alpha x}$, $x\geq 0$	$\dfrac{1}{\alpha}$	$\dfrac{1}{\alpha^2}$	$\dfrac{\alpha}{\alpha-t}$
Normal	$f(x)=\dfrac{1}{\sqrt{2\pi}\,\sigma}\exp\left(-\dfrac{(x-\mu)^2}{2\sigma^2}\right)$, $-\infty<x<\infty$	μ	σ	$\exp(t\mu+\tfrac{1}{2}t^2\sigma^2)$
Cauchy	$f(x)=\dfrac{1}{\pi\alpha\left(1+\left(\dfrac{x-\beta}{\alpha}\right)^2\right)}$, $-\infty<x<\infty$	—	—	—
Gamma	$f(x)=\dfrac{\gamma^\alpha}{\Gamma(\alpha)}x^{\alpha-1}e^{-\gamma x}$, $\alpha,\gamma>0,\ x>0$	$\dfrac{\alpha}{\gamma}$	$\dfrac{\alpha}{\gamma^2}$	$\dfrac{1}{(1-t/\gamma)^\alpha}$, $|t/\gamma|<1$
Student's t	$f(t)=\dfrac{\Gamma((n+1)/2)}{\sqrt{n\pi}\,\Gamma(n/2)}\cdot\dfrac{1}{(1+t^2/n)^{(n+1)/2}}$, $-\infty<t<\infty$	0	$\dfrac{n}{n-2},\ n>2$	—

discuss other important types of random variables and find some of their properties. These random variables and their density functions may appear somewhat artificial but they arise quite naturally in statistics. Ways in which some of them can arise are discussed in the examples of chapters 7 and 8.

5.3.1 CAUCHY DISTRIBUTION

A random variable X with pdf

$$f(x) = \frac{1}{\pi\alpha\left(1 + \left(\frac{x-\beta}{\alpha}\right)^2\right)}, \quad -\infty < x < \infty, \ \alpha > 0 \tag{5.13}$$

is said to have a Cauchy distribution or pdf. The general shape of this pdf is similar to that of a normal pdf except that it decreases more slowly for large values of x. This slow drop off causes problems as will be shown. The cdf of this function is

$$F(x) = \int_{-\infty}^{x} \frac{1}{\pi\alpha\left(1 + \left(\frac{x-\beta}{\alpha}\right)^2\right)} dx$$

$$= \frac{1}{2} + \frac{1}{\pi} \arctan\left(\frac{x-\beta}{\alpha}\right)$$

The effect of the slow convergence to zero by this pdf is that the random variable X does not possess any finite moments $E(X^n)$ for $n \geq 1$. The moment generating function of X in this case does not exist for $t \neq 0$. Consider $m_X(t)$, $t > 0$ for the parameters $\alpha = 1$, $\beta = 0$:

$$m_X(t) = E(e^{tX})$$

$$= \int_{-\infty}^{\infty} e^{tx} \frac{1}{\pi(1+x^2)} dx \geq \int_{0}^{\infty} e^{tx} \frac{1}{\pi(1+x^2)} dx$$

and using the exponential expansion $e^{tx} = 1 + tx + r(t)$, where $r(t) > 0$ for $t > 0$,

$$m_X(t) \geq \int_{0}^{\infty} (1 + tx) \frac{1}{\pi(1+x^2)} dx = \infty$$

It turns out, although we will not show it, that the characteristic function of X does exist and this shows up one advantage of using the characteristic function rather than the moment generating function. Further properties of Cauchy random variables are considered in Chapters 7 and 8.

EXAMPLE 5.12

The parameters β and α in the Cauchy pdf are analogous, in some sense, to the mean and variance. Demonstrate the slow convergence of this density to zero for large values of x by calculating (i) $P(|X - \beta| < \alpha)$, (ii) $P(|X - \beta| < 2\alpha)$,

144 *Applied Probability*

and (iii) $P(|X-\beta|<3\alpha)$ and compare results with the analogous expressions for the normal pdf found earlier.

The cdf of the Cauchy pdf of Eq. (5.13) is

$$F(x)=\frac{1}{2}+\frac{1}{\pi}\arctan\left(\frac{x-\beta}{\alpha}\right)$$

(i) $P(|X-\beta|<\alpha)=F(\beta+\alpha)-F(\beta-\alpha)$

$$=\frac{1}{2}+\frac{1}{\pi}\arctan(1)-\frac{1}{2}-\frac{1}{\pi}\arctan(-1)$$

$$=\frac{1}{2}$$

(ii) $P(|X-\beta|<2\alpha)=F(\beta+2\alpha)-F(\beta-2\alpha)$

$$=\frac{1}{2}+\frac{1}{\pi}\arctan(2)-\frac{1}{2}-\frac{1}{\pi}\arctan(-2)$$

$$=.706$$

(iii) $P(|X-\beta|<3\alpha)=F(\beta+3\alpha)-F(\beta-3\alpha)$

$$=\frac{1}{2}+\frac{1}{\pi}\arctan(3)-\frac{1}{2}-\frac{1}{\pi}\arctan(-3)$$

$$=.795$$

These three values are to be compared to the values .6826, .9544, and .9974, respectively, for the normal pdf. The higher values for the normal pdf indicate a larger concentration of area closer to the mean for the normal than for the Cauchy with implications as to their rate of fall off. The situation is represented in Figure 5.3.

The Cauchy pdf has a unique maximum (at $x=\beta$) and it has two points of inflection at $\beta\pm\alpha/\sqrt{3}$.

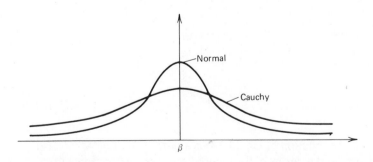

FIGURE 5.3

5.3.2 *GAMMA DISTRIBUTION*

A random variable X is said to have a gamma distribution if its pdf is of the form

$$f(x) = \frac{\gamma^{\alpha}}{\Gamma(\alpha)} x^{\alpha-1} \exp(-\gamma x), \quad \alpha, \gamma > 0, x > 0 \tag{5.14}$$

In this expression $\Gamma(\alpha)$ is the gamma function defined earlier by

$$\Gamma(\alpha) = \int_0^\infty x^{\alpha-1} e^{-x}\, dx$$

Integrating by parts yields

$$\Gamma(\alpha) = x^{\alpha-1} e^{-x} \Big|_0^\infty + \int_0^\infty (\alpha-1) e^{-x} x^{\alpha-2}\, dx$$

$$= (\alpha-1)\Gamma(\alpha-1)$$

For any value of α, we can use this relation successively to write $\Gamma(\alpha)$ as

$$\Gamma(\alpha) = (\alpha-1)(\alpha-2)\cdots(\alpha-(n-1))\Gamma(1+x)$$

where $\alpha = n + x$, n an integer, and $0 \le x < 1$. Thus we need only values of $\Gamma(y)$ for $1 \le y < 2$ and tables for this function exist. Of course, when α is an integer, $\Gamma(\alpha) = (\alpha-1)!$, the familiar factorial function. With this function it is relatively easy to show that $f(x)$ is indeed a pdf. The moment generating function of X is

$$m_X(t) = E(e^{tX})$$

$$= \int_0^\infty e^{tx} \frac{\gamma^{\alpha}}{\Gamma(\alpha)} x^{\alpha-1} \exp(-\gamma x)\, dx$$

and notice that for $t = 0$, the integrand is just the pdf $f(x)$.

Making the change of variable $y = \gamma x$ yields

$$E(e^{tX}) = \int_0^\infty e^{t(y/\gamma)-y} \frac{y^{\alpha-1}}{\Gamma(\alpha)}\, dy$$

or

$$m_X(t) = \frac{1}{(1-t/\gamma)^{\alpha}} \quad \text{for} \quad |t/\gamma| < 1 \tag{5.15}$$

and, since $m_X(0) = 1$, the function of Eq. (5.14) is a pdf. The cdf of X can also be found, but it is expressed in terms of the incomplete gamma function and we omit it. The mean of X can be found from the moment generating function of

Eq. (5.15) as

$$E(X) = \frac{d}{dt}\left(\frac{1}{(1-t/\gamma)^\alpha}\right)\Big|_{t=0}$$

$$= -\frac{\alpha(-1/\gamma)}{(1-t/\gamma)^{\alpha+1}}\Big|_{t=0}$$

$$= \frac{\alpha}{\gamma} \tag{5.16}$$

and similarly

$$E(X^2) = \frac{d^2}{dt^2}\left(\frac{1}{(1-t/\gamma)^\alpha}\right)\Big|_{t=0} = \frac{\alpha(\alpha+1)}{\gamma^2}$$

and

$$V(X) = \frac{\alpha}{\gamma^2} \tag{5.17}$$

If $\alpha = 1$, then the pdf of X is

$$f(x) = \gamma \exp(-\gamma x), \quad x > 0, \gamma > 0$$

and this is simply an exponential random variable with parameter γ. When $\alpha = n/2$, for a positive integer n, $\gamma = 1/2\sigma^2$, then the pdf is

$$f(x) = \frac{x^{n/2-1}}{2^{n/2}\sigma^n \Gamma(n/2)}\exp\left(-\frac{x}{2\sigma^2}\right), \quad x \geq 0, n = 1, 2, \ldots \tag{5.18}$$

and, when $\sigma = 1$, this is referred to as the central chi-square density function with n degrees of freedom. Much as we did for the standard normal pdf, it is useful to define a parameter $\chi^2_{\alpha;n}$ such that

$$\int_0^{\chi^2_{\alpha;n}} f(x)\, dx = \alpha$$

where $f(x)$ is the chi-square density function with n degrees of freedom, as in Eq. (5.18). That is, $\chi^2_{\alpha;n}$ is the ordinate such that the area under the pdf to the left of it is α. Notice that the cdf of this pdf depends on n and it is not feasible to tabulate one for each value of n. Instead only certain values of α are chosen that occur most frequently in practice, and the values of $\chi^2_{\alpha;n}$ are given for these values for various values of n. This is done in Appendix E, which will be used in later chapters. If X has the pdf of Eq. (5.18), then $Y = \sqrt{X/n}$ has the pdf

$$f_Y(y) = \begin{cases} \dfrac{2(n/2)^{n/2}}{\sigma^n \Gamma(n/2)}y^{n-1}\exp\left(-\dfrac{ny^2}{2\sigma^2}\right), & y > 0 \\[2ex] 0, & \text{otherwise} \end{cases} \tag{5.19}$$

and this is referred to as the central chi density function with n degrees of freedom when $\sigma = 1$. The means and variances of these density functions are easily calculated from Eqs. (5.16) and (5.17) by appropriate identification of parameters. The chi and chi-square random variables are of considerable importance in statistics and the reason for this will be shown in Chapter 8.

The density function of Eq. (5.19) with $n = 2$ and $\sigma^2 = 2$ is referred to as the Rayleigh probability density function

$$f(y) = y\, e^{-y^2/2}, \quad y \geq 0$$

which appears frequently in communications and other engineering problems.

The gamma pdf takes on a wide variety of shapes depending on its two parameters α and γ, and thus is a useful pdf in terms of modeling random phenomena. Let us examine the shape of this pdf a little further. The first two derivatives of the pdf are given by

$$f'(x) = \frac{\gamma^\alpha}{\Gamma(\alpha)} x^{\alpha-1}\, e^{-\gamma x}\left(\frac{(\alpha-1)}{x} - \gamma\right)$$

$$= \frac{\gamma^{\alpha+1}}{\Gamma(\alpha)} x^{\alpha-2}\, e^{-x/\gamma}\left(\frac{(\alpha-1)}{\gamma} - x\right) \tag{5.20}$$

$$f''(x) = \frac{\gamma^{\alpha+1}}{\Gamma(\alpha)} x^{\alpha-3}\, e^{-\gamma x}\left[\frac{(\alpha-1)(\alpha-2)}{\gamma} - x(2(\alpha-1)) + \gamma x^2\right] \tag{5.21}$$

The maximum of the function, from Eq. (5.20), appears at $x = (\alpha-1)/\gamma$ except that if $\alpha < 1$, then the maximum appears outside the range of definition ($x \geq 0$) and in this case the pdf is a monotonically decreasing function to 0 as $x \to \infty$. Notice that the slope of $f(x)$ at $x = 0$ can range in value between $-\infty$ and $+\infty$ depending on the values of α and γ. For $\alpha > 2$, the slope at the origin is zero. If $\alpha < 1$, then $f(x) \to \infty$ as $x \to 0$. For $\alpha = 1$ the gamma pdf is simply the exponential law with parameter γ. For $\alpha > 1$, the pdf has a unique maximum, as mentioned, at $x = (\alpha-1)/\gamma$. The quadratic in the brackets of Eq. (5.21) has the solutions

$$x_{1,2} = [(\alpha-1) \pm (\alpha-1)^{1/2}]/\gamma$$

and, for $\alpha > 2$, the pdf has two points of inflection at these solutions. Again, tables for the gamma distribution can be found although they have not been included in this text. The general shape of the gamma density for certain values of its parameters is shown in Figure 5.4.

EXAMPLE 5.13

Consumer demand for milk in a certain locality, per week, is known to be a gamma random variable. If the average demand is 24,000 liters and the most likely demand is 22,000 liters, what is the variance of the demand?

From Eq. (5.16) the mean of a gamma random variable is $\alpha/\gamma = 24{,}000$ and from Eq. (5.20) its most likely value (the value at which its pdf is maximum)

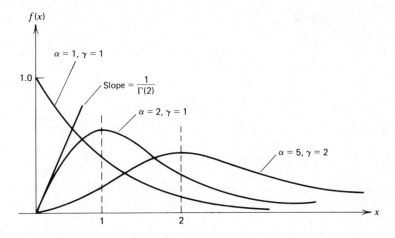

f(x)

α = 1, γ = 1

1.0

Slope = $\dfrac{1}{\Gamma(2)}$

α = 2, γ = 1

α = 5, γ = 2

1   2

x

FIGURE 5.4

is $(\alpha - 1)/\gamma = 22{,}000$. Consequently, $1/\gamma = 2000$ and $\alpha = 12$. From Eq. (5.17) the variance of the demand is $\alpha/\gamma^2 = 12/(4 \times 10^6) = 3 \times 10^{-6}$.

EXAMPLE 5.14

The lifetime T in hours of a certain mechanical part is modeled by a Rayleigh random variable with parameter α, that is, the pdf of T is

$$f(t) = \frac{t}{\alpha^2} \exp\left(-\frac{t^2}{2\alpha^2}\right), \quad t > 0$$

If 10% of those parts that have lasted 100 hours fail before 110 hours, determine the parameter α.

It is useful to first determine the cdf of T:

$$F_T(t) = \int_0^t \frac{x}{\alpha^2} \exp\left(-\frac{x^2}{2\alpha^2}\right) dx$$

$$= -\exp\left(-\frac{x^2}{2\alpha^2}\right)\Big|_0^t = 1 - e^{-t^2/2\alpha^2}, \quad t > 0$$

From the problem statement we are given that $P(T < 110 | T > 100) = 0.10$ and from the notion of conditional probability,

$$P(T < 110 | T > 100) = \frac{P(100 < T < 110)}{P(T > 100)}$$

$$= \frac{F_T(110) - F_T(100)}{1 - F_T(100)}$$

$$= \frac{e^{-100^2/2\alpha^2} - e^{-110^2/2\alpha^2}}{e^{-100^2/2\alpha^2}} = 0.1$$

Rearranging this equation gives

$$.9 = \exp\left(-\frac{110^2}{2\alpha^2} + \frac{100^2}{2\alpha^2}\right)$$

or

$$\ln(.9) = -2100/2\alpha^2$$

which has the solution $\alpha^2 \cong 9{,}962$ or $\alpha \cong 99.8$.

5.3.3 *STUDENT'S t-DISTRIBUTION*

The student's *t*-distribution with n degrees of freedom (which was first introduced by W. Gossett who published his work under the name of Student) is given by the pdf

$$f(t) = \frac{\Gamma((n+1)/2)}{\sqrt{n\pi}\,\Gamma(n/2)} \cdot \frac{1}{(1+t^2/n)^{(n+1)/2}}, \quad -\infty < t < \infty$$

As the density function is symmetric about the origin, it has a mean of zero. If $n = 1$, it is a Cauchy pdf with parameters $\alpha = \sqrt{n}$ and $\beta = 0$, and in this case it has no variance. Similarly, the variance will not exist for $n = 2$. It will, however, have a finite second moment when $n > 2$ that can be calculated as follows:

$$V(T) = E(T^2)$$

$$= \int_{-\infty}^{\infty} \frac{\Gamma((n+1)/2)}{\sqrt{n\pi}\,\Gamma(n/2)} \frac{t^2}{(1+t^2/n)^{(n+1)/2}}\, dt$$

$$= \frac{\Gamma((n+1)/2)}{\sqrt{n\pi}\,\Gamma(n/2)}\left\{ -t \cdot \frac{n}{n-1} \cdot \frac{1}{(1+t^2/n)^{(n-1)/2}}\bigg|_{-\infty}^{\infty}\right.$$

$$\left. + \frac{n}{n-1}\int_{-\infty}^{\infty} \frac{1}{(1+t^2/n)^{(n-1)/2}}\, dt\right\}$$

$$= \frac{\Gamma((n+1)/2)}{\sqrt{n\pi}\,\Gamma(n/2)}\left\{\frac{n}{n-1}\int_{-\infty}^{\infty} \frac{(1+t^2/n)}{(1+t^2/n)^{(n+1)/2}}\, dt\right\}$$

$$= \frac{n}{n-1} + \frac{1}{n-1}V(T)$$

where use has been made of integration by parts. Rearranging this equation gives

$$V(T) = n/(n-2) \quad \text{for } n > 2$$

The curve has a single maximum at the origin and its points of inflection are easily obtained as

$$\left(\frac{n+3}{2}\right) \pm \frac{(n^2+2n+9)^{1/2}}{2}$$

The general appearance of the pdf for large values of n is quite similar to the normal pdf.

As for the standard normal and chi-square densities, it is convenient to define a parameter $t_{\alpha;n}$ such that if T is a random variable with Student's t-distribution with n degrees of freedom, then

$$P(T \le t_{\alpha;n}) = \alpha = F_T(t_{\alpha;n})$$

As with the chi-square density, this is a function of both α and n and so values of $t_{\alpha;n}$ are tabulated in Appendix F for various values of n and those values of α most often appearing in practice. This table will be useful in later chapters.

A summary of the pdf's and their properties appears in Table 5.1 as a convenient reference.

Problems

1. Find the mean and the median of the pdf
$$f(x) = \begin{cases} 4(x - x^3), & 0 \le x \le 1 \\ 0, & \text{otherwise} \end{cases}$$

2. An electronic component has a lifetime T that has the pdf
$$f(t) = \frac{t}{a^2} \exp\left(-\left(\frac{t}{a}\right)^2\right), \quad t > 0, \ a = 10^4$$
Find the mean and variance of the pdf.

3. Show that the function
$$f(x) = \frac{x^n}{n!} e^{-x}, \quad x > 0, \ n \text{ a positive integer}$$
is a pdf and find its mean and variance.

4. For what value of the constant A in terms of a and b is the function
$$f(x) = \begin{cases} A(x - a)(b - x), & a \le x \le b \\ 0, & \text{elsewhere} \end{cases}$$
a pdf and what is its mean and variance.

5. The lifetime of a lightbulb is an exponentially distributed random variable with parameter $\lambda = .001$ hours. What fraction of the bulbs will last longer than 1100 hours?

6. If in Problem 5 five bulbs are chosen, what is the probability that three of them will still be operating after 1200 hours?

7. The number of cars through a toll gate during any period of length T in minutes is a Poisson random variable with parameter $(.5)T$. Starting at some arbitrary time, what is the probability that it will take longer than two minutes for the first car to arrive at the gate?

8. Under what conditions will the median of a distribution coincide with the mean? Give an example of a distribution that has a median but not a mean.

9. The length of nails is assumed to have a pdf $f(x) = a(x - 9)(10 - x)$, $9 \le x \le 10$. Find the mean, median, and mode of this pdf.

10. Show that if a pdf is symmetric about $x = a$, then the mean of the pdf is just a.

11. For what value of the constant c will the expression $E((e^X - c)^2)$ be minimized if $X \sim N(\mu, \sigma^2)$.

12. The error Y in a certain process is related to the error in a certain measurement X by $Y = 5 + 4X + X^2$. If $X \sim N(0, 1)$, find the mean and variance of Y.

13. The error of a measuring instrument is assumed to have a normal distribution with mean zero. If it is known that the probability that the absolute value of this error exceeds .01 is .90, what is the standard deviation of the error.

14. Using the tables of Appendix D, sketch the cdf of the normal random variable $N(4, 9)$.

15. Let X be the weight of a box of cereal. If $X \sim N(250, 100)$ (grams), and boxes weighing less than 245 grams must be repackaged, out of every ten boxes how many would have to be repackaged on the average?

16. It has been observed that a certain measurement is a normal random variable of which 25% are less than 10.2 cm and 10% are greater than 12.0 cm. What are the mean and variance of the measurements.

17. Suppose the lifetime of a system is modeled as a normal random variable with mean 2500 hours. If 95% of the systems are to last at least 2000 hours, what is the largest value that the standard deviation can have?

18. In Problem 17, if $\sigma = 400$ hours, what is the probability that a system which has lasted 2300 hours will last a further 500 hours? Compare this with the probability that a system will last 2800 hours.

19. If X is a normal random variable with mean μ and variance σ^2, express the third and fourth central moments of X, $E((X - \mu)^3)$, and $E((X - \mu)^4)$ in terms of μ and σ^2 only.

20. It is known that 5% of the weighings of an automatic meat packaging machine differ from the nominal value set by more than 2%. If the nominal value is 500 grams and the actual weight is a normal random variable, what fraction of the packages will weigh less than 495 grams?

21. It is unknown whether a certain length measurement X has either a normal pdf $N(20, .01)$ (in meters) or a Cauchy pdf

$$f(x) = \frac{1}{\pi(.1)\left(1 + \left(\frac{x-20}{.1}\right)^2\right)}$$

If we choose to model X by the pdf for which $P(|X - 20| < .15)$ is maximum, which pdf would be chosen?

22. A reading X taken from a chemical process is known to be a normal random variable with mean 4.0 gm and 8% of the readings exceed 4.2 gm. If it is possible to adjust the scales in a linear manner to read $Y = aX + b$, how should a and b be chosen in order that Y have the same mean but fewer than 2% of the readings of Y exceeding 4.2 gm?

23. If X is a normal random variable with mean μ and variance σ^2 for an interval of length τ, for what value of x is the probability $P(x \le X \le x + \tau)$ a maximum?

24. The IQ's of 6 year olds is assumed to be a normal random variable. If it is known that 15% of the children have IQ's under 90 and 2% exceed 135, what percentage of the children have IQ's between 100 and 120?

25. A machine, when operating correctly, produces items that have weight which is normally distributed with mean 20 gm and variance 4 gm². If the machine goes out of adjustment, it produces items that have a weight which is normally distributed

with mean 23 gm with variance 9 gm^2. If an item has weight 21 gm, is it more likely to have been produced when the machine is operating correctly or not?

26. The number of machine breakdowns over a period of T hours in a certain plant is a Poisson random variable with parameter $(.2)T$. Find the pdf of the time between breakdowns.

27. Show that an exponential random variable X with parameter λ has the property that $P(X > t + s \mid X > t) = P(X > s)$.

28. A random variable X is said to have a log-normal distribution if it has the pdf

$$f(x) = \frac{1}{\sqrt{2\pi}\sigma x} \exp\left\{-\frac{1}{2\sigma^2}(\ln(x) - \mu)^2\right\}, \quad x > 0, \, \sigma > 0$$

Using the properties of the normal pdf, find an expression for the kth moment of X and the variance of X.

29. A random variable X is said to have a Weibull distribution with parameters α and λ if it has the density function

$$f(x) = \lambda \alpha x^{\alpha - 1} \exp(-\lambda x^\alpha), \quad \lambda, \alpha > 0, x \geq 0$$

Find the cdf of this pdf. Find an expression for the kth moment of this pdf. (This pdf commonly occurs in reliability studies and will be used in Chapter 12.)

30. The total time taken to complete a certain job T is a gamma random variable with pdf

$$f(t) = \frac{\gamma^\alpha}{\Gamma(\alpha)} t^{\alpha - 1} e^{-\gamma t}, \quad t \geq 0$$

Where $\alpha = 4$ and $\gamma = 1$ hours. What fraction of jobs will take longer than 5 hours to complete?

CHAPTER 6

Bivariate Random Variables and Their Distributions

To this point we have only attempted to deal with a single random variable at a time. It is of course possible to define many random variables on the one sample space, and it is in fact of interest to do so in many situations. For example, in a certain experiment we might be interested in measuring the speed and direction of atomic particles, being emitted randomly from a surface, reporting the outcome as a pair (v, θ), where θ is the direction angle with respect to some reference and v the speed. The speed and direction may or may not be related in a probabilistic sense. The study of the individual random variables V and θ as well as their relative behavior is of interest. In a communication system the amplitude A and phase Φ might be random, depending on the medium through which they were transmitted, and an accurate probabilistic description of their joint behavior may be crucial to the design of an efficient receiver. We study here some of the problems associated with considering more than one random variable at a time. By and large the techniques developed will be extensions of the single variable techniques already considered.

6.1 Bivariate Random Variables and Their Joint Distributions

Definition   Let S be the sample space of some experiment E. Let X and Y be two mappings from S to the real numbers. Then the pair (X, Y) that assigns a point in the real (x, y) plane to each point $s \in S$ is called a bivariate or two-dimensional random variable. Similarly, if $X_1, X_2, \ldots, X_n$ are n real-valued functions on S the n-tuple $(X_1, X_2, \ldots, X_n)$ is called an n-dimensional or n-variate random variable or vector.

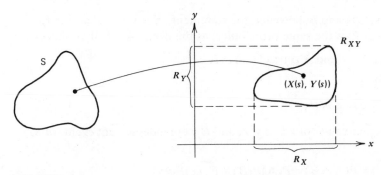

FIGURE 6.1

A picture of the situation is shown in Figure 6.1. The range space of the bivariate random variable (X, Y) is denoted by R_{XY} and defined by

$$R_{XY} = \{(x, y) | \text{there is an } s \in S \text{ such that } X(s) = x, Y(s) = y\}$$

The study of bivariate or n-variate random variables proceeds much as for the single variate case. Similar functions will be defined, except the reader will be required to think in terms of higher dimensions rather than just the real line.

If the random variables X and Y are each, by themselves, discrete random variables, we will call (X, Y) a *discrete bivariate random variable*. Similarly, if X and Y are each, by themselves, continuous random variables, we will call (X, Y) a *continuous bivariate random variable*. It is possible that one of X and Y is discrete while the other is continuous, but such cases seldom arise in practice and can be treated as special cases when they do. The discussion of this section is valid for either discrete or continuous bivariate random variables.

Our first task is to define cumulative distribution functions for bivariate random variables and relate the concept of independence of events, introduced in Chapter 4, to independence of random variables.

Let X and Y be two random variables, both defined on the sample space S. Their individual cdf's, denoted by appropriate subscripts, are

$$F_X(x) = P(X \le x) \quad \text{and} \quad F_Y(y) = P(Y \le y)$$

If A and B are events of S defined by

$$A = \{s \in S \,|\, X(s) \le x\} \quad \text{and} \quad B = \{s \in S \,|\, Y(s) \le y\}$$

then

$$F_X(x) = P(A) \quad \text{and} \quad F_Y(y) = P(B) \tag{6.1}$$

where $P(A)$ and $P(B)$ are the probabilities that the outcome of one trial of the experiment is in A or B, respectively. With two random variables we can define a third function, called the joint cumulative distribution function (joint cdf) of X and Y, which will be denoted by $F_{XY}(x, y)$ and defined by the equation

$$F_{XY}(x, y) = P(X \le x \text{ and } Y \le y) \tag{6.2}$$

In terms of the sets A and B, however, the statement $\{X \le x \text{ and } Y \le y\}$ in Eq. (6.2) is equivalent (has the same probability) as the event $A \cap B$ in S, that is,

$$\{s \in S \,|\, X(s) \le x \text{ and } Y(s) \le y\} = A \cap B$$

It must follow that

$$F_{XY}(x, y) = P(A \cap B) \tag{6.3}$$

If, for two particular values of x and y, A and B were independent events of S, we could write

$$F_{XY}(x, y) = P(A \cap B) = P(A)P(B) = F_X(x)F_Y(y)$$

which follows from Eqs. (6.1) and (6.3).

Definition   Two random variables X and Y will be called *independent* if

$$F_{XY}(x, y) = F_X(x)F_Y(y) \tag{6.4}$$

for every value of x and y.

Independence of random variables is a simple extension of the idea of independence of events, that is, X and Y are independent random variables if

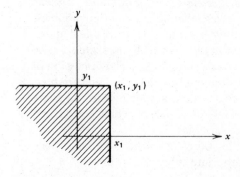

FIGURE 6.2

Applied Probability

and only if the events

$$\{s \in S \,|\, X(s) \leq x_1\} \quad \text{and} \quad \{s \in S \,|\, Y(s) \leq y_1\}$$

are independent for all values of x_1 and y_1. Notice that $F_{XY}(x_1, y_1)$ is a *probability*, the probability that $X(s) \leq x_1$ and $Y(s) \leq y_1$ or, equivalently, the probability that (X, Y) lies in the shaded area shown in Figure 6.2, where x_1 and y_1 are arbitrary values of x and y, respectively. The following example expands on this idea. We will drop the subscripts on $F_{XY}(x, y)$ when the meaning is clear.

EXAMPLE 6.1

The bivariate random variable (X, Y) has the joint cdf $F(x, y)$. Express the probability that the outcome (X, Y) on any particular trial, will fall in the shaded region of Figure 6.3 in terms of the function $F(x, y)$ only.

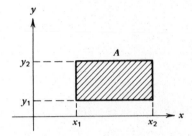

FIGURE 6.3

Denote the shaded area by A. By simple reasoning we can break the probability that (X, Y) lies in A into four terms:

$$P((X, Y) \in A) = P(X \leq x_2, Y \leq y_2) - P(X \leq x_2, Y \leq y_1)$$
$$- P(X \leq x_1, Y \leq y_2) + P(X \leq x_1, Y \leq y_1) \qquad (6.5)$$

where each term on the right-hand side of this equation corresponds to one of the regions in Figure 6.4. But each term in Eq. (6.5) is easily expressed in terms

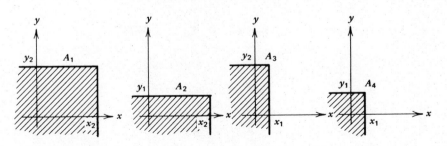

FIGURE 6.4

of $F(x, y)$ and so

$$P((X, Y) \in A) = F(x_2, y_2) - F(x_2, y_1) - F(x_1, y_2) + F(x_1, y_1)$$

The last term in Eq. (6.5) corresponds to area A_4 in Figure 6.4 and must be added in as the probability of a point in A_4 has been subtracted twice (once because A_4 is contained in A_2 and once because A_4 is contained in A_3), while they have only been added once (because A_4 is contained in A_1).

Joint cdf's have many properties analogous to the single variable cdf's. We list the important properties here and leave their verification to the reader.

(1) $\lim\limits_{\substack{x \to \infty \\ y \to \infty}} F_{XY}(x, y) = 1, \quad \lim\limits_{\substack{x \to -\infty \\ y \to -\infty}} F_{XY}(x, y) = 0$

(2) $0 \le F_{XY}(x, y) \le 1$

(3) If $x_1 \le x_2$ and $y_1 \le y_2$, then

(a) $F_{XY}(x_1, y_1) \le F_{XY}(x_2, y_1) \le F_{XY}(x_2, y_2)$

(b) $F_{XY}(x_1, y_1) \le F_{XY}(x_1, y_2) \le F_{XY}(x_2, y_2)$

(6.6)

(4) $\lim\limits_{x \to \infty} F_{XY}(x, y) = F_Y(y), \quad \lim\limits_{y \to \infty} F_{XY}(x, y) = F_X(x)$

where $F_X(x)$ and $F_Y(y)$ are the cdf's of X and Y, respectively.

The next two examples illustrate some of these aspects of joint cdf's.

EXAMPLE 6.2

An item is drawn from a large lot and two random variables X and Y defined as follows: If the item drawn has a defect of Type 1, $X = 1$ and otherwise $X = 0$; if the item drawn has a defect of Type 2, $Y = 1$ and otherwise $Y = 0$. The joint cdf of the discrete bivariate random variable is known to be

$$F_{XY}(x, y) = \begin{cases} 0, & x < 0 \text{ or } y < 0 \\ p_1, & 0 \le x < 1, 0 \le y < 1 \\ p_2, & 0 \le x < 1, y \ge 1 \\ p_3, & x \ge 1, 0 \le y < 1 \\ 1, & x \ge 1, y \ge 1 \end{cases}$$

Under what conditions on these probabilities will X and Y be independent random variables?

The marginal cdf's of X and Y are, using the properties of Eq. (6.6),

$$F_X(x) = \begin{cases} 0, & x < 0 \\ p_2, & 0 \le x < 1 \\ 1, & x \ge 1 \end{cases} \qquad F_Y(y) = \begin{cases} 0, & y < 0 \\ p_3, & 0 \le y < 1 \\ 1, & y \ge 1 \end{cases}$$

respectively. For X and Y to be independent we must have $F_{XY}(x, y) =$

$F_X(x)F_Y(y)$. By considering the marginal cdf's, it is clear that for $0 \le x < 1$, $0 \le y < 1$ we must have $p_1 = p_2 p_3$ and this is a necessary and sufficient condition for X and Y to be independent.

EXAMPLE 6.3

Let (X, Y) be a bivariate random variable with joint cdf

$$F(x, y) = \begin{cases} (1 - e^{-x})(1 - e^{-y}), & 0 \le x < \infty, \ 0 \le y < \infty \\ 0, & \text{elsewhere} \end{cases}$$

Show that X and Y are independent random variables and evaluate $P(X < 1)$.

In order to verify that X and Y are independent random variables, we must show that their joint cdf $F(x, y)$ satisfies Eq. (6.4). There is a subtle point involved here. It is obvious that $F(x, y)$ is a product of a function of x, $(1 - e^{-x})$, and a function of y, $(1 - e^{-y})$. This alone is not enough to guarantee the independence of X and Y. These two functions must be the cdf's of X and Y, respectively, in order that Eq. (6.4) be satisfied. From Property 4 of Eq. (6.6)

$$F_X(x) = \lim_{y \to \infty} F(x, y)$$

$$= \lim_{y \to \infty} (1 - e^{-x})(1 - e^{-y}) = (1 - e^{-x}), \quad x \ge 0$$

and similarly

$$F_Y(y) = \lim_{x \to \infty} F(x, y)$$

$$= (1 - e^{-y}), \quad y \ge 0$$

Since $F(x, y)$ is indeed the product of these two functions, X and Y are independent random variables. Having computed $F_X(x)$, the final part of the question is simple since

$$P(X < 1) = F_X(1) = (1 - e^{-1})$$

6.2 Joint and Conditional Distributions

Joint cdf's are defined for both bivariate discrete and continuous random variables, although the emphasis in the preceding section was on the continuous case. As with the single variable case it is often convenient to define joint probability distribution and density functions.

Let (X, Y) be a bivariate discrete random variable, where X and Y are each discrete random variables and (X, Y) can take on the values (x_i, y_j) for certain allowable sets of integers i and j. Denote $P(X = x_i, Y = y_j)$ by $p_{XY}(x_i, y_j)$, which we call the *joint probability distribution function* of the discrete bivariate random variable (X, Y). This joint probability distribution function has two properties:

(1) $0 \le p_{XY}(x_i, y_j) \le 1$
(2) $\sum_i \sum_j p_{XY}(x_i, y_j) = 1$

and any collection of numbers $p_{XY}(x_i, y_j)$ satisfying these two properties is a joint probability distribution function. The joint cdf of the bivariate random variable (X, Y) is

$$F(x, y) = \sum_{x_i < x} \sum_{y_j < y} p_{XY}(x_i, y_j) \qquad (6.7)$$

but this tends to be an awkward function to use. It is also possible to "invert" Eq. (6.7) to express $p_{XY}(x_i, y_j)$ in terms of the joint cdf, but a simple expression results only in particular cases.

The random variables X and Y each have a probability distribution of their own and, since the joint probability distribution function contains all the information on (X, Y), these must be obtainable from it. To avoid confusion we denote the various probability distribution functions by $p_{XY}(x_i, y_j)$, $p_X(x_i)$, and $p_Y(y_j)$. Suppose that for a fixed value $X = x_i$ the random variable Y can only take on the possible values $y_{i_1}, y_{i_2}, \ldots, y_{i_n}$. To calculate $P(X = x_i) = p_X(x_i)$ we merely sum

$$P(X = x_i) = p_X(x_i)$$

$$= p_{XY}(x_i, y_{i_1}) + p_{XY}(x_i, y_{i_2}) + \cdots + p_{XY}(x_i, y_{i_n})$$

since the only way $X = x_i$ can occur is in conjunction with the values $Y = y_{i_1}, y_{i_2}, \ldots, y_{i_n}$. More generally we simply write

$$p_X(x_i) = \sum_{y_j} p_{XY}(x_i, y_j), \qquad p_Y(y_j) = \sum_{x_i} p_{XY}(x_i, y_j) \qquad (6.8)$$

where the first summation is taken over all possible pairs (x_i, y_j) with x_i fixed and the second summation over all possible pairs (x_i, y_j) with y_j fixed. The probability distributions $p_X(x_i)$, $p_Y(y_j)$, when obtained via Eq. (6.8), are referred to as the *marginal distributions* of X and Y, respectively. These functions are the ordinary probability distribution functions of X and Y, but when derived from the joint probability distribution function the adjective marginal is added.

If X and Y are independent random variables, then

$$F_{XY}(x, y) = F_X(x)F_Y(y) \qquad (6.9)$$

It can be shown that as a consequence of this fact

$$p_{XY}(x_i, y_j) = p_X(x_i)p_Y(y_j) \qquad (6.10)$$

and that if Eq. (6.10) is true then so is Eq. (6.9). Equation (6.10) is often taken as the *definition* of the random variables X and Y being independent (if it is true for all values of x_i, y_j) as the joint cdf for bivariate discrete random variables is seldom used or available. Equation (6.10) in words says that the discrete random variables X and Y are independent if and only if their joint probability distribution function equals the product of their marginal distributions. We will drop the subscripts on the joint probability distribution function when the meaning is clear. The following example further demonstrates these ideas.

EXAMPLE 6.4

Table 6.1 gives the joint probability distribution function of the bivariate random variable (X, Y). (i) Find the marginal distribution functions $p_X(x)$ and $p_Y(y)$ and show that X and Y are independent random variables. (ii) Find $P(Y>X)$.

TABLE 6.1

	Y			
X	$y_1=-2$	$y_2=0$	$y_3=1$	$y_4=3$
$x_1=-1$	1/24	1/12	1/12	1/24
$x_2=0$	1/12	1/6	1/6	1/12
$x_3=1$	1/24	1/12	1/12	1/24

(i) The marginal distribution function of X, $p_X(x_i)$, $i=1, 2, 3$ is obtained by summing along rows:

$$p_X(-1)=1/24+1/12+1/12+1/24=1/4, \quad p_X(0)=1/2, \quad p_X(1)=1/4$$

The marginal distribution function of Y, $p_Y(y_j)$, $j=1, 2, 3, 4$ is obtained by summing up columns:

$$p_Y(-2)=1/6, \quad p_Y(0)=1/3, \quad p_Y(1)=1/3, \quad p_Y(3)=1/6$$

For all pairs (x_i, y_j), $i=1, 2, 3$, $j=1, 2, 3, 4$ it is easily verified that

$$p(x_i, y_j)=p_X(x_i)p_Y(y_j)$$

and X and Y are independent random variables.

(ii) To find $P(Y>X)$ we simply sum the probabilities of all points (x_i, y_j) such that $y_j>x_i$:

$$\begin{aligned} P(Y>X)=&p(-1, 0)+p(-1, 1)+p(-1, 3)\\ &+p(0, 1)+p(0, 3)\\ &+p(1, 3)\\ =&7/12 \end{aligned}$$

The case where (X, Y) is a continuous bivariate random variable is mathematically easier to deal with than the discrete case. Suppose (X, Y) has the joint cdf $F(x, y)$ such that the function

$$f(x, y) = \frac{\partial^2 F(x, y)}{\partial x \, \partial y} \qquad (6.11)$$

is a continuous function, except for possibly a finite number of points. In this case we will call $f(x, y)$ the joint *probability density function* (joint pdf) of (X, Y). Since, from the definition

$$F(x, y) = P(X \le x, \, Y \le y)$$

and since, by integrating Eq. (6.11),

$$F(x, y) = \int_{-\infty}^{y} \int_{-\infty}^{x} f(u, v) \, du \, dv$$

$$= P(X \le x, \, Y \le y) \qquad (6.12)$$

From this equation we can obtain an interpretation of the joint pdf $f(x, y)$. Consider a rectangular region A in the (x, y) plane, as shown in Figure 6.5.

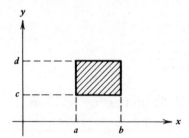

FIGURE 6.5

From Eq. (6.6)

$$P((X, Y) \in A) = P(a \le X \le b, \, c \le Y \le d)$$

$$= F(b, d) - F(b, c) - F(a, d) + F(a, c)$$

$$= \int_{-\infty}^{b'} \int_{\infty}^{d} f(u, v) \, du \, dv - \int_{-\infty}^{b} \int_{-\infty}^{c} f(u, v) \, du \, dv$$

$$- \int_{-\infty}^{a} \int_{-\infty}^{d} f(u, v) \, du \, dv + \int_{-\infty}^{a} \int_{-\infty}^{c} f(u, v) \, du \, dv$$

$$= \int_{a}^{b} \int_{c}^{d} f(u, v) \, du \, dv \qquad (6.13)$$

If the rectangle A is infinitesimally small and $b - a = \Delta x$, $d - c = \Delta y$, then this double integration can be approximated by

$$P((X, Y) \in A) = P(a \le X \le a + \Delta x, \, c \le Y \le c + \Delta y)$$

$$\cong f(a, c) \, \Delta x \, \Delta y$$

Since this is a probability and $\Delta x > 0$, $\Delta y > 0$, it must be that $f(x, y) \geq 0$ for all (x, y). Furthermore, if we expand the rectangle A in Figure 6.5 to encompass the whole plane, Eq. (6.13) becomes

$$P(-\infty < X < \infty, -\infty < Y < \infty) = 1 = \int_{-\infty}^{\infty} \int_{-\infty}^{\infty} f(u, v) \, du \, dv$$

Thus the joint pdf of any continuous bivariate random variable (X, Y), $f(x, y)$ satisfies the two properties:

(1) $f(x, y) \geq 0$

(2) $\int_{-\infty}^{\infty} \int_{-\infty}^{\infty} f(x, y) \, dx \, dy = 1$

and any function with these two properties (along with the necessary continuity condition mentioned) is the joint pdf of some bivariate random variable (X, Y).

A simple extension of Eq. (6.13) yields the fact that for any area A in the (x, y) plane

$$P((X, Y) \in A) = \int_{A} \int f(x, y) \, dx \, dy \tag{6.14}$$

where the integration is over the set A in the (x, y) plane. This probability is just the volume under the curve $f(x, y)$ supported by the area A.

It is to be emphasized again that $f(x, y)$ is a joint density function and it only yields a probability when integrated over a region of its range space R_{XY}. This probability can be thought of as the frequency with which independent observations of the bivariate random variable (X, Y) fall into this region.

As for the discrete case, the individual pdf's of the random variables X and Y can be obtained from their joint pdf $f(x, y)$ and, again, when they are obtained in this manner they are referred to as the marginal pdf's. To obtain the marginal pdf of X, $f_X(x)$, from $f(x, y)$ notice that the cdf of X is given by the equation

$$F_X(x) = P(X \leq x, Y < \infty)$$
$$= F(x, \infty)$$
$$= \int_{-\infty}^{x} \int_{-\infty}^{\infty} f(u, v) \, du \, dv$$

and differentiating both sides of this equation with respect to x gives

$$f_X(x) = \frac{d}{dx} F_X(x) = \int_{-\infty}^{\infty} f(x, v) \, dv$$

Similarly, it can be shown that the marginal pdf of Y is

$$f_Y(y) = \int_{-\infty}^{\infty} f(u, y) \, du$$

If X and Y are independent random variables, by Eq. (6.4) we can write

$$F(x, y) = F_X(x)F_Y(y)$$

and taking partial derivatives of this equation with respect to x and y

$$\frac{\partial^2}{\partial x\, \partial y}F(x, y) = \frac{\partial}{\partial x}F_X(x) \cdot \frac{\partial}{\partial y}F_Y(y)$$

or

$$f(x, y) = f_X(x)f_Y(y) \tag{6.15}$$

in analogy with Eq. (6.10) for the discrete case. In words, this equation says that the random variables X and Y are independent random variables if and only if their joint pdf factors into the product of the marginal pdf's of X and Y. Again it should be noted that it is not sufficient for $f(x, y)$ merely to factor into a product of a function of x and a function of y in order to have independence. The following series of examples develops the ideas of this section further.

EXAMPLE 6.5

The joint pdf of the bivariate random variable (X, Y) is

$$f(x, y) = \begin{cases} \frac{1}{8}(x+y), & 0 \le x, y \le 2 \\ 0, & \text{elsewhere} \end{cases}$$

(i) Are X and Y independent random variables? (ii) Evaluate $P(|X - Y| > 1)$.

(i) To determine whether X and Y are independent, we first find their marginal pdf's and apply the test of Eq. (6.15):

$$f_X(x) = \int_0^2 \tfrac{1}{8}(x+y)\, dy$$

$$= \tfrac{1}{8}\left(xy + \frac{y^2}{2} \right)\Big|_0^2$$

$$= \begin{cases} \frac{1}{4}(x+1), & 0 \le x \le 2 \\ 0, & \text{elsewhere} \end{cases}$$

and since $f(x, y)$ is symmetric as a function and in its region of definition

$$f_Y(y) = \begin{cases} \frac{1}{4}(y+1), & 0 \le y \le 2 \\ 0, & \text{elsewhere} \end{cases}$$

and since $f(x, y) \ne f_X(x)f_Y(y)$, they are not independent random variables.

(ii) To evaluate $P(|X - Y| > 1)$ we must first describe the region, which will be denoted by A, in the (x, y) plane, and apply Eq. (6.14). The region is shown shaded in Figure 6.6. Since $f(x, y)$ is symmetric with respect to x and y, we can find the probability of (X, Y) lying in one of the shaded areas A' and

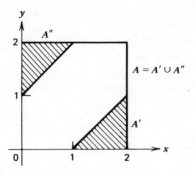

FIGURE 6.6

double it to obtain the answer:

$$P(|X - Y| > 1) = 2 \int_1^2 \int_0^{x-1} f(x, y) \, dx \, dy$$

$$= 2 \int_1^2 \left(\int_0^{x-1} \tfrac{1}{8}(x + y) \, dy \right) dx$$

$$= 2 \int_1^2 \left(\tfrac{1}{8} \left(xy + \frac{y^2}{2} \right) \Big|_0^{x-1} \right) dx$$

$$= 2 \int_1^2 \tfrac{1}{8}(x(x-1) + \tfrac{1}{2}(x-1)^2) \, dx$$

$$= \tfrac{1}{4}$$

Notice in particular from this example that if a function represents a joint pdf only over a region of the (x, y) plane, then this region must be clearly specified. Notice also the manner in which the integral over the area A' is accomplished.

Definition The bivariate random variable (X, Y) is said to be uniformly distributed over the region R_{XY} if its joint pdf $f(x, y)$ is

$$f(x, y) = \begin{cases} \text{constant } C, & (x, y) \in R_{XY} \\ 0, & \text{elsewhere.} \end{cases} \tag{6.16}$$

Since

$$\int_{R_{XY}} \int f(x, y) \, dx \, dy = 1 = \int_{R_{XY}} \int C \cdot dx \, dy = C \times \text{Area} \, (R_{XY})$$

it follows immediately that the constant C in Eq. (6.16) must be

$$C = 1/\text{Area of } R_{XY}$$

EXAMPLE 6.6

Let X be the crushing strength of a concrete beam and Y the load applied. If the joint pdf of the bivariate random variable (X, Y) is assumed uniformly distributed over the region $\{2000 \le x \le 2400, 2200 \le y \le 2600\}$, what is the probability that a given beam will be able to withstand the load applied?

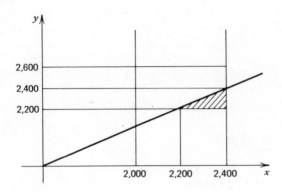

FIGURE 6.7

The probability in question is simply the probability that X exceeds Y and this region is shown shaded in Figure 6.7. The pdf is

$$f(x, y) = \begin{cases} \dfrac{1}{16 \times 10^4}, & (x, y) \in R_{XY} \\ 0, & \text{elsewhere} \end{cases}$$

and integrating this function over the shaded region gives

$$P(X > Y) = \int_{2200}^{2400} \int_{2200}^{x} f(x, y) \, dy \, dx$$

$$= \frac{1}{16 \times 10^4} \times \text{Area of shaded region}$$

$$= \frac{2 \times 10^4}{16 \times 10^4} = \frac{1}{8}$$

EXAMPLE 6.7

The bivariate random variable (X, Y) has the joint pdf

$$f(x, y) = \begin{cases} 4xy, & 0 < x < 1, 0 < y < 1 \\ 0, & \text{elsewhere} \end{cases}$$

(i) Find the probability that exactly one of the random variables is greater than 0.9. (ii) Find the probability that $X + Y > 1$.

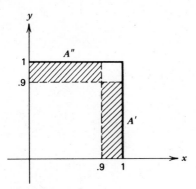

FIGURE 6.8

(i) *Method 1.* The area of the (x, y) plane where exactly one of the random variables is greater than .9 is shown shaded in Figure 6.8. From symmetry we can write

P(exactly one of the random variables is greater than .9)

$$= 2 \int_{A'} \int f(x, y)\, dx\, dy$$

$$= 2 \int_{.9}^{1} \left(\int_{0}^{.9} 4xy\, dy \right) dx$$

$$= 2 \int_{.9}^{1} 4x \left(\frac{y^2}{2} \right) \Big|_{0}^{.9} dx$$

$$= 2 \times 4 \frac{(.81)}{2} \frac{x^2}{2} \Big|_{.9}^{1}$$

$$= 2 \times .81(1 - .81) = .3078$$

(i) *Method 2.* It is easy to show that X and Y are independent random variables with marginal pdf's,

$$f_X(x) = \begin{cases} 2x, & 0 < x < 1 \\ 0, & \text{elsewhere} \end{cases}$$

$$f_Y(y) = \begin{cases} 2y, & 0 < y < 1 \\ 0, & \text{elsewhere} \end{cases}$$

Since X and Y are independent random variables, we can write

P(exactly one of X or Y greater than .9)

$$= P(X > .9 \text{ and } Y < .9) + P(X < .9 \text{ and } Y > .9)$$

$$= 2P(X > .9)P(Y < .9)$$

$$= 2 \left(\int_{.9}^{1} 2x\, dx \right) \left(\int_{0}^{.9} 2y\, dy \right) = .3078$$

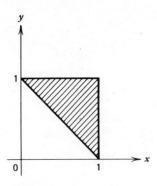

FIGURE 6.9

(ii) The region in the (x, y) plane corresponding to the event $\{X + Y > 1\}$ is shown in Figure 6.9. It follows that

$$P(X + Y > 1) = \int_0^1 \left(\int_x^1 4xy \, dy \right) dx$$

$$= \int_0^1 4x \cdot \left(\frac{y^2}{2} \right) \Big|_x^1 dx$$

$$= \int_0^1 2x(1 - x^2) \, dx$$

$$= 1/2$$

EXAMPLE 6.8

Show that if (X, Y) is a bivariate random variable with pdf

$$f(x, y) = \begin{cases} 8xy, & 0 < y < x < 1 \\ 0, & \text{elsewhere} \end{cases} \tag{6.17}$$

then X and Y are not independent random variables. The region where $f(x, y)$ is defined as shown in Figure 6.10. The marginal pdf's are easily found as

$$f_X(x) = \begin{cases} 4x^3, & 0 < x < 1 \\ 0, & \text{elsewhere} \end{cases}$$

$$f_Y(y) = \begin{cases} 4y(1 - y^2), & 0 < y < 1 \\ 0, & \text{elsewhere} \end{cases}$$

and X and Y are clearly not independent since $f(x, y) \neq f_X(x) f_Y(y)$.

Notice in Example 6.8 that the form of the joint pdf in Eq. (6.17) is similar to that in Example 6.7 (where the random variables were independent). The difference between the two examples is the region where the joint pdf is defined. A little reflection will reveal that if a bivariate random variable (X, Y)

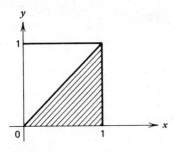

FIGURE 6.10

has a range space R_{XY} that depends functionally on x or y, then X and Y cannot be independent random variables.

The next example introduces the important notion of jointly normal random variables. The two marginal densities will be normal pdf's.

EXAMPLE 6.9

The bivariate random variable (X, Y) is said to be a bivariate normal random variable if its joint pdf is given by

$$f(x, y) = \frac{1}{2\pi\sigma_X\sigma_Y(1-\rho_{XY}^2)^{1/2}} \exp\left[-\left(\left(\frac{x-\mu_X}{\sigma_X}\right)^2\right.\right.$$
$$\left.\left.-2\rho_{XY}\left(\frac{x-\mu_X}{\sigma_X}\right)\left(\frac{y-\mu_Y}{\sigma_Y}\right)+\left(\frac{y-\mu_Y}{\sigma_Y}\right)^2\right)\Big/2(1-\rho_{XY}^2)\right] \quad (6.18)$$

(i) Find the marginal pdf's of X and Y. (ii) Under what conditions on ρ_{XY} will X and Y be independent?

Expressions for the two marginal pdf's are

$$f_X(x) = \int_{-\infty}^{\infty} f(x, y)\, dy \quad \text{and} \quad f_Y(y) = \int_{-\infty}^{\infty} f(x, y)\, dx$$

These can be evaluated for the given joint pdf $f(x, y)$ in Eq. (6.18) by completing the square in the exponent. The details are omitted but the reader should check that

$$f_Y(y) = \frac{\exp\left(-\left(\frac{y-\mu_Y}{\sigma_Y}\right)^2\right)}{\sqrt{2\pi}\sigma_Y} \int_{-\infty}^{\infty} \frac{1}{\sqrt{2\pi}\sigma_X(1-\rho_{XY}^2)^{1/2}}$$
$$\times \exp\left[-\left(\left(\frac{x-\mu_X}{\sigma_X}\right)-\rho_{XY}\left(\frac{y-\mu_Y}{\sigma_Y}\right)\right)^2\Big/2(1-\rho_{XY}^2)\right] dx$$

and applying a modified form (i) of Eq. (5.8) shows that this integral must be unity (show that it is a normal pdf with mean

$$\rho_{XY}\frac{\sigma_X}{\sigma_Y}(y-\mu_Y)+\mu_X$$

and variance $\sigma_X^2(1-\rho_{XY}^2)$). The marginal pdf of Y is then

$$f_Y(y)=\frac{1}{\sqrt{2\pi}\sigma_Y}\exp\left[-\frac{1}{2}\left(\frac{y-\mu_Y}{\sigma_Y}\right)^2\right], \quad -\infty<y<\infty$$

and, similarly, the marginal pdf for X is

$$f_X(x)=\frac{1}{\sqrt{2\pi}\sigma_X}\exp\left[-\frac{1}{2}\left(\frac{x-\mu_X}{\sigma_X}\right)^2\right], \quad -\infty<x<\infty$$

(ii) It is not difficult to show that when $\rho_{XY}=0$, $f(x,y)=f_X(x)f_Y(y)$ and in this case X and Y are independent. The parameter ρ_{XY} will be discussed in the next section. It is called the correlation coefficient of X and Y and it will be shown that $-1\le\rho_{XY}\le1$.

Chapter 4 introduced the notion of conditional probability. With bivariate random variables, either continuous or discrete, we can extend this notion to conditional probability distributions and density functions. Such functions will answer the question of "what is the pdf of the random variable Y, given that $X=x$."

Definition (i) If (X,Y) is a discrete bivariate random variable with joint probability distribution function $p(x_i,y_j)$, the *conditional probability distribution function* of Y, given that $X=x_i$, is

$$p_{Y|X}(y_j|x_i)=\frac{p(x_i,y_j)}{p(x_i)} \tag{6.19}$$

provided that $p(x_i)>0$.

(ii) If (X,Y) is a continuous bivariate random variable with joint pdf $f(x,y)$, the *conditional probability density function* (conditional pdf) of Y given that $X=x$ is

$$f_{Y|X}(y|x)=\frac{f(x,y)}{f_X(x)} \tag{6.20}$$

provided that $f_X(x)>0$, where $f_X(x)$ is the marginal pdf of X.

We can similarly define the pdf's $p(x_i|y_j)$ and $f(x|y)$. We could also define conditional cdf's since conditional random variables are proper random variables. The functions $p_{Y|X}(y_j|x_i)$ and $f_{Y|X}(y|x)$ of Eqs. (6.19) and (6.20) are perhaps best thought of as functions of the variable y with the x variable arbitrary, but fixed. As functions of the y variable, these functions satisfy all the requirements of ordinary distribution and density functions. Thus, in the discrete case

(i) $p_{Y|X}(y_j|x_i)\ge0$ and (ii) $\sum_j p_{Y|X}(y_j|x_i)=1$

since

$$\sum_j p_{Y|X}(y_j|x_i) = \sum_j \frac{p(x_i, y_j)}{p_X(x_i)} = \frac{p_X(x_i)}{p_X(x_i)} = 1$$

For the continuous case

(i) $f_{Y|X}(y|x) \geq 0$ and   (ii) $\displaystyle\int_{-\infty}^{\infty} f_{Y|X}(y|x)\,dy = 1$

since

$$\int_{-\infty}^{\infty} f_{Y|X}(y|x)\,dy = \int_{-\infty}^{\infty} \frac{f(x, y)}{f_X(x)}\,dy = \frac{f_X(x)}{f_X(x)} = 1$$

Subscripts will be omitted when the meaning is clear from the context, which pdf is being used.

The function $f(y|x)$ can be considered in several lights. First note that the range space of Y given that $X = x$ may depend on x. This will be the case when R_{XY} is functionally dependent on x and y. Now imagine the joint pdf $f(x, y)$ as a surface above the (x, y) plane and drop a vertical plane parallel to the y axis (and perpendicular to the x axis). The intersection of the surface $f(x, y)$ with this plane defines a function of y, for a fixed x. This function is "almost" the function $f(y|x)$. Since we want $f(y|x)$ to be a probability density function it should integrate to unity, that is,

$$f(y|x) = Cf(x, y), \quad x \text{ fixed}$$

where C is chosen so that

$$\int_{-\infty}^{\infty} f(y|x)\,dy = 1 = \int_{-\infty}^{\infty} Cf(x, y)\,dy = Cf_X(x)$$

or

$$C = \frac{1}{f_X(x)} \quad \text{and} \quad f(y|x) = \frac{f(x, y)}{f_X(x)}$$

For fixed x, $f(y|x)$ is simply the scalar multiple $1/f_X(x)$ times $f(x, y)$.

A probabilistic interpretation of the function $f(y|x)$ as a conditional pdf can be obtained as follows. Consider the conditional expression

$$P(y \leq Y \leq y + \Delta y | x \leq X \leq x + \Delta x) = \frac{P(x \leq X \leq x + \Delta x, y \leq Y \leq y + \Delta y)}{P(x \leq X \leq x + \Delta x)}$$

which for small Δx and Δy can be approximated by

$$\frac{f(x, y)\,\Delta x\,\Delta y}{f_X(x)\,\Delta x} = \frac{f(x, y)}{f_X(x)}\,\Delta y = f(y|x)\,\Delta y$$

This interpretation justifies referring to $f(y|x)$ as a conditional pdf.

Bivariate Random Variables and Their Distributions   171

EXAMPLE 6.10

The bivariate random variable (X, Y) has the joint pdf

$$f(x, y) = \begin{cases} Cx^2(8-y), & x < y < 2x, \ 0 \le x \le 2 \\ 0, & \text{elsewhere} \end{cases}$$

where $C = 5/112$. Find the marginal pdf's of X and Y and the conditional pdf's $f_{X|Y}(x|y)$ and $f_{Y|X}(y|x)$.

The marginal pdf of X is given by

$$f_X(x) = \int_x^{2x} Cx^2(8-y) \, dy$$

$$= C\left(8x^2 y - \frac{x^2 y^2}{2}\right)\Big|_x^{2x}$$

$$= \begin{cases} C(8x^3 - \frac{3}{2}x^4), & 0 \le x \le 2 \\ 0, & \text{elsewhere} \end{cases} \tag{6.21}$$

and that of Y by

(i) $0 \le y \le 2$, $f_Y(y) = \int_{y/2}^{y} Cx^2(8-y) \, dx = C(8-y)\dfrac{7y^3}{24}$

(ii) $2 \le y \le 4$, $f_Y(y) = \int_{y/2}^{2} Cx^2(8-y) \, dx = C(8-y)\dfrac{1}{3}\left(8 - \dfrac{y^3}{8}\right)$ $\qquad$ (6.22)

The conditional pdf of X given $Y = y$ is:

(i) $0 \le y \le 2$, $f_{X|Y}(x|y) = \dfrac{f(x, y)}{f_Y(y)}$

$$= \frac{Cx^2(8-y)}{C(8-y) \cdot 7y^3/24} = \frac{24x^2}{7y^3}, \quad \frac{y}{2} < x < y$$

(ii) $2 \le y \le 4$,

$$= \begin{cases} \dfrac{Cx^2(8-y)}{C(8-y)\dfrac{1}{3}\left(8 - \dfrac{y^3}{8}\right)} = \dfrac{24x^2}{(64 - y^3)}, & \dfrac{y}{2} < x < 2 \\ 0, & \text{elsewhere} \end{cases} \tag{6.23}$$

and the conditional pdf of Y given $X = x$ is

$$f_{Y|X}(y|x) = \frac{f(x, y)}{f_X(x)}$$

$$= \frac{Cx^2(8-y)}{Cx^3(8 - \frac{3}{2}x)}$$

$$= \begin{cases} \dfrac{2(8-y)}{x(16-3x)}, & x < y < 2x \\ 0, & \text{elsewhere} \end{cases} \tag{6.24}$$

172 *Applied Probability*

This example contains some important points. First notice that the marginal pdf $f_Y(y)$ is defined by different functions in different regions of the y axis, a consequence of the fact that $f(x, y)$ is only defined as the function $Cx^2(8-y)$ in the shaded region of Figure 6.11. Everywhere else, $f(x, y)$ is zero. Secondly, the manner in which the regions of definition of $f_{Y|X}(y|x)$ and $f_{X|Y}(x|y)$ are obtained is important. Consider $f_{Y|X}(y|x)$: When $X = x$, the random variable Y can only assume values between x and $2x$ as seen from Figure 6.11. This functional dependence of R_{XY} on x and y and the regions of definition for the various functions must be computed carefully.

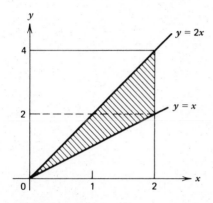

FIGURE 6.11

EXAMPLE 6.11

It has been observed that the joint distribution of the fractional part of a number Y and the number itself X, where $X \geq 1$, which is the result of a computer operation involving a random number generator, have the joint pdf

$$f(x, y) = \frac{4}{5} \frac{(x+y)}{x^3}, \quad 0 \leq y \leq 1, \ 1 \leq x < \infty$$

(i) Find $P(0 < Y < \frac{1}{2}|X = 2)$. (ii) For what value of A is it true that $P(0 < Y < \frac{1}{2}|X > A) = 5/16$?

(i) The conditional pdf of Y given X is given by $f(x, y)/f_X(x)$ and we first find the marginal pdf of X:

$$f_X(x) = \int_0^1 \frac{4}{5} \frac{(x+y)}{x^3} \, dy$$

$$= \frac{4}{5}\left(\frac{1}{x^2}y + \frac{y^2}{2x^3}\right)\Big|_0^1$$

$$= \frac{4}{5}\left(\frac{1}{x^2} + \frac{1}{2x^3}\right), \quad 1 \leq x < \infty$$

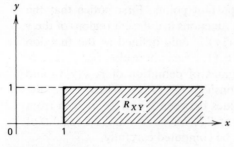

FIGURE 6.12

The conditional pdf of Y given X is

$$f_{Y|X}(y|x) = \frac{\frac{4}{5}(x+y)/x^3}{\frac{4}{5}(1/x^2 + 1/2x^3)} = \frac{2(x+y)}{(2x+1)}, \quad 0 \le y < 1$$

The probability $P(0 < Y < \frac{1}{2} | X = 2)$ is then

$$\int_0^{1/2} f_{Y|X}(y|x=2)\, dy = \int_0^{1/2} \frac{2(2+y)}{5}\, dy$$

$$= \frac{2}{5}\left(2y + \frac{y^2}{2}\right)\Big|_0^{1/2}$$

$$= \frac{2}{5}\left(1 + \frac{1}{8}\right) = \frac{9}{20}$$

(ii) The conditional probability $P(0 < Y < \frac{1}{2} | X > A)$ is found by computing the probabilities $P(\{0 < y < \frac{1}{2}\} \cap \{X > A\})$ and $P(X > A)$. From the marginal density of X:

$$P(X > A) = \int_A^\infty \frac{4}{5}\left(\frac{1}{x^2} + \frac{1}{2x^3}\right) dx$$

$$= \frac{4}{5}\left(\frac{1}{A} + \frac{1}{4A^2}\right)$$

and from the joint density

$$P(\{0 < Y < \frac{1}{2}\} \cap \{X > A\}) = \int_A^\infty \int_0^{1/2} \frac{4}{5}\frac{(x+y)}{x^3}\, dy\, dx$$

$$= \frac{4}{5}\int_A^\infty \left(\frac{1}{x^2}y + \frac{y^2}{2x^3}\right)\Big|_0^{1/2} dx$$

$$= \frac{4}{5}\int_A^\infty \left(\frac{1}{2x^2} + \frac{1}{8x^3}\right) dx$$

$$= \frac{4}{5}\left(\frac{1}{2A} + \frac{1}{16A^2}\right)$$

The conditional probability is

$$P(0<Y<\tfrac{1}{2}|X>A)=\frac{\tfrac{4}{5}(1/2A+1/16A^2)}{\tfrac{4}{5}(1/A+1/4A^2)}$$

$$=\frac{1}{4}\left(\frac{8A+1}{4A+1}\right)$$

and the value of A for which this is $5/16$ is $A=1/12$.

EXAMPLE 6.12

The joint pdf of the temperature in degrees Centigrade at two localities is, at a particular time of day on a particular day, assumed to be a joint normal pdf

$$f(x,y)=\frac{1}{4\pi\sqrt{3}}\exp\left\{-\frac{2}{3}\left[\left(\frac{x-20}{2}\right)^2-\frac{(x-20)(y-20)}{4}+\left(\frac{y-20}{2}\right)^2\right]\right\}$$

$$-\infty<x,\ y<\infty$$

Find the conditional pdf $f_{Y|X}(y|x)$. If the temperature at location 1, X, is 21°C, what is the probability that the temperature at location 2, Y, is between 20 and 22°C?

Comparing the exponent of this pdf with that in Eq. (6.18) it is seen that $\rho_{XY}=1/2$, $\sigma_X=\sigma_Y=2$, and $\mu_X=\mu_Y=20$. The marginal pdf is, by the results of Example 6.9, a normal pdf $N(20,4)$ and the conditional pdf is

$$f_{Y|X}(y|x)=\frac{f(x,y)}{f_X(x)}$$

$$=\frac{1}{\sqrt{6\pi}}\exp\left[-\frac{1}{6}\left(y-\frac{x}{2}-10\right)^2\right],\quad -\infty<y<\infty$$

(after some computation). If $X=21$°C, the probability that Y is between 20 and 22°C is

$$P(20<Y<22|X=21)=\int_{20}^{22}f_{Y|X}(y|21)\,dy$$

$$=\frac{1}{\sqrt{6\pi}}\int_{20}^{22}\exp\left[-\frac{1}{6}\left(y-\frac{21}{2}-10\right)^2\right]dy$$

$$=\int_{-1/2}^{3/2}\frac{1}{\sqrt{2\pi}}\exp\left(-\frac{z^2}{2}\right)dz$$

by using the substitution $z=(y-41/2)/\sqrt{3}$. This last expression is just $\Phi(\tfrac{3}{2})-\Phi(-\tfrac{1}{2})=.6247$ using the tables of Appendix D.

It is useful to have the general conditional pdf of two jointly normal random variables. Let X and Y have the joint normal pdf of Eq. (6.18). The marginal pdf is given in Example 6.9 and the conditional pdf is

$$f_{Y|X}(y|x) =$$

$$\frac{\dfrac{1}{2\pi\sigma_X\sigma_Y(1-\rho_{XY}^2)^{1/2}}\exp\left\{-\left[\left(\dfrac{x-\mu_X}{\sigma_X}\right)^2 -2\rho_{XY}\left(\dfrac{x-\mu_X}{\sigma_X}\right)\left(\dfrac{y-\mu_Y}{\sigma_Y}\right)+\left(\dfrac{y-\mu_Y}{\sigma_Y}\right)^2\right]\Big/2(1-\rho_{XY}^2)\right\}}{\dfrac{1}{\sqrt{2\pi}\sigma_X}\exp\left[-\dfrac{1}{2}\left(\dfrac{x-\mu_X}{\sigma_X}\right)^2\right]}$$

which, after some cancellation and rearranging, is

$$f_{Y|X}(y|x)=\frac{1}{\sqrt{2\pi}\sigma_Y(1-\rho_{XY}^2)^{1/2}}\exp\left\{-\frac{1}{2(1-\rho_{XY}^2)}\left[\left(\frac{y-\mu_Y}{\sigma_Y}\right)-\rho_{XY}\left(\frac{x-\mu_X}{\sigma_X}\right)\right]^2\right\}$$

$$=\frac{1}{\sqrt{2\pi}\sigma_Y(1-\rho_{XY}^2)^{1/2}}\exp\left[-\frac{1}{2\sigma_Y^2(1-\rho_{XY}^2)}\right.$$

$$\left.\times\left(y-\rho_{XY}\frac{\sigma_Y}{\sigma_X}(x-\mu_X)-\mu_Y\right)^2\right] \tag{6.25}$$

and the random variable Y, conditioned on the value $X = x$, has the density

$$N\left(\rho_{XY}\frac{\sigma_Y}{\sigma_X}(x-\mu_X)+\mu_Y,\ \sigma_Y^2(1-\rho_{XY}^2)\right)$$

Just as with ordinary pdf's, we can take the expectation of a conditional pdf; which we call the conditional expectation.

Definition (i) If (X, Y) is a discrete bivariate random variable, then

$$E(Y|X=x_i)=\sum_j y_j p_{Y|X}(y_j|x_i) \tag{6.26}$$

(ii) If (X, Y) is a continuous bivariate random variable, then

$$E(Y|X=x)=E(Y|x)=\int_{-\infty}^{\infty} y f_{Y|X}(y|x)\,dy \tag{6.27}$$

If X and Y are independent random variables, then the conditional pdf's reduce to marginal pdf's; in the discrete case, for X and Y independent we have

$$p_{Y|X}(y_j|x_i)=\frac{p(x_i, y_j)}{p_X(x_i)}=\frac{p_X(x_i)p_Y(y_j)}{p_X(x_i)}=p_Y(y_j)$$

while in the continuous case

$$f_{Y|X}(y|x)=\frac{f(x, y)}{f_X(x)}=\frac{f_X(x)f_Y(y)}{f_X(x)}=f_Y(y)$$

This result agrees with intuition since if X and Y are independent, then knowledge of the outcome of X can in no way affect the distribution of Y.

It follows from these observations that if X and Y are independent, then

$$E(Y|x)=E(Y)$$

The function $E(Y|x)$ is, in general, a function of x, say $H(x)$. In some situations we will want to view this function as a function, not of the ordinary variable x, but of the random variable X, $H(X)$. Application of how such a case might arise is given in Chapter 14 where estimation and regression problems are introduced. For the moment we content ourselves with evaluating $E(H(X))$:

$$E(H(X))=E(E(Y|X))$$

$$=\int f_X(x)\left(\int yf_{Y|X}(y|x)\,dy\right)dx$$

$$=\int\int y\frac{f(x,y)}{f_X(x)}\cdot f_X(x)\,dx\,dy$$

$$=\int y\left(\int f(x,y)\,dx\right)dy$$

$$=\int yf_Y(y)\,dy=E(Y)$$

To further interpret the conditional mean $E(Y|X=x)$, recall that we viewed $f_{Y|X}(y|x)$ as the pdf of the random variable Y when the value $X=x$ has been observed. The conditional mean or expectation is just the mean of this density.

EXAMPLE 6.13

Determine the conditional expectation $E(Y|x)$ of the bivariate random variable of Example 6.10.

The conditional pdf $f_{Y|X}(y|x)$ was found to be

$$f_{Y|X}(y|x)=\begin{cases}\dfrac{2(8-y)}{x(16-3x)}, & x<y<2x\\[2mm]0, & \text{elsewhere}\end{cases}$$

and the conditional expectation is

$$E(Y|x)=\int_x^{2x}y\frac{2(8-y)}{x(16-3x)}\,dy$$

$$=\frac{2}{x(16-3x)}\left(4y^2-\frac{y^3}{3}\right)\Bigg|_x^{2x}$$

$$=\frac{2x(36-7x)}{3(16-3x)},\quad 0<x<2$$

EXAMPLE 6.14

Determine $E(Y|x)$ for the bivariate normal random variable of Example 6.9.

The conditional pdf of Y given $X = x$ was found in Eq. 6.25 to have the pdf

$$N\left(\rho_{XY}\frac{\sigma_Y}{\sigma_X}(x-\mu_X)+\mu_Y, \sigma_Y^2(1-\rho_{XY}^2)\right)$$

and from this it follows *without calculation* that

$$E(Y|x)=\rho_{XY}\frac{\sigma_Y}{\sigma_X}(x-\mu_X)+\mu_Y \tag{6.28}$$

and the reader should verify this.

6.3 *The Correlation Coefficient and Regression of the Mean*

The notion of central moments of single variate random variables was defined in previous chapters. With two random variables other moments can be of interest. Let (X, Y) be a continuous random bivariate random variable with joint pdf $f(x, y)$. Only the continuous case will be considered as the equations are analogous for the discrete case. The (i, j)th moment of the bivariate random variable (X, Y) is defined by

$$E(X^iY^j)=\int_{-\infty}^{\infty}\int_{-\infty}^{\infty} x^iy^jf(x, y)\,dx\,dy$$

and the (i, j)th central moment by

$$E((X - E(X))^i(Y - E(Y))^j)$$
$$=\int_{-\infty}^{\infty}\int_{-\infty}^{\infty} (x - E(X))^i(y - E(Y))^jf(x, y)\,dx\,dy$$

By setting either i or j equal to zero in these expressions, the ordinary or central moments are obtained. If X and Y are independent random variables, then for any two functions $G(\cdot)$ and $H(\cdot)$ it is clear that

$$E(G(X)H(Y))=\iint G(x)H(y)f(x, y)\,dx\,dy$$
$$=\int G(x)f_X(x)\,dx \int H(y)f_Y(y)\,dy$$
$$=E(G(X))E(H(Y))$$

In particular, if X and Y are independent,

$$E((X - E(X))^i(Y - E(Y))^j)=E((X - E(X))^i)E((Y - E(Y))^j)$$

A simple and convenient characterization of a joint pdf is given by the $(1, 1)$ central moment that is called the covariance of X and Y and written $Cov(X, Y)$:

$$Cov(X, Y) = E((X - E(X))(Y - E(Y))) = E(XY) - E(X)E(Y)$$

Definition The correlation coefficient of the random variables X and Y is given by

$$\rho_{XY} = \frac{E((X - E(X))(Y - E(Y)))}{\sqrt{V(X)V(Y)}} = \frac{Cov(X, Y)}{\sqrt{V(X)V(Y)}}$$

Definition The random variables X and Y are said to be uncorrelated if $Cov(X, Y) = 0 = E(XY) - E(X)E(Y)$.

Notice immediately that if X and Y are independent, then

$$E((X - E(X))(Y - E(Y))) = E(X - E(X))E(Y - E(Y)) = 0$$

and so the independence of X and Y implies that X and Y are uncorrelated. The converse is *not* true in general as later examples will demonstrate, that is, the fact that X and Y are uncorrelated does not, in general, imply that they are independent.

The correlation coefficient is just a normalized version of the covariance. In fact, we can show that $-1 \le \rho_{XY} \le 1$. Consider the expression $E((tU - V)^2)$ for any two random variables U and V and a real variable t. This expression when viewed as a quadratic in t is greater than or equal to zero, that is,

$$E((tU - V)^2) = t^2 E(U^2) - 2tE(UV) + E(V^2) \ge 0 \qquad (6.29)$$

But a quadratic is positive only if it has complex roots, that is, only if its discriminant is less than zero or

$$(2E(UV))^2 - 4E(U^2)E(V^2) \le 0$$

or

$$E^2(UV) \le E(U^2)E(V^2) \qquad (6.30)$$

These equations are valid for any random variables and in particular we choose $U = X - E(X)$ and $V = Y - E(Y)$. Equation (6.30) then yields

$$E^2((X - E(X))(Y - E(Y))) \le E((X - E(X))^2)E((Y - E(Y))^2) = \sigma_X^2 \sigma_Y^2$$

and dividing this expression by $\sigma_X^2 \sigma_Y^2$ we obtain

$$\frac{(Cov(X, Y))^2}{\sigma_X^2 \sigma_Y^2} = \rho_{XY}^2 \le 1$$

Since ρ_{XY} is a real number, this implies $-1 \le \rho_{XY} \le 1$.

EXAMPLE 6.15

The joint pdf of the random variable (X, Y) is

$$f_{XY}(x, y) = \begin{cases} 6xy(2-x-y), & 0 \le x \le 1, 0 \le y \le 1 \\ 0, & \text{elsewhere} \end{cases}$$

Determine: (i) $\text{Cov}(X, Y)$; (ii) ρ_{XY}.

As we require $E(X)$, $V(X)$, $E(Y)$, and $V(Y)$ we first determine the marginals:

$$f_X(x) = \int_0^1 6xy(2-x-y)\, dy = \begin{cases} x(4-3x), & 0 \le x \le 1 \\ 0, & \text{elsewhere} \end{cases}$$

$$f_Y(y) = \int_0^1 6xy(2-x-y)\, dx = \begin{cases} y(4-3y), & 0 \le y \le 1 \\ 0, & \text{elsewhere} \end{cases}$$

From these marginals we determine that

$$E(X) = E(Y) = \int_0^1 x \cdot x(4-3x)\, dx = 7/12$$

$$V(X) = V(Y) = \int_0^1 (x - 7/12)^2 x(4-3x)\, dx = 43/720$$

$$E(XY) = \int_0^1 \int_0^1 xy \cdot 6xy(2-x-y)\, dx\, dy = 1/3$$

Using these parameters we can determine:

(i) $\text{Cov}(X, Y) = E(XY) - E(X)E(Y) = -1/144$.

(ii) $\rho_{XY} = \dfrac{\text{Cov}(X, Y)}{\sqrt{V(X)V(Y)}} = \dfrac{-1/144}{43/720} = -\dfrac{5}{43}$.

Normal random variables play a prominent role in many problems and applications of probability and statistics. The bivariate normal pdf is further investigated in the following example.

EXAMPLE 6.16

From Eq. (6.18) the joint pdf of a bivariate normal random variable is

$$f_{XY}(x, y) = \frac{1}{2\pi\sigma_X\sigma_Y(1-\rho_{XY}^2)^{1/2}} \exp\left\{-\left[\left(\frac{x-\mu_X}{\sigma_X}\right)^2 - 2\rho_{XY}\left(\frac{x-\mu_X}{\sigma_X}\right)\left(\frac{y-\mu_Y}{\sigma_Y}\right)\right.\right.$$

$$\left.\left. + \left(\frac{y-\mu_Y}{\sigma_Y}\right)^2\right]\Big/ 2(1-\rho_{XY}^2)\right\} \tag{6.31}$$

where μ_X, μ_Y, σ_X^2, σ_Y^2 are the means and variances of X and Y, respectively. Show that ρ_{XY} is the correlation coefficient of (X, Y).

By definition, the correlation coefficient is

$$E\left(\left(\frac{X-\mu_X}{\sigma_X}\right)\left(\frac{Y-\mu_Y}{\sigma_Y}\right)\right)=\int\int_{-\infty}^{\infty}\left(\frac{x-\mu_X}{\sigma_X}\right)\left(\frac{y-\mu_Y}{\sigma_Y}\right)f_{XY}(x,y)\,dx\,dy \quad (6.32)$$

and by changing variables

$$u=\left(\frac{x-\mu_X}{\sigma_X}\right) \quad \text{and} \quad v=\left(\frac{y-\mu_Y}{\sigma_Y}\right)$$

Eq. (6.32) can be written as

$$\text{LHS}=\int\int_{-\infty}^{\infty} uv\frac{1}{2\pi\sigma_X\sigma_Y(1-\rho_{XY}^2)^{1/2}}\exp\left(\frac{-(u^2-2\rho_{XY}uv+v^2)}{2(1-\rho_{XY}^2)}\right)\sigma_X\sigma_Y\,dx\,dy$$

$$=\int_{-\infty}^{\infty}\frac{v}{\sqrt{2\pi}}\left[\int_{-\infty}^{\infty}\frac{u}{\sqrt{2\pi}(1-\rho_{XY}^2)^{1/2}}\exp\left(-\frac{(u-\rho_{XY}v)^2}{2(1-\rho_{XY}^2)}\right)\,du\right]\exp\left(-\frac{v^2}{2}\right)dv$$

$$=\int_{-\infty}^{\infty}\frac{v}{\sqrt{2\pi}}\cdot\rho_{XY}v\cdot\exp(-v^2/2)\,dv=\rho_{XY}$$

where liberal use has been made of the properties of normal pdf's.

EXAMPLE 6.17

The joint pdf of the random variables X and Y is given by

$$f_{XY}(x,y)=\frac{1}{\pi\sqrt{3}}\exp[-\tfrac{2}{3}(x^2-xy+y^2)], \quad -\infty<x,\ y<\infty$$

Find the correlation coefficient ρ_{XY}.

The point of this example is to realize that the joint pdf is, with the right parameters, a bivariate normal pdf. Rather than answer the question by integrating, it is simpler to solve equations for the various parameters. Clearly both X and Y have zero means. The remaining parameters must satisfy the equations:

$$2\pi\sigma_X\sigma_Y(1-\rho_{XY}^2)^{1/2}=\pi\sqrt{3}$$

$$2(1-\rho_{XY}^2)\sigma_X^2=3/2=2(1-\rho_{XY}^2)\sigma_Y^2$$

$$2\rho_{XY}/2\sigma_X\sigma_Y(1-\rho_{XY}^2)=2/3$$

These equations can be solved to give

$$\sigma_X^2=\sigma_Y^2=1 \quad \text{and} \quad \rho_{XY}=1/2$$

A question arises as to what the physical interpretation of the correlation coefficient is. Consider the following situation: An experiment with two random outcomes X and Y is performed and a large number of (X, Y)

readings are taken. The precise relationship between the two random variables is unknown. However, we would like to approximate Y by a linear combination $aX + b$. In particular, we would like to obtain the best approximation in the sense of minimizing the quantity

$$\varepsilon = E((Y - aX - b)^2)$$

which is known as the mean square error of the approximation. The minimization is with respect to the constants a and b. Differentiating ε with respect to a and b, consecutively, and equating the expressions to zero gives:

(i) $E((Y - aX - b)X) = 0 = E(XY) - aE(X^2) - bE(X)$

(ii) $E(Y - aX - b) = 0 = E(Y) - aE(X) - b$

These equations can be solved to yield

$$a = \sqrt{\frac{V(Y)}{V(X)}}\rho_{XY}, \quad b = E(Y) - aE(X) \tag{6.33}$$

If $V(X) = V(Y)$, then ρ_{XY} is the slope of the best linear mean square approximation of X to Y. This is a useful intuitive interpretation of the correlation coefficient.

Now suppose that X and Y are functionally related by the linear equation $Y = cX + d$. It is interesting to calculate ρ_{XY} in this case. By straightforward calculation:

$$\rho_{XY} = \frac{E(XY) - E(X)E(Y)}{\sqrt{V(X)V(Y)}}$$

$$= \frac{E(X(cX + d)) - E(X)(cE(X) + d)}{\sqrt{V(X)c^2 V(X)}}$$

$$= \frac{cE(X^2) + dE(X) - cE^2(X) - dE(X)}{|c|V(X)}$$

$$= \frac{cV(X)}{|c|V(X)} = \pm 1$$

and $\rho_{XY} = +1$ if $c > 0$ and $\rho_{XY} = -1$ if $c < 0$. Thus if X and Y are linearly related, then $\rho_{XY} = \pm 1$. Surprisingly, this result has a converse in that if $\rho_{XY} = \pm 1$, then X and Y are linearly related in the sense that the variance of the random variable $(Y - cX - d)$ is zero for certain values of c and d, We omit the proof of this fact.

It follows that a value of $|\rho_{XY}|$ very close to unity may indicate a linear relationship between X and Y, while other values merely indicate a lack of such a relationship (although not precluding a nonlinear relationship). In this regard the following example is of interest.

EXAMPLE 6.18

The tensile strength T of a wire cable is assumed to be proportional to the square of its diameter D and $T = kD^2$ for some proportionality constant k. If D is assumed to be uniformly distributed over $(8, 10)\,\text{mm}$, find ρ_{DT}.

The calculation of all quantities is straightforward:

$$E(D)=\tfrac{1}{2}\int_8^{10} x\,dx = 9 \qquad\qquad E(T)=kE(D^2)=k\cdot 81\tfrac{1}{3}$$

$$E(D^2)=\tfrac{1}{2}\int_8^{10} x^2\,dx = 81\tfrac{1}{3} \qquad E(T^2)=k^2E(D^4)=k^2\tfrac{1}{2}\int_8^{10} x^4\,dx$$

$$= k^2\cdot\frac{67{,}232}{10}$$

$$E(DT)=\frac{k}{2}\int_8^{10} x^3\,dx = 738k$$

$$V(D)=E(D^2)-E(D)=\tfrac{1}{3} \qquad V(T)=E(T^2)-E^2(T)=108\tfrac{4}{45}k^2$$

$$\rho_{DT}=\frac{738k-9\times 81\tfrac{1}{3}k}{\sqrt{\tfrac{1}{3}}\cdot(108\tfrac{4}{45}k^2)}=\frac{6}{\sqrt{36.03}}\cong .999$$

The value of ρ_{DT} is quite close to unity and yet the relationship between D and T is that of a square law, not linear, and thus the previous statements on the interpretation of the correlation coefficient must be treated with some care. Notice that in this example the joint pdf of T and D is nonzero only above the curve $t = kd^2$ in the (t, d) plane.

The correlation coefficient will be a useful and important statistical parameter in later chapters where more motivation for its use will appear.

Returning to the problem of estimating Y by a linear function of X, suppose the constants a and b are as given in Eq. (6.33) and consider the expression

$$E((Y-aX-b)Y)=E(Y^2)-aE(XY)-bE(Y)$$

$$=E(Y^2)-\sqrt{\frac{V(Y)}{V(X)}}\rho_{XY}E(XY)$$

$$-\left(E(Y)-\sqrt{\frac{V(Y)}{V(X)}}\rho_{XY}E(X)\right)E(Y)$$

$$=V(Y)-\sqrt{\frac{V(Y)}{V(X)}}\rho_{XY}(E(XY)-E(X)E(Y))$$

$$=V(Y)(1-\rho_{XY}^2) \tag{6.34}$$

while

$$E((Y-aX-b)X) = E(XY)-aE(X^2)-bE(X)$$

$$= E(XY)-\rho_{XY}\sqrt{\frac{V(Y)}{V(X)}}E(X^2)$$

$$-\left(E(Y)-\rho_{XY}\sqrt{\frac{V(Y)}{V(X)}}E(X)\right)E(X)$$

$$= E(XY)-E(X)E(Y)$$

$$-\rho_{XY}\sqrt{\frac{V(Y)}{V(X)}}(E(X^2)-E^2(X))$$

$$= \text{Cov}(X, Y)-\rho_{XY}\sqrt{V(X)V(Y)}=0 \qquad (6.35)$$

This last equation is usually interpreted by saying that the data (X) is "orthogonal" to the error $(Y-aX-b)$. Using Eqs. (6.34) and (6.35) we see that the minimum mean square error possible in estimating Y by a linear function of X is

$$\varepsilon^2 = E((Y-aX-b)^2) = E((Y-aX-b)Y) = V(Y)(1-\rho_{XY}^2) \qquad (6.36)$$

Before proceeding further it should be stressed that the problem of estimating one random variable by another (usually with a mean square error criterion) is very important in practice. For example, in a chemical process it may be critical to control a certain parameter Y that because of its nature cannot be measured directly. Instead, a parameter X that may be functionally related to Y, or perhaps is a "noisy" version of Y, is measured. Based on X and what we know about its relationship with Y we would like to use X to estimate in some optimum manner the value of Y. This estimate of Y is then used to control the process in an efficient manner. Another application would be in a communication system where we identify Y as the actual signal at some time t and X as the value of the signal after it has been transmitted through a medium and been distorted by various types of random perturbations. Based on the observed X we want an estimate of Y.

Suppose now that (X, Y) is a bivariate random variable with joint pdf $f_{XY}(x, y)$ and we want that function $H(X)$ which provides the best mean square estimate of Y possible (not just the best linear mean square estimate discussed earlier), that is, we want to determine the $H(X)$ such that

$$E((Y-H(X))^2)$$

is minimized. We first rewrite this equation as

$$E((Y-H(X))^2) = E(((Y-E(Y|X)+(E(Y|X)-H(X)))^2)$$

$$= E((Y-E(Y|X))^2)$$

$$+2E((Y-E(Y|X))(E(Y|X)-H(X)))$$

$$+E((E(Y|X)-H(X))^2) \qquad (6.37)$$

and consider the second term. For any function $G(\cdot)$ consider the expression

$$E((Y - E(Y|X))G(X)) = \int\int yG(x)f_{XY}(x, y)\, dx\, dy$$

$$- \int G(x)f(x) \cdot \left(\int yf(y|x)\, dy\right) dx = 0$$

and, since $E(Y|X)$ is a function of X, the second term in Eq. (6.37) is zero and so

$$E((Y - H(X))^2) = E((Y - E(Y|X))^2) + E((E(Y|X) - H(X))^2)$$

The first term on the right-hand side of this equation does not involve $H(X)$ and so any choice of $H(X)$ will not reduce the "error" of this term. On the other hand, the second term on the right-hand side can be reduced to zero simply by choosing $H(X) = E(Y|X)$. Consequently, this choice for the function $H(X)$ minimizes the mean square error. Choosing this function results in a minimum mean square error

$$\varepsilon_1^2 = E((Y - E(Y|X))^2) \tag{6.38}$$

It can be shown that ε_1^2 is always less than or equal to the mean square error ε^2 (Eq. (6.36)) for the best linear estimator. To summarize, if the best linear estimate of Y in terms of X is wanted, then choose the coefficients as in Eq. (6.33) and the resulting mean square error is given by Eq. (6.36). The best nonlinear estimate of Y is $E(Y|X)$ and the resulting mean square error is given by Eq. (6.38). The following example expands on these notions. The function $E(Y|X) = H(X)$ is called the regression of the mean of Y on X. As described previously, the conditional mean $E(Y|X = x)$ is to be viewed as a function $H(x)$. The function $E(Y|X)$ is then simply $H(X)$.

EXAMPLE 6.19

The bivariate random variable (X, Y) is uniformly distributed over the area shown in Figure 6.13. It is required to estimate Y by (i) a constant, (ii) a

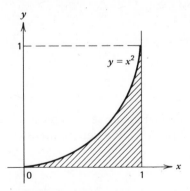

FIGURE 6.13

linear function of X, and (iii) a nonlinear function of X. In each case find the best estimator and determine the minimum mean square error.

The area where $f(x, y)$ is defined is

$$\text{Area} = \int_0^1 \left(\int_0^{x^2} dy \right) dx = \frac{1}{3}$$

and

$$f(x, y) = \begin{cases} 3, & (x, y) \in A \\ 0, & \text{elsewhere} \end{cases}$$

It is convenient to first determine all the parameters that will be required:

$$f_X(x) = \int_0^{x^2} 3 \, dy \qquad\qquad f_Y(y) = \int_{\sqrt{y}}^1 3 \, dx$$

$$= \begin{cases} 3x^2, & 0 \le x \le 1, \\ 0, & \text{elsewhere} \end{cases} \qquad = \begin{cases} 3(1-\sqrt{y}), & 0 \le y \le 1 \\ 0, & \text{elsewhere} \end{cases}$$

$$E(X) = \int_0^1 x \cdot 3x^2 \, dx = \frac{3}{4}, \qquad E(Y) = \int_0^1 y \cdot 3(1-\sqrt{y}) \, dy = \frac{3}{10}$$

$$E(X^2) = \int_0^1 x^2 \cdot 3x^2 \, dx = \frac{3}{5}, \qquad E(Y^2) = \int_0^1 y^2 \cdot 3(1-\sqrt{y}) \, dy = \frac{1}{7}$$

$$V(X) = \frac{3}{5} - \left(\frac{3}{4} \right)^2 = \frac{3}{80}, \qquad V(Y) = \frac{37}{700}$$

$$E(XY) = \int_0^1 \left(\int_0^{x^2} xy \cdot 3 \, dy \right) dx$$

$$= \frac{1}{4}$$

$$\rho_{XY} = \frac{E(XY) - E(X)E(Y)}{\sqrt{V(X)V(Y)}}$$

$$= \frac{\frac{1}{4} - \frac{3}{4} \cdot \frac{3}{10}}{\sqrt{\frac{3}{80} \cdot \frac{37}{700}}} = \sqrt{\frac{35}{111}}$$

(i) The constant C that best approximates Y in the mean square sense is given by $C = E(Y)$ and the resulting mean square error is $V(Y)$. We denote these two quantities by

$$\hat{Y}_0 = E(Y) = \frac{3}{10} \quad \text{and} \quad \varepsilon_0^2 = V(Y) = \frac{37}{700}$$

(ii) From Eqs. (6.33) the best linear mean square estimate of Y by X is

$$\hat{Y}_1 = \sqrt{\frac{V(Y)}{V(X)}}\rho_{XY}X + \left(E(Y) - \sqrt{\frac{V(Y)}{V(X)}}\rho_{XY}E(X)\right)$$

$$= \frac{2}{3}X - \frac{1}{5}$$

and the resulting mean square error of this estimate is, by Eq. (6.36),

$$\varepsilon_1^2 = V(Y)(1 - \rho_{XY}^2) = \tfrac{37}{700}(1 - \tfrac{35}{111})$$

(iii) The best nonlinear estimate of Y by X is just the regression of the mean of Y on X, which we denote by $\hat{Y}_2$:

$$\hat{Y}_2 = E(Y|X) = H(X)$$

The function $H(X)$ is given by the equation

$$E(Y|x) = \int yf(y|x)\,dy$$

$$= \int_0^{x^2} y \cdot \frac{3}{3x^2}\,dy = \frac{1}{x^2}\frac{x^4}{2} = \frac{x^2}{2}$$

and so our third estimate of Y is

$$\hat{Y}_2 = X^2/2$$

The mean square error ε_2^2 that results from using this estimate is from Eq. (6.38),

$$\varepsilon_2^2 = E\left[\left(Y - \frac{X^2}{2}\right)^2\right]$$

$$= \int_0^1\left[\int_0^{x^2}\left(y - \frac{x^2}{2}\right)^2 \cdot 3\,dy\right]dx$$

$$= 1/28$$

Thus, as we allow more general estimators of Y by X, the resulting mean square errors decrease in the following manner:

(i) estimating Y by a constant: $\varepsilon_0^2 = \tfrac{37}{700} \cong .0529$;
(ii) estimating Y by a linear function of X: $\varepsilon_1^2 = \tfrac{37}{700}(1 - \tfrac{35}{111}) \cong .0363$;
(iii) estimating Y by an arbitrary function of X: $\varepsilon_2^2 = \tfrac{1}{28} \cong .0357$,

The three estimates are sketched in Figure 6.14. The estimate $\hat{Y}_1$ has the peculiar property that, for $X < .3$, it yields a negative estimate for Y, even though Y is always positive. This is a result of the linearity constraint on this estimate. The results of this example can be interpreted in the following manner. Suppose we are repeatedly observing the outcomes of a bivariate experiment, that is, an outcome of the experiment is a pair of numbers (X, Y).

Their distribution is uniform over the area shown in Figure 6.13. For some reason, we are now only able to observe X, not Y, and we want to estimate Y using what we know about X. The three estimates discussed here are the usual ones considered (particularly $\hat{Y}_1$ and $\hat{Y}_2$) and their resulting mean square errors give an indication of their performance.

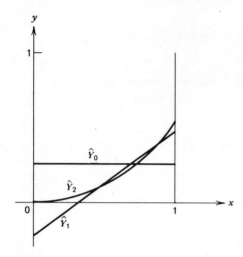

FIGURE 6.14

If (X, Y) is a bivariate normal random variable, then from example (6.14), Eq. (6.28), we saw that

$$E(Y\,|\,X) = \frac{\sigma_Y}{\sigma_X}\rho_{XY}X + \left(\mu_Y - \frac{\sigma_Y}{\sigma_X}\rho_{XY}\mu_X\right) \tag{6.39}$$

In this case the best linear mean square estimate and the best mean square estimate coincide and this turns out to be a very useful and important property of normal random variables. This says that the best estimate of Y is a linear function of X and in practice it is much easier to analyze and build linear systems than nonlinear. This result carries over to a broad class of estimation problems involving normal random variables. It can be shown that if, for any two random variables X and Y, not necessarily normal, $E(Y\,|\,X)$ is equal to a linear function of X, $aX + b$, then the coefficients a and b must be identical with the coefficients in Eq. (6.39).

The following example once again stresses the important properties of normal random variables.

EXAMPLE 6.20

A large company estimates the yield of a planting project by assuming that the base width of the trees at maturity, X, and the height at maturity, Y, measured in meters has a jointly normal pdf

$$f(x, y) = \frac{2}{\pi\sqrt{7}} \exp\left[-\frac{8}{7}\left(4x^2 + \frac{y^2}{4} + 26x - \frac{17}{2}y - \frac{3}{2}xy + 74\right)\right]$$

Find the best mean square estimate of the height using the base width and the mean square error involved with using this estimate.

Since the pdf is jointly normal, we can determine all the parameters needed to find the regression line of Y on X (Eq. (6.39)) by comparing coefficients in this pdf with that of Eq. (6.18). The computations are:

coefficient of the exponential: $\quad \dfrac{1}{2\pi\sigma_X\sigma_Y(1-\rho^2)^{1/2}} = \dfrac{2}{\pi\sqrt{7}}$

coefficient of x^2: $\quad \dfrac{1}{2\sigma_X^2(1-\rho^2)} = \dfrac{32}{7}$

coefficient of y^2: $\quad \dfrac{1}{2\sigma_Y^2(1-\rho^2)} = \dfrac{2}{7}$

coefficient of xy: $\quad \dfrac{2\rho}{2\sigma_X\sigma_Y(1-\rho^2)} = \dfrac{12}{7}$

coefficient of x: $\quad \left(\dfrac{2\mu_X}{\sigma_X^2} - \dfrac{2\rho\mu_Y}{\sigma_X\sigma_Y}\right)\bigg/2(1-\rho^2) = -\dfrac{176}{7}$

coefficient of y: $\quad \left(\dfrac{2\mu_Y}{\sigma_Y^2} - \dfrac{2\rho\mu_X}{\sigma_X\sigma_Y}\right)\bigg/2(1-\rho^2) = \dfrac{68}{7}$

coefficient of constant: $\quad \left(\dfrac{\mu_X^2}{\sigma_X^2} - \dfrac{2\rho\mu_X\mu_Y}{\sigma_X\sigma_Y} + \dfrac{\mu_Y^2}{\sigma_Y^2}\right)\bigg/2(1-\rho^2) = \dfrac{592}{7}$

These equations (seven equations in five unknowns) can be solved in a variety of ways. The solution is: $\sigma_X = \frac{1}{2}$, $\sigma_Y = 2$, $\rho = +\frac{3}{4}$, $\mu_X = 1$, and $\mu_Y = 20$. Using these values in the regression line of Eq. (6.39) gives

$$E(Y|X) = \frac{2}{(\frac{1}{2})}\left(\frac{3}{4}\right)X + \left(20 - \frac{2}{(\frac{1}{2})}\cdot\frac{3}{4}\cdot 1\right)$$

$$= 3X + 17$$

Since this is a linear estimate the associated mean square error is given by Eq. (6.36):

$$\text{mean square error} = V(Y)(1-\rho^2)$$

$$= 4\left(1 - \frac{9}{16}\right) = \frac{7}{4}$$

6.4 Uncorrelatedness and Independence

It is important to have a clear idea of the differences inherent in the assumption of two random variables being uncorrelated and in the assumption of two random variables being independent. Two random variables X and Y are independent if and only if

$$F_{XY}(x, y) = F_X(x)F_Y(y)$$

or equivalently if and only if

$$f_{XY}(x, y) = f_X(x)f_X(y)$$

Two random variables are uncorrelated if and only if

$$E(XY) = E(X)E(Y) \qquad (6.40)$$

and we observed in the previous section that two independent random variables are always uncorrelated. In fact, as has been seen, for any two functions $G(\cdot)$, $H(\cdot)$ it is true that if X and Y are independent random variables, then

$$E(G(X)H(Y)) = E(G(X))E(H(Y)) \qquad (6.41)$$

It can be shown that if Eq. (6.41) holds for any two functions $G(\cdot)$ and $H(\cdot)$, then it implies that X and Y are independent. This equation can be used to gain insight into the difference between uncorrelatedness and independence. Comparing Eqs. (6.40) and (6.41), it is seen that for two random variables to be uncorrelated it is only necessary for the expectations of the identity functions of X and Y to factor into the product of expectations. For the random variables to be independent it is necessary that the expectation of the product of any two functions, one of X and the other of Y, be the product of the expectations of the individual functions. The following example shows one case of how it is possible for two random variables to be uncorrelated but not independent.

EXAMPLE 6.21

Let X and Y be two random variables with the joint probability distribution function:

		Y		
X	-2	-1	1	2
-1	1/16	1/8	1/8	1/16
0	1/16	1/16	1/16	1/16
1	1/16	1/8	1/8	1/16

Show that X and Y are uncorrelated but not independent.

The marginal probability distribution functions of X and Y are

$$p_X(-1) = \frac{3}{8}, \quad p_X(0) = \frac{1}{4}, \quad p_X(1) = \frac{3}{8}$$

and

$$p_Y(-2) = \frac{3}{16}, \quad p_Y(-1) = \frac{5}{16}, \quad p_Y(1) = \frac{5}{16}$$

$$p_Y(2) = \frac{3}{16}$$

It is easy to see that for these distributions, $E(X) = E(Y) = 0$. The calculation of $E(XY)$ results in $E(XY) = 0$. On the other hand, since

$$p(-1, -2) = \frac{1}{6} \neq p_X(-1)p_Y(-2) = \frac{3}{8} \cdot \frac{3}{16}$$

it is clear that X and Y are not independent.

The example shows that uncorrelatedness of random variables does not imply independence, in general. There is one important exception to this rule, treated in the following example.

EXAMPLE 6.22

Show that if (X, Y) is a bivariate normal random variable, then X and Y are independent if and only if they are uncorrelated.

It is only necessary to show that uncorrelated normal random variables are also independent. The joint pdf of two normal random variables is given in Eq. (6.18) and it was shown in Example 6.16 that the parameter ρ_{XY} in this expression is the correlation coefficient of X and Y. If ρ_{XY} is set equal to zero in Eq. (6.18), it reduces to

$$f_{XY}(x, y) = \frac{1}{2\pi\sigma_X\sigma_Y} \exp\left(-\frac{(x-\mu_X)^2}{2\sigma_X^2} - \frac{(y-\mu_Y)^2}{2\sigma_Y^2}\right)$$

$$= f_X(x)f_y(y)$$

and X and Y are independent.

It should be stressed again, however, that uncorrelatedness does not, in general, imply independence. A continuous analog of Example 6.21 is given in the final example.

EXAMPLE 6.23

Let (X, Y) be a bivariate random variable with the joint pdf

$$f(x, y) = \frac{(x^2 + y^2)}{4\pi} \exp\left[-\left(\frac{x^2 + y^2}{2}\right)\right] \quad -\infty < x, y < \infty$$

Show that X and Y are not independent but are uncorrelated.

The marginal pdf of X is

$$f_X(x) = \frac{1}{4\pi} \int_{-\infty}^{\infty} (x^2 + y^2) \exp\left[-\left(\frac{x^2+y^2}{2}\right)\right] dy$$

$$= \frac{e^{-x^2/2}}{2\sqrt{2\pi}} \left(x^2 \int_{-\infty}^{\infty} \frac{e^{-y^2/2}}{\sqrt{2\pi}} dy + \int_{-\infty}^{\infty} y^2 \frac{e^{-y^2/2}}{\sqrt{2\pi}} dy \right)$$

and using the integrals of Eq. (5.8) this reduces to

$$f_X(x) = \frac{(x^2+1)e^{-x^2/2}}{2\sqrt{2\pi}}, \quad -\infty < x < \infty$$

Since $f(x, y)$ is symmetric in x and y

$$f_Y(y) = \frac{(y^2+1)e^{-y^2/2}}{2\sqrt{2\pi}}$$

and as $f(x, y) \neq f_X(x) f_Y(y)$ the random variables X and Y are not independent. Since the marginals are symmetric about the origin, $E(X) = E(Y) = 0$, and to show that X and Y are uncorrelated it is sufficient to show that $E(XY) = 0$:

$$E(XY) = \frac{1}{4\pi} \int_{\infty}^{\infty} \int_{-\infty}^{\infty} xy(x^2+y^2) \exp\left(-\frac{(x^2+y^2)}{2}\right) dx\, dy$$

$$= 0$$

as, for each integral, the integrand is an odd function about the origin.

Problems

1. Two measurements are performed on manufactured items and their deviation from the nominal values, X and Y, are assumed to be independent normal random variables with means of zero and variances σ^2, respectively. Determine the quantities (i) $E(|X|)$; (ii) $E((X^2 + Y^2)^{1/2})$; (iii) $V(X - Y)$.
2. On any particular measurement of the random variables in Problem 1, what is the probability that the Y value exceeds the X value?
3. If X and Y are two independent random variables representing the length and width of a piece of wood from a cutting process, with the pdf's

$$f_X(x) = \begin{cases} 1, & 9 \leq x \leq 10 \\ 0, & \text{elsewhere} \end{cases} \qquad f_Y(y) = \begin{cases} \frac{1}{2}, & 8 \leq y \leq 10 \\ 0, & \text{elsewhere} \end{cases}$$

find (i) the variance of the area of the wood, XY, (ii) the variance of the volume of a box made from such a piece of wood, XY^2.
4. The bivariate random variable (X, Y) has the joint cdf $F(x, y)$. Find the probability that the outcome of any particular realization will fall into the shaded region shown in Figure 6.15. Express your answer in terms of $F(x, y)$ only.

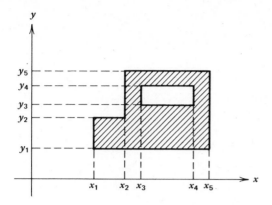

FIGURE 6.15

5. The incomes X and Y above \$10,000 of workers in two different trades are assumed to have the joint pdf

$$f(x, y) = \frac{10^8}{x^2 y^2}, \quad 10{,}000 \le x < \infty, \ 10{,}000 \le y < \infty$$

(i) Find the joint cdf of this joint pdf.
(ii) Find the probability that two workers chosen at random, one from each trade, each have incomes greater than \$25,000.

6. The tensile strengths in kgm/cm² of two kinds of nylon fiber X and Y are assumed to have the bivariate pdf

$$f(x, y) = \left(\frac{e^2}{10^6}\right) \cdot e^{-(x+y)/1000}, \quad 1000 \le x < \infty, \ 1000 \le y < \infty$$

Three pairs of fibers (one of each kind) are tested.
(i) What is the probability that in all three tests the same kind of fiber proves stronger?
(ii) What is the probability that in all three tests the tensile strengths were within 10 kgm/cm² of each other?

7. Let X and Y be independent random variables with pdf's

$$f_X(x) = \begin{cases} \frac{1}{2}, & 40 \le x \le 42 \\ 0, & \text{elsewhere} \end{cases} \qquad f_Y(y) = \begin{cases} \frac{1}{10}, & 90 \le y \le 100 \\ 0, & \text{elsewhere} \end{cases}$$

Determine the probability $P(Y < 95 \mid X + Y > 135)$.

8. Two die, one green and one red, are thrown and X is the sum of the two faces and Y is the number on the green face. Find the probability distribution of X and Y. Find the correlation coefficient of X and Y.

9. Let $F(x, y)$ be the joint cdf of the bivariate random variable (X, Y). Since $P((X, Y) \in A)$ is positive, it follows from Figure 6.16 that

$$F(x_2, y_2) - F(x_2, y_1) - F(x_1, y_2) + F(x_1, y_1) \ge 0$$

Extend this property to three dimensions.

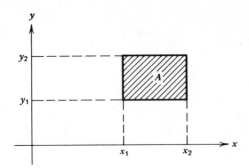

FIGURE 6.16

10. The discrete bivariate random variable (X, Y) has the probability distribution:

		X	
Y	1	2	4
2	.1	.2	.2
3	.04	.05	.08
5	.06	.15	.12

Find the probability distribution of $Z = \max(X, Y)$.

11. If X and Y are jointly distributed, as in Problem 10, find $E(X | Y = 3)$.

12. The bivariate random variable (X, Y) is uniformly distributed over the square $\{-2 \le x \le 2, -2 \le y \le 2\}$. Find the probability of the event $\{|Y| > |X| + 1\}$.

13. The bivariate random variable (X, Y) has the pdf

$$f(x, y) = \begin{cases} Cy^2, & 0 < |y| < x, \ 0 < x < 1 \\ 0, & \text{elsewhere} \end{cases}$$

   (i) Find $P(0 < Y < \frac{1}{8} | X = \frac{1}{4})$.
   (ii) For what value of a is it true that $P(-a < Y < a | X = \frac{1}{2}) = \frac{1}{2}$.

14. If the random variable Z has mean zero and variance unity, find $E(XY)$ and ρ_{XY}, where $X = Z - 1$ and $Y = Z + 1$.

15. The nut diameter X and bolt diameter Y of a nut–bolt pair chosen at random from large lots are assumed to be uniformly distributed over a circle of radius 1 mm centered at $x = y = 1$ cm. What is the probability that the nut diameter exceeds that of the bolt diameter. What is the probability $P(X > 1.025 \text{ cm} | \sqrt{X^2 + Y^2} < 1.05 \text{ cm})$.

16. The random variables X and Y represent the peak signal strengths, in volts, arriving at two physically separated antennas. If their joint pdf is

$$f(x, y) = \begin{cases} \dfrac{xy}{3} \exp\left(-\dfrac{1}{2}\left(x^2 + \dfrac{y^2}{3}\right)\right), & 0 \le x, \ y < \infty \\ 0, & \text{elsewhere} \end{cases}$$

find (i) the mean and variance of X; (ii) $P(X < Y)$; (iii) $P(X < Y < 2X)$.

17. The errors in the measurement of an attribute of a manufactured item are known to be due to three major causes. If the errors due to these causes are normal random variables X, Y, and Z each with mean 0 and variance 1, find the pdf of $X + Y + Z$ if $\text{Cov}(X, Y) = 0.5$, $\text{Cov}(X, Z) = 0.1$, and $\text{Cov}(Y, Z) = -0.4$.

18. The number of customers X and Y at two branches of a chain store are assumed to be independent Poisson random variables, each with parameter λ. Determine the conditional probability distribution $P(X = j \mid X + Y = k)$ and find its mean $E(X \mid X + Y = k)$.

19. Repeat Problem 18 if X and Y each have the geometric probability distribution with parameter θ, that is, $P(X = j) = \theta(1 - \theta)^{j-1}, j = 1, 2, \ldots$.

20. The number of customers that enter a store in a day is assumed to be a Poisson random variable with parameter λ. If each customer has a probability p of making a purchase, what is the probability distribution of the number of customers that make a purchase in a given day?

21. Let X and Y be the average marks obtained by a student in his freshman and sophomore year, respectively. If X and Y are assumed to have a bivariate normal pdf

$$f(x, y) = \frac{2}{\pi 9\sqrt{7}} \exp\left\{ -\frac{8}{7}\left[\left(\frac{x-70}{3}\right)^2 - \frac{2}{27}(x-70)(y-68) + \left(\frac{y-68}{3}\right)^2 \right] \right\},$$

$$0 < x \le 100, \ 0 < y \le 100$$

(neglecting the finite range of X and Y), find an expression for the probability that a student improves his average mark in his sophomore year over his freshman year. What is the correlation coefficient of these two marks?

22. Let (X, Y) be a bivariate normal random variable, where X and Y each have mean zero and variance σ^2 and the correlation coefficient of X and Y is ρ_{XY}. Verify directly that the conditional pdf of Y given X is normal and find its mean $E(Y \mid X)$. (This is a special case of Example 6.14 and Eq. (6.28).)

23. The bivariate random variable (X, Y) is jointly normal, with $X \sim N(0, \sigma_X^2)$ and $Y \sim N(0, \sigma_Y^2)$, and the correlation coefficient is ρ_{XY}. Show that

$$E(X^2 Y^2) = \sigma_X^2 \sigma_Y^2 + 2\rho_{XY}^2 \sigma_X^2 \sigma_Y^2$$

24. Find the variance of the random variable $Z = XY$, where X and Y are as given in Problem 23.

25. If X, Y, and Z are uncorrelated random variables, each with variance σ^2, determine the correlation coefficient of the random variables $U = X + 2Y + Z$ and $V = X + Y - Z$.

26. Find an expression for $V(aX + bY)$, where the correlation coefficient of X and Y is ρ_{XY}.

27. The radius of a ball bearing is assumed to be a random variable uniformly distributed over the range $(0, 2)$ cm. Find the correlation coefficient between the radius and the volume of the bearing.

28. A manufactured item can have two kinds of defects denoted by D_1 and D_2. The fraction of items containing defect 1 is .15, defect 2 is .20, and both D_1 and D_2 is .10. If X is the number of times D_1 was found in an inspection of two items and Y is the number of times D_2 was found, determine the joint probability distribution of X and Y and their covariance.

29. Let (X, Y) be a bivariate random variable with the joint pdf

$$f(x, y) = \frac{1}{6\pi\sqrt{3}} \exp\left[-\frac{2}{3}\left(\frac{x^2}{9} - \frac{5}{6}x + \frac{xy}{6} + \frac{y^2}{4} - y + \frac{7}{4} \right) \right]$$

Find the regression of the mean of Y on X.

30. If T_1 and T_2 are the times to failure of two supposedly identical batteries and they are assumed to be independent random variables with Rayleigh pdf's

$$f_{T_1}(t_1) = t_1 e^{-t_1^2/2}, \quad t_1 > 0$$

$$f_{T_2}(t_2) = t_2 e^{-t_2^2/2}, \quad t_2 > 0$$

respectively, find the quantities $E(T_1)$, $V(T_1 + T_2)$, $V(T_1 - T_2)$, and $\rho_{T_1 T_2}$.

31. The random variables X and Y have a bivariate pdf that is uniformly distributed over the semicircle shown in Figure 6.17.
    (i) Find the marginal pdf of X.
    (ii) Evaluate the quantity $P(Y \geq \frac{3}{2} | X \leq 3)$
    (iii) Find the correlation coefficient ρ_{XY}.

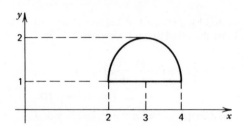

FIGURE 6.17

32. The bivariate random variable (X, Y) has the joint pdf

$$f(x, y) = \begin{cases} \frac{3}{32}(x^2 + y^2), & 0 \leq x, \ y \leq 2 \\ 0, & \text{elsewhere} \end{cases}$$

    (i) Find the correlation coefficient ρ_{XY}.
    (ii) Find the regression line of Y on X.

33. Repeat Problem 32 if the joint pdf is

$$f(x, y) = \begin{cases} \frac{3}{16}(x^2 + y^2), & 0 \leq y \leq x \leq 2 \\ 0, & \text{elsewhere} \end{cases}$$

34. In a psychic research test n people are each asked to pick a number from 1 to 5. If their choices are random and X is the number of people choosing 3 and Y the number of people choosing 4, find an expression for the joint probability distribution function of X and Y. What is $E(X)$. Find the correlation coefficient ρ_{XY}.

35. The outcomes of three experiments X, Y, and Z have been standardized and are assumed to be independent random variables each uniformly distributed over the interval $(0, 1)$.
    (i) Find the probability that exactly two of the outcomes are greater than 0.75.
    (ii) Find the correlation coefficient of the random variables $U = X + Y$ and $V = Z - X$.
    (iii) What is the probability that the sum of the first two outcomes $X + Y$ exceeds the third, Z?

36. The random variables X and Y are independent random variables, uniformly distributed over the intervals $(0, 1)$ and $(1, 2)$ respectively. Find the correlation coefficient between X and $U = X/Y$.

37. Let X and Y be two random variables with a joint pdf

$$f(x, y) = \begin{cases} \frac{1}{2} e^{-y}, & y > |x|, \ -\infty < x < \infty \\ 0, & \text{elsewhere} \end{cases}$$

(i) Are X and Y uncorrelated? Are they independent?
(ii) Find the regression curve of Y on X.

38. If X is uniformly distributed over the interval $(0, 2\pi)$, determine whether the random variables $U = \sin(X)$ and $V = \cos(X)$ are (i) uncorrelated; (ii) independent.

CHAPTER 7

Functions of Random Variables

In many instances it is not the random variable presented to us, either through experimental data or by some other means, that is of interest to us, but rather some function of it. This was encountered in Chapters 2 and 5 where expectations of functions of random variables were computed. The problem treated in this chapter is that of finding the pdf's of functions of single and bivariate random variables. The emphasis is on continuous random variables as the problem of transforming discrete random variables tends to be best treated from basic principles. The results and techniques of this chapter find wide practical application and illustrate certain aspects of probability. In addition, these results will indicate how many of the pdf's considered in Chapter 5 arise in practice.

7.1   The Probability Distribution of a Function of a Random Variable

Let X be a continuous random variable with probability density function $f_X(x)$, cumulative distribution function $F_X(x)$, and suppose $Y = G(X)$, where $G(\cdot)$ is a continuous single-valued function. The problem is to find the pdf $f_Y(y)$ when the pdf of X, $f_X(x)$ is known. The general technique for solving this problem consists of three steps:

(i) Express the event $\{Y \le y\}$ in terms of an event in R_X.
(ii) Find $F_Y(y)$.
(iii) Differentiate $F_Y(y)$ to give $f_Y(y)$, and determine its range of definition.

There are certain classes of functions where an explicit solution can be written down. The above procedure, however, works in principle for all functions $G(x)$.

  One class of functions for which an explicit solution can be found is the class of continuous monotonic functions. A function $y = G(x_2)$ is a continuous monotonically increasing function if $G(x_1) > G(x_2)$ for $x_1 > x_2$ and is continuous. Similarly, $G(x)$ is monotonically decreasing if $G(x_1) < G(x_2)$ for $x_1 > x_2$. If equalities are allowed in these definitions the functions are referred to as monotonically nondecreasing and nonincreasing functions, respectively.

  Assume for the moment that $y = G(x)$ is a monotonically increasing function. An example of such a function is shown in Figure 7.1. Since the

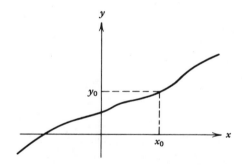

FIGURE 7.1

function is assumed to be strictly increasing (it does not stay horizontal over any finite range), it has an inverse that we denote by $x = G^{-1}(y)$. Each point on the x axis is mapped into exactly one point on the y axis by $y = G(x)$ and each point on the y axis is mapped into exactly one point on the x axis by $x = G^{-1}(y)$. For such a function it is easy to verify that the event corresponding to $\{Y \le y_0\}$ in R_X is $\{X \le x_0\}$, where $y_0 = G(x_0)$. It is this simple manner in which events in R_Y can be translated to events in R_X that allows a general solution to the problem. The three steps of our general technique for monotonic functions are contained in the following development:

$$F_Y(y) = P(Y \le y) = P(X \le G^{-1}(y))$$
$$= F_X(G^{-1}(y)) \tag{7.1}$$

differentiating this expression with respect to y gives

$$f_Y(y) = \frac{d}{dy} F_Y(y)) = \frac{d}{dy}(F_X(G^{-1}(y)))$$

Applying the chain rule of differentiation to this expression yields

$$f_Y(y) = f_X(G^{-1}(y))\frac{d}{dy}G^{-1}(y), \quad x = G^{-1}(y)$$

and this equation can be written

$$f_Y(y) = f_X(x)\frac{dx}{dy} \tag{7.2}$$

on the understanding that, since the left-hand side is a function of y only, we must replace x by $G^{-1}(y)$ every place it appears on the right-hand side.

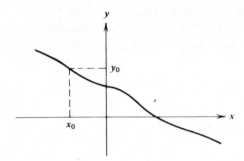

FIGURE 7.2

If the function $y = G(x)$ were monotonically decreasing (as shown in Figure 7.2), then the event $\{Y \le y\}$ in R_Y corresponds to $\{X \ge G^{-1}(y)\}$ implying that

$$F_Y(y) = 1 - F_X(G^{-1}(y))$$

which leads to

$$f_Y(y) = -f_X(x)\frac{dx}{dy} \tag{7.3}$$

instead of Eq. (7.2). In Eq. (7.3), however, since $y = G(x)$ is monotonically decreasing dy/dx is negative. We can combine Eqs. (7.2) and (7.3) into the equation

$$f_Y(y) = f_X(x)\left|\frac{dx}{dy}\right|, \quad y = G(x) \tag{7.4}$$

which is valid for any continuous monotonic (increasing or decreasing) function, $y = G(x)$. Actually, the function need only be monotonic in R_X, rather than on the whole x axis, as can be seen from the development of Eq. (7.4). The quantity $|dx/dy|$ is called the Jacobian of the transformation $y = G(x)$.

EXAMPLE 7.1

The random variable X has the pdf

$$f_X(x) = \begin{cases} 1/x^2, & x \geq 1 \\ 0, & x < 1 \end{cases}$$

Find the pdf of the random variable $Y = e^{-X}$.

Since this function is monotonically decreasing in R_X, we can apply Eq. (7.4) using the fact that

$$\frac{dy}{dx} = \frac{d}{dx} e^{-x} = -e^{-x}, \quad \frac{dx}{dy} = \frac{-1}{e^{-x}} = -e^x$$

and so

$$f_Y(y) = \frac{1}{x^2}|-e^x| = \frac{1}{(\ln y)^2} \cdot \frac{1}{y}, \quad y = -\ln x$$

This function describes the pdf of Y as long as $x \geq 1$, which corresponds to $y \leq e^{-1}$. The complete description of the pdf of Y is

$$f_Y(y) = \begin{cases} \dfrac{1}{y(\ln y)^2}, & 0 \leq y \leq \dfrac{1}{e} \\ 0, & \text{elsewhere} \end{cases}$$

Equation (7.4) is very useful when dealing with functions monotonic in R_X. It is also possible to develop such formulas for specific transformations. For example, consider the transformation $y = x^2$. This function is not one-to-one since two values of x are mapped onto the same value of y. It is, however, possible to develop a formula for $f_Y(y)$ from $f_X(x)$ using the three steps given at the start of this section. It is instructive to go through this exercise for the function $y = x^2$.

(i) The event $A = \{Y \leq y\}$ in R_Y is equivalent to the event $B = \{-\sqrt{y} \leq X \leq \sqrt{y}\}$ in R_X; that is, points in B are mapped into points of A by the function $y = x^2$ and no point not in B is mapped into a point in A as seen in Figure 7.3. As A and B are equivalent events,

$$F_Y(y) = P(Y \leq y) = P(-\sqrt{y} \leq X \leq \sqrt{y})$$

$$= F_X(\sqrt{y}) - F_X(-\sqrt{y})$$

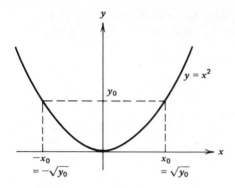

FIGURE 7.3

and differentiating this expression gives

$$f_Y(y) = \frac{d}{dy} F_Y(y)$$

$$= \frac{d}{dy} (F_X(\sqrt{y}) - F_X(-\sqrt{y}))$$

$$= f_X(\sqrt{y}) \frac{1}{2\sqrt{y}} + f_X(-\sqrt{y}) \frac{1}{2\sqrt{y}}$$

$$= \frac{f_X(\sqrt{y}) + f_X(-\sqrt{y})}{2\sqrt{y}} \qquad (7.5)$$

and this equation is valid for all y in the image of R_X under the map $y = x^2$, that is, R_Y.

EXAMPLE 7.2

The random variable X has the pdf

$$f_X(x) = \tfrac{1}{2} e^{-|x|}, \quad -\infty < x < \infty$$

Find the pdf of $Y = X^2$.

From Eq. (7.5) we have

$$f_Y(y) = \frac{\tfrac{1}{2} e^{-\sqrt{y}} + \tfrac{1}{2} e^{-\sqrt{y}}}{2\sqrt{y}} = \begin{cases} \dfrac{1}{2\sqrt{y}} e^{-\sqrt{y}}, & 0 \le y < \infty \\ 0, & y < 0 \end{cases}$$

Notice that if $y > 0$, then it has two possible square roots, one positive and one negative. By convention, $\sqrt{y}$ is always taken as the positive square root.

If, in this example, the random variable X had the pdf $f_X(x) = e^{-x}, x > 0$, then the pdf of $Y = X^2$ would have been

$$f_Y(y) = \frac{e^{-\sqrt{y}} + 0}{2\sqrt{y}} = \begin{cases} \dfrac{1}{2\sqrt{y}} e^{-\sqrt{y}}, & 0 \leq y < \infty \\ 0, & \text{elsewhere} \end{cases}$$

The function $Y = X^2$ is monotonic and increasing in the range R_X, and for this reason we could as well apply Eq. (7.4) to obtain the same answer, that is, in order to apply Eq. (7.4) it is only necessary that the transformation be monotonic in R_X.

EXAMPLE 7.3

The random variable X is uniformly distributed over the interval $(-1, 3)$. Find the pdf of the random variable $Y = X^2$.

In this case R_Y is the interval $(0, 9)$ and

$$f_X(x) = \begin{cases} \frac{1}{4}, & -1 \leq x \leq 3 \\ 0, & \text{elsewhere} \end{cases}$$

A little care must now be exercised in applying Eq. (7.5). For y in the range $(0, 1)$ both $\sqrt{y}$ and $-\sqrt{y}$ are in R_X and so

(i) $0 \leq y \leq 1$, $\quad f_Y(y) = \dfrac{\frac{1}{4} + \frac{1}{4}}{2\sqrt{y}} = \dfrac{1}{4\sqrt{y}}$

For y in the range $(1, 9)$, however, $\sqrt{y}$ is in the range $(1, 3)$ but $-\sqrt{y}$ is less than -1 and so

(ii) $1 \leq y \leq 9$, $\quad f_Y(y) = \dfrac{\frac{1}{4} + 0}{2\sqrt{y}} = \dfrac{1}{8\sqrt{y}}$

A description of the pdf $f_Y(y)$ is

$$f_Y(y) = \begin{cases} 1/4\sqrt{y}, & 0 \leq y \leq 1 \\ 1/8\sqrt{y}, & 1 \leq y \leq 9 \\ 0, & \text{elsewhere} \end{cases}$$

See Figure 7.4.

EXAMPLE 7.4

The velocity of an electronic particle, V, with respect to a reference direction, is assumed to be a normally distributed random variable with mean 0 and variance σ^2. Find the pdf of its kinetic energy $K = \frac{1}{2}mV^2$, where m is its mass, a constant.

The pdf of the particle velocity V is given by the normal pdf

$$f_V(v) = \frac{1}{\sqrt{2\pi}\sigma} \exp\left(-\frac{v^2}{2\sigma^2}\right), \quad -\infty < v < \infty$$

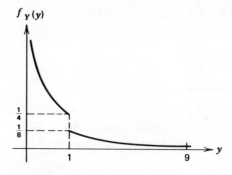

FIGURE 7.4

In order to use Eq. (7.5), let us define a new variable $U = \sqrt{(m/2)}V$ that, as a linear transformation of a normal pdf, has the pdf

$$f_U(u) = \frac{1}{\sqrt{\pi m \sigma}} \exp\left(-\frac{u^2}{m\sigma^2}\right), \quad -\infty < u < \infty$$

Since $K = U^2$, we can now apply Eq. (7.5) to obtain

$$f_K(k) = \frac{1}{2\sqrt{k}}\left(\frac{1}{\sqrt{\pi m \sigma}} e^{-k/m\sigma^2} + \frac{1}{\sqrt{\pi m \sigma}} e^{-k/m\sigma^2}\right)$$

$$= \frac{1}{\sqrt{\pi m k \sigma}} e^{-k/m\sigma^2}, \quad 0 \leq k < \infty$$

and this is a chi-square pdf with one degree of freedom.

A legitimate question to ask is if X is a random variable with pdf $f_X(x)$ and Y has a pdf $f_Y(y)$, is it possible to find a transformation $y = G(x)$ that relates these two random variables? The following example will be useful in clarifying this question.

EXAMPLE 7.5

Let X be a continuous random variable with pdf $f_X(x)$ and cdf $F_X(x)$. Find the pdf of the random variable $Y = F_X(X)$.
This transformation might seem strange at first, but the function $y = F_X(x)$ is a monotonically nondecreasing function defined for all real x and we wish to determine the effect of transforming the random variable X with it. Notice that since $0 \leq F_X(x) \leq 1$ for all real x, y takes on values only in the

interval $(0, 1)$. Using Eq. (7.4) we find that

$$f_Y(y) = f_X(x) \left| \frac{dx}{dy} \right|$$

$$= f_x(x) \frac{1}{\left| \frac{dF_X(x)}{dx} \right|}$$

$$= f_x(x) \cdot \frac{1}{f_X(x)} = \begin{cases} 1, & 0 \le y \le 1 \\ 0, & \text{elsewhere} \end{cases}$$

or, in other words, Y is uniformly distributed over the interval $(0, 1)$ for any random variable X.

This example contains the solution of the question asked previously. If X is a random variable with pdf $f_X(x)$ and cdf $F_X(x)$, then the random variable $Z = F_X(X)$ is uniformly distributed over the interval $(0, 1)$. If Y is a random variable with pdf $f_Y(y)$ and cdf $F_Y(y)$, then the random variable $W = F_Y(Y)$ is uniformly distributed over $(0, 1)$. By the inverse reasoning, if W is uniformly distributed over $(0, 1)$, then $Y = F_Y^{-1}(W)$ has the pdf $f_Y(y)$ assuming this inverse function exists. Composing these two lines of reasoning we find that the transformation $Y = F_Y^{-1}(F_X(X))$ transforms the pdf $f_X(x)$ into $f_Y(y)$. This result is useful in some computer simulation techniques where sequences of random variables with a given distribution are required. The following simple example illustrates.

EXAMPLE 7.6

Let X and Y be random variables with the respective pdf's

$$f_X(x) = \begin{cases} e^{-x}, & x \ge 0 \\ 0, & x < 0 \end{cases} \qquad f_Y(y) = \begin{cases} 1/2\sqrt{y}, & 0 \le y \le 1 \\ 0, & \text{elsewhere} \end{cases}$$

Find a transformation $Y = G(X)$ that relates these two random variables.
From the previous example, since the cdf of X is

$$F_X(x) = \begin{cases} 1 - e^{-x}, & x \ge 0 \\ 0, & x < 0 \end{cases}$$

the random variable $Z = 1 - e^{-X}$ is uniformly distributed over $(0, 1)$. Similarly, as the cdf of Y is

$$F_Y(y) = \int_0^y \frac{1}{2\sqrt{u}} \, du$$

$$= \sqrt{u} \Big|_0^y = \begin{cases} \sqrt{y}, & 0 \le y \le 1 \\ 0, & \text{elsewhere} \end{cases}$$

the random variable $W = \sqrt{Y}$ is uniformly distributed over $(0, 1)$. Composing the first function $Z = (1 - e^{-X})$ with the inverse of the second function $(Y = W^2)$ and changing W to Z yields the relationship $Y = (1 - e^{-X})^2$. In other words, if X has the pdf $f_X(x)$, then the random variable $Y = (1 - e^{-X})^2$ will have the pdf $f_Y(y)$.

EXAMPLE 7.7

Let X be a chi-square random variable with n degrees of freedom with pdf

$$f_X(x) = \begin{cases} \dfrac{x^{n/2-1}}{2^{n/2}\Gamma(n/2)} \exp\left(-\dfrac{x}{2}\right), & x > 0,\ n > 0 \\ 0, & \text{elsewhere} \end{cases}$$

Find the pdf $f_Y(y)$ of the random variable $Y = \sqrt{X/n}$.

Since X is a positive random variable $(f_X(x) = 0,\ x < 0)$, the transformation is one-to-one for $x > 0$. Using Eq. (7.4) gives, for $x = ny^2$,

$$f_Y(y) = \frac{n^{n/2-1}y^{n-2}}{2^{n/2}\Gamma(n/2)} \exp\left(-\frac{ny^2}{2}\right) 2ny$$

$$= \frac{n^{n/2}y^{n-1}}{2^{n/2-1}\Gamma(n/2)} \exp\left(-\frac{ny^2}{2}\right)$$

which is the chi density function of Eq. (5.19), with $\sigma = 1$.

7.2 The Joint Probability Distribution of a Function of a Bivariate Random Variable

Before considering transformations of two variables let us reinterpret a simple case of a single variable transformation, namely, the case when $y = G(x)$ is a monotonic increasing function. Equation (7.4), rewritten, says that

$$f_X(x)\, dx = f_Y(y)\, dy \tag{7.6}$$

where, since $dx/dy > 0$, we have removed the absolute value signs. As a particular case, suppose $y_0 = G(x_0)$ and so

$$f_X(x_0)\, dx = f_Y(y_0)\, dy \tag{7.7}$$

This equation says that the probability of X being in the interval $(x_0, x_0 + dx)$ is the same as the probability of Y being in the interval $(y_0, y_0 + dy)$. To see why this is so we find the event in R_Y equivalent to the event $\{x_0 \le X \le x_0 + dx\}$. Under the transformation $y = G(x)$ the interval $(x_0, x_0 + dx)$ is taken into the interval $(G(x_0), G(x_0 + dx))$ as $G(x)$ is continuous, monotonically increasing, and one-to-one in this example. But we can approximate $G(x_0 + dx)$ by the first-order term

$$G(x_0) + \frac{dG(x_0)}{dx} dx = y_0 + dy$$

where

$$dy = \frac{dG(x_0)}{dx} dx$$

In other words, the event $\{x_0 \le X \le x_0 + dx\}$ in R_X equivalent to the event $\{y_0 \le Y \le y_0 + dy\}$ in R_Y when dy is interpreted correctly. This is the meaning of Eq. (7.7) and hence of (7.6). When the transformation $y = G(x)$ is not one-to-one, then the event $\{x_0 \le X \le x_0 + dx\}$ is equivalent to a union of disjoint events in R_Y that complicates the discussion although it is conceptually the same.

Consider now the problem of transforming a bivariate random variable (X, Y) into another bivariate random variable (U, V) by the transformations

$$U = G_1(X, Y) \qquad V = G_2(X, Y)$$

We are given the joint pdf of (X, Y), $f_{XY}(x, y)$ and wish to compute the joint pdf of (U, V), $f_{UV}(u, v)$. As in the single variable case, we restrict our attention to transformations that yield relatively simple answers. We first assume that the transformations are one-to-one (i.e., they take one point of R_{XY} to one and only one point of R_{UV}) and have the inverse relationships

$$X = H_1(U, V) \qquad Y = H_2(U, V) \tag{7.8}$$

Furthermore, we will assume that all the partial derivatives that we require exist and are continuous, well-behaved functions.

We give an informal derivation of $f_{UV}(u, v)$ in terms of $f_{XY}(x, y)$ and proceed in analogy with the single variable case. Consider a small rectangle A in the x, y plane of area A and suppose this is mapped into the (also small) area B in the u, v plane by the transformation of Eq. (7.7). See Figure 7.5. The two

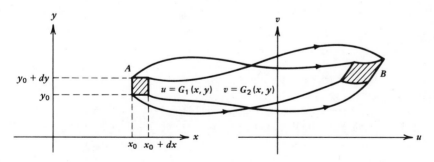

FIGURE 7.5

areas A and B represent equivalent events and we can write

$$P((X, Y) \in A) = P((U, V) \in B)$$

or

$$f_{XY}(x_0, y_0)\, dx\, dy = f_{UV}(u_0, v_0) \times (\text{area of } B) \qquad (7.9)$$

where $u_0 = G_1(x_0, y_0)$ and $v_0 = G_2(x_0, y_0)$. The remaining problem is to determine the area of B and, as with the single variable case, we resort to approximations.

An enlarged version of the area B appears in Figure 7.6a and for small dx and dy we can make the following approximations:

$$G_1(x_0 + dx, y_0) \cong G_1(x_0, y_0) + \frac{\partial}{\partial x} G_1(x_0, y_0)\, dx$$

$$G_1(x_0, y_0 + dy_0) \cong G_1(x_0, y_0) + \frac{\partial}{\partial y} G_1(x_0, y_0)\, dy$$

and similarly for the second function $G_2(x, y)$. A linearized version of the area B appears in Figure 7.6b along with the approximate coordinates. This

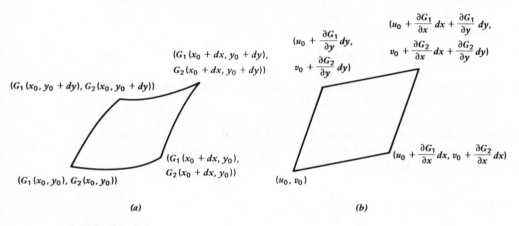

(a)   (b)

FIGURE 7.6

approximate area is assumed to be a parallelogram and it is an interesting exercise to show that its area is

$$\left| \frac{\partial G_1(x_0, y_0)}{\partial x} \cdot \frac{\partial G_2(x_0, y_0)}{\partial y} - \frac{\partial G_2(x_0, y_0)}{\partial x} \cdot \frac{\partial G_1(x_0, y_0)}{\partial y} \right| dx\, dy$$

The term within the absolute value signs is called the *Jacobian* of the transformation and can be expressed as the determinant

$$J(x, y) = \begin{vmatrix} \dfrac{\partial G_1(x, y)}{\partial x} & \dfrac{\partial G_1(x, y)}{\partial y} \\[2ex] \dfrac{\partial G_2(x, y)}{\partial x} & \dfrac{\partial G_2(x, y)}{\partial y} \end{vmatrix}$$

$$= \begin{vmatrix} \dfrac{\partial u}{\partial x} & \dfrac{\partial u}{\partial y} \\[2ex] \dfrac{\partial v}{\partial x} & \dfrac{\partial v}{\partial y} \end{vmatrix} \tag{7.10}$$

It is set as Problem 7.17 to verify that if

$$J(u, v) = \begin{vmatrix} \dfrac{\partial H_1(u, v)}{\partial u} & \dfrac{\partial H_1(u, v)}{\partial v} \\[2ex] \dfrac{\partial H_2(u, v)}{\partial u} & \dfrac{\partial H_2(u, v)}{\partial v} \end{vmatrix}$$

$$= \begin{vmatrix} \dfrac{\partial x}{\partial u} & \dfrac{\partial x}{\partial v} \\[2ex] \dfrac{\partial y}{\partial u} & \dfrac{\partial y}{\partial v} \end{vmatrix}$$

then

$$|J(u, v)| = \frac{1}{|J(x, y)|} \quad \text{(absolute values of determinants)}$$

Returning to Eq. (7.9) with these facts gives

$$f_{XY}(x_0, y_0) \, dx \, dy = f_{UV}(G_1(x_0, y_0), G_2(x_0, y_0))|J(u, v)| \, dx \, dy$$

or, inversely,

$$f_{UV}(u, v) = f_{XY}(H_1(u, v), H_2(u, v)) \frac{1}{|J(x, y)|}$$

or

$$f_{UV}(u, v) = f_{XY}(H_1(u, v), H_2(u, v))|J(u, v)| \tag{7.11}$$

Thus, to determine $f_{UV}(u, v)$, we find $|J(x, y)|$ as a function of u and v and multiply this by $f_{XY}(x, y)$ with x replaced by $H_1(u, v)$ and y by $H_2(u, v)$. The use of Eq. (7.11) will be illustrated with numerous examples. Before doing so we mention that we assumed the functions $u = G_1(x, y)$ and $v = G_2(x, y)$ were one-to-one in the sense that they took one point of R_{XY} to one point of R_{UV}. This is not quite necessary. It is actually only necessary that any small area of

R_{XY} is not mapped onto a single point of R_{UV}. An example of a mapping where the y axis is mapped onto the origin in the u, v plane is given later, and the above theory is used. The only remaining problem in applying Eq. (7.11) is to determine R_{UV} from R_{XY} and for some problems this is nontrivial.

EXAMPLE 7.8

The errors in two measurements X and Y are assumed to be independent random variables, each uniformly distributed over $(0, 1)$. Find the joint and marginal pdf's of the sum and difference of these two errors.

Let $U = X + Y$ and $V = X - Y$ and the inverse functions to these transformations are $X = \frac{1}{2}(U + V)$, $Y = \frac{1}{2}(U - V)$. The Jacobian of the transformation is

$$J(x, y) = \begin{vmatrix} \dfrac{\partial u}{\partial x} & \dfrac{\partial u}{\partial y} \\[2ex] \dfrac{\partial v}{\partial x} & \dfrac{\partial v}{\partial y} \end{vmatrix} = \begin{vmatrix} 1 & 1 \\ 1 & -1 \end{vmatrix} = -2$$

and, since

$$f_{XY}(x, y) = \begin{cases} 1, & 0 \le x, y \le 1 \\ 0, & \text{elsewhere} \end{cases}$$

the joint pdf of U and V is

$$f_{UV}(u, v) = 1 \cdot \frac{1}{|-2|} = \begin{cases} \frac{1}{2}, & (u, v) \in R_{UV} \\ 0, & \text{elsewhere} \end{cases}$$

The problem remains to find R_{UV}. The range R_{XY} is shown in Figure 7.7a. The problem is to determine how this region is transformed under the given transformation, and there are a variety of techniques to accomplish this. In

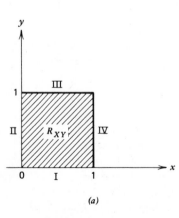

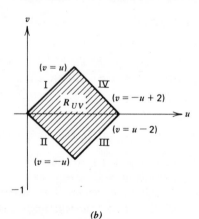

(a)   (b)

FIGURE 7.7

Figure 7.7 it is shown how the boundary lines of R_{XY} are transformed. For example, the line I is described by the equation $y = 0, 0 \le x \le 1$ and, under the mapping $u = x + y, v = x - y$ this becomes $u = x, v = x, 0 \le x \le 1$ or $u = v, 0 \le u, v \le 1$ and this is shown as line I in Figure 7.7b. The other lines are found in a similar manner. The joint random variable (U, V) is uniformly distributed over the shaded region shown in Figure 7.7b (it has area 2).

To find the marginal pdf of U, notice that U can take on values in the interval $(0, 2)$ and that for a value u in $(0, 1)$ the range of values that V may assume is $(-u, u)$ while for $u \in (1, 2)$ the range of values of V is $(u - 2, -u + 2)$. Consequently, the marginal of U is

(i) $0 \le u \le 1$, $\quad f_U(u) = \int_{-u}^{u} \tfrac{1}{2}\, dv = \tfrac{1}{2}(u - (-u)) = u$

(ii) $1 \le u \le 2$, $\quad f_U(u) = \int_{u-2}^{-u+2} \tfrac{1}{2}\, dv = \tfrac{1}{2}(-u + 2 - (u - 2)) = 2 - u$

The marginal pdf of V is found in a similar manner to be

$$f_V(v) = \begin{cases} v + 1, & -1 \le v \le 0 \\ -v + 1, & 0 \le v \le 1 \\ 0, & \text{elsewhere} \end{cases}$$

These pdf's are sketched in Figure 7.8.

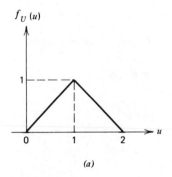

(a) (b)

FIGURE 7.8

EXAMPLE 7.9

The waiting times X and Y of two customers entering a bank at different times are assumed to be independent random variables with the pdf's

$$f_X(x) = \begin{cases} e^{-x}, & x \ge 0 \\ 0, & \text{elsewhere} \end{cases} \qquad f_Y(y) = \begin{cases} e^{-y}, & y \ge 0 \\ 0, & \text{elsewhere} \end{cases}$$

Find the joint pdf of the sum of their waiting times, $U = X + Y$ and the fraction of this time that the first customer spends waiting, $V = X/(X+Y)$. Find the marginal pdf's of U and V and show that they are independent.

We first find the inverse transformations, a process that often proceeds more by intuition than by any set method. Clearly $X = UV$ and $Y = U - X = U - UV$. The Jacobian of the transformation is

$$J(x, y) = \begin{vmatrix} \dfrac{\partial u}{\partial x} & \dfrac{\partial u}{\partial y} \\[2mm] \dfrac{\partial v}{\partial x} & \dfrac{\partial v}{\partial y} \end{vmatrix}$$

$$= \begin{vmatrix} 1 & 1 \\[2mm] \dfrac{y}{(x+y)^2} & -\dfrac{1}{(x+y)^2} \end{vmatrix} = -\frac{(x+y)}{(x+y)^2} = -\frac{1}{(x+y)}$$

and, as a function of u and v, $|J(x, y)| = 1/u$. The joint pdf of U and V is given by

$$f_{UV}(u, v) = e^{-u}u, \quad (u, v) \in R_{UV}$$

To determine the region R_{UV}, we map the positive quadrant in the x, y plane by the transformation $u = x + y$, $v = x/(x+y)$. The line $x = 0$ maps into the line $u = y$, $v = 0$ as y varies from 0 to ∞. The line $y = 0$ maps into the line $u = x$ and $v = 1$ as x varies from 0 to ∞. The complete description of $f_{UV}(u, v)$ is

$$f_{UV}(u, v) = \begin{cases} u e^{-u}, & (u, v) \in R_{UV} \\ 0, & \text{elsewhere} \end{cases}$$

where R_{UV} is shown in Figure 7.9. The marginal pdf's of U and V are given by

$$f_U(u) = \int_0^1 f_{UV}(u, v)\, dv = \int_0^1 u e^{-u}\, dv = u e^{-u}, \quad 0 \le u < \infty$$

$$f_V(v) = \int_0^\infty f_{UV}(u, v)\, dv = \int_0^\infty u e^{-u}\, du = 1, \quad 0 \le v \le 1$$

and U and V are independent random variables.

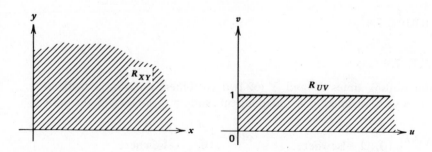

FIGURE 7.9

EXAMPLE 7.10

The random variables X and Y, assumed independent, are the lifetimes of two components chosen from among those that survive 1000 hours. The pdf's of these random variables, in units of 1000 hours, are

$$f_X(x)=\begin{cases}1/x^2, & x\geq 1\\0, & \text{elsewhere}\end{cases} \qquad f_Y(y)=\begin{cases}1/y^2, & y\geq 1\\0, & \text{elsewhere}\end{cases}$$

As a measure of the quality of the manufacturing process the geometric mean $U=\sqrt{XY}$ of the lifetime of the two components is used. Find the joint pdf of U and $V=X$ and the marginal pdf of U.

The Jacobian of the transformation is

$$J(x,y)=\begin{vmatrix}\tfrac{1}{2}\sqrt{y/x} & \tfrac{1}{2}\sqrt{x/y}\\1 & 0\end{vmatrix}=-\frac{1}{2}\sqrt{\frac{x}{y}}=-\frac{1}{2}\frac{V}{U}$$

and the joint pdf of U and V is

$$f_{UV}(u,v)=\frac{1}{x^2y^2}\frac{1}{\left|-\tfrac{1}{2}\sqrt{x/y}\right|}$$

$$=\frac{1}{u^4}\cdot\frac{2u}{v}=\frac{2}{u^3v}\quad(u,v)\in R_{UV}$$

The regions R_{XY} and R_{UV} are shown in Figures 7.10a and 7.10b, respectively.

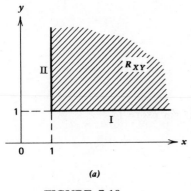

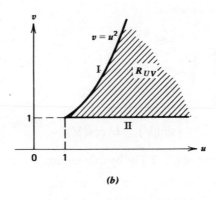

(a)   *(b)*

FIGURE 7.10

The marginal pdf of u is

$$f_U(u)=\int_1^{u^2}\frac{2}{u^3v}\,dv$$

$$=\frac{2}{u^3}\cdot\ln(v)\Big|_1^{u^2}=\frac{2\ln(u^2)}{u^3},\quad u\geq 1$$

In this example we might only have required the pdf of $U = \sqrt{XY}$. Clearly the random variable $V = X$, for such a computation, would be a dummy variable. This, in fact, is a common technique for finding the pdf of a single function of two random variables $U = G(X, Y)$. We define a dummy random variable V (usually $V = X$ or $V = Y$, but not always) in such a way as to make the two-dimensional transformation one-to-one. The joint pdf of U and V is found and the desired marginal of U is then computed. This turns out to be a good method for treating such problems, and several examples of it follow. Notice in the previous example that the line $x = 0$ with y varying from 0 to 1 is transformed into the point $u = 0$, $v = 0$ and so the transformation is not one-to-one. The theory, however, still applies as any finite area is still transformed in a one-to-one manner.

EXAMPLE 7.11

The latitude distance X, and longitude distance Y, in meters, of a missile landing from its target are assumed to be independent normal random variables each with mean zero and variance σ^2. Find the pdf of the random variable $U = \sqrt{X^2 + Y^2}$.

In order to define a dummy variable, define $V = \arctan(Y/X)$ as this is the usual transformation from rectangular to polar coordinates and ensures that it is a one-to-one transformation. The inverse transformation is

$$X = U \cos(V) \quad \text{and} \quad Y = U \sin(V)$$

and has the Jacobian

$$J(u, v) = \begin{vmatrix} \dfrac{\partial x}{\partial u} & \dfrac{\partial x}{\partial v} \\ \dfrac{\partial y}{\partial u} & \dfrac{\partial y}{\partial v} \end{vmatrix}$$

$$= \begin{vmatrix} \cos(V) & -U \sin(V) \\ \sin(V) & U \cos(V) \end{vmatrix} = U \cos^2(V) + U \sin^2(V) = U$$

The joint pdf of (X, Y) is, by assumption,

$$f_{XY}(x, y) = \frac{1}{2\pi\sigma^2} \exp\left(-\frac{(x^2 + y^2)}{2\sigma^2}\right)$$

and using Eq. (7.11)

$$f_{UV}(u, v) = \left[\frac{1}{2\pi\sigma^2} \exp\left(-\frac{u^2}{2\sigma^2}\right)\right] u$$

Since U and V have the interpretation of radius and angle, respectively, in polar coordinates, $R_{UV} = \{0 \le u < \infty, \ 0 \le v \le 2\pi\}$. The marginal pdf of U is given by

$$f_U(u) = \int_0^{2\pi} \frac{u}{2\pi\sigma^2} \exp\left(-\frac{u^2}{2\sigma^2}\right) dv$$

$$= \frac{u}{2\pi\sigma^2} \exp\left(-\frac{u^2}{2\sigma^2}\right) \cdot 2\pi$$

$$= \frac{u}{\sigma^2} \exp\left(-\frac{u^2}{2\sigma^2}\right), \quad 0 \le u < \infty$$

This is the Rayleigh pdf mentioned in connection with Eq. (5.19) with $n = 2$ and $\sigma = \sqrt{2}$. In this derivation, the random variables U and V are seen to be independent, that is,

$$f_{UV}(u, v) = f_U(u) f_V(v)$$

since the marginal of V, $f_V(v)$ is simply

$$f_V(v) = \frac{1}{2\pi}, \quad 0 \le v \le 2\pi$$

In an engineering context it is sometimes possible to resolve a signal into the two components X and Y and in such a problem U is the amplitude of the wave, which has a Rayleigh density, given in Eq. (5.19), and V is the phase of the signal, which has a uniform density and U and V are independent random variables (if X and Y are).

We will use the technique of the previous example to find the pdf of certain generic types of transformations that occur frequently in practice. These include $U = X + Y$, $U = XY$, and $U = X/Y$. The pdf's of these three functions will be found as the next three examples. The range R_U will depend on R_{XY}, which is not assumed to be known, and so all mention of R_U is omitted in these examples. In a particular use of these equations, it will, of course, have to be determined. In all the examples we choose the dummy variable $V = X$ and take the joint pdf of (X, Y) to be $f_{XY}(x, y)$.

EXAMPLE 7.12

Determine the pdf of $U = X + Y$.
The bivariate transformation $U = X + Y$, $V = X$ has the inverse $X = V$, $Y = U - V$ and Jacobian

$$J(u, v) = \begin{vmatrix} \dfrac{\partial x}{\partial u} & \dfrac{\partial x}{\partial v} \\ \dfrac{\partial y}{\partial u} & \dfrac{\partial y}{\partial v} \end{vmatrix} = \begin{vmatrix} 0 & 1 \\ 1 & -1 \end{vmatrix} = -1$$

The joint pdf of (U, V) is

$$f_{UV}(u, v) = f_{XY}(v, u - v) \cdot |-1|$$

and the marginal of U is

$$f_U(u) = \int f_{XY}(v, u - v)\, dv \tag{7.12}$$

If X and Y are independent random variables, then

$$f_U(u) = \int f_X(v) f_Y(u - v)\, dv \tag{7.13}$$

An integral of this form is called a convolution integral and occurs frequently in engineering, physics, and other areas.

EXAMPLE 7.13

Determine the pdf of $U = XY$.
The inverse transformation is $X = V$, $Y = U/V$ and has the Jacobian

$$J(u, v) = \begin{vmatrix} 0 & 1 \\ 1/v & -u/v^2 \end{vmatrix} = -1/v$$

giving the joint pdf

$$f_{UV}(u, v) = f_{XY}(v, u/v)|1/v|$$

and the marginal pdf

$$f_U(u) = \int f_{XY}(v, u/v)|1/v|\, dv \tag{7.14}$$

EXAMPLE 7.14

Determine the pdf of $U = X/Y$.
The inverse transformation is $X = V$, $Y = .V/U$ and the Jacobian is

$$J(u, v) = \begin{vmatrix} 0 & 1 \\ -v/u^2 & 1/u \end{vmatrix} = \frac{v}{u^2}$$

and the marginal pdf of U is

$$f_U(u) = \int f_{XY}(v, v/u)|v/u^2|\, dv \tag{7.15}$$

EXAMPLE 7.15

Show that if X and Y are standard normal random variables, then $U = X/Y$ is a Cauchy random variable.

216 *Applied Probability*

Using Eq. (7.15), for X and Y independent,

$$f_U(u) = \int_{-\infty}^{\infty} \frac{e^{-v^2/2}}{\sqrt{2\pi}} \frac{e^{-\frac{1}{2}(v/u)^2}}{\sqrt{2\pi}} \left|\frac{v}{u^2}\right| dv$$

$$= \int_{-\infty}^{\infty} \left|\frac{v}{u^2}\right| \frac{1}{2\pi} \exp\left[-\frac{v^2}{2}\left(1+\frac{1}{u^2}\right)\right] dv$$

$$= \int_0^{\infty} \frac{v}{u^2 2\pi} \exp\left[-\frac{v^2}{2}\left(1+\frac{1}{u^2}\right)\right] dv$$

$$- \int_{-\infty}^0 \frac{v}{u^2 2\pi} \exp\left[-\frac{v^2}{2}\left(1+\frac{1}{u^2}\right)\right] dv$$

$$= -\frac{1}{2\pi(1+u^2)} \exp\left[-\frac{v^2}{2}\left(1+\frac{1}{u^2}\right)\right]\Big|_0^{\infty}$$

$$+ \frac{1}{2\pi(1+u^2)} \exp\left[-\frac{v^2}{2}\left(1+\frac{1}{u^2}\right)\right]\Big|_{-\infty}^0$$

$$= \frac{1}{\pi(1+u^2)}, \quad -\infty < u < \infty$$

which is just the Cauchy density function of Eq. (5.13) with $\beta = 0$ and $\alpha = 1$.

Equations (7.12) through (7.15) will be useful in several of the problems. For the present we apply Eq. (7.13) to a particular situation.

EXAMPLE 7.16

Determine the pdf of the random variable $U = X + Y$, where X and Y are independent random variables with pdf's

$$f_X(x) = \begin{cases} \frac{1}{4}, & -2 \le x \le 2 \\ 0, & \text{elsewhere} \end{cases} \qquad f_Y(y) = \begin{cases} y+1, & -1 \le y \le 0 \\ -y+1, & 0 \le y \le 1 \\ 0, & \text{elsewhere} \end{cases}$$

In order to use Eq. (7.13) we have first to determine R_{UV}, where $U = X + Y$ and $V = X$. The region is shown in Figure 7.11. For the marginal pdf of U

$$f_U(u) = \int f_X(v) f_Y(u-v) \, dv$$

and the range of integration depends on the value of u. We consider the various ranges

(i) $-3 \le u \le -1$: $f_U(u) = \int_{-2}^{u+1} \frac{1}{4} \cdot f_Y(u-v) \, dv$

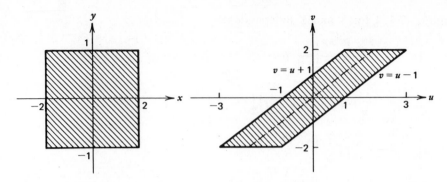

FIGURE 7.11

and the situation is complicated by the definition of $f_Y(y)$. When $-1 \leq u - v \leq 0$, then $u \leq v \leq u + 1$ and $f_Y(u - v) = u - v + 1$. When $0 \leq u - v \leq 1$, then $u - 1 \leq v \leq u$ and $f_Y(u - v) = -(u - v) + 1$. This is indicated on the dotted line in Figure 7.11, and implies that the region $-3 \leq u \leq -1$ must be split into two regions:

(i) (a) $-3 \leq u \leq -2$: $f_U(u) = \frac{1}{4} \int_{-2}^{u+1} (u - v + 1)\, dv = \frac{1}{8}(u + 3)^2$

(b) $-2 \leq u \leq -1$: $f_U(u) = \frac{1}{4} \int_{-2}^{u} (v - u + 1)\, dv + \frac{1}{4} \int_{u}^{u+1} (u - v + 1)\, dv$

$$= \frac{1}{8}(-u^2 - 2u + 1) = \frac{1}{4}\left(1 - \frac{(u+1)^2}{2}\right)$$

(ii) $-1 \leq u \leq 1$: $f_U(u) = \frac{1}{4} \int_{u-1}^{u} (v - u + 1)\, dv + \frac{1}{4} \int_{u}^{u+1} (u - v + 1)\, dv$

$$= \frac{1}{4}$$

(iii) (a) $1 \leq u \leq 2$: $f_U(u) = \frac{1}{4} \int_{u-1}^{u} (v - u + 1)\, dv + \frac{1}{4} \int_{u}^{2} (u - v + 1)\, dv$

$$= \frac{1}{4}\left(1 - \frac{(u-1)^2}{2}\right)$$

(b) $2 \leq u \leq 3$: $f_U(u) = \frac{1}{4} \int_{u-1}^{2} (v - u + 1)\, dv = \frac{1}{8}(u - 3)^2$

The job of verifying that these five equations do indeed describe a valid pdf (i.e., is positive with unit area) is left as an exercise to the reader.

The point of this example is to illustrate the care that must be taken in evaluating the integrals and regions encountered in some of these transformation problems.

In Chapter 5 it was observed that a linear transformation $Y = aX + b$ of a normal random variable is again a normal random variable and that if $X \sim N(\mu, \sigma^2)$, then $Y \sim N(a\mu + b, a^2\sigma^2)$. The final example of this section shows that if X and Y are normal random variables, then their sum is also a normal random variable. Thus, if $X \sim N(\mu_x, \sigma_x^2)$ and $Y \sim N(\mu_y, \sigma_y^2)$, then $Z = X + Y$ has the pdf $N(\mu_x + \mu_y, V(X + Y))$, where $V(X + Y) = \sigma_x^2 + \sigma_y^2 + 2\,\mathrm{Cov}(X, Y)$. This is another useful property of normal random variables to keep in mind. It is only proven in the example for the case where X and Y are independent as in the correlated case the proof tends to be very involved.

EXAMPLE 7.17

Show that if (X, Y) is a jointly normal random variable with X and Y independent, then the random variable $Z = X + Y$ is also a normal random variable.

Using the joint pdf of Eq. (6.18) in Eq. (7.13) yields

$$f_Z(x) = \frac{1}{2\pi\sigma_x\sigma_y} \int_{-\infty}^{\infty} \exp\left\{-\frac{1}{2}\left[\left(\frac{v - \mu_x}{\sigma_x}\right)^2 + \left(\frac{z - v - \mu_y}{\sigma_y}\right)^2\right]\right\} dv$$

Expanding the exponent and gathering terms in v we obtain

$$f_Z(z) = \frac{1}{2\pi\sigma_x\sigma_y} \int_{-\infty}^{\infty} \exp\left\{-\frac{1}{2}\left[v^2\left(\frac{1}{\sigma_x^2} + \frac{1}{\sigma_y^2}\right)\right.\right.$$
$$\left.\left. -2v\left(\frac{\mu_x}{\sigma_x^2} + \frac{2(z - \mu_y)}{\sigma_y^2}\right) + \frac{(z - \mu_y)^2}{2\sigma_y^2} + \frac{\mu_x^2}{\sigma_x^2}\right]\right\} dv$$

and completing the square in v and integrating gives

$$f_Z(z) = \frac{1}{\sqrt{2\pi}(\sigma_x + \sigma_y)} \exp\left(-\frac{1}{2}\frac{(v - \mu_x - \mu_y)^2}{(\sigma_x^2 + \sigma_y^2)}\right)$$

and $Z \sim N(\mu_x + \mu_y, \sigma_x^2 + \sigma_y^2)$. There is a simpler way to show this result, and this method will be used extensively in the next chapter. In Chapter 5 it was shown that the mgf of X is

$$m_X(t) = E(e^{tX}) = \exp(t\mu_x + \tfrac{1}{2}t^2\sigma_x^2)$$

In the case when X and Y are independent we can write

$$m_Z(t) = E(e^{tZ}) = E(e^{t(X+Y)}) = E(e^{tX})E(e^{tY})$$
$$= m_X(t)m_Y(t)$$
$$= \exp(t\mu_x + \tfrac{1}{2}t^2\sigma_x^2)\exp(t\mu_y + \tfrac{1}{2}t^2\sigma_y^2)$$
$$= \exp[t(\mu_x + \mu_y) + \tfrac{1}{2}t^2(\sigma_x^2 + \sigma_y^2)]$$

which is the mgf of a normal random variable $N(\mu_x + \mu_y, \sigma_x^2 + \sigma_y^2)$.

7.3 Other Types of Transformations

There are other types of transformations that arise in statistical and other problems which must be solved by applying basic principles rather than resorting to general techniques. These problems are instructive and we consider a few of them here.

EXAMPLE 7.18

Let X and Y be two random variables with joint pdf $f_{XY}(x, y)$ and joint cdf $F_{XY}(x, y)$. Find the cdf of the random variable $U = \max(X, Y)$.

Clearly this function does not satisfy the differentiability conditions imposed in the previous section and another approach is required. In trying to find the cdf of the random variable U, notice that the region corresponding to

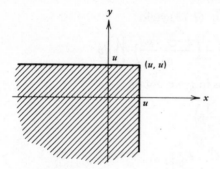

FIGURE 7.12

the event $\{\max(X, Y) \le u\}$ is shown in Figure 7.12 and the probability of this event is

$$P(U \le u) = F_{XY}(u, u) = F_U(u)$$

If X and Y are independent random variables, then

$$F_U(u) = F_X(u)F_Y(u)$$

and differentiating to find the pdf gives

$$f_U(u) = \left(\frac{d}{du}F_X(u)\right)F_Y(u) + F_X(u)\left(\frac{d}{du}F_Y(u)\right)$$

$$= f_X(u)F_Y(u) + F_X(u)f_Y(u)$$

$$= f_X(u)\int_{-\infty}^{u} f_Y(y)\, dy + f_Y(u)\int_{-\infty}^{u} f_X(x)\, dx \qquad (7.16)$$

In the case where X and Y have the same pdf $f(x)$ and cdf $F(x)$, the pdf of

$U = \max(X, Y)$ is

$$f_U(u) = 2f(u)F(u)$$

We will give another derivation of this equation in Example 7.20.

EXAMPLE 7.19

Find the cdf of the random variable $V = \min(X, Y)$, where X and Y are as in Example 7.18.

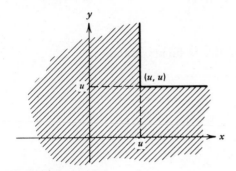

FIGURE 7.13

The region in the x, y plane corresponding to the event $\{\min(X, Y) \le v\}$ is shown in Figure 7.13 and the probability of this event is

$$P(V \le v) = F_X(v) + F_Y(v) - F_{XY}(v, v)$$

$$= F_V(v)$$

If X and Y are independent random variables, then

$$F_V(v) = F_X(v) + F_Y(v) - F_X(v)F_Y(v)$$

and differentiating with respect to v gives

$$f_V(v) = f_X(v) + f_Y(v) - f_X(v)F_Y(v) - F_X(v)f_Y(v) \qquad (7.17)$$

Again, a slightly different approach to this result is given in the next example.

In certain sampling problems of statistics, n observations, usually assumed to be independent, of a single random variable with pdf $f(x)$ are made. Since these observations, which we denote by $X_1, X_2, \ldots, X_n$, are from the same population (i.e., they are independent but have the same pdf $f(x)$) certain questions on the amount of variation in the sample can be of interest. Some of these are dealt with in the next example.

EXAMPLE 7.20

Let $X_1, X_2, \ldots, X_n$ be n independent random variables each with pdf $f(x)$. Find the pdf of the following random variables:

(i) $U = \max(X_1, X_2, \ldots, X_n)$

(ii) $V = \min(X_1, X_2, \ldots, X_n)$

(iii) $R = \max(X_1, X_2, \ldots, X_n) - \min(X_1, X_2, \ldots, X_n)$

(i) The probability that U falls in the interval $(u, u + du)$ is simply the probability that one of the random variables falls in this interval and that the others are all less than u. The probability that X_1 falls in $(u, u + du)$ and that $X_2, \ldots, X_n$ are all less than u is

$$f(u)\, du\left(\int_{-\infty}^{u} f(x)\, dx\right)^{n-1}$$

Since there are n ways of choosing the variable to be maximum,

$$f_U(u) = nf(u)\left(\int_{-\infty}^{u} f(x)\, dx\right)^{n-1}$$

For $n = 2$ this is just

$$f_U(u) = 2f(u)\int_{-\infty}^{u} f(x)\, dx = 2f(u)F(u)$$

where $F(x)$ is the cdf of $f(x)$. This expression is just Eq. (7.16) for the case where $F_X(u) = F_Y(u)$.

(ii) The probability that V falls in the interval $(v, v + dv)$ is the probability that one of the variables falls in $(v, v + dv)$ and all others are greater than v, that is,

$$f_V(v) = nf(v)\left(\int_{v}^{\infty} f(x)\, dx\right)^{n-1}$$

For $n = 2$ this reduces to

$$f_V(v) = 2f(v)(1 - F(v))$$

which is Eq. (7.17) for $f_X(v) = f_Y(v)$.

(iii) To find the pdf of R, $f_R(r)$, we first find the joint pdf of U and V. Using arguments similar to those in (i) and (ii), $f_{UV}(u, v)\, du\, dv$ is the probability of one random variable falling in the interval $(u, u + du)$ (and this random variable can be chosen in n ways) times the probability of another random variable falling in the interval $(v, v + dv)$, $v < u$ (and this random variable can be chosen in $(n-1)$ ways), times the probability that the remaining $(n-2)$ random variables fall in the range (v, u), that is,

$$f_{UV}(u, v) = n(n-1)f(u)f(v)\left(\int_{v}^{u} f(x)\, dx\right)^{n-2} \tag{7.18}$$

To find the pdf of $R = U - V$, we apply arguments similar to those used in

222 *Applied Probability*

deriving Eq. (7.12) of the previous section to arrive at the expression

$$f_R(r) = \int f_{UV}(y, y-r)\, dy$$

$$= n(n-1) \int f(y) f(y-r) \left(\int_{y-r}^{y} f(x)\, dx \right)^{n-2} dy \qquad (7.19)$$

To indicate the difficulties encountered in applying the rather complicated looking expression of Eq. (7.19), we conclude the section with a brief application of it.

EXAMPLE 7.21

If n points are chosen at random on the interval $(0, 1)$, find the pdf of the maximum distance between any two of them.

The problem here is to determine the ranges of integration in Eq. (7.19) for the pdf

$$f(x) = \begin{cases} 1, & 0 \le x \le 1 \\ 0, & \text{elsewhere} \end{cases}$$

It is clear that for the interior integral to be nonzero we must have $0 < y - r <$ $y < 1$ and

(i) $r > 0$.

The functions $f(y)$ and $f(y-r)$ are nonzero only on the intervals

(ii) $0 < y < 1$ and
(iii) $0 < y - r < 1$ or $y - 1 < r < y$

and all three of these conditions must be met to obtain a nonzero integrand. These conditions define the region shown in Figure 7.14 and Eq. (7.19) reduces to

$$f_R(r) = n(n-1) \int_{r}^{1} r^{n-2}\, dy$$

$$= \begin{cases} n(n-1) r^{n-2}(1-r), & 0 \le r \le 1 \\ 0, & \text{elsewhere} \end{cases}$$

By reasoning similar to that employed in the previous examples, it is a straightforward matter to show that the pdf $f_M(m)$ of the median of a sample $X_1, X_2, \ldots, X_n$, each X_i having the pdf $f_X(x)$, is

$$f_M(m) = \frac{n!}{\{[(n-1)/2]!\}^2} \left(\int_{-\infty}^{m} f_X(x)\, dx \right)^{(n-1)/2} \left(\int_{m}^{\infty} f_X(x)\, dx \right)^{(n-1)/2} f_X(m)$$

where n is assumed to be odd.

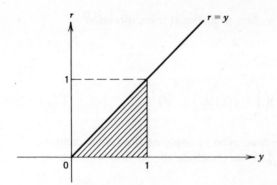

FIGURE 7.14

Problems

1. If X is a standard normal random variable, find the pdf of $Y = \sqrt{|X|}$.
2. If $X = A \cos \phi$, where A is a constant and ϕ is uniformly distributed between 0 and 2π, find the pdf of X.
3. Let X be a random variable with an exponential pdf with parameter 1. Find the pdf of the random variable $Y = 1 - e^{-X}$, $X \geq 0$.
4. A square law rectifier has the characteristic

$$y = \begin{cases} kx^2, & x > 0 \\ 0, & x \leq 0 \end{cases}$$

   where x and y are the input and output voltages, respectively. If the input to the rectifier is noise having a Rayleigh pdf

$$f(x) = \begin{cases} (2x/\beta) \exp(-x^2/\beta), & x \geq 0, \quad \beta > 0 \\ 0, & x < 0 \end{cases}$$

   find the pdf of the output.
5. The input into a control system X undergoes a saturation type of linear transformation shown in Figure 7.15. If X is an exponential random variable with parameter λ, find the pdf of Y.

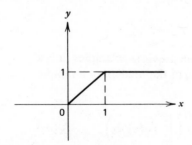

FIGURE 7.15

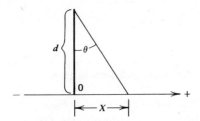

FIGURE 7.16

6. A light emitting source emits a light ray at an an angle θ from the vertical as shown in Figure 7.16. If θ is a random variable uniformly distributed over $(-\pi/2, \pi/2)$, what is the pdf of the random variable X? What are the mean and variance of this pdf?

7. Find the pdf of the random variable $Y = 1/X$ if the random variable X has the pdf

$$f(x) = \begin{cases} \dfrac{6x}{(1+x)^4}, & 0 \le x < \infty \\ 0, & x < 0 \end{cases}$$

8. The errors in a weighing process are known to have a normal pdf with mean μ and variance σ^2. What is the pdf of the absolute value of the errors?

9. The demand D for a certain brand of coffee, per week, is assumed to be uniformly distributed over the interval $(10{,}000, 12{,}000)$ kg. The cost per kg is inversely proportional to cost in that cost $= K/(10{,}000 + D)$. Find the pdf of the cost as a function of the constant K.

10. Find the transformation $Y = H(X)$ such that if X has a pdf

$$f(x) = \begin{cases} 6x(1-x), & 0 \le x \le 1 \\ 0, & \text{elsewhere} \end{cases}$$

then Y has the pdf

$$g(y) = \begin{cases} 3(1-\sqrt{y}), & 0 \le y \le 1 \\ 0, & \text{elsewhere} \end{cases}$$

11. The diameter X of a coin is a random variable with $E(X) = 1$ cm and $V(X) = 2$ mm^2. The coin is 3 mm thick. What is the expected volume of a coin?

12. Suppose the random variable X has a probability density function

$$f(x) = \begin{cases} \tfrac{1}{2}, & 0 \le x \le 1 \\ 1/(2x^2), & x > 1 \\ 0, & x < 0 \end{cases}$$

Find the pdf of $Y = 1/X$.

13. The radius R of a tennis ball is, approximately, a normal random variable with mean 4.50 cm and variance .04 cm^2. Find the pdf of the volume of the tennis ball. What is the expected value of the volume?

14. If X is a standard normal random variable, what transformation $Y = H(X)$ would yield a pdf of the form

$$g(y) = \begin{cases} e^{-y}, & y \ge 0 \\ 0, & y < 0 \end{cases}$$

for Y?

15. If X is a random variable with a Cauchy pdf

$$f(x) = \frac{1}{\pi(1+x^2)}, \quad -\infty < x < \infty$$

find the pdf of the random variable $Y = 1/X$. Under what conditions, in general, will the pdf of Y be the same as that for X?

16. If $Y = H(X)$, find the joint cdf $F(x, y)$ of X and Y in terms of the cdf of X, $F_X(x)$.

17. If $U = G_1(X, Y)$ and $V = G_2(X, Y)$, and, for the inverse transformations $X = H_1(U, V)$ and $Y = H_2(U, V)$, show that $|J(x, y)| \cdot |J(u, v)| = 1$, where

$$J(x, y) = \begin{vmatrix} \dfrac{\partial u}{\partial x} & \dfrac{\partial u}{\partial y} \\ \dfrac{\partial v}{\partial x} & \dfrac{\partial v}{\partial y} \end{vmatrix} \quad \text{and} \quad J(u, v) = \begin{vmatrix} \dfrac{\partial x}{\partial u} & \dfrac{\partial x}{\partial v} \\ \dfrac{\partial y}{\partial u} & \dfrac{\partial y}{\partial v} \end{vmatrix}$$

18. In the manufacture of hockey pucks, the radius R and thickness T are assumed to be independent random variables uniformly distributed over the intervals $(5.0, 5.1)$ and $(2, 2.1)$ cm, respectively. Find the pdf of the volume of a puck.

19. The random variables X and Y have the pdf's

$$f(x) = \begin{cases} x \exp(-x^2/2), & 0 \le x < \infty \\ 0, & \text{elsewhere} \end{cases} \qquad g(y) = \begin{cases} y \exp(-y^2/2), & y \ge 0 \\ 0, & y < 0 \end{cases}$$

Find the pdf of the quotient $Z = X/Y$.

20. The times between store holdups in a town are assumed to be independent random variables, each with an exponential pdf with parameter $\lambda = .1 \text{ day}^{-1}$. If T_1 and T_2 are two successive time intervals between holdups, find the pdf of $T_1/(T_1 + T_2)$.

21. Let X and Y be independent normal random variables each with mean 0 and variance 1. Under what conditions on the constants a, b, c, and d will the random variables $U = aX + bY$ and $V = cX + dY$ be independent? In particular, are $U = \cos(\theta)X + \sin(\theta)Y$ and $V = \sin(\theta)X - \cos(\theta)Y$ independent?

22. Let X, Y, and Z be independent random variables each with the exponential pdf $f(x) = \lambda e^{-\lambda x}$, $x > 0$. What is the probability that $X + Y > Z$?

23. Let X and Y be normal random variables each with mean zero and variance σ^2 and correlation coefficient ρ. Find the pdf of the random variable $Z = X + Y$.

24. If $X \sim N(\mu_x, \sigma_x^2)$, $Y \sim N(\mu_y, \sigma_y^2)$ and the correlation coefficient between X and Y is ρ_{xy}, for what values of a, b, c, and d are the random variables

$$U = aX + bY \quad \text{and} \quad V = cX + dY$$

independent?

25. Let X and Y be independent random variables with pdf's

$$f_X(x) = \begin{cases} \frac{1}{10}, & 100 < x < 110 \\ 0, & \text{elsewhere} \end{cases} \qquad f_Y(y) = \begin{cases} \frac{1}{2}, & 10 \le y \le 12 \\ 0, & \text{elsewhere} \end{cases}$$

representing the length and width of a rectangle. Find the pdf of the perimeter of the rectangle.

26. In Problem 25 find the pdf of the area of the rectangle.

27. If X and Y are independent standard normal random variables, find the pdf of $Z = X/Y$.

28. Find the pdf of the random variable $Z = X_1 + X_2 + X_3$, where each X_i is independent and uniformly distributed over $(-1, 1)$.

29. The random variables X and Y are independent and both are uniformly distributed over the interval (2, 4). Find the pdf of the random variable $Z = |X - Y|$.

30. If the random variables X and Y are independent standard normal random variables, show that the random variables

$$Z = \frac{X+Y}{2} \quad \text{and} \quad W = \frac{(X-Z)^2 + (Y-Z)^2}{2}$$

are independent and find their marginal pdf's.

31. Extending Problem 30 show that the sample mean and sample variance of a random sample of size n of a standard normal population are independent, where

$$\bar{X} = \frac{1}{n} \sum_i^n X_i \quad \text{and} \quad S^2 = \frac{1}{n-1} \sum_i^n (X_i - \bar{X})^2$$

32. If (X, Y) is a bivariate random variable uniformly distributed over the circle $x^2 + y^2 \leq 1$, find the joint pdf of the random variables $R = (X^2 + Y^2)^{1/2}$ and $\theta = \arctan(X/Y)$.

33. Let X and Y be two independent random variables, each with an exponential pdf with parameter 1. Find the pdf of the random variable $Z = \sqrt{X + Y}$.

34. A manufacturer has two machines turning out copper wire. If the machines are operating correctly, the tensile strength, in kilograms, of the wire is a random variable X that has a pdf

$$f(x) = \begin{cases} e^{-x}, & x \geq 0 \\ 0, & x < 0 \end{cases}$$

The manufacturer takes a sample from each machine at random. What is the probability that the tensile strengths of the wires will be within 10 kg of one another?

35. If X and Y are independent random variables, show that $aX + b$ and $cY + d$ are also independent random variables for any finite constants a, b, c, and d, $a \neq 0$, $c \neq 0$.

36. The random variables X and Y are independent and have the pdf's

$$f_X(x) = \begin{cases} x e^{-x^2/2}, & x \geq 0 \\ 0, & x < 0 \end{cases} \qquad f_Y(y) = \begin{cases} y e^{-y^2/2}, & y \geq 0 \\ 0, & y < 0 \end{cases}$$

Find the probability that X is less than or equal to kY. Find also the probability that $X < Y < 2X < 1$.

37. The sum of two exponentially distributed random variables with the same parameter has a gamma pdf. What would the pdf of the sum be if they did not have the same parameter?

38. A random variable X is considered to be a signal and is uniformly distributed over the range (0, 1). Another random variable N, independent of X, is considered to be noise and has a triangular density function over the range $(-1, 1)$. If the random variable $Y = X + N$ is observed, find the pdf of Y.

39. Let X and Y be independent random variables each with a Cauchy pdf, that is,

$$f_X(x) = \frac{1}{\pi(1+x^2)}, \quad -\infty < x < \infty$$

$$f_Y(y) = \frac{1}{\pi(1+y^2)}, \quad -\infty < y < \infty$$

Find the pdf of the sum $Z = X + Y$.

Functions of Random Variables 227

40. Let $X_1, X_2, \ldots, X_n$ be a random sample from the Cauchy pdf of Problem 39. Extend the results of that problem by finding the pdf of $Z = X_1 + X_2 + \cdots + X_n$.

41. In a traffic survey it was found that the number of cars X passing a certain point on a road in any one-hour period has an exponential distribution

$$P(X = j) = ar^i, \quad j = 0, 1, 2, \ldots$$

where $a = \frac{3}{4}$ and $r = \frac{1}{4}$. Let X_1 and X_2 be the number of cars passing in two separate one-hour periods and assume these are independent random variables. What is the probability that $X_2 \geq X_1$?

42. Let X, Y, and Z be independent random variables, each with an exponential pdf with parameter 1. Find the pdf of their sum. Find the probability of the event $X < Y < Z$.

CHAPTER 8

Sums of Random Variables

In the previous chapters problems involving transformations of two or more random variables were considered. Of particular importance in applications is the ability to deal with sums of independent random variables. Such problems arise frequently in statistical inference, that is, drawing conclusions from a set of observations on a random variable. In the next section such problems are encountered and it will be shown that many of the density functions introduced in Chapter 5 arise naturally in this context.

The law of large numbers, introduced in Section 8.2, is a reassuring result of probability theory. Simply stated it says that under certain conditions the average of a sample of observations drawn from a distribution tends toward the distribution mean as the number of observations in the sample increases. This formalizes an intuitive feeling that was introduced along with the axioms of probability in the first chapter.

A result of a slightly different nature, the central limit theorem, emphasizes, and in some part helps to explain, the importance of the normal density function. This theorem appears frequently in applications of probability and statistics and has been proved under a variety of conditions. In essence it says that a standardized sum of independent random variables has a cumulative distribution function that approaches the function of a standardized normal random variable. It is therefore not surprising that many distributions can be approximated using the normal distribution. We explore the many relationships that exist between the normal and other distributions. In many instances the approximations developed here give accurate answers and can be of importance in applications.

8.1 Sums of Random Variables and Sampling Distributions

As an introduction, suppose we are interested in how well a manufacturing process is doing in producing an item according to a specification on one of its characteristics, say weight. One approach would be to choose n items at random from the production and weigh them. In this situation the set of all items produced is referred to as the population and, if $X_1, X_2, \ldots, X_n$ are the random variables, X_i giving the weight of the ith item sampled, then $(X_1, X_2, \ldots, X_n)$ is called a sample of size n. Now since we have n random variables we would need an n-variable pdf, $f(x_1, x_2, \ldots, x_n)$ in order to describe the sample probabilistically. However, in all cases considered in this book we will assume that the random variables $X_1, X_2, \ldots, X_n$ are independent random variables and each random variable has the common pdf $f(x)$. In this case we refer to the set of random variables $X_1, X_2, \ldots, X_n$ as a random sample of X, a random variable with this pdf, and

$$f(x_1, x_2, \ldots, x_n) = f(x_1)f(x_2) \cdots f(x_n)$$

This will greatly simplify our analyses. A particular set of outcomes of the random sample, $x_1, x_2, \ldots, x_n$ is called a set of observations.

In order to deduce information about the population, it is invariably necessary to form functions of the sample. For example, intuitively an estimate of the mean of the population μ, where

$$\mu = \int_{R_x} xf(x)\, dx$$

is given by

$$\bar{X} = \frac{1}{n} \sum_{i=1}^{n} X_i$$

and this quantity, referred to as the sample mean, is itself a random variable. A knowledge of the pdf of $\bar{X}$ would enable us to answer questions such as how close $\bar{X}$ is likely to be to μ, the population mean (which is a constant, not a random variable). Similarly we might estimate the variance of $f(x)$, σ^2, by

$$S^2 = \frac{1}{(n-1)} \sum_{i=1}^{n} (X_i - \bar{X})^2$$

(the reason for the factor $(n-1)$ rather than n is explored in the next chapter), and this quantity is called an estimator for the population variance. We will refer to any real-valued function $G(X_1, X_2, \ldots, X_n)$ of the sample as a statistic. These ideas are gathered in the following more formal definition:

Definition   Let X be a random variable with pdf $f(x)$ and $X_1, X_2, \ldots, X_n$ a set of n independent random variables each with pdf $f(x)$. The set of random variables $X_1, X_2, \ldots, X_n$ is called a random sample of size n of X. A real-valued function of a random sample is called a statistic.

     Many of the statistics that will be of interest to us in later applications can be expressed as a sum of independent random variables and for this reason we will look at various aspects of such sums in this chapter. For the remainder of this section we consider some pdf's that arise as the sum of identically distributed, independent random variables and some curious "reproducing properties" that some of them have.

     To begin with, recall that if $Y = a_1 X_1 + \cdots + a_n X_n$, then

$$E(Y) = E(a_1 X_1 + \cdots + a_n X_n)$$
$$= a_1 E(X_1) + \cdots + a_n E(X_n)$$

The variance of Y can also be found:

$$V(Y) = E((Y - E(Y))^2)$$
$$= E\left[\left(\sum_{i=1}^{n} [a_i X_i - a_i E(X_i)] \right)^2 \right]$$
$$= \sum_{i=1}^{n} \sum_{j=1}^{n} a_i a_j E\{[X_i - E(X_i)][X_j - E(X_j)]\}$$
$$= \sum_{i=1}^{n} \sum_{j=1}^{n} a_i a_j \, \text{Cov}(X_i, X_j)$$

If X_i and X_j are pairwise uncorrelated, that is, $\text{Cov}(X_i, X_j) = 0$, $i \neq j$, then, since $\text{Cov}(X_i, X_i) = V(X_i)$,

$$V(Y) = \sum_{i=1}^{n} a_i^2 V(X_i)$$

and, in particular, this is true if $X_1, X_2, \ldots, X_n$ are independent random variables. An example of the use of this equation is in calculating the mean and variance of a binomial random variable. This was done in Chapter 3 in two ways, one being by tedious calculation and the other (Problem 3.16) by differentiating the moment generating function.

     Let X_i be the random variable representing the outcome of an experiment on the ith trial where

$$X_i = \begin{cases} 1, & \text{if the } i\text{th experiment is a success} \\ 0, & \text{otherwise} \end{cases}$$

If the experiment is repeated n times independently and $P(X_i = 1) = \theta$ for each experiment, then find the mean and variance of the random variable $Y = X_1 + X_2 + \cdots + X_n$.

The random variable Y is just the number of successes in n Bernoulli trials, that is, a binomial random variable with parameters n and θ. The random variable X_i has mean

$$E(X_i) = 0 \cdot (1 - \theta) + 1 \cdot (\theta) = \theta$$

and since

$$E(X_i^2) = 0 \cdot (1 - \theta) + 1^2 \cdot (\theta) = \theta$$

its variance is

$$V(X_i) = \theta - \theta^2 = \theta(1 - \theta) = \theta\eta, \quad \eta = 1 - \theta$$

It follows that

$$E(Y) = \sum_{i=1}^{n} E(X_i) = n\theta$$

and

$$V(Y) = \sum_{i=1}^{n} V(X_i) = n\theta\eta$$

which agrees with Eqs. (3.6) and (3.7).

If $X_1, X_2, \ldots, X_n$ is a set of independent random variables and the pdf of X_i is $f_i(x)$ and its mgf is $m_i(t)$ (i.e., it is not necessarily a random sample since the pdf's are not assumed identical), then the mgf of the sum $Y = X_1 + X_2 + \cdots + X_n$ is

$$
\begin{aligned}
m_Y(t) &= E(e^{tY}) \\
&= E(e^{t(X_1 + X_2 + \cdots + X_n)}) \\
&= E(e^{tX_1}) \cdots E(e^{tX_n}) \\
&= m_1(t)m_2(t) \cdots m_n(t)
\end{aligned}
\tag{8.1}
$$

and we use this fact to examine the reproducing properties of certain pdf's mentioned earlier. Notice that if each $m_i(t)$ is the same, then $m_Y(t) = m_1(t)^n$.

8.1.1 BINOMIAL RANDOM VARIABLES

Let $X_1, X_2, \ldots, X_n$ be independent binomial random variables where X_i has parameters n_i, θ_i. From Chapter 2 the mgf of X_i is

$$m_i(t) = (\theta_i e^t + \eta_i)^{n_i}, \quad \eta_i = 1 - \theta_i, \quad i = 1, \ldots, n$$

and using Eq. (8.1) the mgf of Y is

$$m_Y(t) = \prod_{i=1}^{n} (\theta_i e^t + \eta_i)^{n_i}$$

If $\theta_i = \theta$, $i = 1, 2, \ldots, n$, then

$$m_Y(t) = \prod_{i=1}^{n} (\theta e^t + \eta)^{n_i} = (\theta e^t + \eta)^m$$

where $m = n_1 + n_2 + \cdots + n_n$. Since this is the mgf of a binomial random variable with parameters m and p, we have shown that:

If $X_1, X_2, \ldots, X_n$ is a set of independent binomial random variables, X_i having parameters n_i and θ, then their sum $Y = X_1 + X_2 + \cdots + X_n$ is a binomial random variable with parameters m and θ, $m = n_1 + n_2 + \cdots + n_n$.

It is, of course, important that the parameter θ be the same for each random variable since the result is not true if they differ.

8.1.2 POISSON RANDOM VARIABLES

From Table 2.1 the mgf of a Poisson random variable X_i with parameter λ_i is

$$m_i(t) = e^{\lambda_i(e^t - 1)}$$

and so the mgf of $Y = X_1 + X_2 + \cdots + X_n$, where $X_1, X_2, \ldots, X_n$ are independent random variables, is

$$m_Y(t) = \prod_{i=1}^{n} e^{\lambda_i(e^t - 1)}$$
$$= e^{\lambda(e^t - 1)}$$

where $\lambda = \lambda_1 + \lambda_2 + \cdots + \lambda_n$. As this is again the pdf of a Poisson random variable with parameter λ, we have shown:

If $X_1, X_2, \ldots, X_n$ are independent Poisson random variables, X_i having parameter λ_i, then $Y = X_1 + X_2 + \cdots + X_n$ is also a Poisson random variable with parameter $\lambda = \lambda_1 + \lambda_2 + \cdots + \lambda_n$.

8.1.3 GAMMA RANDOM VARIABLES

We first prove an important and useful fact on a relationship between exponential and gamma distributions. Let $X_1, X_2, \ldots, X_n$ be independent exponential random variables, each with parameter λ, and recall from Table 5.1 that the mgf of each X_i is

$$m_i(t) = \frac{\lambda}{\lambda - t}$$

Using Eq. (8.1), the mgf of $Y = X_1 + X_2 + \cdots + X_n$ is

$$m_Y(t) = \left(\frac{\lambda}{\lambda - t}\right)^n$$

which is the mgf of a gamma pdf with parameters λ and n, that is, the pdf corresponding to this mgf is

$$f_Y(y) = \frac{\lambda^n}{\Gamma(n)} y^{n-1} e^{-\lambda y}, \quad y > 0$$

The sum of n independent, exponentially distributed random variables, each with parameter λ, is a gamma random variable with parameters λ and n.

Now let $X_1, X_2, \ldots, X_n$ be a set of n independent exponentially distributed random variables, each with parameter λ. The random variable $Y = X_1 + X_2 + \cdots + X_n$ has a gamma pdf with parameters λ and n. Rather than summing all random variables at once, define, for a given set of positive integers $n_1, n_2, \ldots, n_m$, $n_1 + n_2 + \cdots + n_m = n$, the random variables

$$U_i = X_{k_i+1} + X_{k_i+2} + \cdots + X_{k_i+n_i}$$

$$k_i = n_1 + n_2 + \cdots + n_{i-1}, \quad i = 2, 3, \ldots, m, \ k_1 = 0$$

By exactly the same argument as before, U_i, as a sum of n_i independent exponentially distributed random variables each with parameter λ, is a gamma distributed random variable with parameters λ and n_i with mgf

$$m_{U_i}(t) = \left(\frac{\lambda}{\lambda - t}\right)^{n_i}$$

Since $Y = X_1 + X_2 + \cdots + X_n = U_1 + U_2 + \cdots + U_m$ and

$$m_Y(t) = \prod_{i=1}^{m} m_{U_i}(t) = \left(\frac{\lambda}{\lambda - t}\right)^n$$

we have shown:

If $U_1, U_2, \ldots, U_m$ are m independent gamma random variables, U_i having parameters λ and n_i, then their sum $Y = U_1 + U_2 + \cdots + U_m$ also has a gamma density function, with parameters λ and $n = n_1 + n_2 + \cdots + n_m$. A direct proof of this fact using the gamma mgf is also simple.

8.1.4 NORMAL RANDOM VARIABLES

Let $X_1, X_2, \ldots, X_n$ be n independent normally distributed random variables, $X_i \sim N(\mu_i, \sigma_i^2)$ and $Y = X_1 + X_2 + \cdots + X_n$. From Table 5.1 the mgf of X_i, $m_i(t)$, is

$$m_i(t) = \exp(\mu_i t + \tfrac{1}{2}\sigma_i^2 t)$$

and hence the mgf of Y is

$$m_Y(t) = \prod_{i=1}^n \exp(\mu_i t + \tfrac{1}{2}\sigma_i^2 t)$$

$$= \exp(\mu t + \tfrac{1}{2}\sigma^2 t)$$

where $\mu = \mu_1 + \mu_2 + \cdots + \mu_n$ and $\sigma^2 = \sigma_1^2 + \sigma_2^2 + \cdots + \sigma_n^2$. It follows that Y is a normal random variable $N(\mu, \sigma^2)$ and this is an extension of the result of Example 7.17.

8.1.5 *CHI-SQUARE RANDOM VARIABLES*

If X is a standard normal random variable, then, using Eq. (7.5), the pdf of $Y = X^2$ is

$$f(y) = \frac{1}{\sqrt{2\pi y}} e^{-y/2}, \quad y > 0$$

and, from Table 5.1, Y is a gamma random variable with parameter $\gamma = \tfrac{1}{2}$ and $\alpha = \tfrac{1}{2}$. Comparing this pdf with Eq. (5.18), Y is also a chi-square random variable with parameters $n = 1$ and $\sigma^2 = 1$. The mgf of this density is

$$m_Y(t) = \frac{1}{(1-2t)^{1/2}}$$

Now let $X_1, X_2, \ldots, X_n$ be a set of n independent normal random variables each with mean zero and variance σ^2, and let $Y = X_1^2 + X_2^2 + \cdots + X_n^2$. The mgf of X_i^2 is

$$m_i(t) = \frac{1}{(1-2t\sigma^2)^{1/2}}$$

and the mgf of Y is

$$m_Y(t) = \prod_{i=1}^n \left(\frac{1}{1-2t\sigma^2}\right)^{1/2} = \frac{1}{(1-2t\sigma^2)^{n/2}} \tag{8.2}$$

Comparing this with the mgf of a gamma random variable, we conclude that the pdf of Y is

$$f_Y(y) = \left(\frac{1}{2\sigma^2}\right)^{n/2} \frac{y^{(n-2)/2}}{\Gamma(n/2)} e^{-y/2\sigma^2} \tag{8.3}$$

and, as observed in Eq. (5.18) when $\sigma = 1$, this is referred to as the central chi-square density with n degrees of freedom. It is important to recognize that the sum of the squares of n independent standard normal random variables is a chi-square random variable with n degrees of freedom.

Assume now that $U_1, U_2, \ldots, U_m$ are m independent random variables, U_i having a chi-square density with n_i degrees of freedom. Using Eq. (8.2), the mgf of $Z = U_1 + U_2 + \cdots + U_m$ is

$$m_Z(t) = E(e^{tZ})$$
$$= E(e^{t(U_1 + U_2 + \cdots + U_n)})$$
$$= E(e^{tU_1})E(e^{tU_2}) \cdots E(e^{tU_n})$$
$$= \prod_{i=1}^{m} \left(\frac{1}{1-2t}\right)^{n_i/2}$$
$$= \left(\frac{1}{1-2t}\right)^{n/2}$$

where $n = n_1 + n_2 + \cdots + n_m$. This is the mgf of a chi-square density with n degrees of freedom, and we have shown:

If $U_1, U_2, \ldots, U_m$ is a set of independent chi-square random variables, U_i having n_i degrees of freedom, then their sum also has a chi-square density with $n = n_1 + n_2 + \cdots + n_m$ degrees of freedom.

For the remainder of the section we apply these results to determining the density functions of the sample mean and sample variance under various conditions. Let $X_1, X_2, \ldots, X_n$ be a random sample of $X \sim N(\mu, \sigma^2)$ and consider

$$\bar{X} = \frac{1}{n} \sum_{i=1}^{n} X_i$$

As a linear sum of normal random variables, $\bar{X}$ is itself normal with mean

$$E(\bar{X}) = \frac{1}{n} \sum_{i=1}^{n} E(X_i) = \mu$$

and variance

$$V(\bar{X}) = V\left(\frac{1}{n} \sum_{i=1}^{n} X_i\right) = \frac{1}{n^2} V\left(\sum_{i=1}^{n} X_i\right)$$
$$= \frac{1}{n^2} \sum_{i=1}^{n} V(X_i) = \frac{n\sigma^2}{n^2} = \frac{\sigma^2}{n}$$

that is,

$$\bar{X} \sim N(\mu, \sigma^2/n)$$

Now the random variable $(X_i - \mu)/\sigma$ is a standard normal random variable and so

$$Y = \sum_{i=1}^{n} \frac{(X_i - \mu)^2}{\sigma^2} \tag{8.4}$$

is, from previous results, a chi-square random variable with n degrees of freedom.

If the mean of the population was not known, we might replace μ in Eq. (8.4) by the sample mean and try to find the pdf of the random variable

$$Z = \sum_{i=1}^{n} \frac{(X_i - \bar{X})^2}{\sigma^2} \qquad (8.5)$$

It is convenient to write this random variable in a slightly different form as follows:

$$\sum_{i=1}^{n} (X_i - \mu)^2 = \sum_{i=1}^{n} (X_i - \bar{X} + \bar{X} - \mu)^2$$

$$= \sum_{i=1}^{n} (X_i - \bar{X})^2 + n(\bar{X} - \mu)^2 + 2(\bar{X} - \mu) \sum_{i=1}^{n} (X_i - \bar{X})$$

$$= \sum_{i=1}^{n} (X_i - \bar{X})^2 + n(\bar{X} - \mu)^2$$

and so

$$Z = \sum_{i=1}^{n} \frac{(X_i - \mu)^2}{\sigma^2} - \frac{n}{\sigma^2}(\bar{X} - \mu)^2$$

$$= Y - \frac{n}{\sigma^2}(\bar{X} - \mu)^2 \qquad (8.6)$$

Now it is claimed that $\bar{X}$ and $(X_i - \bar{X})$ are independent random variables for $i = 1, 2, \ldots, n$ and to see this note that they are uncorrelated:

$$E[\bar{X}(X_i - \bar{X})] = E\left[X_i\left(\frac{1}{n}\sum_{j=1}^{n} X_j\right)\right] - E\left[\left(\frac{1}{n^2}\sum_{i=1}^{n}\sum_{j=1}^{n} X_i X_j\right)\right]$$

$$= \frac{(n-1)}{n}\mu^2 + \frac{1}{n}(\sigma^2 + \mu^2) - \left(\frac{n^2 - n}{n^2}\right)\mu^2 - \frac{1}{n}(\sigma^2 + \mu^2)$$

$$= 0 = E(\bar{X})E(X_i - \bar{X})$$

It follows that $\bar{X}$ and

$$Z = \frac{1}{\sigma^2}\sum_{i=1}^{n} (X_i - \bar{X})^2$$

are independent since $\bar{X}$ is independent with every term in the sum. The random variable $(\bar{X} - \mu)/(\sigma/\sqrt{n})$ is a standard normal random variable and letting $U = ((\bar{X} - \mu)/(\sigma/\sqrt{n}))^2$, the mgf of U is

$$m_U(t) = \frac{1}{(1 - 2t)^{1/2}}$$

Since Y is a chi-square random variable with n degrees of freedom, its mgf is

$$m_Y(t) = \frac{1}{(1 - 2t)^{n/2}}$$

Since U and Z are independent and $Y = Z + U$,

$$m_Y(t) = m_Z(t)m_U(t)$$

or

$$m_Z(t) = \frac{m_Y(t)}{m_U(t)} = \frac{1}{(1-2t)^{(n-1)/2}}$$

showing that Z is a chi-square random variable with $(n-1)$ degrees of freedom, that is, the pdf of Z is

$$f_Z(z) = \frac{(z)^{(n-3)/2}}{2^{(n-1)/2}\Gamma((n-1)/2)} e^{-z/2}, \quad 0 \leq z < \infty \qquad (8.7)$$

The percentage points of this distribution $\chi^2_{\alpha;n}$, mentioned in the previous chapter, are given in Appendix E and their use is illustrated in the examples later in the section.

We will need one more density function for the next chapter and, as preparation for it, we prove that if V is a standard normal random variable and W is a χ^2 random variable with n degrees of freedom, then the pdf of $T = V/\sqrt{W/n}$ is the t-distribution with n degrees of freedom if V and W are independent. The respective densities of V and W are

$$f_V(v) = \frac{e^{-v^2/2}}{\sqrt{2\pi}}$$

and

$$f_W(w) = \frac{w^{(n-2)/2} e^{-w/2}}{2^{n/2}\Gamma(n/2)}$$

Define a dummy random variable $S = V$ and the Jacobian of the bivariate transformation $S = V$, $T = V/\sqrt{W/n}$ is

$$J(v,w) = \begin{vmatrix} \dfrac{\partial S}{\partial v} & \dfrac{\partial S}{\partial w} \\[2mm] \dfrac{\partial T}{\partial v} & \dfrac{\partial T}{\partial w} \end{vmatrix}$$

$$= \begin{vmatrix} 1 & 0 \\[2mm] \dfrac{1}{\sqrt{w/n}} & -\dfrac{v\sqrt{n}}{2w^{3/2}} \end{vmatrix} = -\frac{v\sqrt{n}}{2w^{3/2}}$$

The joint pdf of S and T is ($V = S$, $W = nS^2/T^2$):

$$f_{ST}(s,t) = f_V(s)f_W\left(\frac{ns^2}{t^2}\right) \cdot \frac{1}{(t^3/2ns^2)}$$

$$= \frac{e^{-s^2/2}}{\sqrt{2\pi}} \cdot \frac{1}{2^{n/2}\Gamma(n/2)} \cdot \frac{n^{(n-2)/2}s^{n-2}}{t^{n-2}} e^{-ns^2/2t^2} \cdot \frac{2ns^2}{t^3}$$

and the marginal pdf of T is:

$$f_T(t) = \frac{n^{n/2}}{\sqrt{2\pi}2^{(n-2)/2}\Gamma(n/2)t^{n+1}} \int_0^\infty s^n e^{-s^2(\frac{1}{2}+n/2t^2)} \, ds$$

Changing the variable of integration to $r = s^2\sigma^2$, where $\sigma^2 = (\frac{1}{2}+n/2t^2)$, noting that $dr = 2\sigma^2 s \, ds$,

$$f_T(t) = \frac{n^{n/2}}{\sqrt{2\pi}2^{(n-2)/2}\Gamma(n/2)t^{n+1}2\cdot\sigma^{n+1}} \int_0^\infty r^{(n-1)/2} e^{-r} \, dr$$

$$= \frac{\Gamma((n+1)/2)}{\sqrt{n\pi}\Gamma(n/2)} \frac{1}{(1+t^2/n)^{(n+1)/2}}$$

which is just the t-distribution with n degrees of freedom discussed in Chapter 5. The percentage points for this distribution, $t_{\alpha;n}$, are tabulated for certain values of t and n in Appendix F.

To return to our sampling problem suppose $X_1, X_2, \ldots, X_n$ is a random sample of $X \sim N(\mu, \sigma^2)$ and let

$$S^2 = \frac{1}{(n-1)} \sum_{i=1}^n (X_i - \bar{X})^2$$

be the sample variance estimator. From Eq. (8.5) and the previous discussion the random variable $(n-1)S^2/\sigma^2$ has a chi-square distribution with $(n-1)$ degrees of freedom, while $(\bar{X}-\mu)/(\sigma/\sqrt{n}) \sim N(0,1)$. From our derivation above (this time using $(n-1)$ rather than n), the random variable

$$\frac{\bar{X}-\mu}{(\sigma/\sqrt{n})} \Bigg/ \sqrt{\frac{(n-1)S^2}{\sigma^2(n-1)}} = \frac{\bar{X}-\mu}{S/\sqrt{n}}$$

has a t-distribution with $(n-1)$ degrees of freedom, and this is the result required in Chapter 9.

For the remainder of the section we consider examples that illustrate the use of this material and of the tables in Appendices E and F.

EXAMPLE 8.1

From experience it is known that the service time of a bank teller is an exponentially distributed random variable with parameter $\lambda = .25 \text{ min}^{-1}$ (mean service time is four minutes). What is the time t_0 for which there is 10% chance that the length of time the fifth customer will have to wait to complete his business?

If T_i is the time the teller takes to service the ith customer, the total time before the fifth customer leaves the bank is $T = T_1 + T_2 + T_3 + T_4 + T_5$. Since T_i is an exponential random variable with parameter $\lambda = .25$, the random variable T has the gamma pdf

$$f_T(t) = \frac{\lambda^5}{4!} t^4 e^{-\lambda t}, \quad t \geq 0$$

as was shown earlier in the section. The probability that the customer has to wait more than t_0 minutes is

$$P(T \geq t_0) = \int_{t_0}^{\infty} \frac{(\frac{1}{4})^5}{4!} t^4 e^{-t/4} \, dt$$

Making the substitution $s = t/2$ gives the equation

$$P(T \geq t_0) = \int_{t_0/2}^{\infty} \frac{1}{2^5 \Gamma(5)} s^4 e^{-3/2} \, ds$$

The integrand is a chi-square density with 10 degrees of freedom and, from Appendix E,

$$.1 = \int_{15.987}^{\infty} \frac{1}{2^5 \Gamma(5)} s^4 e^{-s/2} \, ds$$

and so $t_0 = 31.974$ min. There is a probability .10 that the fifth customer will take more than 31.974 min. to complete service.

EXAMPLE 8.2

The number of people absent from a factory due to sickness is assumed to be a Poisson random variable with parameter $\lambda = 5$. If the number of absentees is assumed independent from day to day, what is the probability that the number of employee-days lost to sickness exceeds 30 in a five-day week?

If N_i is the number of absences on the ith day, the total number of absences is $N = N_1 + N_2 + N_3 + N_4 + N_5$. It follows that N is a Poisson random variable with parameter $\lambda = 25$. From Appendix C

$$\sum_{j=31}^{\infty} \frac{(25)^j}{j!} e^{-25} = .1367$$

which is the probability that more than 30 employee-days are lost in a week.

EXAMPLE 8.3

The weight of drill ore samples is assumed to be a normal random variable with variance 156 g^2 and unknown mean. Ten samples are taken. What is the probability that the actual mean is greater than the sample mean by at least 10 g?

It is known that $(\bar{X} - \mu)/(\sigma/\sqrt{n})$ is a standard normal random variable, where $\bar{X} = (1/n) \sum_1^n X_i$. To calculate $P(\mu > \bar{X} + 10)$, note that

$$P(\mu > \bar{X} + 10) = P(\bar{X} - \mu < -10)$$

$$= P((\bar{X} - \mu)/(\sigma/\sqrt{n}) < -10/(\sigma/\sqrt{n}))$$

$$= \Phi\left(\frac{-10}{\sigma/\sqrt{10}}\right)$$

$$= \Phi\left(\frac{-10}{\sqrt{15.6}}\right) = \Phi(-2.53) = .0057$$

240 *Applied Probability*

EXAMPLE 8.4

It is desired to test the sample variance of Example 8.3 to see how close it is to the (known) actual variance. In particular, find the constant a such that $P(\sigma^2(1-a)\leq S^2\leq\sigma^2(1+a))=.95$.

It was shown that the random variable $Z=(\sum_1^n(X_i-\bar{X})^2)/\sigma^2$ is a chi-square random variable with $(n-1)$ degrees of freedom and notice that $(\sigma^2/(n-1))Z=S^2$, the sample variance. To determine a such that

$$P(\sigma^2(1-a)\leq S^2\leq\sigma^2(1+a))$$

$$=.95$$

$$=P((n-1)(1-a)\leq Z\leq(n-1)(1+a))$$

we use the table of Appendix E. Unfortunately, these tables do not permit an exact solution but notice that, when $n=10$ (nine degrees of freedom)

$$P(Z\leq16.9190)=.950$$

and, if we equate $9(1+a)=16.9190$, then $a=.879$. The quantity $9(1-a)=1.089$ and, using the table again,

$$P(Z\leq1.089)\leq.005$$

With these equations we can state that

$$P(1.089\leq Z\leq16.9190)\cong.95$$

and, with probability .95, the range $(\sigma^2\pm.879\sigma^2)$ includes the sample variance. This method of solution is approximate but quite accurate for this case since the upper limit was the controlling factor.

The techniques of these last two examples are illustrative of the types to be used in the next two chapters.

To conclude the section we consider the concept of a random sum of random variables. Such sums appear in population studies, atomic reactions, and so forth. The following example begins consideration of them and their study is continued in the problems at the end of the chapter.

EXAMPLE 8.5

In an atomic reaction, N particles are fired at a surface. Each particle releases a random number X_i of particles. If the random variables $X_1, X_2, \ldots, X_N$ are independent, each having a mean μ and variance σ^2 and if N is a random variable with mean m and variance s^2, find the mean and variance of the random sum $X=X_1+X_2+\cdots+X_N$.

Given that $N=n$, conditional means and variances can be calculated as

$$E(X|N=n)=E(X_1+X_2+\cdots+X_N|N=n)=n\mu$$

and

$$E(X^2|N=n)=n(\sigma^2+\mu^2)+(n^2-n)\mu^2$$

In Chapter 6 it was shown that $E(E(X|N))=E(X)$ and, similarly, $E(E(X^2|N))=E(X^2)$. It follows immediately that

$$E(E(X|N))=E(N\mu)=m\mu$$

and

$$
\begin{aligned}
V(X) &= E(X^2)-E(X)^2 \\
&= E(N(\sigma^2+\mu^2)+(N^2-N)\mu^2)-m^2\mu^2 \\
&= m(\sigma^2+\mu^2)+(s^2+m^2)\mu^2-m\mu^2-m^2\mu^2 \\
&= m\sigma^2+\mu^2 s^2
\end{aligned}
$$

8.2 The Law of Large Numbers and the Central Limit Theorem

The subject of the approximation of distributions has, for practical reasons, received much attention. Two fundamental results in this direction, the law of large numbers and the central limit theorem, are discussed in this section. In the literature these results are proved under a variety of conditions and assumptions. We only consider simple versions of them.

Before presenting these two results we prove a useful inequality (The Chebyshev inequality) of probability theory:

Chebyshev Inequality   If X is a random variable with mean μ and variance σ^2, then for any positive number k

$$P(|X-\mu|\ge k\sigma)\le 1/k^2 \tag{8.8}$$

Proof   By definition

$$\sigma^2 = E((X-\mu)^2)$$

$$= \int_{-\infty}^{\infty}(x-\mu)^2 f_X(x)\,dx$$

$$\ge \int_{|x-\mu|\ge k\sigma}(x-\mu)^2 f_X(x)\,dx$$

that is, rather than integrate over the whole real line, we only integrate over those values of x for which $|x-\mu|\ge k\sigma$—hence the inequality since we have not increased and may have decreased the value of the integral by doing this. In this new range of integration, we further decrease the value of the integral by replacing $(x - \mu)^2$ by $k^2\sigma^2$, a lesser quantity in this range by definition:

$$\sigma^2 \geq \int_{|x-\mu| \geq k\sigma} k^2\sigma^2 f_X(x)\, dx$$

$$= k^2\sigma^2 \int_{|x-\mu| \geq k\sigma} f_X(x)\, dx$$

$$= k^2\sigma^2 P(|X - \mu| \geq k\sigma)$$

and finally

$$P(|X - \mu| \geq k\sigma) \leq 1/k^2$$

as required.

The inequality was only proved for the case where X is a continuous random variable, but it is equally valid for discrete random variables.

EXAMPLE 8.6

The rainfall X in a certain locality is a normally distributed random variable with mean 40 cm and variance 4 cm². Find a simple upper bound on the probability that the rainfall in a particular year will be more than 5 cm from the mean.

From the inequality of Eq. (8.8),

$$P(|X - \mu| \geq k\sigma) = P(|X - 40| \geq 2k) \leq 1/k^2$$

Choosing $k = 5/2$ gives the equation

$$P(|X - 40| \geq 5) \leq \frac{1}{(5/2)^2} = \frac{4}{25} = .16$$

The actual value for this probability can be calculated as

$$P(|X - 40| \geq 5) = 2P(X < 35) = 2\Phi((35 - 40)/2)$$

$$= 2\Phi(-5/2)$$

which, from the table in Appendix D, is $2 \times .0062 = .0124$. The agreement in this case is not very good but in many practical situations even such a loose upper bound can be surprisingly useful.

Certain variations of the Chebyshev inequality follow immediately:

(i) $P(|X - \mu| \geq \varepsilon) \leq \sigma^2/\varepsilon^2$
(ii) $P(|X - \mu| < k\sigma) \geq 1 - 1/k^2$
(iii) $P(|X - \mu| < \varepsilon) \geq 1 - \sigma^2/\varepsilon^2$

Also, since

$$P(X-\mu \geq \varepsilon) \leq P(|X-\mu| \geq \varepsilon) \leq \sigma^2/\varepsilon^2 \tag{8.9}$$

we can use this equation as a one-sided version of Chebyshev's inequality.

EXAMPLE 8.7

If X is the number of cars passing a point on a street in some fixed period of time, and is modeled as a Poisson random variable with parameter 4, find a simple upper bound for the probability that more than 10 cars passed this point in this time interval.

Using the one-sided Chebyshev inequality of Eq. (8.9) gives

$$P(X-4 \geq \varepsilon) \leq 4/\varepsilon^2$$

and choosing $\varepsilon = 6$ this can be written as

$$P(X \geq 10) \leq 4/36 \cong .111$$

From tables, the actual value of this probability is about .008 and again the bound is very loose. This is not very surprising since it would not be reasonable to expect a single, simple bound to be strong for all types of random variables and all possible parameters of them.

We will use the Chebyshev inequality in proving a law of large numbers. Let $X_1, \ldots, X_n$ be a set of random variables that we assume are independent, each having the same pdf $f(x)$ (we will deal with the continuous case, the argument being the same for the discrete case). The sample mean was defined in Section 8.1 as

$$\bar{X} = S_n = \frac{1}{n} \sum_{i=1}^{n} X_i$$

and a sample variance can be defined in an analogous manner. Since we assumed the random variables are independent,

$$E(S_n) = \mu \qquad V(S_n) = \sigma^2/n$$

although it was not necessary to assume the X_i independent for this result.

The Weak Law of Large Numbers If $X_1, X_2, \ldots, X_n$ is a sequence of independent random variables with identical pdf's and $E(X_i) = \mu$, $V(X_i) = \sigma^2 < \infty$, then

$$P(|S_n - \mu| \geq \varepsilon) \leq \sigma^2/n\varepsilon^2, \quad \varepsilon > 0 \tag{8.10}$$

Proof Apply Chebyshev's inequality to the random variable S_n and the result follows immediately.

It is not really necessary in this law to assume that the random variables X_i have the same distribution since it can be modified easily to take care of the case for different distributions. There is also a strong law of large numbers that gives more information on precisely how S_n converges to μ, but it will not be discussed here. The following example indicates one use for the weak law.

EXAMPLE 8.8

How many samples of a random variable should be taken if we want to have a probability of at least .95 that the sample mean will not deviate by more than $\sigma/10$ from the true mean?

From Eq. (8.10)

$$P\left(|S_n - \mu| \geq \frac{\sigma}{10}\right) = 1 - P\left(|S_n - \mu| < \frac{\sigma}{10}\right) \leq \frac{\sigma^2}{n \cdot (\sigma^2/100)}$$

or

$$P\left(|S_n - \mu| < \frac{\sigma}{10}\right) \geq 1 - 100/n$$

Since we want this probability to be at least .95, n must be at least $n \geq 100/.05 = 2000$.

The Chebyshev bound, and the weak law of large numbers, usually lead to very weak bounds. It is not surprising since the bound involves only the mean and variance of the pdf and not the pdf itself. To illustrate this point, suppose in Example 8.8 we additionally assumed that the samples of the random variable are independent and each have the normal pdf $N(\mu, \sigma^2)$. In this case S_n has the pdf $N(\mu, \sigma^2/n)$ and we can write

$$P\left(|S_n - \mu| < \frac{\sigma}{10}\right) = P\left(-\frac{\sqrt{n}}{10} < \frac{S_n - \mu}{\sigma/\sqrt{n}} < \frac{\sqrt{n}}{10}\right)$$

$$= \Phi\left(\frac{\sqrt{n}}{10}\right) - \Phi\left(-\frac{\sqrt{n}}{10}\right)$$

For this quantity to be .95, $\sqrt{n}/10$ must have the value 1.96 or n must be 385, a substantial saving from the 2000 samples required in the Example 8.8. In any case, where the form of the pdf involved is not known, however, there is no alternative in such problems but to use the weak law of large numbers.

A weak law of large numbers for discrete random variables can also be formulated and this is done in the next example for the particular case of Bernoulli trials.

EXAMPLE 8.9

A page of a large book is chosen at random and it is assumed that there is a probability θ that it contains at least one typographical error. If n pages are chosen and S_n is the fraction of these containing errors to n, how large should n be if it is desired that this ratio differs from the actual θ (assumed known) by less than .01 with probability at least .95?

Let us do the general case of repeating a Bernoulli trial n times where the outcome on the ith trial X_i is a 1 with probability θ and 0 with probability $1-\theta$. The random variable $Y = X_1 + \cdots + X_n$ is binomially distributed with parameters n and θ. We also know that $E(X_i) = \theta$ and $V(X_i) = \theta(1-\theta)$. Applying the weak law of large numbers to the random variable $S_n = (1/n)(X_1 + \cdots + X_n)$ yields

$$P(|S_n - \theta| \geq \varepsilon) \leq \frac{\sigma^2}{n\varepsilon^2} = \frac{\theta(1-\theta)}{n\varepsilon^2}$$

or equivalently

$$P(|S_n - \theta| < \varepsilon) \geq 1 - \frac{\theta(1-\theta)}{n\varepsilon^2}$$

For the case at hand, the probability of obtaining a page with at least one error on any given trial is θ, and $\varepsilon = .01$ and the probability is to exceed 0.95, that is

$$1 - \frac{\theta(1-\theta)}{n\varepsilon^2} \geq 0.95$$

or

$$n \geq \frac{\theta(1-\theta)}{(0.05)\varepsilon^2} = 2\theta(1-\theta) \times 10^5$$

Another very important result of probability theory is the Central Limit Theorem, which we state here without proof. Again only a very restricted version of the theorem is presented. It has been proved in the literature under a variety of conditions and assumptions.

THE CENTRAL LIMIT THEOREM

Let $X_1, X_2, \ldots, X_n$ be n independent random variables, each with the same cdf $F(x)$. Suppose that the mean $E(X_i) = \mu$ and variance $V(X_i) = \sigma^2$ are both finite. Then for any real x

$$\lim_{n \to \infty} P\left[\frac{1}{\sqrt{n}} \sum_{i=1}^{n} \left(\frac{X_i - \mu}{\sigma} \right) \leq x \right] = \Phi(x) \tag{8.11}$$

where $\Phi(x)$ is the cdf of a standard normal random variable.

In words, the theorem says that the sum of a properly standardized sequence of random variables has a cdf which converges toward that of a standard normal random variable. The theorem has many generalizations. For example, it is not necessary that the random variables X_i have the same cdf. However, the above form is satisfactory for our purpose. Notice how much stronger this theorem is than the law of large numbers which simply states that under certain conditions the sample mean converges to the actual mean. The

Central Limit Theorem actually gives a cdf toward which the cdf of a standardized sum converges.

In practice, whenever an observed random variable is known to be a sum of a large number of "well-behaved" random variables, then we have some justification for assuming that their sum is normally distributed. For example, atmospheric noise in space is made up of emissions from various radio sources in the universe. Their emissions are independent and their net effect at any antenna in space is to produce normally distributed noise.

It is possible to obtain an estimate of the error involved with using Eq. (8.11) when n is some finite number. However, such calculations will not be of interest to us.

EXAMPLE 8.10

At a bank, the interest is calculated on 100 accounts and rounded up or down to the nearest cent. If it is assumed that the round-off error is uniformly distributed between $(-\frac{1}{2}, \frac{1}{2})$ and that the round-off errors are independent, find the probability that the sum of the errors does not exceed 2 cents in magnitude.

Let X_i be the round-off error on the ith account, and note that since X_i is uniformly distributed over the interval $(-\frac{1}{2}, \frac{1}{2})$, $E(X_i) = 0$, $V(X_i) = 1/12$. From the central limit theorem we have the approximation

$$P\left(\frac{1}{10\sigma} \sum_{i=1}^{100} X_i \le x\right) \cong \Phi(x)$$

and consequently

$$P\left(\sum_{i=1}^{100} X_i \le 10\sigma x\right) \cong \Phi(x) \tag{8.12}$$

We want an expression for the probability

$$P\left(-2 \le \sum_{i=1}^{100} X_i \le 2\right)$$

and by Eq. (8.12) this can be approximated by

$$P\left(-2 \le \sum_{i=1}^{100} X_i \le 2\right) \cong \Phi(x) - \Phi(-x)$$

where $10\sigma x = 2$ or $x = 2/10 \times 1/\sqrt{12} = .693$. From tables then the required probability is .512.

The following example serves mainly to indicate once again the loose bounds that one generally obtains from the law of large numbers in the form that we have stated it.

EXAMPLE 8.11

In a dice rolling game a player can either win 6 dollars or lose either 1, 2, or 3 dollars, and each of these outcomes is equally likely. How many times should the game be played if the player wishes to keep his average loss or gain per game to less than 2 dollars with probability at least 0.95?

(i) Using the law of large numbers we get a lower bound on n as follows. Let X_i be his loss or gain on the ith game and note that $E(X_i) = 0$, $V(X_i) = 50$. From Eq. (8.10)

$$P\left(\left|\frac{1}{n}\sum_{i=1}^{n} X_i\right| \ge \varepsilon\right) \le \frac{\sigma^2}{n\varepsilon^2}$$

or equivalently

$$P\left(\left|\frac{1}{n}\sum_{i=1}^{n} X_i\right| < \varepsilon\right) \ge 1 - \frac{2}{n\varepsilon^2}$$

As we require $\varepsilon = 2$ and the right-hand side of this equation is to be at least .95, we have

$$1 - \frac{50}{n \cdot 4} = .95$$

or n is to be at least $50/.05 = 100$.

(ii) Using the central limit theorem, Eq. (8.11) gives

$$P\left(\frac{1}{\sqrt{n}}\sum_{i=1}^{n} \frac{X_i}{\sigma} \le x\right) \cong \Phi(x) \tag{8.13}$$

or, by some transposition,

$$P\left(\frac{1}{n}\sum_{i=1}^{n} X_i \le \frac{x\sigma}{\sqrt{n}}\right) = \Phi(x)$$

From the problem statement we require n such that

$$P\left(-2 \le \frac{1}{n}\sum_{i=1}^{n} X_i \le 2\right) = .95 = \Phi(x) - \Phi(-x)$$

which, from Appendix D, implies that $x = 1.96$. Consequently, as $x\sigma/\sqrt{n}$ is equal to 2, we have $\sqrt{n} = 1.96\sqrt{50}/2$ or $n = 48$. It is seen from this development that the bound $n = 48$ is significantly tighter than the bound $n = 100$ obtained by using the law of large numbers, as expected.

8.3 The Approximation of Distributions

The Central Limit Theorem states that a sum of independent random variables is approximately normally distributed. This fact plays an important role in approximating certain distributions. In this section a few of the more important approximations are considered, concentrating mainly on the relationship among the binomial, Poisson, and normal distributions. The use and availability of computers gives a more direct way of obtaining values for certain probabilities. Nonetheless, the approximation techniques developed here are still useful and interesting.

The binomial distribution has arisen in many problems of the text and although we were satisfied with an expression rather than a number, the ability to calculate such probabilities as

$$P(Y = k) = \binom{n}{k} \theta^k (1 - \theta)^{n-k}$$

is important. For large values of n and k, however, the calculations become awkward and we must invariably resort to approximation techniques. To use the results of the previous section, we use the fact that a binomial random variable Y can be written as

$$Y = X_1 + X_2 + \cdots + X_n$$

where X_i is a Bernoulli random variable taking on the value 1 with probability θ and 0 with probability $(1 - \theta)$. Since the X_i are also independent, we can apply the Central Limit Theorem to the random variable

$$S_n = \frac{1}{\sqrt{n}} \sum_{i=1}^{n} \left(\frac{X_i - E(X_i)}{\sqrt{V(X_i)}} \right) = \frac{1}{\sqrt{n}} \sum_{i=1}^{n} \frac{(X_i - \theta)}{\sqrt{\theta(1 - \theta)}} \tag{8.14}$$

For large n, S_n is normally distributed and

$$P(S_n \leq x) \cong \Phi(x) = \frac{1}{\sqrt{2\pi}} \int_{-\infty}^{x} e^{-t^2/2} \, dt \tag{8.15}$$

Substituting Eq. (8.14) in (8.15) gives

$$P\left[\frac{1}{\sqrt{n\theta(1 - \theta)}} \left(\sum_{i=1}^{n} (X_i - \theta) \right) \leq x \right] = P(Y \leq x\sqrt{n\theta(1 - \theta)} + n\theta) \cong \Phi(x) \tag{8.16}$$

or, perhaps more conveniently,

$$P(Y \leq \beta) \cong \Phi\left(\frac{\beta - n\theta}{\sqrt{n\theta(1 - \theta)}} \right) \tag{8.17}$$

Due to the fact that we are approximating a discrete density by a continuous one, a slightly better approximation is given by

$$P(Y \leq \beta) \cong \Phi\left(\frac{\beta + \frac{1}{2} - n\theta}{\sqrt{n\theta(1 - \theta)}} \right) \tag{8.18}$$

and such an adjustment is referred to as a continuity correction. These

Sums of Random Variables 249

expressions would be more useful if they were accompanied by some estimate or upper bound of the error involved with their use. Such estimates are readily available in the literature but take us too far from our purpose and we omit them. From Eq. (8.17) we can write

$$P(k \leq Y \leq l) = \sum_{j=k}^{l} \binom{n}{j} \theta^j (1-\theta)^{n-j}$$

$$= \Phi\left(\frac{l-n\theta}{\sqrt{n\theta(1-\theta)}}\right) - \Phi\left(\frac{k-n\theta}{\sqrt{n\theta(1-\theta)}}\right) \qquad (8.19)$$

or adjusting this equation with continuity corrections as with Eq. (8.18),

$$P(k \leq Y \leq l) \cong \Phi\left(\frac{l+\frac{1}{2}-n\theta}{\sqrt{n\theta(1-\theta)}}\right) - \Phi\left(\frac{k-\frac{1}{2}-n\theta}{\sqrt{n\theta(1-\theta)}}\right) \qquad (8.20)$$

For the individual binomial probabilities it can be shown that

$$P(Y = k) = \binom{n}{k} \theta^k (1-\theta)^{n-k}$$

$$\cong \frac{1}{\sqrt{2\pi n\theta(1-\theta)}} \exp\left(-\frac{(k-n\theta)^2}{2n\theta\eta}\right), \quad \eta = 1-\theta, \qquad (8.21)$$

or, using Eq. (8.20), we can obtain another approximation,

$$P(Y = k) \cong \Phi\left(\frac{k+\frac{1}{2}-n\theta}{\sqrt{n\theta(1-\theta)}}\right) - \Phi\left(\frac{k-\frac{1}{2}-n\theta}{\sqrt{n\theta(1-\theta)}}\right) \qquad (8.22)$$

The significant feature of these approximations is that we can calculate binomial probabilities for large values of n from a table of the standardized normal cdf as given in Appendix D. The values of n and θ for which these approximations are valid depend, of course, on the accuracy required. Generally they are better for θ close to $\frac{1}{2}$ and this is reasonable since the normal density is always symmetric about its mean while the binomial is only symmetric for $\theta = \frac{1}{2}$. It is reported in Parzen (1960) that Eq. (8.20) achieves at least two decimal place accuracy for $n\theta > 37$ and further suggests that reasonable accuracy is achieved with these equations, for the purpose of textbook problem solving, if $n\theta(1-\theta) > 10$. For θ not close to $\frac{1}{2}$, the approximation gets better as n increases. The following examples illustrate the uses of these approximations. Some comparisons using Eqs. (8.21) and (8.22) are given in Table 8.1 along with another approximation given later.

EXAMPLE 8.12

What is the probability of obtaining more than 520 heads in 1000 tosses of a fair coin?

From Eq. (8.17) with $n = 1000$ and $\theta = \frac{1}{2}$, and letting Y be the number of heads obtained in 1000 tosses of the coin,

$$P(Y \le 520) \cong \Phi\left(\frac{520 - 500}{\sqrt{1000(\frac{1}{2})(1 - \frac{1}{2})}}\right)$$

$$= \Phi\left(\frac{40}{31.62}\right) = \Phi(1.26) = 0.8962$$

and, consequently,

$$P(Y > 520) = 1 - 0.8962 = 0.1038$$

Using Eq. (8.18) involving the continuity correction gives

$$P(Y < 520) \cong \Phi\left(\frac{520 + \frac{1}{2} - 500}{\sqrt{1000(\frac{1}{2}) \cdot \frac{1}{2}}}\right)$$

$$= \Phi(1.296) = .9026$$

EXAMPLE 8.13

How many experiments should be performed so that the probability of obtaining at least 40 successes is at least .95 if the experiments are independent and the probability of a success on any trial is .2?

The probability of 40 or fewer successes is, by Eq. (8.17),

$$P(Y \le 40) \cong \Phi\left(\frac{40 - .2n}{\sqrt{n(.2)(.8)}}\right)$$

and we would like this to be $1 - .95 = .05$. From tables, this condition implies that

$$\frac{40 - (.2)n}{\sqrt{n(.2)(.8)}} = -1.645$$

Letting $x = \sqrt{n}$, this equation yields the quadratic equation

$$.2x^2 - .658x - 40 = 0$$

which has a positive solution $x \cong 15.65$ or $n \cong 245$, implying that n should be at least 245 to meet the conditions of the problem. Again, using Eq. (8.18) would yield a slightly more accurate answer.

EXAMPLE 8.14

When n is large and $n\theta$ is an integer, find the most probable value for the binomial random variable X with parameters n and θ. Estimate the probability that X takes on this value.

From Problem 3.25 a binomial random variable with parameters n and θ has its most probable values at the smallest integer greater than $n\theta - (1 - \theta)$.

If $n\theta$ is an integer, this value is just $n\theta$. From Eq. (8.21),

$$P(X = k) \cong \frac{1}{\sqrt{2\pi n\theta(1-\theta)}} \exp\left(-\frac{(k-n\theta)^2}{2n\theta(1-\theta)}\right)$$

When $k = n\theta$ this takes on the value

$$P(X = n\theta) = \frac{1}{\sqrt{2\pi n\theta(1-\theta)}}$$

In the first section of this chapter it was shown that the sum of two independent Poisson random variables with parameters λ_1 and λ_2, respectively, is a Poisson random variable with parameter $\lambda_1 + \lambda_2$. Using this fact, we can view a Poisson random variable Y with parameter λ as a sum of the independent Poisson random variables X_i, $i = 1, 2, \ldots, n$, where X_i has parameter λ/n, that is,

$$Y = X_1 + X_2 + \cdots + X_n \qquad E(X_i) = \lambda/n = V(X_i), \quad i = 1, 2, \ldots, n$$

The Central Limit Theorem then implies that the random variable

$$Z = \frac{Y - E(Y)}{\sqrt{V(Y)}} = \frac{Y - \lambda}{\sqrt{\lambda}}$$

is approximately normal in the sense that

$$P(Z \le z) = \Phi(z)$$

It follows that, by appropriate substitution,

$$P(Y \le k) = P\left(Z \le \frac{k-\lambda}{\sqrt{\lambda}}\right) \cong \Phi\left(\frac{k-\lambda}{\sqrt{\lambda}}\right)$$

Using a continuity correction again, a more accurate approximation would be

$$P(Y \le k) \cong \Phi\left(\frac{k + \frac{1}{2} - \lambda}{\sqrt{\lambda}}\right) \tag{8.23}$$

and this approximation tends to improve as λ increases. It is a good approximation if λ is moderately large, say $\lambda \ge 20$.

EXAMPLE 8.15

A radioactive source can be modeled as a Poisson random variable with parameter 30 particles per hour. What is the probability that fewer than 10 particles are emitted in a 10-minute period?

If Y is the number of particles emitted during a 10-minute period, then it is a Poisson random variable with parameter $\lambda = 30/6 = 5$ particles per 10 minutes. Using Eq. (8.23) gives

$$P(Y \leq 10) \cong \Phi\left(\frac{10+\frac{1}{2}-5}{\sqrt{5}}\right)$$

$$= \Phi\left(\frac{5.5}{\sqrt{5}}\right) = \Phi(2.46) = .9931$$

The actual value of this quantity from tables is

$$P(Y \leq 10) = .9863$$

indicating that the approximation is within 1% of the true value.

EXAMPLE 8.16

The number of people entering a store is Poissonly distributed with parameter 100 people per hour. How long should you wait in order to have a probability of .90 that more than 200 people have entered the store?

If Y is the number of people entering the store in the interval $(0, t)$, then it is Poissonly distributed with parameter $100t$, where t is expressed in hours. To determine t so that $P(Y > 200) = .90$ or $P(Y \leq 200) = .10$, use Eq. (8.23) and set

$$.10 = P(Y \leq 200) \cong \Phi\left(\frac{200+\frac{1}{2}-100t}{\sqrt{100t}}\right)$$

and from tables this equation implies that

$$\frac{200+\frac{1}{2}-100t}{10\sqrt{t}} = -1.28$$

or, setting $x = \sqrt{t}$,

$$100x^2 - 12.8x - 200.5 = 0$$

which has the solution $x \cong 1.48$ or $t = 2.19$ hours.

From Eq. (8.23) an approximation for $P(Y = k)$, with continuity corrections, can be obtained as

$$P(Y = k) \cong \Phi\left(\frac{k+\frac{1}{2}-\lambda}{\sqrt{\lambda}}\right) - \Phi\left(\frac{k-\frac{1}{2}-\lambda}{\sqrt{\lambda}}\right) \tag{8.24}$$

and when λ is an integer the maximum of the approximation occurs at $k = \lambda$ (as opposed to λ and $\lambda + 1$ for the actual distribution, as found in Example 2.8. In this manner the Poisson distribution can be approximated by using the normal pdf. Table 8.1 shows the exact value and approximate values for $\lambda = 100$, using Eq. (8.24). The approximation tends to be better for larger values of λ and for values of the random variable close to λ.

TABLE 8.1

k	$P(X=k)$ EXACT[a]	$P(X=k)$ APPROXIMATION USING Eq. (8.24)[b]
70	.000218	.000445
80	.003285	.005406
90	.022535	.024197
100	.039861	.039878
110	.025765	.024197
120	.006674	.005406
130	.000758	.000445

[a] *Using tables of E. C. Molina from which Appendix C was taken.*
[b] *Using tables of Appendix D.*

So far in this section we have been able to express both binomial and Poisson random variables as the sums of independent random variables and apply the Central Limit Theorem to obtain good approximations. In certain circumstances we can also use the Poisson distribution to approximate the binomial distribution. Before tackling the mathematical proof we argue intuitively as follows. Let Y be a Poisson random variable with parameter λ. Express Y as a sum of n independent random variables X_i,

$$Y = X_1 + X_2 + \cdots + X_n$$

where X_i is a Poisson random variable with parameter λ/n. Choose n in such a way that $P(X_i > 1)$ is negligible, which implies λ/n must be small. In this case the random variable X_i is essentially a Bernoulli random variable with

$$P(X_i = 0) = e^{-\lambda/n} \cong 1 - \frac{\lambda}{n} \quad \text{and} \quad P(X_i) = \frac{\lambda}{n}(e^{-\lambda/n}) \cong \frac{\lambda}{n}$$

With this interpretation Y can be viewed as a binomial random variable with parameters n and λ/n:

$$P(Y = k) = \binom{n}{k}\left(\frac{\lambda}{n}\right)^k \left(1 - \frac{\lambda}{n}\right)^{n-k} \tag{8.25}$$

But Y is a Poisson random variable with

$$P(Y = k) = e^{-\lambda}\frac{\lambda^k}{k!} \tag{8.26}$$

and we have the approximation

$$e^{-\lambda}\frac{\lambda^k}{k!} \cong \binom{n}{k}\left(\frac{\lambda}{n}\right)^k \left(1 - \frac{\lambda}{n}\right)^{n-k}$$

which will be good when λ/n is small, n large, and λ "moderate." To approximate a given binomial distribution

$$P(Y=k)=\binom{n}{k}\theta^k(1-\theta)^{n-k}$$

for n large and θ small ($\theta < .10$), and $n\theta$ "moderate" we set $\lambda = n\theta$ and use the equation

$$P(Y=k)\cong e^{-\lambda}\frac{\lambda^k}{k!}$$

as long as $\lambda = n\theta$ is a moderate value.

We can achieve the same result by mathematical manipulation and it is instructive to do so. For each n, $n \geq 1$, let X_n be a binomial random variable with parameters n and θ_n. We now make an assumption on the behavior of n and θ_n, namely, that $\lim_{n\to\infty} n\theta_n = \lambda_n \to \lambda = $ a constant. Since X_n for some fixed n has a binomial distribution, we have

$$P(X_n=k)=\binom{n}{k}\theta_n^k(1-\theta_n)^{n-k}$$

$$=\frac{n!}{k!(n-k)!}\frac{n^k\theta_n^k}{n^k}(1-\theta_n)^n(1-\theta_n)^{-k}$$

$$=\frac{1}{k!}\frac{n(n-1)\cdots(n-k+1)}{nn\cdots n}\lambda_n^k\left(1-\frac{\lambda_n}{n}\right)^n(1-\theta_n)^{-k}$$

$$=\frac{1}{k!}(1)\cdot\left(1-\frac{1}{n}\right)\cdot\left(1-\frac{2}{n}\right)$$

$$\cdots\left(1-\frac{(k-1)}{n}\right)\cdot\lambda_n^k\cdot\left(1-\frac{\lambda_n}{n}\right)^n(1-\theta_n)^{-k}$$

where each factor has been fixed in a form convenient for further manipulations. Let us fix k and consider the first k brackets as $n \to \infty$. Any term of the form $(1-i/n)$, $0 \leq i \leq (k-1)$ will tend to unity. Since, by assumption, $\lim_{n\to\infty} \lambda_n = \lambda$, then $\lambda_n^k \to \lambda^k$. The fact that

$$\lim_{n\to\infty}\left(1-\frac{\lambda}{n}\right)^n=e^{-\lambda}$$

is usually encountered in a first calculus course. We can thus use the approximation

$$\lim_{n\to\infty}\left(1-\frac{\lambda_n}{n}\right)^n=e^{-\lambda}$$

Finally, since θ_n tends to zero as $n \to \infty$, each term $(1-\theta_n)^{-k}$ tends to unity as n becomes large. Putting these facts together in the above equation yields the approximation

$$\lim_{n\to\infty}\binom{n}{k}\theta_n^k(1-\theta_n)^{n-k}=\frac{\lambda^k}{k!}e^{-\lambda} \tag{8.27}$$

As a result, whenever n is large, θ small, and $n\theta$ a moderate value, this approximation can be used with some confidence.

Some examples of the use of Eqs. (8.21) and (8.22) are shown in Table 8.2. The approximations are not always very good, of course, and some experience is necessary before they can be used with confidence. The probability $P(Y=k)$, where Y is a binomial random variable with parameters $n = 1000$ and $\theta = .1$, is shown in Table 8.2 along with approximations.

TABLE 8.2

k	EXACT VALUE[a]	USING Eq. (8.21)[b]	USING Eq. (8.22)[c]
90	.02484	.03590	.02413
100	.04202	.04204	.04203
110	.02348	.03590	.02413

[a] From Tables.
[b] From Appendix D.
[c] From Appendix D, using interpolation.

EXAMPLE 8.17

If .1% of garments manufactured by a clothing company are flawed, what is the probability that 20 garments among 10,000 will be flawed?

If Y is the random variable giving the number of flawed garments, then clearly it is a binomial random variable with distribution

$$P(Y=k)=\binom{10,000}{k}(.001)^k(1-.001)^{10,000-k}$$

and this is a difficult expression to evaluate exactly. In the approximation of Eq. (8.27), we choose $\lambda = n\theta = 10,000(.001) = 10$, which appears to satisfy the definition of "moderate," and so

$$P(Y=k)=\binom{10,000}{k}(.001)^k(1-.001)^{10,000-k}$$

$$\cong e^{-10}\frac{10^k}{k!}$$

and

$$P(Y=20)\cong e^{-10}\frac{10^{20}}{20!}$$

$$\cong 1.8659\times 10^{-3}$$

EXAMPLE 8.18

In an outbreak of flu it is assumed that 10% of the people in a town will catch it. If 1000 people are chosen at random, what is the most probable number of people that will have had the flu and what is its probability?

The number of people who have had flu, X, is binomially distributed with parameters $n = 1000$ and $\theta = .1$. From Example 8.14 the most probable value of X is $n\theta = 100$. An approximation of the probability that X attains this value is also given in that example as

$$P(X = 100) \cong \frac{1}{\sqrt{2\pi n\theta(1-\theta)}} = \frac{1}{\sqrt{2\pi \times 90}} = .0419$$

We can also use Eq. (8.27) to approximate this quantity, with $\lambda = n\theta = 100$, as

$$P(X = 100) \cong e^{-100} \frac{100^{100}}{100!} = .0399$$

showing reasonably good agreement. The continuity correction of Eq. (8.22) yields the approximation

$$P(X = 100) \cong .0418$$

From Chapter 2 the probability distribution obtained when drawing from a dichotomous population without replacement yields a hypergeometric distribution. For example, drawing k items from a lot containing n items, r of which are of Type I and $(n - r)$ of Type II, drawing without replacement, the probability of drawing j items of Type I was calculated as

$$P(X = j) = \frac{\binom{r}{j}\binom{n-r}{k-j}}{\binom{n}{k}} \tag{8.28}$$

If the number of items n is large and r/n is a moderate value (between 0 and 1, of course), then there is not much difference between drawing with or without replacement; that is, each time an item is drawn, there is approximately the same ratio of the number of Type I items to the total number as before the drawing. In this case, the situation is approximately binomial and we can write

$$P(X = j) \cong \binom{k}{j}\left(\frac{r}{n}\right)^j\left(1-\frac{r}{n}\right)^{n-j} \tag{8.29}$$

and this approximation is generally valid for n large and k small (compared with n).

EXAMPLE 8.19

From a lot of 50 items of which 10 are defective, four are drawn at random. What is the probability that exactly two of them will be defective?

From the approximation of Eq. (8.29),

$$P(X=2)=\frac{\binom{10}{2}\binom{40}{2}}{\binom{50}{4}}\cong\binom{4}{2}\left(\frac{1}{5}\right)^2\left(\frac{4}{5}\right)^2=.154$$

The actual probability is .152 and the agreement in this case is good.

In the case where the binomial distribution can be approximated by either the Poisson or the normal distribution, then we can use these to approximate the hypergeometric distribution. For reference, if X is a hypergeometric random variable given by Eq. (8.28), then the Poisson approximation is

$$P(X=j)=e^{-\lambda}\frac{\lambda^j}{j!},\quad j=1,2,\dots \tag{8.30}$$

where $\lambda = kr/n$ and this approximation will be valid when r/n is small and k large. If k is large while r/n is not small, then we can use the normal approximation given by

$$P(X\le j)\cong\Phi(s) \tag{8.31}$$

where

$$s=\frac{\left(j-\frac{kr}{n}+\frac{1}{2}\right)}{\sqrt{\frac{kr}{n}\left(1-\frac{r}{n}\right)\left(\frac{n-k}{n-1}\right)}}$$

the expression in the denominator being the standard deviation of the hypergeometric distribution.

EXAMPLE 8.20

A bank dispenses 10,000 one dollar bills and is told at the end of the day that 100 of them were counterfeit. If 100 of the bills are recovered, what is the probability that there will be at least two of the counterfeit bills among them?

Let X be the number of counterfeit bills recovered. Since X has a hypergeometric distribution with $n = 10,000$, $r = 100$, $k = 100$, and the probability

$$P(X\ge2)=\sum_{j=2}^{100}\frac{\binom{100}{j}\binom{9900}{100-j}}{\binom{10,000}{100}}\cong1-\Phi(s)$$

where

$$s = \frac{(1-1+\frac{1}{2})}{\sqrt{1\left(1-\dfrac{100}{10,000}\right)\left(\dfrac{10,000-100}{9999}\right)}} \cong .505$$

and

$$P(X \ge 2) = 1 - \Phi(.505) \cong .307$$

Many other approximations for these and other distributions are available. For example, if X is the number of Bernoulli trials to obtain the kth success, then

$$P(X = j) = \binom{j-1}{k-1} \theta^k \eta^{j-k}, \quad j = k, k+1, \ldots, \infty, \ \theta + \eta = 1$$

TABLE 8.3

APPROXIMATING	BY	COMMENTS	EQUATION NUMBER
Y, a binomial random variable with parameters n, θ		Good approximation when $n\theta(1-\theta) > 10$. Accuracy improves as n increases.	
$P(k \le Y \le l)$	$\Phi\left(\dfrac{l-n\theta}{\sqrt{n\theta(1-\theta)}}\right) - \Phi\left(\dfrac{k-n\theta}{\sqrt{n\theta(1-\theta)}}\right)$		(8.19)
	$\Phi\left(\dfrac{l+\frac{1}{2}-n\theta}{\sqrt{n\theta(1-\theta)}}\right) - \Phi\left(\dfrac{k-\frac{1}{2}-n\theta}{\sqrt{n\theta(1-\theta)}}\right)$		(8.20)
$P(Y = k)$	$\dfrac{1}{\sqrt{2\pi n\theta(1-\theta)}} \exp\left(-\dfrac{(k-n\theta)^2}{2n\theta(1-\theta)}\right)$		(8.21)
	$\Phi\left(\dfrac{k+\frac{1}{2}-n\theta}{\sqrt{n\theta(1-\theta)}}\right) - \Phi\left(\dfrac{k-\frac{1}{2}-n\theta}{\sqrt{n\theta(1-\theta)}}\right)$		(8.22)
Y, a Poisson random variable with parameter λ		Good approximation for large λ	
$P(Y \le k)$	$\Phi\left(\dfrac{k+\frac{1}{2}-\lambda}{\sqrt{\lambda}}\right)$		(8.23)
$P(Y = k)$	$\Phi\left(\dfrac{k+\frac{1}{2}-\lambda}{\sqrt{\lambda}}\right) - \Phi\left(\dfrac{k-\frac{1}{2}-\lambda}{\sqrt{\lambda}}\right)$		(8.24)
Y, a binomial random variable with parameters n and p		Good approximation when $\theta < .1$ and $n\theta$ moderate (≤ 10)	
$P(Y = k)$	$e^{-\lambda}\dfrac{\lambda^k}{k!}, \quad \lambda = n\theta$		(8.27)

Now let Y be a binomial random variable with parameters n and θ. It is clear then that the probability that Y is greater than or equal to k is just the same as the probability that X is less than or equal to n, that is, $P(Y \geq k) = P(X \leq n)$. Similarly, the probability that it takes more than n trials to achieve k successes is just the same as the probability that the number of successes in n trials is fewer than k, that is, $P(X > n) = P(Y < k)$.

An interesting, deeper treatment of the approximation of distributions is given in Johnson and Kotz (1969).

The contents of this section are summarized in Table 8.3 where the approximations involving the binomial, Poisson, and normal distributions are listed. Comments on when the approximations are valid are also given although as noted in the text these conditions are necessarily imprecise.

Problems

1. Determine the mean and variance of a chi-square distribution with n degrees of freedom.

2. Determine the mean and variance of the t-distribution with n degrees of freedom.

3. Using the tables of Appendices D and F, compare the cumulative distribution function of a standard normal random variable with that of a t-distribution with 10 degrees of freedom.

4. Let $X_1, X_2, \ldots, X_n$ be independent standard normal random variables. The random variable $Y = \sum_{i=1}^{n} X_i^2$ is a chi-square random variable with n degrees of freedom. Compare the cdf of Y for $n = 8$ with that of the appropriate normal pdf.

5. Let $X_1, X_2, \ldots, X_n$ be a random sample of the random variable X with pdf $f(x)$. If $\mu_i = E((X - E(X))^i)$, $i = 1, 2, 3, 4$ are the first four central moments, find the mean and variance of the sample variance

$$S^2 = \frac{1}{(n-1)} \sum_{i=1}^{n} (X_i - \bar{X})^2$$

6. If $X_1, X_2, \ldots, X_n$ are independent normal random variables each with mean μ and variance σ^2, then

$$Y = \frac{1}{\sigma^2} \sum_{1}^{n} (X_i - \mu)^2$$

is a chi-square random variable with n degrees of freedom. How large can n be in order for the probability that Y exceeds 20 to be less than .1?

7. If, in Problem 6, μ was unknown and the sample mean was used in its place, how many samples would be needed to satisfy the conditions?

8. Let $X_1, X_2, \ldots, X_n$ be a random sample of a normal random variable X whose mean and variance are both unknown. Find the number a, in terms of the sample mean and sample variance, such that the probability that the mean is greater than a is less than .1.

9. Let X be a chi-square random variable with n degrees of freedom. Using the table of Appendix E, sketch the curve of p versus n for values of p and n satisfying $P(X > 5) = p$.

10. The weekly consumption of water in a town is assumed to be a random variable having a gamma density function with parameters $\lambda = 4$ and $n = 7$, where

consumption is measured in millions of liters per week. What is the probability that consumption in a particular week will exceed 1.5 million liters?

11. The number of particles from a radioactive source is a Poisson random variable with parameter $\lambda = .01$ per minute. If there are 30 such sources, estimate the probability that the total number of particles observed in a 1 minute period exceeds 2 by (i) using the Central Limit Theorem; (ii) from Appendix C.

12. If X is a uniformly distributed random variable on the interval $(-\frac{1}{2}, \frac{1}{2})$, compare $P(|X| \geq k\sigma)$ as a function of k obtained from the Chebyshev inequality with the exact value.

13. Let $X_1, X_2, \ldots, X_{20}$ be 20 identically distributed random variables, each with an exponential pdf with parameter $\lambda = 10$. Compare the probability that the sample mean differs from the true mean by less than .1 obtained from the law of large numbers with the true value.

14. Repeat Problem 13 using the Central Limit Theorem instead of the law of large numbers.

15. Obtain an approximation of the gamma distribution by the normal distribution and consider the conditions under which the approximation will be good.

The following nine problems deal with random sums of random variables first introduced in Example 8.5.

16. The generating function (gf) of a discrete probability distribution $p_k = P(X = x_k)$, $k = 0, 1, 2, \ldots$ is defined by the series

$$\phi(s) = \sum_{k=0}^{\infty} p_k s^k$$

Although it is quite a different function from the moment generating function of the distribution, show that $E(X) = \phi'(1)$.

17. Let $X_1, X_2, \ldots, X_N$ be independent random variables each with the same distribution and the same generating function $\phi(s)$. If N itself is an integer valued random variable with probability distribution function q_k, $k = 0, 1, 2, \ldots$ and gf $\psi(s)$, show that the gf of the random variable $Y = X_1 + X_2 + \cdots + X_N$ is $\beta(s) = \psi(\phi(s))$. (This is referred to as a compound distribution, i.e., the gf is a function of a function.)

18. Let $X_1, X_2, \ldots, X_N$ be N independent Bernoulli random variables each with parameter θ and let N be a Poisson random variable with parameter λ that is independent of X_i, $i = 1, 2, \ldots, N$. Determine the probability distribution of $Y = X_1 + X_2 + \cdots + X_N$, and find its mean.

19. In a marketing survey, each person entering a store is asked whether he has a credit card for that store. If it is assumed that each person will have a card with probability $p = .05$ and the number of people entering the store per hour is a Poisson random variable with parameter $\lambda = 100$, find the probability distribution of the number of people with cards.

20. At a given time $t = 0$, the number of bacteria present in a culture is a random variable N that has the geometric distribution

$$P(N = j) = \rho^j(1 - \rho), \quad j = 0, 1, 2, \ldots$$

If, one hour later, each bacterium is still alive (with probability θ) or has died (with probability $(1 - \theta)$), each acting independently of the others, find the probability distribution of the number of bacteria after this one hour period.

21. In Problem 20 what is the expected number of bacteria present after the one hour period? What is the probability the whole culture died?

22. A man who has just learned he has won a lottery decides to give a dollar to everybody in the room and to each member of their families who is not there. If the number of people N in the room has a Poisson distribution with parameter λ and the number of people X_i in the ith persons family not including the person in the room has the geometric distribution

$$P(X = j) = \rho^j(1-\rho), \quad j = 0, 1, 2, \ldots$$

what is the expected amount of money he will pay out?

23. The number of trees N in a pine forest has a Poisson distribution with parameter $\lambda = 10,000$. Experience has shown that next spring each tree will have either (i) died, with probability $\frac{1}{2}$, (ii) survived with probability $\frac{1}{4}$, or (iii) survived and produced a shrub with probability $\frac{1}{4}$. What is the expected number of trees that will be growing next spring?

24. In a particular type of reproduction (i.e., of diploid cells, called meiosis) we consider pairs of chromosomes in a cell where a chromosome may be either of type A or a. Each chromosome splits into two and combines with chromosomes from other cells to give cells with either AA, Aa, or aa chromosome combinations. If the fraction of cells in the first generation of type AA, Aa, and aa is, respectively, r, s, and t $(r+s+t=1)$, what is the distribution of cell types after the first mating? Show that this distribution remains constant for all succeeding generations, a result known as Hardy–Weinberg equilibrium.

25. In an outbreak of measles the probability of a child contracting the disease is .3. In a school of 300 students what is the probability, approximately, that more than 120 students will contract the disease?

26. The number of broken letters appearing on a page in a newspaper is a Poisson random variable with parameter 1. In a 60-page newspaper what is the probability that 40 or more pages will have no broken letters?

27. About 1% of the population of a country have been convicted of a crime. In a random sampling of the population what is the probability that no more than 80 people in a sample of 10,000 have been convicted?

28. The number of telephone calls handled by a local exchange in a 24-hour (workday) period is assumed to be a Poisson random variable with parameter 20,000. What is the probability that it will be required to handle more than 20,500 calls in any given such period?

29. A nail making machine produced about .05% of its output defective, either by malforming its head or point. In a bag of 10,000 such nails what is the probability that more than five of them will be defective?

30. Five percent of voters fill out ballots incorrectly. In a polling station for 5000 voters what is the probability that more than 300 ballots will be spoiled?

31. Compare the Poisson approximation to the normal approximation (Eq. (8.22)) for the binomial distribution with parameters $n = 20$, $\theta = .1$ by comparing the various values of $P(X = k)$, $k = 0, 1, 2, 3, 4, 5$.

32. Find an estimate of the probability $P(X \geq 25)$, where X is a Poisson random variable with parameter $\lambda = 20$.

CHAPTER 9

Estimation

An important application of probability lies in the area of statistical inference, which, briefly stated, may be described as drawing conclusions about certain properties of a random variable based on a random sample of observations of it. For our purposes we will divide the subject matter of statistical inference into estimation and hypothesis testing. We consider problems of estimation in this chapter and problems of hypothesis testing in the next. To help differentiate between the two subjects consider the following example. Suppose in the manufacturing process of steel reinforcing rods the quality of the product is determined by the tensile strength. In order to obtain some information on the strength, we choose samples and destructively test them and obtain values $x_1, x_2, \ldots, x_n$, which we view as realizations of the random variables $X_1, X_2, \ldots, X_n$. Assuming that these random variables all have the same pdf, what can we say about the pdf from these observations? Two parameters that might be useful, for example, are the mean and the variance, and the problem arises as to how we should use these observations to gain estimates of these parameters. Such problems are estimation problems and will be dealt with in this chapter.

As an illustration of an hypothesis testing situation suppose that we have been testing the steel rods for some time and that our experience has shown that when the process is operating correctly the mean of the rod strengths is μ_0, but occasionally the blast furnace temperature control fails and its temperature falls by a certain amount. Rods manufactured at the lower temperature have an average tensile strength of $\mu_1 < \mu_0$. Given a batch of rods, we can ask the question whether they were manufactured at the correct or at the lower temperature, that is, based on the tensile strengths of a sample from the batch we are required to make a decision, and not estimate a parameter. Such problems are treated in the next chapter.

The next section deals with point estimates, that is, based on the observations $x_1, x_2, \ldots, x_n$ we arrive at a single number that is our estimate of the parameter of the population of interest. Section 9.2 takes the different approach of finding an interval for which we can give the probability that it will contain the parameter, that is, we can state how confident we are, in terms of probability, that the parameter will lie in the interval. The last section deals with Bayesian estimation techniques that utilize, in addition to the sample information obtained, prior information about parameter values that may be in our possession.

9.1 Point Estimators and Their Properties

As in the previous chapter, let $X_1, X_2, \ldots, X_n$ be a random sample drawn from a population that we here assume has a pdf $f(x; \theta)$ which depends on an unknown parameter θ. Based on the random sample, it is desired to find a function, say $g(X_1, X_2, \ldots, X_n)$, which may be used as an estimator of θ. For example, if θ is the mean of the pdf, then we might choose

$$g(X_1, X_2, \ldots, X_n) = \frac{1}{n} \sum_{i=1}^{n} X_i$$

and it will be shown later that this is indeed a good estimator for the mean of a population. The probability that the random sample takes on the values $X_1 = x_1, X_2 = x_2, \ldots, X_n = x_n$ is

$$f(x_1; \theta)f(x_2; \theta) \cdots f(x_n; \theta)$$

(i.e., the random variables $X_1, X_2, \ldots, X_n$ are independent).

As a matter of notation we will denote a function $g(X_1, X_2, \ldots, X_n)$ used to estimate the parameter θ as

$$\hat{\theta} = g(X_1, X_2, \ldots, X_n)$$

and refer to it as an *estimator* of θ. For a particular set of observations, $X_1 = x_1, X_2 = x_2, \ldots, X_n = x_n$, the value of the estimator $g(x_1, x_2, \ldots, x_n)$ will be called an *estimate* of θ and denoted by $\hat{\theta}$. Thus an estimator is a random variable and an estimate is a particular realization of it.

There are many different methods of obtaining point estimators, each with their own characteristics, and which one is better than the others will sometimes depend on what is meant by "better." This point will be clarified as we progress.

9.1.1 METHOD OF MOMENTS

This method is based on the fact that, intuitively, the rth sample moment of the sample $X_1, X_2, \ldots, X_n$

$$\bar{X}_r = \frac{1}{n} \sum_{i=1}^{n} X_i^r = g(X_1, X_2, \ldots, X_n)$$

should approximate the rth moments of X about the origin

$$E(X^r; \theta) = \int_{-\infty}^{\infty} x^r f(x; \theta)\, dx$$

where X is assumed to be a continuous random variable with pdf $f(x)$. Formulas for the case where X is a discrete random variable are analogous. When it is desired to denote the dependence of the pdf $f(x)$ on the unknown parameter θ directly it will be written $f(x; \theta)$. If these two quantities are equated

$$E(X^r; \theta) = \overline{X^r} \qquad (9.1)$$

and the equation solved for θ, then we could use this value as an estimator $\hat{\theta}$ of θ. It is only an estimate as Eq. (9.1) will not in general be strictly true. When there is only one unknown parameter, we take $r = 1$.

As an example, suppose it is known that the random sample $X_1, X_2, \ldots, X_n$ is from a population with an exponential pdf with unknown parameter μ (i.e., each X_i has the pdf $f(x; \mu) = \mu e^{-\mu}$). The samples here are not assumed to be independent, although we may need this assumption for other methods. The sample mean is

$$\bar{X} = \frac{1}{n} \sum_{i=1}^{n} X_i$$

and the population mean

$$E(X; \mu) = 1/\mu$$

and equating these two and solving yields the estimator

$$\hat{\mu} = 1 \bigg/ \left(\frac{1}{n} \sum_{i=1}^{n} X_i \right)$$

This method can also be used when there is more than one unknown parameter. If s parameters, $\theta_1, \theta_2, \ldots, \theta_s$ of the pdf are unknown, then the first s population and sample moments are found and equated:

$$E(X; \theta_1, \theta_2, \ldots, \theta_s) = \bar{X}$$
$$E(X^2; \theta_1, \theta_2, \ldots, \theta_s) = \overline{X^2}$$
$$\vdots$$
$$E(X^s; \theta_1, \theta_2, \ldots, \theta_s) = \overline{X^s}$$

and a solution to these equations, $\hat{\theta}_1, \hat{\theta}_2, \ldots, \hat{\theta}_s$ is used for the required estimators.

For example, suppose $X_1, X_2, \ldots, X_n$ are samples taken from a normal population with unknown parameters μ and σ^2. The first and second moments of the population are

$$E(X; \mu, \sigma) = \mu$$
$$E(X^2; \mu, \sigma) = \sigma^2 + \mu^2$$

and equating these with the appropriate sample moments gives

$$\bar{X} = \frac{1}{n} \sum_{i=1}^{n} X_i = \mu$$

$$\overline{X^2} = \frac{1}{n} \sum_{i=1}^{n} X_i^2 = \sigma^2 + \mu^2$$

and, solving these equations gives the estimators

$$\hat{\mu} = \bar{X} \quad \text{and} \quad \hat{\sigma}^2 = \overline{X^2} - (\bar{X})^2$$

The method of moments for estimators is relatively simple and often yields good estimates, but as Example 9.2 will show, it is not without its problems. It is also quite difficult to say anything about the estimators obtained, analytically. A few examples will further illustrate the method.

EXAMPLE 9.1

A sample of four light bulbs burn out at the times 4.1, 4.5, 3.9, and 5.0 days. If it is assumed that the population burn-out time has a gamma pdf

$$f(x; \alpha, \gamma) = \frac{x^{\alpha-1} e^{-x/\gamma}}{\gamma^{\alpha} \Gamma(\alpha)}$$

estimate α and γ by the method of moments.

The first two moments of the population are, from Chapter 5,

$$E(X; \alpha, \gamma) = \alpha\gamma \qquad E(X^2; \alpha, \gamma) = \alpha\gamma^2(1+\alpha)$$

and the first two moments of the sample are

$$\bar{X} = \frac{1}{4} \sum_{i=1}^{4} X_i = 4.375 \quad \text{and} \quad \overline{X^2} = \frac{1}{4} \sum_{i=1}^{4} X_i^2 = 19.3175$$

Equating

$$\alpha\gamma = 4.375 \qquad \alpha\gamma(\alpha + \alpha\gamma) = 19.3175$$

yields the solution $\hat{\gamma} = .063$ and $\hat{\alpha} = 69.4$ and these are the parameter estimates by the method of moments.

EXAMPLE 9.2

It is known that the samples 3.1, 0.2, 1.6, 5.2, and 2.1 are from a random variable that is uniformly distributed over the unknown range (a, b). Find estimates for a and b by the method of moments.

If X is a uniformly distributed random variable over the interval (a, b), then

$$E(X; a, b) = \frac{a+b}{2} \quad \text{and} \quad E(X^2; a, b) = \frac{b^3 - a^3}{3(b-a)} = \frac{b^2 + ab + a^2}{3}$$

The sample mean is 2.44 and the sample second moment is 8.73. The estimates are the solutions to the equations

$$2.44 = \frac{a+b}{2} \qquad 8.73 = \frac{b^2 + ab + a^2}{3}$$

which are easily calculated as $\hat{a} = .44$ and $\hat{b} = 5.33$.

This example shows up one of the weaknesses of the method of moments. The estimate of a is .44 but one of the observations was .2 indicating that a is at most .2. In spite of such contradictions the method of moments is still useful in many applications. The final example of this method simply indicates that it can also be used for discrete distributions.

EXAMPLE 9.3

Every hour 10 items are drawn off an assembly line, inspected, and the number defective noted. If the data for one day's output are

Hour number	1	2	3	4	5	6	7	8
Number defective	1	1	2	0	2	1	0	1

estimate the probability that an item is defective, by the method of moments.

To consider this problem we will assume that each item will be defective with probability θ, independent of other items being defective, and hence that the number of defective items in a sample of size n has a binomial distribution with parameters n and θ. The mean of a binomial population, for $n = 10$, is 10θ and the sample mean is .8 and the estimate of θ by the method of moments is then $\hat{\theta} = .08$.

9.1.2 MAXIMUM LIKELIHOOD ESTIMATION

In discussing maximum likelihood point estimators we will again assume that the samples are drawn from a population characterized by a continuous pdf $f(x; \theta)$ with one (or more) unknown parameters. Discrete random variables are handled in a similar manner and will be encountered in the examples. Consider first the case where only one parameter θ is unknown and assume that a sample $X_1, X_2, \ldots, X_n$ has been drawn. If the sample values are $X_i = x_i, i = 1, 2, \ldots, n$ the value of the n-variate density function at the point $(x_1, x_2, \ldots, x_n)$ is

$$L(\theta) = f(x_1, x_2, \ldots, x_n; \theta)$$

and, if the samples are independent, this can be written as

$$L(\theta) = f(x_1; \theta)f(x_2; \theta) \cdots f(x_n; \theta) \qquad (9.2)$$

The function $L(\theta)$ is referred to as the likelihood function of the sample. The value of θ that maximizes $L(\theta)$ will be taken as the estimate $\hat{\theta}$. If the value of θ

that maximizes $L(\theta)$ can be expressed as a function $g(x_1, x_2, \ldots, x_n)$, then

$$\hat{\theta} = g(X_1, X_2, \ldots, X_n)$$

is the maximum likelihood estimator of θ. Picking that value of θ which maximizes the likelihood function has a great deal of intuitive appeal. We will see later in the section that a maximum likelihood estimator has many other desirable properties.

One way of finding the value of θ that maximizes Eq. (9.2) is by using elementary calculus, that is, letting $dL(\theta)/d\theta = 0$ and solving for θ. This turns out to be a little complicated but a simple observation will help. The function $\ln(x)$ is a monotonically increasing function of x for $x \geq 0$. It is readily seen then that the value of θ that maximizes $L(\theta)$ also maximizes $\ln[L(\theta)]$. But we can write $\ln[L(\theta)]$ as

$$\ln[L(\theta)] = \sum_{i=1}^{n} \ln[f(x_i; \theta)]$$

and differentiating with respect to θ and equating to zero gives

$$\frac{\partial}{\partial \theta}\{\ln[L(\theta)]\} = 0$$

$$= \sum_{i=1}^{n} \frac{1}{f(x_i; \theta)} \frac{\partial}{\partial \theta}[\ln f(x_i; \theta)] \tag{9.3}$$

Solving this equation for θ, and hence $\hat{\theta}$, is often easier than working directly with the likelihood function, particularly when $f(x; \theta)$ is exponential or normal.

Sometimes a maximum of $\ln[L(\theta)]$ will not exist and this situation will be discussed in Example 9.5. When a solution to Eq. (9.3) does exist it must, of course, be checked that it does correspond to a maximum.

If $X_1, X_2, \ldots, X_n$ is a random sample drawn from a population with pdf $f(x; \theta_1, \theta_2)$, the log of the likelihood ratio is

$$\ln[L(\theta_1, \theta_2)] = \sum_{i=1}^{n} \ln f(x_i; \theta_1, \theta_2)$$

and the maximum likelihood estimates of θ_1 and θ_2 in this case correspond to the point (θ_1, θ_2) at which $\ln[L(\theta_1, \theta_2)]$ achieves its maximum value. It is found by solving the equations

$$\frac{\partial}{\partial \theta_1} \ln[L(\theta_1, \theta_2)] = 0$$

$$\frac{\partial}{\partial \theta_2} \ln[L(\theta_1, \theta_2)] = 0$$

The extension to n unknown parameters is clear.

Suppose that $X_1, X_2, \ldots, X_n$ is a sample from a normal population with known variance σ^2 but unknown mean μ. The likelihood function is, from Eq.

(9.2),

$$L(\mu) = \prod_{i=1}^{n} \frac{1}{\sqrt{2\pi}\sigma} \exp\left(-\frac{(x_i - \mu)^2}{2\sigma^2}\right) \tag{9.4}$$

where $X_i = x_i$, $i = 1, 2, \ldots, n$, or, equivalently,

$$\ln[L(\mu)] = -\frac{1}{2\sigma^2} \sum_{i=1}^{n} (x_i - \mu)^2 - n \ln(\sqrt{2\pi}\sigma) \tag{9.5}$$

Differentiating this expression with respect to μ gives

$$\frac{\partial}{\partial \mu}[L(\mu)] = \frac{1}{\sigma^2} \sum_{i=1}^{n} (x_i - \mu) = 0$$

and hence

$$\hat{\mu} = \frac{1}{n} \sum_{i=1}^{n} X_i$$

This is the same estimator for the mean of a population as given by the method of moments. This will not always happen for other density functions or parameters and, when the estimators differ, the maximum likelihood estimator is usually regarded as the most reliable. If it had been assumed that both the mean and the variance were unknown in this problem, then we would have regarded the right-hand side of Eq. (9.4) or (9.5) as a function of μ and σ^2 and solved the equations

$$\frac{\partial}{\partial \mu}\{\ln[L(\mu)]\} = 0 = \frac{1}{\sigma^2} \sum_{i=1}^{n} (x_i - \mu)$$

$$\frac{\partial}{\partial \sigma^2}\{\ln[L(\sigma^2)]\} = 0 = \frac{1}{2\sigma^4} \sum (x_i - \mu)^2 - \frac{n}{2\sigma^2}$$

and the resulting estimators are

$$\hat{\mu} = \frac{1}{n} \sum_{i=1}^{n} X_i = \bar{X} \quad \text{and} \quad \hat{\sigma}^2 = \frac{1}{n} \sum_{i=1}^{n} (X_i - \bar{X})^2$$

and these values do indeed result in a maximum of the function $L(\mu, \sigma^2)$. A few examples will further illustrate the application of maximum likelihood estimation. Notice that before we can use this method the pdf of X must be assumed known, except for some of its parameters.

EXAMPLE 9.4

In analyzing the flow of traffic through a cafeteria checkout, the times between arrivals at the register are, for 10 customers, recorded as 4.2, 1.1, 6.3, 5.2, 2.2, 2.8, 7.4, 1.2, 1.9, and 3.1. If it is assumed that these times are exponentially distributed with parameter λ, find the maximum likelihood estimate of λ and compare it with the estimate obtained from the method of moments.

The population mean for exponentially distributed random variables is simply

$$E(X; \lambda) = 1/\lambda$$

and so the estimate obtained for λ by the method of moments is

$$\hat{\lambda} = 1 \Big/ \left(\frac{1}{10} \sum_{i=1}^{10} x_i \right) = \frac{1}{3.54}$$

The likelihood function for this case is

$$L(\lambda) = \prod_{i=1}^{10} \lambda \exp(-\lambda x_i)$$

$$= \lambda^{10} \exp\left(-\lambda \sum_{i=1}^{10} x_i \right)$$

$$= \lambda^{10} \exp(-35.4\lambda)$$

Differentiating with respect to λ and equating the result to zero yields

$$10\lambda^9 \exp(-35.4\lambda) - 35.4\lambda^{10} \exp(-35.4\lambda) = 0$$

which gives the estimate

$$\hat{\lambda} = 1 \Big/ \left(\frac{1}{10} \sum_{i=1}^{10} x_i \right) = \frac{1}{3.54}$$

as before.

When the likelihood function does not have a maximum, then other reasoning must be used as the following example illustrates.

EXAMPLE 9.5

Find the maximum likelihood estimates for the parameters a and b of Example 9.2.

In this case the likelihood function is

$$L(a, b) = \frac{1}{(b-a)^5}$$

and we note that it is independent of the observations. Differentiating with respect to a and b is of little value in this case since it does not yield a solution. However, since we have observed the values 3.1, 0.2, 1.6, 5.2, and 2.1, it seems that we should choose a and b so that $(b-a)$ is as small as possible, consistent with the given data. Consequently, we choose $\hat{a} = \min_i X_i = 0.2$ and $\hat{b} = \max_i X_i = 5.2$ since these values maximize $L(a, b)$ for the observations given. Arguing in this manner avoids the difficulty encountered with the method of moments.

EXAMPLE 9.6

The crushing strength of concrete samples, in kilograms per square centimeter is modeled as a gamma distributed random variable with pdf

$$f(x;\theta) = \frac{x\,e^{-x/\gamma}}{\gamma^2}, \quad x \geq 0$$

where γ is unknown. Find the maximum likelihood estimate of γ based on the observations 5.4, 7.1, 6.2, 6.4, and 4.9.

The likelihood function for this density is

$$L(\gamma) = \prod_{i=1}^{5} \frac{1}{\gamma^2} x_i \exp\left(-\frac{x_i}{\gamma}\right)$$

$$= \frac{1}{\gamma^{10}} \left(\prod_{i=1}^{5} x_i\right) \exp\left[-\frac{1}{\gamma}\left(\sum_{i=1}^{5} x_i\right)\right]$$

and differentiating with respect to γ and equating the result to zero yields

$$\frac{\partial}{\partial\gamma} L(\gamma) = 0$$

$$= \left(\prod_{i=1}^{5} x_i\right)\left[-\frac{10}{\gamma^9} \exp\left(-\frac{1}{\gamma}\sum_{i=1}^{5} x_i\right)\right.$$

$$\left. - \left(\sum_{i=1}^{5} x_i\right)\frac{1}{\gamma^{10}} \exp\left(-\frac{1}{\gamma}\sum_{i=1}^{5} x_i\right)\right]$$

or

$$\hat{\gamma} = \frac{1}{10}\sum_{i=1}^{5} X_i = \frac{\bar{X}}{2}, \quad \hat{\gamma} = 3.0$$

As was mentioned, precisely the same techniques can be applied to discrete random variables where we define the likelihood function as

$$L(\theta) = P(x_1;\theta)P(x_2;\theta)\cdots P(x_n;\theta)$$

and $P(x_i;\theta)$ is the probability distribution function of the random variable X when the unknown parameter is θ.

As an example we find the maximum likelihood estimate of the parameter θ when X is a binomial random variable with parameters n and (unknown) θ. The likelihood function is

$$L(\theta) = \prod_{i=1}^{N} \binom{n}{x_i} \theta^{x_i}(1-\theta)^{n-x_i}$$

when the sample is of size N and $X_i = x_i$, $i = 1, 2, \ldots, N$. Differentiating the natural log of this function with respect to θ and equating to zero yields

$$\frac{\partial}{\partial\theta}\{\ln[L(\theta)]\}=\frac{\partial}{\partial\theta}\left\{\ln\left[\prod_i\binom{n}{x_i}\right]+\ln[\theta^{(\Sigma x_i)}(1-\theta)^{nN-\Sigma x_i}]\right\}$$

$$=\frac{\partial}{\partial\theta}\left\{(\ln\theta)\left(\sum_{i=1}^N x_i\right)+\left(nN-\sum_{i=1}^N x_i\right)[\ln(1-\theta)]\right\}$$

$$=\left(\sum_{i=1}^N x_i\right)\frac{1}{\theta}+\left(nN-\sum_{i=1}^N \cdot x_i\right)\left(-\frac{1}{1-\theta}\right)=0$$

which with some rearranging gives

$$\hat{\theta}=\frac{1}{nN}\sum_{i=1}^N x_i$$

an intuitively appealing result. The following example requires a similar calculation for the Poisson distribution.

EXAMPLE 9.7

The number of defects found on a roll of carpet has a Poisson distribution with parameter λ. If four rolls of carpet are inspected and found to have 12, 4, 9, and 15 defects, respectively, find the maximum likelihood estimate for λ.

The likelihood function for a Poisson random variable with parameter λ is

$$L(\lambda)=\prod_{i=1}^4 \lambda^{x_i}\frac{e^{-\lambda}}{(x_i)!}=\frac{1}{\prod_i (x_i)!}\lambda^{\Sigma x_i}e^{-4\lambda}$$

and differentiating gives

$$\frac{\partial L(\lambda)}{\partial\lambda}=\frac{1}{\prod_i (x_i)!}[(\Sigma x_i)\lambda^{(\Sigma x_i)-1}e^{-4\lambda}-4\lambda^{\Sigma x_i}e^{-4\lambda}]$$

which gives the estimate

$$\hat{\lambda}=\frac{1}{4}\sum_{i=1}^4 x_i=10.$$

EXAMPLE 9.8

The random sample $X_1, X_2, \ldots, X_n$ is drawn from a normal population with known mean μ and unknown variance σ^2. Compare the estimates of σ^2 obtained by the method of moments and the maximum likelihood method.

When the mean is known, the maximum likelihood estimate of σ^2 is obtained from the likelihood function

$$L(\sigma^2)=\left(\frac{1}{\sqrt{2\pi}\sigma}\right)^n \exp\left(-\frac{1}{2\sigma^2}\sum_{i=1}^n (x_i-\mu)^2\right)$$

and

$$\hat{\sigma}^2 = \frac{1}{n} \sum_{i=1}^{n} (x_i - \mu)^2$$

$$= \frac{1}{n} \sum_{i=1}^{n} x_i^2 - \frac{2\mu}{n} \sum_{i=1}^{n} x_i + \mu^2$$

By the method of moments we would equate

$$E(X^2; \sigma^2) = \sigma^2 + \mu^2 = \frac{1}{n} \sum_{i=1}^{n} x_i^2$$

or

$$\hat{\sigma}^2 = \frac{1}{n} \sum_{i=1}^{n} x_i^2 - \mu^2$$

and the two estimates are not, in general, the same. They would only be the same if

$$\mu = \frac{1}{n} \sum_{i=1}^{n} X_i$$

As a final word on maximum likelihood estimates we note, without proof, that they enjoy what is referred to as the invariant property. Thus if $\hat{\theta}$ is a maximum likelihood estimate of θ, then $h(\hat{\theta})$ is the maximum likelihood estimate for $h(\theta)$, where $h(\cdot)$ is some function. This property can be useful in saving computation since, otherwise, in determining a maximum likelihood estimate for $h(\theta)$ the likelihood function would have to be expressed as a function of $h(\theta)$.

9.1.3 *PROPERTIES OF ESTIMATORS*

For the first two parts of this section various aspects of obtaining a statistic, which is an estimator for an unknown parameter, have been discussed. For the simple parameters used the two statistics obtained by the method of moments and the maximum likelihood technique were often, but not always, the same. In cases where the estimates differ we will have a choice as to which one to use and the question arises as to which one is better. The term "better" cannot be answered in general since it will usually be context dependent. For the remainder of this section we discuss various properties that estimators might have in order to clarify the choice of one estimator over another. Bear in mind that estimators (or statistics) are functions of random variables and hence are themselves random variables with distribution functions, means, and so forth.

Unbiasedness The estimator $\hat{\theta}$ of the parameter θ is said to be unbiased if $E(\hat{\theta}) = \theta$. For example, the sample mean

$$\bar{X} = \frac{1}{n} \sum_{1}^{n} X_i$$

is always unbiased, no matter what the distribution of X_i is, since

$$E(\bar{X}) = \frac{1}{n} \sum_{i=1}^{n} E(X_i) = \frac{1}{n} \sum_{i=1}^{n} \mu = \mu$$

and this is true for both continuous and discrete random variables.

A result that is less obvious than this is the fact that the sample variance

$$S_1^2 = \frac{1}{n} \sum_{i=1}^{n} (X_i - \bar{X})^2 = \hat{\sigma}^2$$

is not an unbiased estimator for the population variance σ^2. To see this, assuming the X_i are uncorrelated at least, we calculate

$$E(S_1^2) = \frac{1}{n} \sum_{i=1}^{n} E[(X_i - \bar{X})^2]$$

$$= \frac{1}{n} \sum_{i=1}^{n} E(X_i^2 - 2X_i\bar{X} + \bar{X}^2)$$

$$= \frac{1}{n} \sum_{i=1}^{n} [E(X_i^2) - 2E(X_i\bar{X}) + E(\bar{X}^2)]$$

$$= \frac{1}{n} \sum_{i=1}^{n} \left(\sigma^2 + \mu^2 - \frac{2}{n}[\sigma^2 + \mu^2 + (n-1)\mu^2] \right.$$

$$\left. + \frac{1}{n^2}[n\sigma^2 + n\mu^2 + n(n-1)\mu^2] \right)$$

$$= \sigma^2 \left(1 - \frac{2}{n} + \frac{1}{n} \right) + \mu^2(1 - 2 + 1)$$

$$= \frac{(n-1)}{n} \sigma^2$$

which is not equal to σ^2. We conclude however that

$$S^2 = \frac{1}{n-1} \sum_{i=1}^{n} (X_i - \bar{X})^2$$

is an unbiased estimator of σ^2 and we will refer to this quantity as the sample variance. Notice from Example 9.8 that the maximum likelihood estimate of σ^2 when μ is known and the population normal, is not an unbiased estimate. The bias in many estimates can be removed by multiplying by an appropriate constant.

Efficiency When there is more than one unbiased estimator of a parameter θ, say $\hat{\theta}_1$ and $\hat{\theta}_2$, it would be useful to have a criterion to choose between them. Clearly we would like to choose the one that is closer to the true value of the parameter more of the time. We discussed in Chapter 2 the fact that the variance of a random variable gives an indication of the spread of the distribution. A random variable with small variance is more likely to assume values closer to its mean than a random variable with larger variance, roughly speaking.

If $\hat{\theta}_1$ and $\hat{\theta}_2$ are unbiased estimators of θ and $V(\hat{\theta}_1) < V(\hat{\theta}_2)$, we say that $\hat{\theta}_1$ is a *more efficient* estimator of θ than $\hat{\theta}_2$. If among all the *unbiased* estimators of θ, $\hat{\theta}_0$ is the one with the smallest variance, we call it the *most efficient* estimator of θ. The quantity $V(\hat{\theta})$ is called the mean square error of the estimator $\hat{\theta}$. It is not always true that the most efficient estimate is the one with the least variance since we require it to be unbiased. A short example will illustrate the point.

EXAMPLE 9.9

Let X_1 and X_2 be independent normal random variables with mean μ and variance σ^2. Find the mean square error of the two estimators

$$\hat{\mu}_1 = \frac{X_1 + X_2}{2} \quad \text{and} \quad \hat{\mu}_2 = \frac{X_1 + X_2}{k}$$

The estimator $\hat{\mu}_1$ is unbiased and has a mean square error of

$$V(\hat{\mu}_1) = \sigma^2/2$$

The estimator $\hat{\mu}_2$ is biased if $k \neq 2$ and its mean square error is

$$V(\hat{\mu}_2) = \frac{\sigma^2}{(k^2/2)}$$

For $k > 2$, $V(\hat{\mu}_2) < V(\hat{\mu}_1)$, and, by letting $k \to \infty$, $V(\hat{\mu}_2) \to 0$, but it is not a better estimator than $\hat{\mu}_1$ as its mean is also going to zero. For large k, the pdf's of the two estimates might be as shown in Figure 9.1. This is the reason for insisting that the most efficient estimator be unbiased.

Suppose in a particular circumstance we insist that the estimator of θ be linear in the sample observations $X_1, X_2, \ldots, X_n$, that is,

$$\hat{\theta} = a_1 X_1 + a_2 X_2 + \cdots + a_n X_n$$

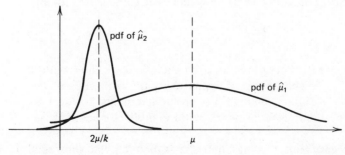

FIGURE 9.1

For example, when estimating the mean we took $a_i = 1/n$, $i = 1, 2, \ldots, n$. We could also use such an estimator to estimate the variance of a population, but it would not be a very good one. Because of simplicity, however, we may be willing to sacrifice some efficiency and use such an estimator. The best linear unbiased estimate of θ is the one of the form $\hat{\theta} = \sum a_i X_i$, $\mu(\sum a_i) = E(\theta)$, which has the minimum variance.

As an example we show that

$$\hat{\mu} = \frac{1}{n} \sum_{i=1}^{n} X_1$$

is the best linear unbiased estimator for μ. Suppose $\hat{\mu}_1 = \sum_{i=1}^{n} a_i X_i$ is an unbiased estimator of μ with lower variance. Since it is unbiased we must have

$$E(\hat{\mu}_1) = \mu = \mu \sum_{i=1}^{n} a_i \quad \text{or} \quad \sum_{i=1}^{n} a_i = 1$$

Assuming that the X_i are uncorrelated, the variance of $\hat{\mu}_1$ is $\sigma^2(\sum a_i^2)$ and $V(\hat{\mu}) = \sigma^2/n$ and, from the assumptions, $\sigma^2(\sum a_i^2) < \sigma^2/n$, or $\sum_i a_i^2 < 1/n$. Consider the sum

$$
\begin{aligned}
0 \le \sum_{i=1}^{n} \left(a_i - \frac{1}{n} \right)^2 &= \sum_{i=1}^{n} \left(a_i^2 - \frac{2a_i}{n} + \frac{1}{n^2} \right) \\
&= \sum_{i=1}^{n} a_i^2 - \frac{2}{n} \sum_{i=1}^{n} a_i + \frac{1}{n} \\
&= \sum_{i=1}^{n} a_i^2 - \frac{1}{n}
\end{aligned}
$$

which, by assumption, is less than 0. This is impossible unless $a_i = 1/n$ implying that $\hat{\mu}$ is the best linear unbiased estimate of μ. Using a linear sum of the observations to estimate some parameters (for example, the variance) will in general give quite bad results.

Consistency   A slightly sharper concept than the variance of an estimate, which gives a better idea of how close the estimate is to the actual parameter, is the notion of consistency. The estimator $\hat{\theta}$ of θ is said to be *consistent* if, for any small $\delta > 0$,

$$\lim_{n \to \infty} P(|\theta - \hat{\theta}| < \delta) = 1$$

where n is the number of observations used by the statistic. Thus, with a consistent estimator, one is guaranteed of approaching the true value of the parameter if the sample size is large enough. The problem remains, however, to determine with more precision what is meant by the sample size being "large enough."

It is a relatively easy matter, using Chebyshev's inequality, to show that the sample mean is a consistent estimator of μ. The variance of $\hat{\mu}$ is σ^2/n, where

σ^2 is the variance of X_i, $i = 1, 2, \ldots, n$ and the samples are uncorrelated. Using a variation of Eq. (8.8) we conclude that

$$P(|\mu - \hat{\mu}| < \varepsilon) = P\left(\left|\mu - \frac{1}{n}\sum_{i=1}^{n} X_i\right| < \varepsilon\right) \geq 1 - \frac{\sigma^2}{n\varepsilon^2}$$

and, as the right-hand side of the inequality tends to unity with n, $\hat{\mu}$ is a consistent estimator of μ. It can, in fact, be shown that most maximum likelihood estimators are consistent.

Fortunately there exists a simple criterion to determine whether or not an estimator is consistent. Using Chebyshev's inequality we can write

$$P(|\hat{\theta}_n - \theta| \geq \varepsilon) \leq \frac{E[(\hat{\theta}_n - \theta)^2]}{2} = \frac{1}{\varepsilon^2} E\{[\hat{\theta}_n - E(\hat{\theta}_n) + E(\hat{\theta}_n) - \theta]^2\}$$

$$= \frac{1}{\varepsilon^2} E\{[\hat{\theta}_n - E(\hat{\theta}_n)]^2\} + E\{[E(\hat{\theta}_n) - \theta]^2\}$$

$$+ 2E\{[\hat{\theta}_n - E(\hat{\theta}_n)][E(\hat{\theta}_n) - \theta]\}$$

$$= \frac{1}{\varepsilon^2}(V(\hat{\theta}_n) + E\{[E(\hat{\theta}_n) - \theta]^2\})$$

where $\hat{\theta}_n$ is the estimator of the parameter θ based on a random sample of size n. If $\lim_{n\to\infty} E(\hat{\theta}_n) = \theta$ and $\lim_{n\to\infty} V(\hat{\theta}_n) \to 0$, then the estimate is consistent. In practice, these relatively simple conditions are easier to apply than the original definition.

Many other properties of estimators are of interest and all are attempts to clarify the nature of the information that is being obtained about the population. These topics however are beyond the scope of this book.

EXAMPLE 9.10

The random variable X is uniformly distributed over the interval $(0, a)$. If $X_1, X_2, \ldots, X_n$ is a random sample of size n, show that

$$\hat{A} = \max(X_1, X_2, \ldots, X_n)$$

is a consistent estimator of the parameter a.

If X is uniformly distributed over $(0, a)$, then, from Chapter 8, the pdf of $Y = \max(X_1, X_2, \ldots, X_n)$ is $n[F(y)]^{n-1}f(y)$ or

$$g(y) = n\left(\frac{y}{a}\right)^{n-1}\frac{1}{a}, \quad 0 \leq y \leq a$$

The expected value of this density, and hence of the estimate $\hat{A}$ is

$$E(\hat{A}) = \int_0^a yg(y)\,dy = \frac{1}{a^n}\int_0^a ny^n\,dy = \frac{n}{n+1}a$$

and so the estimate is not unbiased. Notice that in the limit, however,

$\lim_{n \to \infty} E(\hat{A}) = a$, and so the estimate is asymptotically unbiased. To show that the estimate is consistent requires the variance of the estimate for which we calculate

$$E(\hat{A}^2) = \int_0^a y^2 n \left(\frac{y}{a}\right)^{n-1} \frac{1}{a} \, dy$$

$$= \frac{n}{a^n} \frac{y^{n+2}}{n+2} \bigg|_0^a = \frac{na^2}{(n+2)}$$

The variance is

$$V(A) = E(A^2) - (E(A))^2$$

$$= \frac{na^2}{(n+2)} - \frac{n^2 a^2}{(n+1)^2}$$

$$= a^2 \frac{n}{(n+2)(n+1)^2}$$

and, as $n \to \infty$, the variance does go to zero showing that it is consistent.

9.2 Confidence Intervals

In Section 9.1 we considered point estimators and their properties. To estimate a parameter θ an estimator $\hat{\theta}$ was determined according to some criterion. A random sample was taken and the estimate of θ, $\hat{\theta}$, was found. The result is a single number and there is no indication of how likely it is that this number $\hat{\theta}$ is close to θ, the unknown true value. The method of confidence intervals is an attempt to overcome this problem with point estimators. Here we construct an interval (L, U) whose endpoints are statistics of the random sample (and hence random variables) and give a statement of the likelihood that the true value of the parameter lies in this interval.

Definition Let $X_1, X_2, \ldots, X_n$ be a random sample drawn from a population with pdf $f(x; \theta)$, where θ is an unknown parameter. The statistics L and U determine a $100(1-\alpha)\%$ confidence interval (L, U) for the parameter θ if

$$P(L < \theta < U) \geq 1 - \alpha, \quad 0 < \alpha < 1$$

Rather than a single number $\hat{\theta}$, we seek an interval (L, U) that we will be $100(1-\alpha)\%$ confident contains the true value of θ. Another way of viewing such intervals is to take 100 random samples and for each form the interval (L_i, U_i), $i = 1, 2, \ldots, 100$. These will be different intervals, depending in each case on the random sample drawn, but approximately $100(1-\alpha)\%$ of them will contain the (fixed and unknown) parameter θ. In practice, only one random sample is generally used, the interval (L, U) determined, and the statement that with $100(1-\alpha)\%$ confidence this interval contains θ. The quantity $(1-\alpha)$ is called the confidence coefficient.

Much of the remainder of this section will be concerned with normal populations and we recall from Chapter 8 that if $X_1, X_2, \ldots, X_n$ are independent normal random variables each with mean μ and variance σ^2, then the random variable

$$Y = \frac{1}{\sigma^2} \sum_{i=1}^{n} (X_i - \bar{X})^2$$

is a χ^2 random variable with $(n-1)$ degrees of freedom that is, it has the pdf

$$f_Y(y) = \frac{1}{2^{(n-1)/2}\Gamma((n-1)/2)} \cdot y^{((n-1)/2)-1} e^{-y/2}, \quad y > 0$$

Also recall that if S^2 is the sample variance,

$$S^2 = \frac{1}{(n-1)} \sum_{i=1}^{n} (X_i - \bar{X})^2$$

for the normal population with mean μ and variance σ^2, then $(\bar{X} - \mu)/(S/\sqrt{n})$ has a t-distribution with $(n-1)$ degrees of freedom, that is, its pdf is

$$f(t) = \frac{\Gamma(n/2)}{\Gamma((n-1)/2)\sqrt{(n-1)\pi}} \left(1 + \frac{t^2}{(n-1)}\right)^{-n/2}, \quad -\infty < t < \infty$$

9.2.1 CONFIDENCE INTERVALS FOR THE MEAN OF A NORMAL POPULATION WITH KNOWN VARIANCE

A random sample $X_1, X_2, \ldots, X_n$ is drawn from a normal population that is known to have variance σ^2. From the sample we want to define two limits U and L, which are functions of the sample and hence random variables such that

$$P(L < \mu < U) \geq 1 - \alpha$$

for some α, $0 < \alpha < 1$. The random variable

$$Z = \frac{\bar{X} - \mu}{\sigma/\sqrt{n}}$$

is a standard normal variable and hence for a given α we can find a number $z_{1-\alpha/2} \ (= -z_{\alpha/2})$ such that

$$P(-z_{1-\alpha/2} \leq Z \leq +z_{1-\alpha/2}) = \Phi(z_{1-\alpha/2}) - \Phi(-z_{1-\alpha/2})$$

$$= 1 - \alpha$$

But, rewriting the argument gives

$$P(-z_{1-\alpha/2} \leq Z \leq +z_{1-\alpha/2})$$

$$= P\left[-z_{1-\alpha/2} \leq \left(\frac{\bar{X}-\mu}{\sigma/\sqrt{n}}\right) \leq +z_{1-\alpha/2}\right]$$

$$= P\left(\bar{X} - \frac{\sigma}{\sqrt{n}}z_{1-\alpha/2} \leq \mu \leq \bar{X} + \frac{\sigma}{\sqrt{n}}z_{1-\alpha/2}\right)$$

$$= 1 - \alpha$$

and so

$$L = \bar{X} - \frac{\sigma}{\sqrt{n}}z_{1-\alpha/2} \quad \text{and} \quad U = \bar{X} + \frac{\sigma}{\sqrt{n}}z_{1-\alpha/2} \tag{9.6}$$

where $z_{1-\alpha/2}$ is a constant depending only on α, define the lower and upper confidence limits, and hence the $100(1-\alpha)\%$ confidence interval. Typically α takes on the values .90, .95, and .99 and, from the standard normal cdf the corresponding values of $z_{1-\alpha/2}$ are 1.645, 1.960, and 2.576, respectively.

EXAMPLE 9.11

A surveyor measures the baseline of a property four times and finds that the average of his measurements is 585.145 m. From past experience he knows that the variance of his measurements for such a distance will be .010 m^2.

(a) What is the 99% confidence interval for these measurements?
(b) What is the probability that the true distance will fall within σ of the average?
(c) How many times should he measure the distance in order to have a 90% confidence interval less than 5 cm wide?

(a) To a high degree of accuracy measurements of this type are normally distributed and we make this assumption. From Eq. (9.6) the 99% confidence interval for these measurements is

$$\left(585.145 - \frac{.1}{2}(2.576), 585.145 + \frac{.1}{2}(2.576)\right)$$

$$= (585.016, 585.274)$$

(b) Using Eq. (9.6), the probability that μ lies within σ of $\bar{X}$ is

$$P(\bar{X} - \sigma \leq \mu \leq \bar{X} + \sigma) = P\left(-2 \leq \frac{\bar{X}-\mu}{\sigma/\sqrt{4}} \leq 2\right)$$

which, from the tables for the standard normal cdf is .9545.

(c) The width of the 90% confidence interval is not a random variable and is simply

$$\frac{2\sigma}{\sqrt{n}} \times 1.645 = \frac{2 \times .1 \times 1.645}{\sqrt{n}} \text{ m}$$

In order for this to be less than 5 cm, we require $n = 44$.

At times only upper or lower confidence limits are required. For example in reliability testing of components, as discussed in Chapter 12, we might model the time to failure of a component by a normal random variable. To find a lower confidence limit for μ, since

$$P\left(\frac{\bar{X} - \mu}{\sigma/\sqrt{n}} \le z_{1-\alpha}\right) = 1 - \alpha$$

$$= P\left(\bar{X} - z_{1-\alpha}\frac{\sigma}{\sqrt{n}} \le \mu\right) \tag{9.7}$$

the lower confidence limit for the $100(1-\alpha)\%$ confidence level is $\bar{X} - z_{1-\alpha}\sigma/\sqrt{n}$.

EXAMPLE 9.12

The lifetime of a mechanical relay in a heating system is assumed to be a normal random variable with variance 6.4 days^2. Five items are tested and fail at 104.1, 86.2, 94.1, 112.7, and 98.8 days, respectively. What is the 95% lower confidence limit for $E(T)$, the expected time to failure for the population?

The sample mean is 99.18 days and, from the table for the standard normal cdf, $z_\alpha = 1.645$. Consequently, the lower 95% confidence limit is

$$L = 99.18 - 1.645(2.53/\sqrt{5}) = 97.32$$

and we are 95% confident that the true mean of the population is greater than 97.32 days.

EXAMPLE 9.13

Two coin stamping machines produce coins that have a weight which is normally distributed. The variance of the weight for both machines is $.9 \times 10^{-3}$ g^2. In an examination of 10 coins from each machine the sample means were 4.10 and 4.15 g, respectively. What is the 95% confidence interval for the difference in the means?

If the output of one machine has the normal distribution $N(\mu_X, \sigma^2)$ and the other $N(\mu_Y, \sigma^2)$, then the random variable

$$Z = \frac{\bar{X} - \bar{Y} - \mu_X + \mu_Y}{\sigma\sqrt{2/n}}$$

is $N(0, 1)$, assuming the outputs are independent. Consequently,

$$P(-z_{1-\alpha/2} \leq Z \leq z_{1-\alpha/2})$$

$$= 1 - \alpha$$

$$= P\left(\bar{X} - \bar{Y} - z_{1-\alpha/2}\sigma\sqrt{\frac{2}{n}} \leq \mu_X - \mu_Y \leq \bar{X} - \bar{Y} + z_{1-\alpha/2}\sigma\sqrt{\frac{2}{n}} \right)$$

To determine 95% confidence intervals, $z_{1-\alpha/2} = 1.960$. For the figures given, the limits are

$$(4.10 - 4.15) \pm 1.960\sqrt{\frac{2}{10}} .03 = -.15 \pm .023$$

9.2.2 CONFIDENCE INTERVALS FOR THE MEAN OF A NORMAL POPULATION WITH UNKNOWN VARIANCE

In the case where both μ and σ^2 are both known and confidence limits for the mean are required, we use S^2, the unbiased sample variance instead of σ^2, and proceed as before. The random variable

$$\frac{\bar{X} - \mu}{S/\sqrt{n}}$$

has, as derived in Chapter 8, a t-distribution with $(n - 1)$ degrees of freedom. The point $t_{1-\alpha/2;n-1}$ is defined by the condition

$$P\left[-t_{1-\alpha/2;n-1} \leq \left(\frac{\bar{X} - \mu}{S/\sqrt{n}} \right) \leq t_{1-\alpha/2;n-1} \right]$$

$$= 1 - \alpha$$

$$= P\left(\bar{X} - t_{1-\alpha/2;n-1}\frac{S}{\sqrt{n}} \leq \mu \leq \bar{X} + t_{1-\alpha/2;n-1}\frac{S}{\sqrt{n}} \right) \qquad (9.8)$$

and the upper and lower confidence limits are

$$\text{LHS} = \bar{X} - t_{1-\alpha/2;n-1}\frac{S}{\sqrt{n}} \quad \text{and} \quad L = \bar{X} + t_{1-\alpha/2;n-1}\frac{S}{\sqrt{n}} \qquad (9.9)$$

respectively. The point $t_{1-\alpha/2;n-1}$ is, of course, found from a tabulation of the cdf of the t-distribution with the appropriate number of degrees of freedom as in Appendix F. Notice that as the point $t_{n;n-1}$ is defined for the t-distributed random variable T (with $(n - 1)$ degrees of freedom) by

$$P(T \leq t_{\eta;n-1}) = \eta$$

then

$$P(T \leq t_{1-\alpha/2;n-1}) = 1 - \alpha/2$$

and

$$P(-t_{1-\alpha/2;n-1} \le T \le t_{1-\alpha/2;n-1}) = 1 - \frac{\alpha}{2} - \frac{\alpha}{2} = 1 - \alpha$$

EXAMPLE 9.14

Suppose that the surveyor of Example 9.11 computed the sample variance of his four measurements rather than using the known variance of 0.01 m^2, and suppose that the sample variance is 0.01 m^2. What would his 99% confidence interval be?

For $1 - \alpha = .99$, using Eq. (9.8) and tables for t-distribution with three degrees of freedom indicates that $t_{.995;3} = 5.841$ and consequently the confidence limits are

$$\text{LHS} = \bar{X} - 5.841 \times \left(\frac{.1}{2}\right) \quad \text{and} \quad \bar{L} = \bar{X} + 5.841 \times \left(\frac{.1}{2}\right)$$

and the width of the confidence interval, for this particular sample, is .584 m. This is considerably wider than the .258 m obtained in Example 9.11 and this is in spite of the fact that the same figures were used for sample mean and variance. The fact that the variance was assumed to be a random variable here, however, meant that we had to widen our interval over the known variance case in order to obtain the same level of confidence. Notice also that in the case where the variance was assumed known, the confidence interval is of width $(2\sigma/\sqrt{n})z_{1-\alpha/2}$ and is not a function of the sample, that is, it is not a random variable. The width of the interval when the sample variance is used is $2(S/\sqrt{n})t_{1-\alpha/2;n-1}$ and is a random variable.

9.2.3 CONFIDENCE INTERVALS FOR THE VARIANCE OF A NORMAL POPULATION WITH UNKNOWN MEAN

The unbiased estimator of the sample variance

$$S^2 = \frac{1}{(n-1)} \sum_{i=1}^{n} (X_i - \bar{X})^2$$

is such that $(n-1)S^2/\sigma^2$ has a χ^2 distribution with $(n-1)$ degrees of freedom. Proceeding as in the previous section, we define the point $\chi^2_{\eta;n-1}$ by the equation

$$P\left(\frac{(n-1)S^2}{\sigma^2} \le \chi^2_{\eta;n-1}\right) = \eta$$

which implies that

$$P\left(\chi^2_{\alpha/2;n-1} \le \frac{(n-1)S^2}{\sigma^2} \le \chi^2_{1-\alpha/2;n-1}\right) = 1 - \frac{\alpha}{2} - \frac{\alpha}{2} = 1 - \alpha \qquad (9.10)$$

and this is equivalent to

$$P\left(\frac{(n-1)S^2}{\chi^2_{1-\alpha/2;n-1}} \leq \sigma^2 \leq \frac{(n-1)S^2}{\chi^2_{\alpha/2;n-1}}\right) = 1-\alpha$$

Consequently, a confidence interval for the variance of a normal population is

$$\left(\frac{1}{\chi^2_{1-\alpha/2;n-1}} \sum_{i=1}^{n} (X_i - \bar{X})^2, \quad \frac{1}{\chi^2_{\alpha/2;n-1}} \sum_{i=1}^{n} (X_i - \bar{X})^2\right)$$

What would be the effect on this interval if we assumed that the mean of the population was known?

EXAMPLE 9.15

In a highly sensitive pneumatic control system the gas pressure is assumed to be a normally distributed random variable. In a sequence of eight readings it was found that $\sum (X_i - \bar{X})^2 = 1.02$ (kilopascal)^2. Determine the 99% confidence interval for the variance of the gas pressure.

From Eq. (9.10) for $\alpha = .01$ and from tables of the χ^2 distribution we determine that

$$\chi^2_{.005;7} = .9893 \quad \text{and} \quad \chi^2_{.995;7} = 20.28$$

and the corresponding confidence interval is

$$\left(\frac{1.02}{20.28}, \frac{1.02}{.9893}\right) \quad \text{or} \quad (.050, 1.03) \text{ kilopascal}$$

The large interval is an indication of the high degree of confidence required.

9.2.4 THE LOWER CONFIDENCE INTERVAL FOR THE MEAN OF AN EXPONENTIAL POPULATION

In Chapter 8 it was shown that if $X_1, X_2, \ldots, X_n$ are independent and identically distributed exponential random variables with parameter γ, then their sum $\sum X_i$ has the gamma distribution

$$f(x) = \frac{\gamma^n}{(n-1)!} x^{n-1} \exp(-\gamma x), \quad x \geq 0$$

and so the random variable $Y = 2\gamma(\sum X_i)$ has the pdf

$$g(y) = \frac{1}{2^n(n-1)!} y^{n-1} \exp\left(-\frac{y}{2}\right), \quad y \geq 0$$

This is also the pdf of a χ^2 random variable with $2n$ degrees of freedom. Consequently,

$$P(2\gamma \sum X_i \leq \chi^2_{1-\alpha;2n}) = 1-\alpha$$

or, equivalently,

$$P\left(\frac{2\sum X_i}{\chi^2_{1-\alpha;2n}} \le \frac{1}{\gamma}\right) = 1-\alpha$$

giving the lower confidence interval

$$\left(\frac{2\sum X_i}{\chi^2_{1-\alpha;2n}}, \infty\right)$$

for the mean $1/\gamma$ of the exponential distribution.

EXAMPLE 9.16

A manufacturer wishes to have a warranty on the appliances produced. A lifetime test of 10 appliances produced a mean time to failure of 1192 hours. With what level of confidence can the manufacturer state the mean lifetime is greater than 1000 hours?

Since the true mean of the assumed exponential population $1/\gamma$ is to be greater than 1000 hours, we require

$$\frac{2\sum X_i}{\chi^2_{1-\alpha;20}} \le 1000$$

or, equivalently,

$$\chi^2_{1-\alpha;20} \ge \frac{2\times 1192 \times 10}{1000} = 23.84$$

and from tables we conclude that $1-\alpha \cong .750$ giving a confidence level of about 75%.

9.2.5 CONFIDENCE INTERVALS FOR THE PARAMETER OF THE BINOMIAL DISTRIBUTION

Suppose X is a binomially distributed random variable with parameters n and θ, and the problem is to derive the $100(1-\alpha)\%$ confidence interval for the parameter θ based on an observation. The random variable X has mean $n\theta$ and variance $n\theta(1-\theta)$ and

$$Z = \frac{X-n\theta}{\sqrt{n\theta(1-\theta)}}$$

is a standard random variable. When n is large and θ moderate (see Section 8.3), then the distribution of Z can be approximated by that of a standard normal random variable and we can write

$$P(-z_{\alpha/2} \le Z \le z_{1-\alpha/2}) = \Phi(-z_{\alpha/2}) - \Phi(z_{1-\alpha/2})$$

$$= 1-\alpha$$

The argument in this expression is

$$-z_{\alpha/2} \leq \frac{X - n\theta}{\sqrt{n\theta(1-\theta)}} \leq z_{1-\alpha/2}$$

and expressing this equation as a function of θ and noting that $z_{\alpha/2} = z_{1-\alpha/2}$, we obtain

$$(n^2 + nz^2_{\alpha/2})\theta^2 - (2nX + nz^2_{\alpha/2})\theta + X^2 \leq 0 \qquad (9.11)$$

As a quadratic in θ, the roots of this equation can be solved to give

$$\theta_{U,L} = \frac{2nX + nz^2_{\alpha/2} \pm [(2nX + nz^2_{\alpha/2})^2 - 4X^2(n^2 + nz^2_{\alpha/2})]^{1/2}}{2(n^2 + nz^2_{\alpha/2})}$$

$$= \frac{\left(\dfrac{X}{n}\right) + (z^2_{\alpha/2/2n}) \pm \dfrac{z_{\alpha/2}}{\sqrt{n}}\left[\dfrac{X}{n}\left(1 - \dfrac{X}{n}\right) + \dfrac{z^2_{\alpha/2}}{4n}\right]^{1/2}}{1 + z^2_{\alpha/2}/n}$$

Since the curve of Eq. (9.11) is a quadratic in θ and takes on negative values for $\theta_L \leq \theta \leq \theta_U$, (θ_L, θ_U) is the $100(1-\alpha)\%$ confidence interval for θ.

This derivation is, of course, only valid for the case when n is large and θ moderate. Other cases must be handled with different, and generally more complicated, methods.

For a given α, $z_{\alpha/2}$ is fixed and so for large n the confidence limits can be approximated by

$$\theta_{U,L} = \left(\frac{X}{n}\right) \pm \frac{z_{\alpha/2}}{\sqrt{n}}\left[\frac{X}{n}\left(1 - \frac{X}{n}\right)\right]^{1/2} \qquad (9.12)$$

EXAMPLE 9.17

One thousand items from a production line are tested and 14 are found to be defective. Determine 95% confidence limits for θ, the probability an item is defective.

From Eq. (9.12) the confidence limits are

$$\theta_{U,L} = \left(\frac{14}{1000}\right) \pm \frac{z_{\alpha/2}}{\sqrt{1000}}\left[\frac{14}{1000}\left(\frac{986}{1000}\right)\right]^{1/2}$$

where $z_{\alpha/2} = 1.96$. The limits are $.014 \pm .006$ and the width and center point are random variables.

9.3 Bayesian Estimation

The previous sections of this chapter were concerned with estimating some unknown, but fixed, parameter θ of a distribution. The dependence of the population pdf on θ was shown as $f(x; \theta)$, and only the sample was used to construct an estimate of θ. Moreover, probabilities had a relative frequency interpretation. Such an approach is often referred to as a classical or objective approach to probability.

For Bayesian methods we introduce the notion of subjective probabilities. These are simply ones that reflect our experience, intuition, or belief rather than are the result of experimentation. For this section we will assume that the unknown parameter θ is actually a random variable. The distribution chosen for θ will reflect the state of our knowledge about θ. Thus, for example, if we are quite sure that θ is close to some value, we would choose a pdf for θ that has a small variance that is centered on that value. Interpreting θ as a random variable and then attempting to estimate it represents a radically different approach to estimation from that employed in the previous sections and requires some additional terminology.

As a first step in Bayesian estimation techniques, prior to taking a random sample of the population, some *prior distribution* or pdf $h(\theta)$ of θ is assumed. As mentioned, this is chosen in some manner to represent adequately our beliefs as to the nature of θ. Since θ is now taken to be a random variable, what was viewed as the population pdf, $f(x; \theta)$ in previous sections, is now viewed as a conditional pdf and written $f_{X|\theta}(x|\theta) = f(x|\theta)$. It follows that we can find the joint pdf of the sample $X = (X_1, X_2, \ldots, X_n)$ and θ as

$$f_{X,\theta}(x_1, x_2, \ldots, x_n, \theta) = f_{X|\theta}(x_1, \ldots, x_n|\theta)h(\theta)$$

$$= \prod_{i=1}^{n} f(x_i|\theta)h(\theta)$$

and the marginal pdf of the sample

$$f(x_1, \ldots, x_n) = \int_{R_\theta} f_{X,\theta}(x_1, \ldots, x_n, \theta)\, d\theta$$

$$= \int_{R_\theta} \prod_{i=1}^{n} f(x_i|\theta)h(\theta)\, d\theta$$

The other conditional pdf

$$g(\theta|x_1, \ldots, x_n) = \frac{f_{X|\theta}(x_1, \ldots, x_n|\theta)h(\theta)}{f(x_1, \ldots, x_n)} \tag{9.13}$$

is referred to as the *posterior distribution* or pdf of θ. Thus the prior pdf $h(\theta)$ represents our beliefs about θ prior to sampling and the posterior pdf $g(\theta|x_1, \ldots, x_n)$ takes into account the change in our feelings having observed the sample. The mean of the posterior pdf

$$\hat{\theta} = E(\theta|x_1, x_2, \ldots, x_n)$$

$$= \int \theta g(\theta|x_1, x_2, \ldots, x_n)\, d\theta \tag{9.14}$$

is called the Bayes estimate of θ and is a function of the random sample $x_1, x_2, \ldots, x_n$. The following three examples illustrate the use of Bayes estimates.

EXAMPLE 9.18

The weight of a large metal casting has an unknown mean with variance 4 gm. The weight is assumed to be normally distributed. Prior information indicates that the mean should be close to 100 gm. If it is assumed that the mean is itself normally distributed with a mean of 100 gm and variance of 1 gm, find the Bayes estimate of the mean. If 10 such castings are found to have a sample mean of 99.2 gm, compare the Bayes estimate with the sample mean.

To develop a general formula, let $f(x_i|\mu)$ be a normal pdf $N(\mu, \sigma^2)$, where μ is the unknown mean, and let μ have the pdf $h(\mu) = N(\mu_1, \sigma_1^2)$. The joint conditional sample pdf is

$$f_{X|\mu}(x_1, \ldots, x_n|\mu) = \left(\frac{1}{\sqrt{2\pi}\sigma}\right)^n \exp\left(-\frac{1}{2\sigma^2} \sum_{i=1}^{n} (x_i - \mu)^2\right)$$

The unconditional sample pdf is found by integration:

$$f(x_1, \ldots, x_n) = \int_{-\infty}^{\infty} f_{X|\mu}(x_1, \ldots, x_n|\mu) h(\mu) \, d\mu$$

$$= \int_{-\infty}^{\infty} \left(\frac{1}{\sqrt{2\pi}\sigma}\right)^n \frac{1}{\sqrt{2\pi}\sigma_1} \exp\left(-\frac{1}{2\sigma^2} \sum_{i=1}^{n} (x_i - \mu)^2\right)$$

$$-\frac{1}{2\sigma_1^2}(\mu - \mu_1)^2\right) d\mu$$

This integral can be evaluated by rearranging terms in the exponent and completing the square. The posterior pdf is then found as

$$g(\mu|x_1, \ldots, x_n) = \frac{f_{X|\mu}(x_1, \ldots, x_n|\mu) h(\mu)}{f(x_1, \ldots, x_n)}$$

$$= \frac{1}{\sqrt{2\pi}\sigma'} \exp\left\{-\frac{1}{2\sigma'^2}\left[\mu - \sigma'^2\left(\frac{\mu_1}{\sigma_1^2} + \frac{n\bar{x}}{\sigma^2}\right)\right]^2\right\}$$

where $\sigma'^2 = (1/\sigma_1^2 + n/\sigma^2)^{-1}$. Since

$$g(\mu|x_1, \ldots, x_n) \sim N\left[\sigma'^2\left(\frac{\mu_1}{\sigma_1^2} + \frac{n\bar{x}}{\sigma^2}\right), \sigma'^2\right]$$

the expectation, the Bayes estimate of μ, is

$$\hat{\mu} = E(\mu|x_1, \ldots, x_n) = \left(\frac{1}{\sigma_1^2} + \frac{n}{\sigma^2}\right)^{-1}\left(\frac{\mu_1}{\sigma_1^2} + \frac{n\bar{x}}{\sigma^2}\right)$$

For the figures in the problem statement, $\sigma^2 = 4$, $\mu_1 = 100$, $\sigma_1^2 = 1$ and the Bayes estimate of μ is

$$\hat{\mu} = \frac{4}{n+4}\left(100 + \frac{n\bar{x}}{4}\right) \tag{9.15}$$

From the data given this is

$$\hat{\mu} = \frac{4}{14}\left(\frac{10}{4} \times 99.2 + 100\right) = 99.4$$

while the maximum likelihood estimate for this case is the sample mean 99.2. Our prior information on μ has skewed the estimate toward the mean of the density of μ. Notice that as the number of samples n becomes large, the estimate of Eq. (9.15) tends to

$$\lim_{n \to \infty} \hat{\mu} = \bar{x}$$

which is the maximum likelihood estimate.

EXAMPLE 9.19

In a proficiency test the times taken for participants to complete a given task are noted. It is assumed that these times are independent exponentially distributed random variables with parameter λ. If λ is itself assumed to be an exponentially distributed random variable with parameter η, find the Bayes estimate of λ.

The conditional density of n sample times is

$$f(t_1, t_2, \ldots, t_n | \lambda) = \prod_{i=1}^{n} \lambda \exp(-\lambda t_i)$$

$$= \lambda^n \exp\left(-\lambda \sum_{i=1}^{n} t_i\right)$$

and the assumed prior density for λ is

$$h(\lambda) = \eta e^{-\eta \lambda}, \quad \eta, \lambda > 0$$

The n-variate sample density is found by evaluating the integral

$$f(t_1, t_2, \ldots, t_n) = \int_0^\infty f(t_1, t_2, \ldots, t_n | \lambda) h(\lambda) \, d\lambda$$

$$= \int_0^\infty \lambda^n \exp\left(-\lambda \sum_{i=1}^{n} t_i\right) \eta e^{-\eta \lambda} \, d\lambda$$

$$= \eta \int_0^\infty \lambda^n \exp\left[-\lambda\left(\eta + \sum_{i=1}^{n} t_i\right)\right] d\lambda$$

$$= \eta \frac{n!}{(\eta + \Sigma t_i)^{n+1}}$$

The posterior distribution for λ is then

$$g(\lambda | t_1, t_2, \ldots, t_n) = \eta \frac{\lambda^n \exp[-\lambda(\eta + \Sigma t_i)]}{\eta \dfrac{n!}{(\eta + \Sigma t_i)^{n+1}}}$$

and its mean, the Bayes estimate for λ, is

$$\hat{\lambda} = \int_0^\infty \lambda g(\lambda | t_1, t_2, \ldots, t_n) \, d\lambda$$

$$= \frac{(n+1)!}{(\eta + \Sigma \, t_i)^{n+2}} \cdot \frac{(\eta + \Sigma \, t_i)^{n+1}}{n!}$$

$$= \frac{(n+1)}{(\eta + \Sigma \, t_i)}$$

The maximum likelihood estimate for λ is simply $n/\Sigma \, t_i$ and, as n becomes large, these two estimates will become close.

The final example illustrates the fact that if the prior distribution of the parameter θ is uniform, then the maximum likelihood estimate of θ is the mode of the posterior distribution of θ, that is, that value of θ which maximizes the distribution of θ, given the random sample.

EXAMPLE 9.20

If the pdf of a sample value, given that the unknown value of the mean is μ, is $N(\mu, \sigma^2)$ and if the prior distribution of the parameter μ is assumed to be uniformly distributed over the region (a, b), show that the Bayes estimate of μ is just the sample mean.

The joint conditional pdf of the sample is

$$f_{X|\mu}(x_1, x_2, \ldots, x_n | \mu) = \left(\frac{1}{\sqrt{2\pi}\sigma}\right)^n \exp\left(-\frac{1}{2\sigma^2} \sum_{i=1}^n (x_i - \mu)^2\right)$$

and the prior distribution of the parameter μ is assumed to be

$$h(\mu) = \frac{1}{b-a}, \quad a \le \mu \le b$$

The joint pdf of the sample is

$$f(x_1, x_2, \ldots, x_n) = \int_a^b \left(\frac{1}{b-a}\right) \cdot \left(\frac{1}{\sqrt{2\pi}\sigma}\right)^n \cdot \exp\left(-\frac{1}{2\sigma^2} \sum_{i=1}^n (x_i - \mu)^2\right) d\mu$$

$$= \frac{\exp\left(-\dfrac{n}{2\sigma^2}(\overline{x^2} - \bar{x}^2)\right)}{(b-a)(\sqrt{2\pi}\sigma)^{n-1}\sqrt{n}} \left[\Phi\left(\frac{b-\bar{x}}{\sigma/\sqrt{n}}\right) - \Phi\left(\frac{a-\bar{x}}{\sigma/\sqrt{n}}\right)\right]$$

and the posterior distribution of μ having observed the sample is

$$g(\mu | x_1, x_2, \ldots, x_n) = \frac{f_{X|\mu}(x_1, x_2, \ldots, x_n | \mu) h(\mu)}{f(x_1, x_2, \ldots, x_n)}$$

$$= \frac{\sqrt{n}}{\sqrt{2\pi}\sigma} \left[\Phi\left(\frac{b-\bar{x}}{\sigma/\sqrt{n}}\right) - \Phi\left(\frac{a-\bar{x}}{\sigma/\sqrt{n}}\right)\right] \cdot \exp\left(-\frac{n}{2\sigma^2}(\mu - \bar{x})^2\right)$$

As a function of μ this conditional pdf is symmetric about $\mu = \bar{x}$ and its mean, the Bayes estimate of μ, is $\bar{x}$.

Problems

1. The weight of ball bearings is assumed to be a normally distributed random variable with a known mean $\mu = 10$ and an unknown variance. Fifteen bearings are weighed and found to have weights satisfying

$$\sum_{i=1}^{15} X_i = 145.104 \text{ g} \qquad \sum_{i=1}^{15} X_i^2 = 1407.441 \text{ g}^2$$

   Find the maximum likelihood estimate of σ, the standard deviation. If instead of using the known mean we used the sample mean, what would the maximum likelihood estimate of σ be?

2. Let X be a Bernoulli random variable with a probability θ of success on each trial. If $X_1, X_2, \ldots, X_n$ are outcomes of n trials, show that the maximum likelihood estimate of θ is $\bar{X}$.

3. The lifetime of an appliance is assumed to have an exponential distribution. If 10 appliances are tested and found to have a mean life of 4.52 years, what is the maximum likelihood estimate of the probability that an appliance will last 5 years?

4. The interarrival times of 10 consecutive customers have a sample mean of 14.3 seconds. If these times have an exponential distribution with parameter λ, find an estimate of λ using the method of moments.

5. The number of telephone calls arriving at a switchboard between 10:00 A.M. and 10:10 A.M. during a work week is 11, 5, 8, 7, and 10. If this number is assumed to have a Poisson distribution, use the invariance property of maximum likelihood estimates (mentioned in Section 9.1) to find the maximum likelihood estimate of the probability that there will be no calls during this interval on a given day.

6. A Bernoulli experiment with a probability θ of success on each trial is performed. In five repetitions of the experiment the number of trials to obtain the first success were 4, 7, 6, 6, and 3. Determine the maximum likelihood estimate for the parameter θ. If the above sequence is viewed as 26 repetitions of the single experiment, determine the maximum likelihood estimate of θ and compare it with the first one.

7. The height of five-year-old children is assumed to be a normally distributed random variable. If 4 five year olds, chosen at random, have heights 91 cm, 87 cm, 84 cm, and 92 cm, respectively, find an estimate for the mean and variance of the distribution using the method of moments.

8. The random variable $Y = \min(X_1, X_2, \ldots, X_n)$ is used as an estimate of the lower limit a, where X is uniformly distributed over the interval $(a, a+1)$. For what value C is CY an unbiased estimate of a?

9. The diameter of a metal pipefitting is assumed to be uniformly distributed over the interval $(a, a+10)$ mm. If a sequence of five samples yields the measurements 1.040, 1.046, 1.045, 1.047, and 1.044 m, respectively, find an estimate for a using the method of moments. Can the maximum likelihood method be of any use here?

10. The lifetime of males in a certain country is assumed to be a normally distributed random variable with variance 121 years^2. In a sampling of 100 obituaries it was found that the sample mean was 68.2 years. Find the 95% lower confidence limit on the mean.

11. The yield from an acre of wheat is assumed to be normally distributed. For 10 acres it is found that

$$\sum_{i=1}^{10} X_i = 204.1 \text{ bushels} \qquad \sum_{i=1}^{10} X_i^2 = 4269 \text{ bushels}^2$$

Find the 95% confidence limits for the mean yield per acre.

12. The time of service before repair of a piece of construction equipment has an exponential distribution with parameter λ. In a small sample of 10 it is found that the mean time of service before repair is 2.65 days. Find the 95% lower confidence limit for the actual mean time.

13. Determine the $100(1-\alpha)\%$ upper confidence limit for the variance of a normal random variable based on a random sample of size n, and assuming the mean is unknown.

14. In an opinion poll it was found that out of 1000 people questioned 476 preferred candidate A and the remainder candidate B. Find the 95% confidence interval for the fraction of the population who will vote for candidate A.

15. The weight of a certain volume of landfill is a normally distributed random variable with variance 1600 kg^2. At a particular landfill site 8 trucks are weighed and found to have an average weight of 1042 kg. Find the 95% confidence interval for the average weight of landfill per truck. How many trucks should be weighed to reduce this interval to a width of 10 kg?

16. For what constant C is the estimate

$$\hat{\sigma}^2 = C \sum_{i=2}^{n} (X_i - X_{i-1})^2$$

an unbiased estimate for the variance of a normal population? For what value of C is it the most efficient estimator of the variance?

17. A cigarette manufacturer wishes to be able to estimate to within 1% of its true value with 95% confidence the proportion of the population that prefers its brand over any other. How large a sample is needed?

18. The random variables X_1 and X_2 have the same unknown mean μ but their variances are 2 and 3, respectively. Find the minimum variance linear unbiased estimate of μ and compare its efficiency with $\bar{X}$.

19. The concentration of ore in earth samples of a certain weight is assumed to have a gamma distribution

$$f(x) = \lambda^k \frac{x^{k-1}}{\Gamma(k)} e^{-\lambda x}, \quad x \geq 0$$

If n earth samples are tested and found to have concentrations $X_1, X_2, \ldots, X_n$, use these to find the maximum likelihood estimate of the parameter λ, assuming that k is known.

20. The lifetime of an electronic component is normally distributed with mean μ and variance σ^2. The reliability $R(t)$ of the component is defined (cf. Chapter 12) as the probability that it is still functioning at time t. If $X_1, X_2, \ldots, X_n$ are n samples, and if σ^2 is assumed known, show carefully how to determine a $100(1-\alpha)\%$ confidence interval for $R(t)$.

21. The pdf of a random variable X is assumed to be of the form $f(x) = Cx^\alpha, 0 \leq x \leq 1$ for some number α and constant C. If $X_1, X_2, \ldots, X_n$ is a random sample of size n, find the maximum likelihood estimate of α.

22. Find 95% confidence limits on the parameter λ of a Poisson distribution if, in a random sample of size 100, the sample mean was 46.21.

23. The time to failure of a device has a pdf

$$f(y) = 3a^3 y^{-4}, \quad y \geq a$$

For a random sample of size n, find the maximum likelihood estimate of the parameter a.

24. In a fatigue test of certain metal rods it is assumed that the time to failure is an exponentially distributed random variable with parameter λ. Eight rods are tested and found to have a mean time to failure of 4.62 hours. Find the 95% lower confidence limit on μ. From this derive a 95% lower confidence limit on the probability that a rod will last 4.5 hours.

25. An item that survives a 100-hour factory test is assumed to be nondefective. If 60 items are tested and 4 found to be defective, find a 95% confidence interval for the probability θ that an item is defective.

26. The energy in a reflected sonar signal from a submarine is a normally distributed random variable. If the submarine is friendly, its mean is μ_1 and variance is σ_1^2, while if it is unfriendly its mean is μ_2 and variance σ_2^2. Determine $100(1-\alpha)\%$ confidence limits on the difference of the means $\mu_1 - \mu_2$ assuming that σ_1^2 and σ_2^2 are known.

27. Find the maximum likelihood estimate of the parameter λ of the Weibull distribution

$$f(x) = \lambda \alpha x^{\alpha-1} e^{-\lambda x^{\alpha}}, \quad x > 0$$

using a sample of size n assuming α is known.

28. The $100(1-\alpha)\%$ confidence interval of a normally distributed random variable with known variance σ^2 is $(\bar{X} \pm (\sigma/\sqrt{n})z_{1-\alpha/2})$ for a random sample of size n. What is the probability that this overlaps at least half of the interval $(\mu \pm (\sigma/\sqrt{n})z_{1-\alpha/2})$?

29. In a random sample of size 10, it was found that $\sum_{i=1}^{10} X_i = 41.2$ and $\sum_{i=1}^{10} X_i^2 = 178.7$. Find the maximum likelihood estimate of the variance assuming the population is normal.

30. It is assumed that items coming off an assembly line are acceptable with probability θ or defective with probability $(1-\theta)$, but no information on θ is available. If n items are sampled and k found to be defective, determine a Bayes estimate of θ.

31. The probability that an item is defective is either .05 (when the machine is operating well) with probability .8 or .10 (when the machine falls out of adjustment). Find the Bayes estimate of θ based on a sample of 2 items.

32. The probability of obtaining a head on a toss of the coin is θ. Find the Bayes estimate of θ, based on a single toss of the coin if: (i) the prior density of θ is uniformly distributed over (0, 1); (ii) the prior density of θ is uniformly distributed over (.45, .55).

33. The continuous random variable Y has a beta density function if

$$f(y) = \frac{\Gamma(\alpha + \beta)}{\Gamma(\alpha)\Gamma(\beta)} y^{\alpha-1}(1-y)^{\beta-1}, \quad 0 < x < 1$$

Show that its mean is $\alpha/(\alpha + \beta)$. Show that if X is a binomial random variable with parameters n and θ, and the prior distribution of θ is assumed uniform on the interval (0, 1), then the posterior density of θ is a beta density.

34. It is certain that the parameter θ of the density $f(x, \theta)$ is actually the value θ_0. What is the Bayes estimate for θ in this case? How do you interpret this result?

CHAPTER 10

Testing Hypotheses

There are numerous situations that arise in practice where important decisions have to be made based on limited data. For example, a business manager may have to decide on whether or not to commit a plant to the manufacture of a new product based on inconclusive evidence as to its market acceptability. The choice is twofold—either one goes ahead or one does not. To further complicate matters, there may be certain costs involved. If the product is not successful, the company may go bankrupt.

As another example, suppose a radar technician observes activity on the radar screen. Based on his experience, he is required to make a decision as to a course of action to take, for example, cause an alert, fire missiles, or identify the intruder as friendly. Each decision carries a cost factor with it. If he mistakenly identifies it as friendly, the consequences may be serious. To carry the examples to a less serious level, a poker player is continuously faced with decisions and usually the costs of a wrong decision versus the benefits of a correct decision are clear to him.

These situations are examples where hypothesis testing procedures can be applied. Basically they are probabilistic methods designed to yield information on the consequences of deciding on one course of action over another. The techniques find wide applicability in engineering, social sciences, and management.

10.1 *Testing Hypotheses*

One of the simplest situations in decision theory arises when we are asked to decide on the value of some parameter of a probability density function, given a sequence of observations. Suppose we observe a sequence $X_1, X_2, \ldots, X_n$ of n random variables and are told that each has been obtained independently from a mechanism that has either a pdf $f_0(x)$ or $f_1(x)$. It is our problem to decide which pdf is appropriate. To simplify matters we assume that $f_0(x)$ and $f_1(x)$ both have the same form and differ only in their means, that is, $f_1(x)$ is simply a translation of $f_0(x)$. Assume that $f_0(x)$ has mean μ_0 and $f_1(x)$, μ_1. We call the hypothesis that the mean is μ_0 the null hypothesis, H_0, and the hypothesis that the mean is μ_1 the alternate hypothesis, H_1; this is standard terminology. To write this more explicitly:

H_0: $\mu = \mu_0$ (the null hypothesis)
H_1: $\mu = \mu_1$ (the alternate hypothesis)

Upon observing $X_1, X_2, \ldots, X_n$ we have to decide which of the two hypotheses is valid. The sample space is the set of all real n-tuples, that is, Euclidean n-space, E_n.

We can think of the decision process as dividing the space E_n of observations up into two regions R_0 and R_1. If the observed vector $(X_1, \ldots, X_n)$ lies in R_0, we will decide on H_0; if in R_1, decide on H_1. A test of an hypothesis is then a partitioning of E_n into the two regions R_0 and R_1. The region R_0 is referred to as the acceptance region and R_1 the critical or rejection region (since the null hypotheses is rejected whenever the sample falls in R_1). Notice the difference between this situation and estimation where an actual value was the end result.

There is always the possibility that, even though the null hypothesis is true (i.e., the actual population pdf is $f_0(x)$), the sample lies in the region R_1 in which case H_1 will be accepted. This is referred to as an *error of Type I* and it occurs with a probability P_I, which is also denoted by α. Similarly, if H_1 is in fact true (the population pdf is $f_1(x)$), there is a chance that the sample lies in R_0, in which case H_0 will be accepted as true. This is called *a Type II error* and it occurs with a probability P_{II} also denoted as β. P_I is commonly called the *level of significance* and $1-P_{II}$ *the power of the test* (which is the probability of accepting H_1 when H_1 is true). If for a given test $P_I = \alpha$, its rejection region is called a rejection region of size α.

In the discussion we assumed that the two possibilities were H_0: $\mu = \mu_0$ and H_1: $\mu = \mu_1$. This situation however is quite general. We can formulate the two hypotheses H_0: $\theta = \theta_0$ versus H_1: $\theta = \theta_1$ for any parameter θ of a population. Of course we could also have a test of the form H_0: $\theta^{(1)} = \theta_{01}$, $\theta^{(2)} = \theta_{02}, \ldots, \theta^{(k)} = \theta_{0k}$ versus H_1: $\theta^{(1)} = \theta_{11}$, $\theta^{(2)} = \theta_{12}, \ldots, \theta^{(k)} = \theta_{1k}$. In some cases the parameter values may not be known precisely and we are merely interested in a test of the form H_0: $\theta = \theta_0$ versus H_1: $\theta \neq \theta_0$. The alternate hypothesis does not specify a value for θ explicitly—it only excludes one.

It would appear from the discussion so far that it is irrelevant as to which hypothesis is chosen as the null and which for the alternative. Typically, however, we choose the null hypothesis so that an error of Type I is more important or has more serious consequences than an error of Type II. One reason for doing so is that the Neyman–Pearson lemma, developed in the next section, gives the test which minimizes P_{II} for a given P_I. Thus the control is on P_I and so the hypotheses are chosen accordingly. This usually implies that H_0 is chosen as the hypothesis more likely to be false so that in rejecting it the probability of doing so is known as it is a design parameter of the test.

An hypothesis is called *simple* if all parameters in the test are specified exactly. Otherwise it is called *composite*. For example, a test of the form H_0: $\theta = \theta_0$ versus H_1: $\theta = \theta_1$ is a test of two simple hypotheses. A test such as H_0: $\theta = \theta_0$ versus H_1: $\theta \neq \theta_0$ is a simple null hypothesis versus a composite alternative hypothesis test. Similarly the test H_0: $|\theta| < k$ versus H_1: $|\theta| > k$ is an example of a composite versus composite test. Thus a distribution under a simple hypothesis is completely specified while under a composite hypothesis the distribution is, to a certain extent, undetermined. The study of composite hypotheses is analytically quite difficult and some problems encountered have no solutions. Some examples will be given in Section 10.3. For the remainder of this section we consider some examples and properties of simple hypotheses. The next section will show how to construct the regions R_0 and R_1 for simple hypotheses analytically (the Neyman–Pearson lemma), a problem that we will approach heuristically in this section.

EXAMPLE 10.1

We are told by the bank manager that a number of new pennies that the bank obtained from the mint are not fair pennies. In fact, it is known that the probability of a head with one of these pennies is $\frac{7}{12}$ and of a tail $\frac{5}{12}$. Having received some change from the bank recently we take a penny from the change and want to test whether it is a fair penny. We set up the hypotheses:

H_0:   the penny is fair (the null hypothesis)
H_1:   the penny is not fair (the alternate hypothesis)

Notice that alternative ways of describing these two hypotheses is by letting θ be the probability of getting a head and X be a binomial random variable (i.e., $X = 1$ if a head is obtained and 0 if a tail) and defining H_0: $\theta = \frac{1}{2}$ versus H_1: $\theta = \frac{7}{12}$ or H_0: $E[X] = \frac{1}{2}$ versus H_1: $E[X] = \frac{7}{12}$. We proceed with this example intuitively rather than rigorously. If we threw the coin a large number N, of times, we would expect "about" $N/2$ heads if the coin is fair and $\frac{7}{12}N$ if it is not. Let us, arbitrarily, agree to accept the hypothesis H_0 that the coin is fair if the sample mean

$$\bar{X} = \frac{1}{N} \sum_{i=1}^{N} X_i$$

X_i = outcome of the ith throw, is less than $\frac{13}{24}$ (half way between $\frac{1}{2}$ and $\frac{7}{12}$).

Similarly, accept H_1 if $\bar{X} \ge \frac{13}{24}$. This corresponds to dividing the sample space S of binary n-tuples into two regions R_0 and R_1 defined by

$$R_0 = \left((X_1, \ldots, X_N) \in S \;\middle|\; \frac{1}{N} \sum_{i=1}^{N} X_i < \frac{13}{24} \right)$$

and

$$R_1 = \bar{R}_0 = \left((X_1, \ldots, X_N) \in S \;\middle|\; \frac{1}{N} \sum_{i=1}^{N} X_i \ge \frac{13}{24} \right)$$

To calculate the two types of errors, we note that $E[X_i] = \theta$ and $V(X_i) = \theta(1-\theta)$, where $\theta = \frac{1}{2}$ under H_0 and $\theta = \frac{7}{12}$ under H_1. Suppose the experiment of tossing the coin N times is conducted when H_0 is true. What is the probability that we will make an error of Type I and decide that H_1 is in effect? Let P_I be the probability of a Type I error. Then we have

$$P_I = P\left(\frac{1}{N} \sum_{i=1}^{N} X_i > \frac{13}{24} \;\middle|\; \theta = \frac{1}{2} \right)$$

$$= P\left(\frac{1}{N} \sum_{i=1}^{N} \left(X_i - \frac{1}{2} \right) > \frac{1}{24} \;\middle|\; \theta = \frac{1}{2} \right)$$

$$= P\left(\sum_{i=1}^{N} X_i > \frac{13N}{24} \;\middle|\; \theta = \frac{1}{2} \right)$$

For N large, we can use the normal approximation to the binomial distribution developed in Section 8.3 to give

$$P_I = P\left(\sum_{i=1}^{N} X_i > \frac{13N}{24} \;\middle|\; \theta = \frac{1}{2} \right) \cong \frac{1}{\sqrt{2\pi}} \int_{C}^{\infty} e^{-t^2/2} \, dt$$

where

$$C = \left(\frac{13N}{24} - \frac{N}{2} \right) \Big/ \left(\frac{N}{4} \right)^{1/2} = \frac{\sqrt{N}}{12}$$

Consequently we have

$$P_I \cong 1 - \Phi(\sqrt{N}/12)$$

For $N = 100$, using the tables for the normal distribution gives

$$P_I \cong 1 - .798 = .202$$

A similar calculation for a Type II error, which occurs with probability P_{II}, yields

$$P_{II} = P\left(\frac{1}{N} \sum_{i=1}^{N} X_i < \frac{13}{24} \;\middle|\; p = \frac{7}{12} \right)$$

$$\cong \frac{1}{\sqrt{2\pi}} \int_{-\infty}^{C'} e^{-t^2/2} \, dt = \Phi(C')$$

where

$$C' = \left(\frac{13N}{24} - \frac{7N}{12}\right) \Big/ \left(N \cdot \frac{7}{12} \cdot \frac{5}{12}\right)^{1/2} = \frac{-\sqrt{N}}{2\sqrt{35}}$$

If $N = 100$, then $C' = -.841$ yielding

$$P_{\mathrm{II}} \cong .200$$

For $N = 1000$ these probabilities are $P_{\mathrm{I}} = .0042$ and $P_{\mathrm{II}} = .0037$.

EXAMPLE 10.2

In a binary communication system, during every T seconds one of two possible signals $s_0(t)$ or $s_1(t)$ is transmitted. Such systems are frequently used in a variety of situations such as computer communications and even voice transmission in the telephone network. We will assume for simplicity and convenience that $s_0(t) = +A$ and $s_1(t) = -A$, $0 \le t \le T$ and that the effect of the channel is to add noise $n(t)$, which for each t is a normal random variable, so that the receiver has available only $r(t) = s_i(t) + n(t)$. Based only on his observation of $r(t)$, $0 \le t \le T$ he must decide whether $s_0(t)$ or $s_1(t)$ was sent. We make a further simplifying assumption: Rather than assume $r(t)$ is available, we assume only that samples $r(t_i)$, $i = 1, \ldots, N$ are available, $t_i = i(T/N)$, $1 \le i \le N$. Our two hypotheses now are

H_0: $s_0(t)$ was transmitted
H_1: $s_1(t)$ was transmitted

and, based on observations $r(t_i) = s_i(t_i) + n(t_i)$, $1 \le i \le N$, we have to decide on one of these hypotheses and the probabilities of making the two types of errors. We will assume that the noise random variables $n(t_i)$ are independent normal random variables, $1 \le i \le N$, with zero mean and variance σ^2. A reasonable test under these conditions is to decide on H_0 if $\sum_{i=1}^{N} r(t_i) \ge 0$ and on H_1 otherwise. The joint pdf of $\mathbf{r}$, if the received N-tuple is $\mathbf{r} = (r(t_1), \ldots, r(t_N))$, given that H_0 is in effect is

$$L(\mathbf{r}; H_0) = \prod_{i=1}^{N} \frac{\exp[-(r_i - A)^2/2\sigma^2]}{\sqrt{2\pi}\sigma}$$

while

$$L(\mathbf{r}; H_1) = \prod_{i=1}^{N} \frac{\exp[-(r_i + A)^2/2\sigma^2]}{\sqrt{2\pi}\sigma}$$

If we let $Y = \sum_{i=1}^{N} r(t_i)$, then if H_0 is in effect, Y is a normal random variable with mean NA and variance $N\sigma^2$, that is, $g(y|H_0)$ the pdf of Y under H_0 is given by

$$g(y|H_0) = \frac{\exp[-(y - NA)^2/2N\sigma^2]}{\sqrt{2\pi N}\sigma}$$

while

$$g(y \mid H_1) = \frac{\exp[-(y + NA)^2 / 2N\sigma^2]}{\sqrt{2\pi N}\sigma}$$

It is now a relatively simple matter to determine P_{I} and P_{II}:

$$P_{\mathrm{I}} = (\text{probability } Y < 0 \text{ when } H_0 \text{ is true})$$

$$= \int_{-\infty}^{0} g(y \mid H_0)\, dy$$

$$= \int_{-\infty}^{0} \frac{\exp[-(y - NA)^2 / 2N\sigma^2]\, dy}{\sqrt{2\pi N}\sigma}$$

$$= \int_{-\infty}^{-NA/\sigma\sqrt{N}} \frac{e^{-t^2/2}}{\sqrt{2\pi}}\, dt = \Phi\!\left(\frac{-NA}{\sigma\sqrt{N}}\right)$$

and

$$P_{\mathrm{II}} = (\text{probability } Y > 0 \text{ when } H_1 \text{ is true})$$

$$= \int_{0}^{\infty} g(y \mid H_1)\, dy$$

$$= \int_{0}^{\infty} \frac{\exp[-(y + NA)^2 / 2N\sigma^2]\, dy}{\sqrt{2\pi N}\sigma}$$

$$= \int_{NA/\sigma\sqrt{N}}^{\infty} \frac{e^{-t^2/2}}{\sqrt{2\pi}}\, dt = 1 - \Phi\!\left(\frac{NA}{\sigma\sqrt{N}}\right) = \Phi\!\left(-\frac{NA}{\sigma\sqrt{N}}\right)$$

and the two kinds of errors have the same probability. Figure 10.1 shows the two density functions and the areas of the shaded regions are the probabilities of the two types of errors.

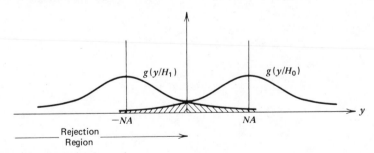

FIGURE 10.1

These two examples have been discussed on an entirely intuitive basis. The aim of any hypothesis testing procedure should be to use the information available to the best possible advantage. The problem is how to interpret the term "best possible advantage."

Suppose the parameter θ appears in a simple hypothesis test of the form H_0: $\theta = \theta_0$ versus H_1: $\theta = \theta_1$. Usually the probability of an error, of either kind, decreases as we take more samples. The probability of accepting H_0 is typically a function $O(\theta)$ of θ and is called the *operating characteristic* (*OC*) of the test. Its complement $P(\theta) = 1 - O(\theta)$ is called the *power function* of the test.

To illustrate these ideas further, and bearing in mind that, at least for the present we are proceeding intuitively, we modify the last example.

EXAMPLE 10.3

Determine the *OC* curve $O(A)$ for Example 10.2 if, with a test of the form

$$Y = \sum_{i=1}^{N} r(t_i) \leq C, \quad C \text{ a constant}$$

it is desired to have $P_1 \leq .01$.

From Example 10.2, the constant C can be determined by the requirement:

$$P_1 = P\left(\sum_{i=1}^{N} r(t_i) < C \text{ when } H_0 \text{ is true}\right)$$

$$= \int_{-\infty}^{C} g(y \mid H_0) \, dy$$

$$= \int_{-\infty}^{C} \frac{\exp[-(y-NA)^2/2N\sigma^2] \, dy}{\sqrt{2\pi N}\sigma}$$

$$= \int_{-\infty}^{(C-NA)/\sigma\sqrt{N}} \frac{e^{-t^2/2}}{\sqrt{2\pi}} \, dt = \Phi\left(\frac{C-NA}{\sigma\sqrt{N}}\right) = .01$$

When N, A, and σ are specified, C can be found. As a function of these variables

$$\frac{C-NA}{\sigma\sqrt{N}} = 2.32 \quad \text{or} \quad C = NA + 2.32\sigma\sqrt{N}$$

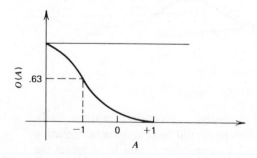

FIGURE 10.2

The probability of accepting H_0: $s_0(t)$ was transmitted:

$$O(A) = P(Y > C | H_0 \text{ is true}) + P(Y > C | H_1 \text{ is true})$$

$$= \int_C^\infty \frac{\exp[-(y - NA)^2 / 2N\sigma^2]}{\sqrt{2\pi N \sigma}} \, dy$$

$$+ \int_C^\infty \frac{\exp[-(y + NA)^2 / 2N\sigma^2]}{\sqrt{2\pi N \sigma}} \, dy$$

$$= 1 - \Phi\left(\frac{C - NA}{\sigma\sqrt{N}}\right) + 1 - \Phi\left(\frac{C + NA}{\sigma\sqrt{N}}\right)$$

$$= 2 - \Phi(2.32) - \Phi\left(\frac{2NA}{\sigma\sqrt{N}} + 2.32\right)$$

$$= 1.01 - \Phi\left(\frac{2NA}{\sigma\sqrt{N}} + 2.32\right)$$

As a function of A, for fixed N and σ, this gives the OC curve. It is shown in Figure 10.2 for $\sigma = 1$, for $N = 1$.

The power of a test is the probability of accepting H_1 when it is true, and if there exists two tests of level α we would choose that test with the greater power. The test that maximizes the power for a given level is called the *most powerful test*. In Section 10.3 we will be mainly concerned with testing a simple null hypothesis against a composite alternative one. In such a case it may turn out that there exists a single test that is the most powerful for any value of the parameter allowable under H_1. In such a case the test is called the *uniformly most powerful test*. Unfortunately, such tests often do not exist, as will be seen in Section 10.3.

For easy reference the terminology introduced in this section for testing the hypothesis H_0 versus H_1 involving the parameter μ is summarized:

An error of Type I occurs when H_0 is rejected when it is true.

An error of Type II occurs when H_1 is rejected when it is true.

P_I, the probability of a Type I error, is also called the level of significance of the test.

$I - P_{II}$ is called the power of the test.

$O(\theta)$, the operating characteristic of the test, is the probability of accepting H_0, regardless of which hypothesis is true, as a function of θ.

$1 - O(\theta)$ is the power function of the test.

A test is said to be the uniformly most powerful test (over a range of alternatives) if it minimizes P_{II} (maximizes the power) over the range for a fixed P_I and sample size.

If the level of the test is α its rejection region is said to be of size α.

10.2 Simple Hypotheses and the Neyman–Pearson Lemma

The previous section introduced the terminology of hypothesis testing and proceeded in a largely intuitive way. In this section we restrict our attention to testing the simple hypotheses $H_0: \theta = \theta_0$ versus $H_1: \theta = \theta_1$. Since both H_0 and H_1 are simple, the distributions under these hypotheses are completely specified and the error probabilities $P_I = \alpha$ and $P_{II} = \beta$ can be uniquely determined. The problem then is to construct that test which, for a given level α, minimizes β, that is, the problem is to find the critical region of size α that gives the most powerful test. The Neyman–Pearson lemma that follows gives the solution to this problem. For this situation we denote the joint pdf of the set of observations $\mathbf{x} = (x_1, x_2, \ldots, x_n)$ under $H_0: \theta = \theta_0$ as $L(\mathbf{x}; \theta_0)$, as in Example 10.2, and similarly, for $H_1: \theta = \theta_1$, $L(\mathbf{x}; \theta_1)$. If the sample is random and x_i has the pdf $f_X(x; \theta_0)$, then

$$L(\mathbf{x}; \theta_0) = \prod_{i=1}^{n} f(x_i; \theta_0)$$

The likelihood ratio is sometimes viewed as a random variable and sometimes as a function of the set of observations, as is any other statistic.

The Neyman–Pearson Lemma To test the two simple hypotheses $H_0: \theta = \theta_0$ versus $H_1: \theta = \theta_1$, the critical region R_1 of size α defined by

$$R_1 = \left\{ \mathbf{x} \in E_n \,\middle|\, \frac{L(\mathbf{x}; \theta_1)}{L(\mathbf{x}; \theta_0)} \geq C \right\}$$

is the most powerful region of size α (i.e., has the smallest probability of Type II error among all regions of size α). The constant C is chosen so that $P_I = \alpha$.

Proof From the assumptions of the lemma,

$$P_I = \alpha = \int_{R_1} L(\mathbf{x}; \theta_0) \, d\mathbf{x}$$

where the integration is n-fold over the region R_1 of E_n. Similarly,

$$P_{II} = \beta = \int_{R_0} L(\mathbf{x}; \theta_1) \, d\mathbf{x} = 1 - \int_{R_1} L(\mathbf{x}; \theta_1) \, d\mathbf{x}$$

since $R_1 = \bar{R}_0$. Now suppose that R_1' is some other region of size α;

$$\alpha = \int_{R_1'} L(\mathbf{x}; \theta_0) \, d\mathbf{x}$$

It is required to show that the probability of a Type II error associated with R_1',

P'_{II} is at least as large as β, that is,

$$P'_{II} = 1 - \int_{R'_1} L(\mathbf{x}; \theta_1)\, d\mathbf{x} \geq 1 - \int_{R_1} L(\mathbf{x}; \theta_1)\, d\mathbf{x}$$

$$= P_{II} = \beta$$

or, equivalently, that

$$\int_{R'_1} L(\mathbf{x}; \theta_1)\, d\mathbf{x} \leq \int_{R_1} L(\mathbf{x}; \theta_1)\, d\mathbf{x}$$

Referring to the Venn diagram of Figure 10.3 it is clear that

$$\int_{R_1} L(\mathbf{x}; \theta_1)\, d\mathbf{x} - \int_{R'_1} L(\mathbf{x}; \theta_1)\, d\mathbf{x} = \int_{R_1 \backslash S} L(\mathbf{x}; \theta_1)\, d\mathbf{x}$$

$$- \int_{R'_1 \backslash S} L(\mathbf{x}; \theta_1)\, d\mathbf{x}$$

$$> C \Bigg(\int_{R_1 \backslash S} L(\mathbf{x}; \theta_0)\, d\mathbf{x}$$

$$- \int_{R'_1 \backslash S} L(\mathbf{x}; \theta_0)\, d\mathbf{x} \Bigg)$$

and the right-hand side of this equation can be written as

$$C \Bigg(\int_{R_1 \backslash S} L(\mathbf{x}; \theta_0)\, d\mathbf{x} + \int_{S} L(\mathbf{x}; \theta_0)\, d\mathbf{x} - \int_{S} L(\mathbf{x}; \theta_0)\, d\mathbf{x} - \int_{R'_1 \backslash S} L(\mathbf{x}; \theta_0)\, d\mathbf{x} \Bigg)$$

$$= C \Bigg(\int_{R_1} L(\mathbf{x}; \theta_0)\, d\mathbf{x} - \int_{R'_1} L(\mathbf{x}; \theta_0)\, d\mathbf{x} \Bigg) = 0$$

since, by assumption, both integrals in this last equation equal α. It follows that

$$\int_{R_1} L(\mathbf{x}; \theta_1)\, d\mathbf{x} - \int_{R'_1} L(\mathbf{x}; \theta_1)\, d\mathbf{x} > 0$$

which was required to prove.

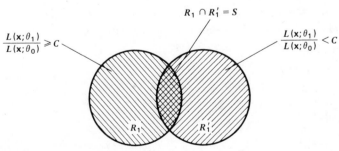

$$R_1 \cap R'_1 = S$$

$$\frac{L(\mathbf{x}; \theta_1)}{L(\mathbf{x}; \theta_0)} \geq C$$

$$\frac{L(\mathbf{x}; \theta_1)}{L(\mathbf{x}; \theta_0)} < C$$

$$R_1 \qquad R'_1$$

FIGURE 10.3

This lemma constructs the most powerful critical region of size α. Its use for normal data is illustrated in two cases before more general examples are considered. The ratio $L(\mathbf{x}; \theta_1)/L(\mathbf{x}; \theta_0)$ is referred to as the likelihood ratio of the sample.

Case i $H_0: \mu = \mu_0$ versus $H_1: \mu = \mu_1 < \mu_0$ (σ^2 known).
The function $L(\mathbf{x}; \mu_0)$ is

$$L(\mathbf{x}; \mu_0) = \prod_{i=1}^{n} \frac{1}{\sqrt{2\pi}\sigma} \exp\left(-\frac{(x_i - \mu_0)^2}{2\sigma^2}\right)$$

$$= \left(\frac{1}{\sqrt{2\pi}\sigma}\right)^n \exp\left(-\frac{1}{2\sigma^2} \sum_{i=1}^{n} (x_i - \mu_0)^2\right)$$

and similarly for $L(\mathbf{x}; \mu_1)$. The Neyman–Pearson lemma then yields the test of the form: Reject H_0 if $\mathbf{x} \in R_1$, where

$$R_1 = \left(\mathbf{x} \left| \frac{L(\mathbf{x}; \mu_1)}{L(\mathbf{x}; \mu_0)} \geq C\right.\right)$$

The likelihood ratio can be expressed as

$$\frac{L(\mathbf{x}; \mu_1)}{L(\mathbf{x}; \mu_0)} = \frac{\left(\dfrac{1}{\sqrt{2\pi}\sigma}\right)^n \exp\left(-\dfrac{1}{2\sigma^2} \sum (x_i - \mu_1)^2\right)}{\left(\dfrac{1}{\sqrt{2\pi}\sigma}\right)^n \exp\left(-\dfrac{1}{2\sigma^2} \sum (x_i - \mu_0)^2\right)}$$

$$= \exp\left(-\frac{n}{2\sigma^2}(\mu_1^2 - \mu_0^2) + \bar{x}n\frac{(\mu_1 - \mu_0)}{\sigma^2}\right)$$

For $\mu_1 < \mu_0$ the argument in the exponent is a monotonically decreasing function of

$$\bar{x} = \frac{1}{n} \sum_{i=1}^{n} x_i$$

and so the test of the form

$$\frac{L(\mathbf{x}; \mu_1)}{L(\mathbf{x}; \mu_0)} \geq C$$

can be reduced to a test of the form: Reject H_0 if $\bar{x} \leq C'$. To find the constant C', note that under H_0, $\bar{X}$ is a normal random variable with mean μ_0 and variance σ^2/n and

$$\alpha = P_\mathrm{I} = P(\bar{X} \leq C' | \mu = \mu_0)$$

$$= \int_{-\infty}^{C'} \frac{1}{\sqrt{2\pi}(\sigma/\sqrt{n})} \exp\left(-\frac{1}{2(\sigma^2/n)}(v - \mu_0)^2\right) dv$$

$$= \Phi\left(\frac{C' - \mu_0}{\sigma/\sqrt{n}}\right)$$

It follows that $(C' - \mu_0)/(\sigma/\sqrt{n}) = z_\alpha$ or $C' = (\sigma/\sqrt{n})z_\alpha + \mu_0$. For this decision constant the probability of a Type II error is

$$\beta = P_{II} = \int_{C'}^{\infty} \frac{1}{\sqrt{2\pi}(\sigma/\sqrt{n})} \exp\left(-\frac{1}{2(\sigma^2/n)}(v - \mu_1)^2\right) dv$$

$$= 1 - \Phi\left(\frac{C' - \mu_1}{(\sigma/\sqrt{n})}\right)$$

$$= 1 - \Phi\left(\frac{\mu_0 - \mu_1}{(\sigma/\sqrt{n})} + z_\alpha\right)$$

These expressions for α and β depend, of course, on the difference between the means being tested and the size of the sample. If it was assumed that the variance was unknown, we could use the fact that $(\bar{X} - \mu_0)/(S/\sqrt{n})$, S the unbiased sample variance, has a t-distribution with $(n-1)$ degrees of freedom to derive the test: Reject H_0 if $\bar{x} < \mu_0 + (s/\sqrt{n})t_{\alpha;n-1}$.

Case ii H_0: $\sigma^2 = \sigma_0^2$ versus H_1: $\sigma^2 = \sigma_1^2 < \sigma_0^2$ (assume μ known).
The ratio of the Neyman–Pearson lemma is

$$\frac{L(\mathbf{x}; \sigma_1)}{L(\mathbf{x}; \sigma_0)} = \frac{\left(\dfrac{1}{\sqrt{2\pi}\sigma_1}\right)^n \exp\left(-\dfrac{1}{2\sigma_1^2}\sum(x_i - \mu)^2\right)}{\left(\dfrac{1}{\sqrt{2\pi}\sigma_0}\right)^n \exp\left(-\dfrac{1}{2\sigma_0^2}\sum(x_i - \mu)^2\right)}$$

$$= \left(\frac{\sigma_0}{\sigma_1}\right)^n \exp\left[-\frac{1}{2}\sum(x_i - \mu)^2\left(\frac{1}{\sigma_1^2} - \frac{1}{\sigma_0^2}\right)\right]$$

which is a monotonically decreasing function of $\sum(x_i - \mu)^2$. The test then is to reject H_0 if $\sum_{i=1}^{n}(x_i - \mu)^2 \le C$. Recalling from Chapter 8 that $\sum_{i=1}^{n}(x_i - \mu)^2/\sigma^2$ is a chi-square random variable with n degrees of freedom,

$$P_I = \alpha = P\left(\sum_{i=1}^{n}\frac{(X_i - \mu)^2}{\sigma_0^2} \le \frac{C}{\sigma_0^2}\,\middle|\,\sigma^2 = \sigma_0^2\right)$$

implies that $C/\sigma_0^2 = \chi_{\alpha;n}^2$ or $C = \sigma_0^2\chi_{\alpha;n}^2$. The probability of a Type II error is

$$\beta = P\left(\sum_{i=1}^{n}(X_i - \mu)^2 > \sigma_0^2\chi_{\alpha;n}^2\,\middle|\,\sigma^2 = \sigma_1^2\right)$$

$$= P\left(\sum_{i=1}^{n}\frac{(X_i - \mu)^2}{\sigma_1^2} > \frac{\sigma_0^2}{\sigma_1^2}\chi_{\alpha;n}^2\,\middle|\,\sigma^2 = \sigma_1^2\right)$$

If the mean μ were unknown, the hypotheses would be composite since the distribution functions associated with them would not be completely specified. Other tests involving normal data can be treated in a similar manner. We conclude this section with a few examples.

EXAMPLE 10.4

A rare disease D_1 that is similar in symptoms to another relatively harmless disease D_0 has struck a community. N persons with the disease are being tested. It is possible to perform certain tests to assist in detecting the diseases but they are not conclusive. These measurements are reflected in a random variable X_i calculated for each patient, $i = 1, \ldots, N$. It is known that if the patient has disease D_0, then X_i has a normal distribution with mean 8 and variance 4. If the patient has disease D_1, then X_i has a normal distribution with mean 8.6 and variance also 4. We want to devise the most powerful test of level α. We assume that all patients have the same disease and that the measurements X_i are independent random variables.

It is required to test H_0: $\mu = 8$ versus H_1: $\mu = 8.6$ ($\sigma^2 = 4$) and, by an analysis similar to Case i earlier in the section, the test is of the form: Reject H_0 if $\bar{X} \geq C$, where $C = \mu_0 - (\sigma/\sqrt{n})z_\alpha = 8 + 3.292/\sqrt{n}$. The probability of a Type II error is

$$P_{II} = P(\bar{X} < C | \mu = 8.6)$$

$$= \int_{8+3.292/\sqrt{n}}^{\infty} \frac{1}{(\sqrt{2\pi} \cdot \sigma/\sqrt{n})} \exp\left(-\frac{1}{2(\sigma^2/n)}(v - 8.6)^2\right) dv$$

$$= 1 - \Phi\left(-\frac{.3}{\sqrt{n}} + 1.646\right)$$

From the Neyman–Pearson lemma, the test is the most powerful.

In the previous example we "fixed" the probability of Type I error and found the test that maximizes the probability of correctly detecting H_1 for a fixed number of samples. Clearly in some situations we can alter the process and, for example, attempt to find the minimum number of samples required to yield a probability of Type I error $\leq \alpha$ and a probability of Type II error $\leq \beta$. The following example illustrates.

EXAMPLE 10.5

The result of a test X_i of radar signals is a normal random variable. Under H_0, X_i has mean 5 and variance 9 while under H_1, X_i has mean 7 and variance 9. Find the number of samples required to result in $P_I \leq .05$ and $P_{II} \leq .02$.

To test the hypotheses H_0: $\mu = 5$ versus H_1: $\mu = 7$ ($\sigma^2 = 9$), the most powerful test will be of the form: Reject H_0 if $\bar{X} \geq C$. Equivalently, we can write this test as: Reject H_0 if

$$\frac{1}{(3/\sqrt{n})} \sum_{i=1}^{n} (X_i - 5) \geq C'$$

where, if H_0 is in effect, the left-hand side of the inequality is a standard normal random variable. For a test of level $\alpha = .05$ it follows that, for any number of samples n, $C' = 1.645$. To calculate P_{II} note that

$$\frac{1}{(3/\sqrt{n})} \sum_{i=1}^{n} (X_i - 5) = \frac{1}{(3/\sqrt{n})} \sum_{i=1}^{n} (X_i - 7) + \frac{2\sqrt{n}}{3}$$

and the test can be reformulated to reject H_0 if

$$\frac{1}{(3/\sqrt{n})} \sum_{i=1}^{n} (X_i - 7) \geq 1.645 - \frac{2\sqrt{n}}{3}$$

where the left-hand side is a standard random variable under H_1. Thus

$$P_{II} = \int_{-\infty}^{1.645 - 2\sqrt{n}/3} \frac{e^{-\mu^2/2}}{\sqrt{2\pi}} \, d\mu$$

$$= \Phi\left(1.645 - \frac{2\sqrt{n}}{3}\right)$$

and from the table of Appendix D, $1.645 - (2\sqrt{n}/3) \leq -2.054$ or $n \geq 31$ in order to attain the limits on both P_I and P_{II}.

The final example of this section considers the application of the Neyman–Pearson lemma to discrete random variables.

EXAMPLE 10.6

Two manufacturers make an electronic component to the same specifications. It turns out that the lifetime of the component is an exponentially distributed random variable with parameter .02 hour^{-1} from the first manufacturer and .015 hour^{-1} from the second. Find the test that minimizes P_{II} for a level of significance of .05, and determine its power function, assuming the null hypothesis tests the parameter $\alpha = .02$ hour^{-1}. What is the power of the test if, in a test of 10 components, the sample mean lifetime was found to be 57.2 hours?

In general, with the two hypotheses

H_0: $\alpha = \alpha_0$
H_1: $\alpha = \alpha_1$

for the exponentially distributed population $f(x) = \alpha e^{-\alpha x}$, $x \geq 0$, the Neyman–Pearson lemma yields the most powerful test that minimizes P_{II} for a given P_I. The test is to reject H_0 if

$$\frac{L(\mathbf{x}; \alpha_1)}{L(\mathbf{x}; \alpha_0)} \geq C$$

For the exponential distribution and a sample of size n,

$$\frac{L(\mathbf{x}; \alpha_1)}{L(\mathbf{x}; \alpha_0)} = \frac{(\alpha_1)^n \exp\left[-\alpha_1\left(\sum_1^n T_i\right)\right]}{(\alpha_0)^n \exp\left[-\alpha_0\left(\sum_1^n T_i\right)\right]}$$

$$= \left(\frac{\alpha_1}{\alpha_0}\right)^n \exp\left((\alpha_0 - \alpha_1)\sum_1^n T_i\right)$$

For fixed values of α_1, α_0, and n, this is a monotonically increasing function of $\sum_1^n T_i = n\bar{T}$ if $\alpha_1 < \alpha_0$. The test can be described as rejecting H_0 if $\sum_1^n T_i \geq C'$ for some constant C' determined by the required level of significance. The rejection or critical region R_1 is then the set of points $(t_1, t_2, \ldots, t_n)$ for which $\sum_1^n T_i \geq C'$.

In Chapter 8 it was shown that the random variable $2\alpha \sum_{i=1}^n T_i$ has a chi-square distribution with $2n$ degrees of freedom and consequently the probability of a Type I error for this test is

$$P_{\mathrm{I}} = P\left(\sum_1^n T_i \geq C' \,|\, \alpha = \alpha_0\right)$$

$$= P\left(2\alpha \sum_1^n T_i \geq 2\alpha C' \,|\, \alpha = \alpha_0\right) = p$$

For a given value of p, if Y is a χ^2 random variable with $2n$ degrees of freedom, then

$$P(Y \geq \chi^2_{1-p;2n}) = p$$

and so

$$2\alpha C' = \chi^2_{1-p;2n} \quad \text{or} \quad C' = \frac{\chi^2_{1-p;2n}}{2\alpha}$$

For the problem, $\alpha_0 = .02$ and $p = .05$, $n = 10$ and

$$C' = \frac{\chi^2_{.95;20}}{.04} = \frac{31.41}{.04} = 785.25$$

The test for the hypotheses H_0: $\alpha = .02$ hour^{-1} versus H_1: $\alpha = .015$ hour^{-1} is to reject H_0 if $\sum_{i=1}^n T_i \geq 785.25$. Notice that $\alpha = .015$ hour^{-1} played no part in determining the regions. The power of the test, which is the probability of rejecting H_0 when H_1 is true, is given by

$$P\left(\sum_1^{10} T_i \geq C' \,|\, \alpha = \alpha_1\right) = P\left(2\alpha_1 \sum_1 T_i \geq 2\alpha_1 C' \,|\, \alpha_1 = .015\right)$$

$$= P\left(2\alpha_1 \sum_1^{10} T_i > 23.56 \,|\, \alpha_1 = .015\right)$$

From a table of the χ^2 distribution ($2\alpha_1 \sum_1^{10} T_i$ is a χ^2 random variable with 20 degrees of freedom), this probability is approximately .735.

308 *Applied Probability*

It should be noted that the aim of an hypothesis test is to arrive at a conclusion regarding the consistency of the data with the given hypothesis. Thus for the same population our conclusion could vary from sample to sample. The techniques of hypothesis testing then are simply useful methods for the interpretation of data.

10.3 *Composite Hypotheses*

The formulation of the Neyman–Pearson lemma of the previous section required computation of the likelihood ratio $L(\mathbf{x}; \theta_1)/L(\mathbf{x}; \theta_0)$ and when both hypotheses are simple this is usually a straightforward matter. This lemma constructed a most powerful test. The construction of the most powerful test when one or both of the hypotheses is composite is a more difficult matter and, as we will see, in many cases such a test will not exist for all possible values of parameters in the composite alternative hypothesis. If the null hypothesis H_0 is composite, then there is not in general a unique value for P_I and the strategy is usually to take a pessimistic point of view and let P_I be the largest Type I probability of error among all values of the parameter in H_0, and similarly for H_1. With this approach in mind we formulate the likelihood ratio test. It usually gives good results, but there is no guarantee that it gives the most powerful test, as was the case in using the Neyman–Pearson lemma.

We will only be interested here in the case where H_0 is simple and H_1 composite, but we state the procedure for the general composite versus composite case. Suppose that under H_0 a collection of parameters $\theta = (\theta^{(1)}, \theta^{(2)}, \ldots, \theta^{(k)})$ are allowed to take values in some region of a k-dimensional space E, and call this region Λ_0. Under H_1 these parameters are allowed to take values in some other region Λ_1 and let $\Lambda = \Lambda_0 \cup \Lambda_1$. As we allow the parameters to vary over Λ_0 for some point or set of points, it takes on its maximum value, $\max L(\mathbf{x}; \theta)$, for the particular sample $\mathbf{X}$. If we allow the parameters to range over Λ, then

$$\max_{\theta \in \Lambda} L(\mathbf{X}; \theta) \ge \max_{\theta \in \Lambda_0} L(\mathbf{X}; \theta)$$

If we form the likelihood ratio

$$L(\mathbf{X}) = \frac{\max_{\Lambda_0} L(\mathbf{X}; \theta)}{\max_{\Lambda} L(\mathbf{X}; \theta)}$$

then our test of H_0 versus H_1 will be to reject H_0 when

$$L(\mathbf{X}) < C$$

where the constant C is chosen to give the desired level of significance.

From its definition, the constant C cannot be greater than unity. The test is different from that given in the Neyman–Pearson lemma, although it reduces to the same test for simple hypotheses. It is to be emphasized that it is not guaranteed to minimize P_{II} for a given P_I.

We will develop tests for normal data for four cases:

(i) H_0: $\mu = \mu_0$ versus H_1: $\mu > \mu_0$ (σ assumed known)
(ii) H_0: $\mu = \mu_0$ versus H_1: $\mu \neq \mu_0$ (σ assumed known)
(iii) H_0: $\mu = \mu_0$ versus H_1: $\mu < \mu_0$ (σ unknown)
(iv) H_0: $\mu^2 = \mu_0^2$ versus H_1: $\mu^2 \neq \mu_0^2$ (σ unknown)

Table 10.1 lists all the tests for normal data and some of these are given as problems at the end of this chapter.

TABLE 10.1 *Composite Hypothesis Tests for Normal Data*

H_0	H_1	CRITICAL REGION OF SIZE α	
$\mu = \mu_0$	$\mu > \mu_0$	$\bar{x} > \mu_0 + \dfrac{\sigma}{\sqrt{n}} z_{1-\alpha}$	σ known
		$\bar{x} > \mu_0 + \dfrac{s}{\sqrt{n}} t_{1-\alpha;n-1}$	σ unknown
$\mu = \mu_0$	$\mu < \mu_0$	$\bar{x} < \mu_0 - \dfrac{\sigma}{\sqrt{n}} z_{1-\alpha}$	σ known
		$\bar{x} < \mu_0 - \dfrac{s}{\sqrt{n}} t_{1-\alpha;n-1}$	σ unknown
$\mu = \mu_0$	$\mu \neq \mu_0$	$\lvert \bar{x} - \mu_0 \rvert > \dfrac{\sigma}{\sqrt{n}} z_{1-\alpha/2}$	σ known
	$\mu \neq \mu_0$	$\lvert \bar{x} - \mu_0 \rvert > \dfrac{s}{\sqrt{n}} t_{1-\alpha/2;n-1}$	σ unknown
$\sigma^2 = \sigma_0^2$	$\sigma^2 > \sigma_0^2$	$\sum (x_i - \bar{x})^2 > \sigma_0^2 \chi^2_{1-\alpha;n-1}$	μ unknown
$\sigma = \sigma_0^2$	$\sigma^2 < \sigma_0^2$	$\sum (x_i - \bar{x})^2 < \sigma_0^2 \chi^2_{\alpha;n-1}$	μ unknown
$\sigma = \sigma_0^2$	$\sigma^2 \neq \sigma_0^2$	$\sum (x_i - \bar{x})^2$ outside the interval $(\sigma_0^2 \chi^2_{1-\alpha/2}, \sigma_0^2 \chi^2_{\alpha/2})$	μ unknown

Case i H_0: $\mu = \mu_0$ versus H_1: $\mu > \mu_0$.

For convenience let the mean under the alternative hypothesis be for the moment $\mu_1 > \mu_0$. The likelihood ratio is

$$L(\mathbf{x}) = \frac{\exp[-\sum (x_i - \mu_0)^2 / 2\sigma^2]}{\max_\mu \exp[-\sum (x_i - \mu)^2 / 2\sigma^2]}$$

$$= \min_\mu \exp\left(\frac{n\bar{x}}{\sigma^2}(\mu_0 - \mu) + \frac{n(\mu_0^2 - \mu^2)}{\sigma^2} \right)$$

For any $\mu > \mu_0$ this is a monotonically decreasing function of $\bar{x}$ and it is easy to see that in this case the critical region is determined by the rule: Reject H_0 when $\bar{X} > C$. To determine the constant C note that

$$P(\bar{X} > C \mid \mu = \mu_0) = \alpha = P\left(\frac{\bar{X} - \mu_0}{\sigma/\sqrt{n}} > \frac{C - \mu_0}{\sigma/\sqrt{n}} \,\bigg|\, \mu = \mu_0\right)$$

from which we conclude that

$$\frac{C - \mu_0}{\sigma/\sqrt{n}} = z_{1-\alpha} \quad \text{or} \quad C = \mu_0 + \frac{\sigma}{\sqrt{n}} z_{1-\alpha}$$

Thus the critical region of size α for this test is determined by the rule: Reject H_0 when $\bar{X} > \mu_0 + (\sigma/\sqrt{n})z_{1-\alpha}$.

Suppose now that the actual mean of the data is $\mu_1 > \mu_0$. The probability of a Type II error, of accepting H_0 when H_1 is true, is

$$P_{\mathrm{II}}(\mu_1) = P\left(\bar{X} < \mu_0 + \frac{\sigma}{\sqrt{n}} z_{1-\alpha} \,\bigg|\, \mu = \mu_1\right)$$

$$= P\left(\frac{\bar{X} - \mu_1}{\sigma/\sqrt{n}} < \frac{\mu_0 - \mu_1}{\sigma/\sqrt{n}} + z_{1-\alpha} \,\bigg|\, \mu = \mu_1\right)$$

$$= \Phi\left(\frac{\mu_0 - \mu_1}{\sigma/\sqrt{n}} + z_{1-\alpha}\right)$$

Notice that the test in this case does not depend on the specific value of μ in effect under the alternative hypothesis although P_{II} does. Thus the test is the uniformly most powerful test for all $\mu > \mu_0$ in H_1. To test H_0: $\mu = \mu_0$ versus H_1: $\mu < \mu_0$ the modifications of the previous argument are simple and yield the test (again, uniformly most powerful) of the form: Reject H_0 if $\bar{X} < \mu_0 - (\sigma/\sqrt{n})z_{1-\alpha}$. The next case is only a little more complicated. Notice the similarities and differences of this test to Case i of the previous section involving only simple hypotheses.

Case ii H_0: $\mu = \mu_0$ versus H_1: $\mu \neq \mu_0$.

Such an alternative hypothesis must lead to a double-sided test, as we will see. The numerator of the likelihood ratio is simply

$$\left(\frac{1}{\sqrt{2\pi}\sigma}\right)^n \exp\left(-\frac{1}{2\sigma^2}\Sigma\,(x_i - \mu_0)^2\right)$$

The denominator of the likelihood ratio is, for $\mu = \mu_1$,

$$\left(\frac{1}{\sqrt{2\pi}\sigma}\right)^n \exp\left(-\frac{1}{2\sigma^2}\Sigma\,(x_i - \mu_1)^2\right)$$

and by differentiating with respect to μ_1 and setting the result to 0, the value of

μ_1 that maximizes this expression is $\mu_1 = \bar{x}$. The likelihood ratio is then

$$L(\mathbf{x}) = \frac{\exp\left(-\dfrac{1}{2\sigma^2}\sum(x_i - \mu_0)^2\right)}{\exp\left(-\dfrac{1}{2\sigma^2}\sum(x_i - \bar{x})^2\right)} = \exp\left(-\frac{(\bar{x} - \mu_0)^2}{2(\sigma^2/n)}\right)$$

Since this is a monotonically decreasing function of the absolute value of the difference between $\bar{x}$ and μ_0, the test is of the form: Reject H_0 if $|\bar{X} - \mu_0| > C$. To determine the constant C for $P_I = \alpha$, we argue as follows:

$$P_I = \alpha = P(|\bar{X} - \mu_0| > C | \mu = \mu_0)$$

$$= P\left(\frac{|\bar{X} - \mu_0|}{\sigma/\sqrt{n}} > \frac{C}{\sigma/\sqrt{n}}\right)$$

$$= 2\Phi\left(\frac{-C}{\sigma/\sqrt{n}}\right) = 2\Phi(z_{\alpha/2})$$

whence

$$C = -\frac{\sigma}{\sqrt{n}}z_{\alpha/2} = \frac{\sigma}{\sqrt{n}}z_{1-\alpha/2}$$

and the critical region of size α is

$$R_1 = \left\{\mathbf{x} \,\Big|\, |\bar{x} - \mu_0| > \frac{\sigma}{\sqrt{n}}z_{1-\alpha/2}\right\}$$

The probability of a Type II error again depends on the particular value $\mu = \mu_1$ in H_1:

$$P_{II} = P\left(|\bar{X} - \mu_0| < \frac{\sigma}{\sqrt{n}}z_{1-\alpha/2} \,\Big|\, \mu = \mu_1\right)$$

$$= P\left(\frac{\mu_0 - \mu_1}{\sigma/\sqrt{n}} - z_{1-\alpha/2} \leq \frac{\bar{X} - \mu_1}{\sigma/\sqrt{n}} \leq \frac{\mu_0 - \mu_1}{\sigma/\sqrt{n}} + z_{1-\alpha/2} \,\Big|\, \mu = \mu_1\right)$$

$$= \Phi\left(\frac{\mu_0 - \mu_1}{\sigma/\sqrt{n}} + z_{1-\alpha/2}\right) - \Phi\left(\frac{\mu_0 - \mu_1}{\sigma/\sqrt{n}} - z_{1-\alpha/2}\right)$$

It turns out that this double-sided test is not a uniformly most powerful test. For example, if μ under H_1 is actually greater than μ_0, then the one-sided test (Case i) is to reject H_0 if $\bar{X} > \mu_0 + (\sigma/\sqrt{n})z_{1-\alpha}$ and this is the most powerful test for this situation. However, the two-sided test is to reject H_0 if $|\bar{x} - \mu_0| > (\sigma/\sqrt{n})z_{1-\alpha/2}$, which is necessarily a less powerful test. A similar argument can be made for $\mu < \mu_0$. Notice that this does not imply that we should use a single-sided test for $H_1: \mu \neq \mu_0$ since, if we used the test for $H_1: \mu > \mu_0$ and, in fact, $\mu < \mu_0$, the probability of a Type II error would be large. It is not uncommon for single-sided tests derived using the likelihood ratio to be most powerful while two-sided tests are not.

The tests for cases where the mean or variance is not assumed to be known follow immediately from the first two cases on recognizing that the random variable $(\bar{X} - \mu)/(S\sqrt{n})$, where S^2 is the unbiased sample variance, has a t-distribution with $(n-1)$ degrees of freedom, while

$$\sum_{i=1}^{n} \frac{(X_i - \bar{X})^2}{\sigma^2}$$

has a χ^2 distribution with $(n-1)$ degrees of freedom. We use these facts in Cases iii and iv.

Case iii $H_0: \mu = \mu_0$ versus $H_1: \mu < \mu_0$ (σ unknown).

It is straightforward to show, using the likelihood ratio, that the test is of the form: Reject H_0 if $\bar{X} < C$ and the only problem is to determine the constant C:

$$P_{\mathrm{I}} = \alpha = P(\bar{X} < C \mid \mu = \mu_0)$$

$$= P\left(\frac{\bar{X} - \mu_0}{S/\sqrt{n}} < \frac{C - \mu_0}{S/\sqrt{n}} \,\bigg|\, \mu = \mu_0\right)$$

which implies that

$$\frac{C - \mu_0}{S/\sqrt{n}} = t_{\alpha;n-1}$$

or

$$C = \mu_0 + \frac{S}{\sqrt{n}} t_{\alpha;n-1} = \mu_0 - \frac{S}{\sqrt{n}} t_{1-\alpha;n-1}$$

The probability of a Type II error can likewise be found as a function of $\mu_0 - \mu$.

Case iv $H_0: \sigma^2 = \sigma_0^2$ versus $H_1: \sigma^2 \neq \sigma_0^2$ (μ unknown).

The numerator of the likelihood ratio is

$$\left(\frac{1}{\sqrt{2\pi}\sigma_0}\right)^n \exp\left(-\frac{1}{2\sigma_0^2}\sum(x_i - \mu)^2\right)$$

and maximizing this over μ, since μ is unknown, yields $\mu = \bar{x}$. Under H_1 (assuming $\mu = \mu_1$ for the moment) the denominator is

$$\left(\frac{1}{\sqrt{2\pi}\sigma}\right)^n \exp\left(-\frac{1}{2\sigma^2}\sum(x_i - \mu)^2\right)$$

The maximum value over σ for this expression is $\sigma^2 = \sum(x_i - \mu)^2/n$. Maximizing the resulting expression over μ then gives $\mu = \bar{x}$ and the likelihood ratio is

$$L(\mathbf{X}) = \left(\frac{(n-1)}{n} \cdot \frac{S^2}{\sigma_0^2}\right)^{n/2} \exp\left(-\frac{(n-1)S^2}{2\sigma_0^2} + \frac{n}{2}\right)$$

This likelihood ratio is not monotonic in S^2. Nonetheless, since $(n-1)S^2/\sigma^2$ has a chi-square distribution with $(n-1)$ degrees of freedom

$$P\left(\chi^2_{\alpha/2;n-1} \leq \frac{(n-1)S^2}{\sigma_0^2} \leq \chi^2_{1-\alpha/2;n-1}\right) = 1 - \alpha$$

and the critical region of size α is to reject H_0 if either

$$\frac{(n-1)S^2}{\sigma_0^2} = \frac{\sum (X_i - \bar{X})^2}{\sigma_0^2} < \chi^2_{\alpha/2;n-1} \quad \text{or} \quad \sum \frac{(X_i - \bar{X})^2}{\sigma_1^2} > \chi^2_{1-\alpha/2;n-1}$$

An expression for the probability of a Type II error can also be found (as a function of σ^2) from these considerations.

For the remainder of the section some examples of the use of these tests are considered.

EXAMPLE 10.7

The length of service of a component is a normally distributed random variable with mean 145 hours and variance 100 hours^2, if it has been manufactured correctly. A defect in the manufacturing process results in shorter lifetimes for the components. From a large lot, which is either all good or all defective, 10 items are chosen, tested, and found to have an average lifetime of 132 hours. Should the lot be accepted if we require the probability of rejecting a good lot to be less than .02? If the lot actually had a mean of 130 hours, what would the probability of accepting it be?

The population is assumed normal with known variance 100 hours^2 and we test the hypotheses $H_0\colon \mu_0 = 145$ versus $H_1\colon \mu = \mu_1 < 145$. For the moment, let us consider the general situation where the population is normal with variance σ^2 and the two hypotheses on its mean are $H_0\colon \mu = \mu_0$ versus $H_1\colon \mu = \mu_1 < \mu_0$. The ratio $L(\mathbf{x}; \mu_0)/\max_\mu L(\mathbf{x}; \mu)$ can be written as

$$\exp\left(-\sum_1^n (x_i - \mu_0)^2/2\sigma^2\right) \Big/ \max_\mu \exp\left(-\sum_1^n (x_i - \mu)^2/2\sigma^2\right)$$

where there is no maximization to be done on the numerator and we have to choose μ, $\mu \leq \mu_0$ to maximize the denominator or minimize the ratio. Rearranging terms, the test becomes, for *any* μ, to reject H_0 if

$$\exp\left[-\frac{1}{2\sigma^2}\left(\sum_1^n (x_i - \mu_0)^2 - (x_i - \mu)^2\right)\right] < C$$

or, equivalently, to reject H_0 if

$$\exp\left(\frac{\mu^2 - \mu_0^2}{2\sigma^2}\right) \exp\left[\left(\sum_1^n x_i\right)\frac{(\mu_0 - \mu)}{2\sigma^2}\right] < k$$

A little reflection on this equation will indicate that, for any value of μ less than μ_0, the left-hand side is an increasing function of $\sum_1^n x_i$. Consequently, the test must be to reject H_0 when $\sum_{i=1}^{n} X_i < C'$ for some new constant C' to be determined by the level of significance. Equivalently we reject H_0 when $\bar{X} < C''$. Since the sample mean under H_0 has mean μ_0 and variance σ^2/n, the level of significance is given by

$$P_{\mathrm{I}} = P(\bar{X} < C'' | \mu = \mu_0) = \int_{-\infty}^{C''} \frac{\exp\left(-\dfrac{(x-\mu_0)^2}{2\sigma^2/n}\right)}{\sqrt{2\pi}\sigma/\sqrt{n}}\, dx$$

$$= \Phi\left(\frac{C'' - \mu_0}{\sigma/\sqrt{n}}\right)$$

The probability of accepting H_0 when H_0 is true is then $1 - P_{\mathrm{I}}$. The probability of accepting H_0 when H_1 is true will obviously depend on the particular value of μ_1:

$$P_{\mathrm{II}} = P(\bar{X} \geq C'' | \mu = \mu_1) = \int_{C''}^{\infty} \frac{\exp\left(-\dfrac{(x-\mu_1)^2}{2\sigma^2/n}\right)}{\sqrt{2\pi}\sigma/\sqrt{n}}\, dx$$

$$= 1 - \Phi\left(\frac{C'' - \mu_1}{\sigma/\sqrt{n}}\right)$$

If the level of significance is α, then z_α is defined by $\Phi(z_\alpha) = \alpha$ and $C'' = z_\alpha \sigma/\sqrt{n} + \mu_0$. The operating characteristic of the test is then

$$O(\mu_1) = 1 - \alpha + 1 - \Phi\left(z_\alpha + \frac{\mu_0 - \mu_1}{\sigma/\sqrt{n}}\right)$$

To return to the problem, if $\bar{X} = 132$ hours, $\mu_0 = 145$ hours, $\sigma = 10$ hours, $n = 10$, and $P_{\mathrm{I}} = .02$, then $z_\alpha = -2.05$ and the test is to reject H_0 ($\mu = 145$ hours) if

$$\bar{X} < \frac{-2.05 \cdot 10}{\sqrt{10}} + 145 = 138.52$$

Since $\bar{X} = 132$, we reject the hypothesis H_0.

If the true mean of the population was 130 hours, then the probability that the lot would be accepted is

$$O(130) = 2 - .02 - \Phi\left(-2.05 + \frac{15}{10/\sqrt{10}}\right)$$

$$= 1.98 - \Phi(2.69) = .984$$

Notice that the operating characteristic was expressed as $O(\mu_1)$ in this last example. Since the probability varies with both μ_0 and μ_1 and, in fact, depends only on $\mu_0 - \mu_1$, it would be reasonable to plot it as a function of $\mu_0 - \mu_1$, in this case.

If we had assumed that $\mu_1 > \mu_0$, then it is clear from this example that the test would have been of the form: Reject H_0: $\mu = \mu_0$ (versus H_1: $\mu = \mu_1 > \mu_0$) if $\bar{X} > C$ for some constant C determined by the level of significance required. These are examples of one-sided tests since the critical region is determined by a single one-sided inequality. The following example illustrates a two-sided inequality.

EXAMPLE 10.8

A certain grade ore A has a density that is a normally distributed random variable with mean 8.40 g/cm^3 and standard deviation 1.50 g/cm^3. Ten samples of some ore are taken and found to have an average density of 9.05 g/cm^3. Should the ore from which the samples were taken be assumed to be of type A with a level of significance of 5%?

The feature of this problem is that we are only interested in whether or not the ore is of type A. This requires hypotheses of the form H_0: $\mu = \mu_0$ versus H_1: $\mu \neq \mu_0$ and the trouble arises in dealing with the composite hypothesis H_1. The test suggested at the beginning of the section requires calculating the likelihood ratio $L(\mathbf{x})$. For any $\mu = \mu_1$ under H_1 we can write the ratio

$$\frac{L(\mathbf{x}; \mu_0)}{L(\mathbf{x}; \mu_1)} = \exp\left[-\frac{1}{2\sigma^2}\left(\sum_{i=1}^{n}[(x_i - \mu_0)^2 - (x_i - \mu_1)^2]\right)\right]$$

$$= \exp\left\{-\frac{1}{2\sigma^2}\left[\left(\sum_{i=1}^{n} 2(\mu_1 - \mu_0)x_i\right) + n(\mu_0^2 - \mu_1^2)\right]\right\}$$

Maximizing the denominator of the left-hand side of this expression with respect to μ_1 is equivalent to minimizing the whole expression and this can be accomplished by differentiating with respect to μ_1. The result is that μ_1 should be chosen as $\bar{x}$ and substituting this value in the expression gives

$$L(\mathbf{x}) = \exp\left(-\frac{n}{2\sigma^2}(\bar{x} - \mu_0)^2\right)$$

and the test is to reject H_0 when $L(\mathbf{x}) < C$ or, equivalently, to reject H_0 when $(\bar{x} - \mu_0)^2 > C'$. The probability of a Type I error is

$$P_{\mathrm{I}} = \int_{-\infty}^{\mu_0 - C''} + \int_{\mu_0 + C''}^{\infty} \frac{\exp\left(-\dfrac{(y - \mu_0)^2}{2\sigma^2/n}\right) dy}{\sqrt{2\pi}(\sigma/\sqrt{n})}$$

$$= 2\int_{\mu_0 + C''}^{\infty} \frac{\exp\left(-\dfrac{(y - \mu_0)^2}{2\sigma^2/n}\right) dy}{\sqrt{2\pi}(\sigma/\sqrt{n})}$$

$$= 2\int_{C''/(\sigma/\sqrt{n})}^{\infty} \frac{\exp(-z^2/2) \, dz}{\sqrt{2\pi}}$$

$$= 2\Phi\left(-\frac{C''}{(\sigma/\sqrt{n})}\right)$$

For the level of significance .05 with $n = 10$, $\sigma = 1.225$ we have

$$.05 = 2\Phi\left(-\frac{C''}{\sigma/\sqrt{n}}\right) \quad \text{or} \quad -\frac{C''}{\sigma/\sqrt{n}} = -1.96$$

$$C'' = .759$$

Thus, the test is to accept H_0 if $7.64 \le \bar{X} \le 9.16$ with the level of significance of .05. Since the sample average is 9.05, we accept H_0 at the .05 significance level.

The final example of the section indicates that the same arguments are easily applied to the discrete case.

EXAMPLE 10.9

The probability that an item is defective is θ. The following experiment is run to decide whether $\theta = .40$ or $\theta < .40$. Items are tested until the first defective is found and the number of tests noted. The experiment is repeated twice with 3 and 5 tests needed, respectively. If the level of significance is .10, should the hypothesis that $\theta = .40$ be accepted?

The situation can be regarded as drawing n samples from a discrete population with a geometric distribution function

$$P(X = k) = (1 - \theta)\theta^{k-1}, \quad k = 1, 2, \ldots$$

Based on the random sample of size n, it is necessary to decide between the two hypotheses H_0: $\theta = \theta_0$ versus H_1: $\theta = \theta_1 < \theta_0$. Under H_0,

$$L(\mathbf{x}; \theta_0) = (1 - \theta_0)^{\sum_1^n k_i} \theta_0^{\sum_1^n k_i - n}$$

while under H_1,

$$L(\mathbf{x}; \theta_1) = (1 - \theta_1)^{\sum_1^n k_i} \theta_1^{\sum_1^n k_i - n}$$

where k_i is the number of items inspected in the ith experiment. For a fixed $\theta_1 \le \theta_0$, the likelihood ratio is

$$L(\mathbf{x}) = \left(\frac{1 - \theta_0}{1 - \theta_1}\right)^{\sum_1^n k_i} \left(\frac{\theta_0}{\theta_1}\right)^{\sum_1^n k_i - n}$$

$$= \left(\frac{(1 - \theta_0)\theta_0}{(1 - \theta_1)\theta_1}\right)^{\sum_1^n k_i} \left(\frac{\theta_0}{\theta_1}\right)^{-n}$$

and, for $\theta_1 + \theta_0 < 1$,

$$\frac{(1 - \theta_0)\theta_0}{(1 - \theta_1)\theta_1} > 1$$

In this case $L(\mathbf{x})$ is an increasing function of $\sum_{i=1}^n k_i$, and the test is to reject H_0 if $L(\mathbf{x}) < C$ or, equivalently, to reject H_0 if

$$\sum_{i=1}^n X_i < C'$$

The sum of n geometrically distributed random variables has a Pascal distribution since it has the interpretation of the number of trials required to achieve the nth defective item. Thus if $Z = \sum_{i=1}^{n} X_i$,

$$P(Z = k) = \binom{k-1}{n-1}(1-\theta)^k \theta^{n-k}, \quad k = n, n+1, \ldots$$

The constant C' is chosen so that

$$\sum_{k=n}^{C'} \binom{k-1}{n-1}(1-\theta_0)^k \theta_0^{n-k} = P_{\mathrm{I}}$$

where the upper limit is adjusted to be the largest integer for which the left-hand side is less than P_{I}. Some calculation shows that, when $n = 2$ and $\theta_0 = .4$, then C' must be 5 and the test is to reject H_0 when $X_1 + X_2 < 5$. Since $X_1 + X_2 = 8 > 5$, the hypothesis that $\theta_0 = .40$ should be accepted.

This last example, the first where nonnormal data were used, introduces the problem of how to treat such cases in general. So far the form of the test and the evaluation of any constants in them to yield $P_{\mathrm{I}} = \alpha$ was straightforward. For nonnormal data this is often not true. It can be shown however that for H_0 simple the statistic $-2 \ln[L(\mathbf{X})]$ is approximately a χ^2 random variable with one degree of freedom under the assumption that H_0 is true. This fact can then be used to determine (approximate) expressions for the necessary test constants and error probabilities.

Finally, it should be noted that there is a clear connection between one- and two-sided hypothesis tests and one- and two-sided confidence intervals. For example, to test H_0: $\mu = \mu_0$ versus H_1: $\mu \neq \mu_0$ (σ known), the test is of the form: Reject H_0 if

$$|\bar{X} - \mu_0| > \frac{\sigma}{\sqrt{n}} z_{1-\alpha/2}$$

or, equivalently, accept H_0 if

$$\mu_0 - \frac{\sigma}{\sqrt{n}} z_{1-\alpha/2} < \bar{X} < \mu_0 + \frac{\sigma}{\sqrt{n}} z_{1-\alpha/2}$$

or if

$$\bar{X} - \frac{\sigma}{\sqrt{n}} z_{1-\alpha/2} < \mu_0 < \bar{X} + \frac{\mu}{\sqrt{n}} z_{1-\alpha/2}$$

This last region is just the $100(1-\alpha)\%$ confidence interval for μ, assumed unknown.

10.4 *Goodness of Fit Tests*

In most of the previous work of this chapter, it was assumed that the distribution function of the random variable under test was known, and the test involved certain parameters of the distribution. Often in a practical situation the distribution is not known and, as a first step in the analysis, it is desired to model it, that is, based on the measurements it is required to choose a distribution that "fits" the data. This could be set up as an hypothesis testing situation where the null hypothesis is represented by a distribution that we feel might fit the observations reasonably well and the alternative hypothesis is that the distribution is not of that form. Obviously we can no longer form a likelihood ratio since there is no assumed form for the distribution under the alternative hypothesis for this case. Another approach is required. One possible approach, considered in this section, is the χ^2 goodness of fit test.

As an introduction to this test suppose that X is a random variable with pdf $f(x)$ and let $I_1, I_2, \ldots, I_m$ be m disjoint intervals such that their union is the range of X, R_X. Only the case of a continuous random variable will be analyzed as the modifications required for a discrete random variable are simple. Define the probabilities $\theta_1, \theta_2, \ldots, \theta_m$ by

$$\theta_j = P(X \in I_j) = \int_{I_j} f(x)\, dx, \quad j = 1, 2, \ldots, m, \quad \sum_{i=1}^{m} \theta_i = 1$$

If $X_1, X_2, \ldots, X_n$ is a random sample of X, then if Y_i is the number of these samples that fall into the ith interval, then

$$P(Y_1 = n_1, Y_2 = n_2, \ldots, Y_m = n_m) = \frac{n!}{n_1! n_2! \cdots n_m!} \theta_1^{n_1} \theta_2^{n_2} \cdots \theta_m^{n_m}$$

where $\sum_{i=1}^{m} n_i = n$; that is, the random variables $Y_1, Y_2, \ldots, Y_m$ have a multinomial distribution.

Since the population X has the pdf $f(x)$, the ratios n_i/n, as n becomes large, should approach the probabilities θ_i, $i = 1, \ldots, m$. If the population is only assumed to have this pdf, but in fact has some other pdf $g(x)$, then some deviation between the observed frequencies n_i and expected frequencies $n\theta_i$ should appear. This simple observation is the basis of the χ^2 goodness of fit test.

Our hypothesis testing situation is now to test H_0 versus H_1, where

H_0: the observed frequencies n_i are consistent with $n\theta_i$, $i = 1, \ldots, m$
H_1: the observed frequencies are not consistent

The approximate equivalence between this hypothesis test and the original test concerning the distributions is clear. We have reduced the test H_0: the distribution is $f(x)$ versus H_1: the distribution is not $f(x)$ to one of testing the multinomial probabilities θ_i obtained from $f(x)$ against the observed frequencies.

Although we cannot form the likelihood ratio for the test directly, we can form the ratio

$$L = \frac{\left(\dfrac{n!}{n_1! n_2! \cdots n_m!}\right) \theta_1^{n_1} \theta_2^{n_2} \cdots \theta_m^{n_m}}{\left(\dfrac{n!}{n_1! n_2! \cdots n_m!}\right) \left(\dfrac{n_1}{n}\right)^{n_1} \left(\dfrac{n_2}{n}\right)^{n_2} \cdots \left(\dfrac{n_m}{n}\right)^{n_m}}$$

where the ratio n_i/n is used as the estimate of θ_i. If we let e_i be the number of observations out of n expected to fall into the interval I_i, then $e_i = n\theta_i$ and we can write L as

$$L = \left(\frac{e_1}{n_1}\right)^{n_1} \left(\frac{e_2}{n_2}\right)^{n_2} \cdots \left(\frac{e_m}{n_m}\right)^{n_m}$$

or, more conveniently,

$$\ln(L) = \sum_{i=1}^{m} n_i \ln\left(\frac{e_i}{n_i}\right)$$

It can be shown that, as n tends to ∞, the random variables $Y_1, Y_2, \ldots, Y_m$, where Y_i is the number of the observations in the interval I_i, have an m-variate normal pdf and that the statistic

$$\Lambda = \sum_{i=1}^{m} \frac{(Y_i - e_i)^2}{e_i} \tag{10.1}$$

tends to a chi-square random variable with $(m-1)$ degrees of freedom. It can also be shown that the sum $\sum_1^m (n_i - e_i)^2/e_i$ approximates $-2 \ln(L)$ and so the statistic Λ is approximately proportional to the log of a likelihood ratio. It will be used below in the goodness of fit test.

An intuitive, nonrigorous way of understanding the above approximation is to note that Y_i is a binomial random variable with parameters n and θ_i. We observed in Chapter 8 that for large n and moderate θ the random variable

$$W_i = \frac{Y_i - n\theta_i}{\sqrt{n\theta_i(1 - \theta_i)}}$$

is approximately a standard normal random variable. Thus the sum of squares

$$\sum_{i=1}^{m} \frac{(Y_i - n\theta_i)^2}{n\theta_i(1 - \theta_i)}$$

is approximately a χ^2 random variable with $(m-1)$ degrees of freedom. If the θ_i are all relatively small, we can make the further approximation $\theta_i(1 - \theta_i) \cong \theta_i$ and conclude that

$$\sum_{i=1}^{m} \frac{(Y_i - n\theta_i)^2}{n\theta_i}$$

is approximately a chi-square random variable with $(m-1)$ degrees of freedom, which is the result of Eq. (10.1).

The χ^2 goodness of fit test is to reject H_0, that the true density is $f(x)$, if

$$\Lambda = \sum_{i=1}^{m} \frac{(Y_i - e_i)^2}{e_i} > \chi^2_{1-\alpha;m-1} \qquad (10.2)$$

This implies that the probability of a Type I error, of rejecting H_0 when in fact it is true, is α, that is, the test is at the $100\alpha\%$ level of significance.

In order to use the test of Eq. (10.2) we should have $e_i \geq 5$, and the intervals I_i should be chosen accordingly. For discrete random variables the choice of I_i is often dictated by n and the range of the random variable in question. For continuous random variables the choice of the intervals is far more arbitrary. A guideline is to choose them approximately the same length, and choose as many as possible consistent with the requirement that $e_i = n\theta_i > 5$. If the data are given in finer intervals, and some of them do not meet the requirement $n\theta_i \geq 5$, they should be combined with others until the requirement is satisfied.

Sometimes we merely wish to test whether or not a population has a distribution function of a certain form without knowing some of its parameters. For example, we may want to test whether or not X has a normal distribution without knowing its mean or variance. To calculate the probabilities θ_i required for Eq. (10.2) these parameters are needed. It can be shown however that if r parameters of $f(x)$ are unknown, then estimates of them obtained from the sample can be used in $f(x)$ in order to calculate the probabilities θ_i. In this case, however, the random variable Λ is better approximated as a χ^2 random variable with $(m - r - 1)$ degrees of freedom; for each unknown parameter the number of degrees of freedom is decreased by 1.

EXAMPLE 10.10

The lifetime of a transformer is thought to be an exponentially distributed random variable with parameter .45 years^{-1}. Fifty of them are tested with the results

Time interval (years)	0–1	1–2	2–3	>4
No. of failures	21	16	9	4

If $\alpha = .05$ is the hypothesis to be accepted?

The exponential pdf with parameter .45 yields the probabilities

$$\theta_1 = \int_0^1 (.45) e^{-.45t} \, dt = .362, \quad \theta_2 = .232, \quad \theta_3 = .145, \quad \theta_4 = .259$$

With these parameters the statistic Λ is calculated as

$$\Lambda = \frac{(21-18.1)^2}{18.1} + \frac{(16-11.6)^2}{11.6} + \frac{(9-7.25)^2}{7.25} + \frac{(4-12.95)^2}{12.95}$$

$$= 8.74$$

From Appendix E, $\chi^2_{.95;3} = 7.81$ and, since $\lambda = 8.74 > \chi^2_{.95;3}$, we reject the hypothesis of the exponential distribution.

EXAMPLE 10.11

The number of particles emitted from a radioactive source is thought to have a Poisson distribution. The number of particles emitted in 100 consecutive 10-sec intervals are as follows:

Number of particles	0	1	2	3	4	>5
Number of 10-sec intervals	11	30	25	20	10	4

Would you accept the hypothesis that the distribution is Poisson at the $\alpha = .01$ level? At the .05 level?

Since the mean is not given, we use the data to approximate it finding $\hat{\lambda} = 2.00$. For this value, the Poisson distribution gives the frequencies

$$\theta_0 = .135, \quad \theta_1 = .270, \quad \theta_2 = .270, \quad \theta_3 = .180, \quad \theta_4 = .090, \quad \theta_5 = .055$$

and the variable is

$$\Lambda = \frac{(11-13.5)^2}{13.5} + \frac{(30-27)^2}{27} + \frac{(25-27)^2}{27} + \frac{(20-18)^2}{18} + \frac{(10-9)^2}{9}$$
$$+ \frac{(4-5.5)^2}{5.5}$$
$$= 1.687$$

The variable $\chi^2_{.99;4}$ is, from Appendix E, 13.77 and since $\Lambda = 1.687 < 13.277$, we accept the hypothesis. At the 5% level of significance $\chi^2_{.95;4} = 9.487$ and the hypothesis is still accepted.

EXAMPLE 10.12

The times between arrivals at a toll booth on a turnpike, during a particular time of day, are thought to be exponentially distributed with parameter 3.0 minute^{-1}. The result of a particular set of 20 readings gave inter arrival times (in seconds) of 14.6, 9.8, 29.8, 25.3, 11.0, 8.7, 30.6, 22.4, 4.6, 19.7, 12.1, 29.0, 5.2, 15.6, 32.1, 8.6, 42.1, 7.8, 21.7, and 7.4. Would you support this hypothesis at the $\alpha = .01$ level of significance? At the .05 level of significance? If the parameter of the exponential distribution used had to be estimated, would this change the results?

With only 20 readings it is difficult to attain much accuracy since if $20\theta_i$ is to be greater than 5, θ_i for each interval must be greater than .25. We choose three intervals, somewhat arbitrarily, and after experimentation, as (0–7.5), (7.5–20), and (>20) seconds. The intervals are not of equal length since the scarcity of data required some accommodation. The probabilities for these intervals are

$$\theta_0 = \int_0^{7.5} \left(\frac{1}{20}\right) \exp\left(-\frac{1}{20}t\right) dt$$

$$= -e^{-1/20t}\big|_0^{7.5} = .314$$

$$\theta_1 = .318$$

$$\theta_2 = .368$$

and note that $20\theta_i \geq 5$ for all i. The observed frequencies for these intervals are:

Interval	0–7.5	7.5–20	>20
Number of times	7	5	8

The variable Λ is then calculated as

$$\Lambda = \frac{(7-6.28)^2}{6.28} + \frac{(5-6.36)^2}{6.36} + \frac{(8-7.36)^2}{7.36}$$

$$= .429$$

From Appendix E $\chi^2_{.99;2} = 9.210$ and $\chi^2_{.95;2} = 5.991$ and both hypotheses are accepted.

If the parameter of the exponential density was unknown, we would have to estimate it. From the data given, the mean interarrival time is 18.4 sec and the estimate for the parameter λ is 1/18.4. The interval probabilities are now

$$\theta_0 = .335, \quad \theta_1 = .328, \quad \theta_2 = .337$$

and the variable Λ is

$$\Lambda = \frac{(7-6.79)^2}{6.70} + \frac{(5-6.56)^2}{6.56} + \frac{(8-6.74)^2}{6.74}$$

$$= .620$$

Since the exponential parameter had to be estimated, this quantity is compared to $\chi^2_{.99;1} = 6.635$ and $\chi^2_{.95;1} = 3.841$ and in both cases we again accept the hypothesis.

Extensions of these ideas involving two-way classification schemes and contingency tables can also be considered, but the simple χ^2 goodness of fit test is sufficient to illustrate the ideas here.

Problems

1. A normal population has variance $\sigma^2 = 4$. Find the probability P_{II} for the hypotheses H_0: $\mu = 0$ versus H_1: $\mu = 1$ at the 5% level for a sample of size 10.

2. Construct a likelihood ratio test for a normal population with variance $\sigma^2 = 25$ for the hypotheses H_0: $\mu = 100$ versus H_1: $\mu = 120$. How many samples are needed at the 5% level of significance so that the probability of deciding on H_1 when it is true is at least .95?

3. In a test of the tensile strength of a nylon thread it is known that the mean is 100 kg. Assuming that the actual strength is a normal random variable it is desired to test H_0: $\sigma^2 = 4.0$ kg^2 versus H_1: $\sigma^2 = 6.4$ kg^2. If 10 samples are taken and $\sum_{i=1}^{10} (x_i - 100)^2 = 45.2$, should H_0 be accepted at the 5% level of significance?

4. The random variable X is uniformly distributed on the interval $(0, a)$. The hypotheses H_0: $a = 2$ versus H_1: $a = 2.5$ is tested by rejecting H_0 if max $(X_1, X_2) > 1.9$, where (X_1, X_2) is a random sample of size 2. What is P_I and P_{II} for this test?

5. The failure law of a class of light bulbs is normal with standard deviation $\sigma = 20$ hours. A sample of size 20 is tested and found to have an average of 644 hours. Would you accept the hypothesis H_0: $\mu = 620$ versus H_1: $\mu = 680$ at the 10% level of significance?

6. The time to failure of a relay is a normally distributed random variable with a standard deviation of 100 hours. Devise a test to decide between the two hypotheses H_0: $\mu = 800$ hours versus H_1: $\mu = 1000$ hours if P_I is to be less than .02 and P_{II} less than .05.

7. The number of absentee workers per day at a large plant is assumed to be a Poisson random variable with mean $\lambda = 40$. A new incentive program is instituted and its effect is to be discussed at the next board meeting. Over a five-day period the average number of absentee workers is found to be 17. What is the level of significance for the result for the hypotheses H_0: $\lambda = 4.0$ versus H_1: $\lambda = 3.5$?

8. The time to failure of an electronic component is an exponentially distributed random variable with parameter either $\alpha_0 = .0010$ hours^{-1} or $\alpha_1 = .0008$ hours^{-1}. How large a sample should be taken from a lot to have the power of the test .75 when the level of significance of the test is 5%?

9. The weight of cereal actually put into a box is a normally distributed random variable with mean 300 g. When the loading machine is working correctly the variance of this weight is 25 g^2. What is the 5% level of significance test for the hypothesis H_0: $\sigma^2 = 25$ versus H_1: $\sigma^2 = 30$ when only 20 samples are used?

10. A population is exponentially distributed with parameter λ. Use the Neyman–Pearson lemma to derive an optimum test for the hypotheses H_0: $\lambda = \lambda_0$ versus H_1: $\lambda = \lambda_1 < \lambda_0$.

11. A normal population has a known mean μ and a variance either σ_0^2 or σ_1^2. Derive the most powerful test for a level of significance α, based on a random sample of size n, for the simple hypotheses H_0: $\sigma = \sigma_0$ versus H_1: $\sigma = \sigma_1$.

12. A light bulb manufacturer claims that a certain bulb will last more than 1000 hours. If the lifetime is exponentially distributed and 25 bulbs tested were found to have an average lifetime of 1014.7 hours, would you accept the hypothesis H_0: $\alpha = .001$ versus H_1: $\alpha > .001$ at the 5% level?

13. The number of accidents per week in a plant was found to be a Poisson random variable with parameter $\lambda = 3.0$. After a campaign to increase safety consciousness it was found that an average of 2.5 accidents occurred per week during a four-week interval. Is this a significant improvement (say at the 5% level of significance)?

14. The number of errors a typesetter makes per page is a Poisson random variable with parameter λ. In the morning this parameter is known to be $\lambda = .1$ while in the afternoon he makes many more errors. One hundred pages are chosen at random and the total number of mistakes found is 14. If these pages are known to have

been all set during either a morning or an afternoon, would you accept the hypothesis at the 5% level of significance that they were set in the morning?

15. Type A fertilizer gives a normally distributed yield, with mean 21 bushels per acre, of a certain crop. Type B fertilizer is used on a 25 acre experimental plot with a resulting average yield 22.2 bushels per acre. Devise a test for the hypotheses H_0: $\mu_0 = 21$ versus H_1: $\mu_1 > 21$ for a new fertilizer for a level of significance of .05 assuming a known variance of 4 bushels^2 per acre.

16. Examples in Section 10.3 derived a test for the composite hypothesis H_0: $\mu = \mu_0$ versus H_1: $\mu < \mu_0$ (or $\mu > \mu_0$) when the population is normally distributed with variance σ^2. What would this test be if the variance was unknown?

17. As in Problem 16 devise the test for the hypotheses H_0: $\mu = \mu_0$ versus H_1: $\mu \neq \mu_0$ when the population is normally distributed with mean μ and unknown variance.

18. Players in a crap game wish to test the fairness of a die by checking to see how frequently a 1 appears. If they test only the hypotheses H_0: $\theta = \frac{1}{6}$ versus H_1: $\theta < \frac{1}{6}$, where θ is the true probability of obtaining a one, determine the test for a .05 level of significance.

19. Suppose X is a normal random variable that has a mean μ and standard deviation 1. The test of the hypotheses H_0: $\mu = 6.0$ versus H_1: $\mu \neq 6.0$ is taken as: Reject H_0 if $|\bar{X} - 6.0| > 0.4$, $\bar{X}$ the sample mean. Sketch P_I plotted against the number of samples taken.

20. As in Problem 18 it is desired to test the fairness of a die, only this time it is to be accomplished by observing the number of rolls needed to get a 6. Would you accept or reject the hypothesis that the die is fair at the 10% level of significance if on three experiments the average number of rolls needed to get a 6 was $7\frac{1}{3}$?

21. The length of time that a stamp press remains in service is an exponentially distributed random variable with parameter λ. To test the hypotheses H_0: $\lambda = .2$ days^{-1} versus H_1: $\lambda > .2$ days^{-1}, the failure time T of one machine is observed. The critical region is defined as $T < 4.1$. Calculate P_I for this test.

22. In a cattle breeding experiment the size of calves at one month is a normally distributed random variable with mean $\mu = 100$ kg and variance $\sigma^2 = 16$ kg^2 on a certain feeding regimen. A different regimen assumed to have the same variance is to be tested for its effectiveness and 10 calves on this one are found to have an average weight of 104.4 kg at one month. At the 5% level of significance should the hypotheses H_0: $\mu = 100$ versus H_1: $\mu = 110$ be accepted?

23. Returning to Problems 18 and 20, the fairness of a particular die is to be tested. It is rolled 40 times with the results:

Face showing	1	2	3	4	5	6
Frequency	8	7	6	4	7	8

Would you accept the hypothesis at the 5% level that the die is fair?

24. A discrete population has a geometric distribution with parameter θ (i.e., $P(X = k) = (1 - \theta)\theta^{k-1}$, $k = 1, 2, \ldots$). Construct a test for the hypotheses H_0: $\theta = \theta_0$ versus H_1: $\theta > \theta_0$, at the $100(1 - \alpha)$% level of significance.

25. A die is rolled 48 times with the following observations:

Face showing	1	2	3	4	5	6
Frequency	4	10	5	7	14	8

In addition, sixes appear on rolls 8, 13, 21, 27, 32, 38, 43, and 48. Of the tests mentioned in Problems 20 and 23 do either support the hypothesis that the die is fair at the 5% level of significance? What conclusions do you draw from these results?

26. Using the Neyman–Pearson lemma show that the test for the hypotheses H_0: $\sigma^2 = \sigma_0^2$ versus H_1: $\sigma^2 > \sigma_0^2$ when the population is normally distributed with unknown mean is to reject H_0 when $\sum (x_i - \bar{x})^2 > \sigma_0^2 \chi_{1-\alpha;n-1}^2$.

27. Measurements of an electronic instrument are assumed to be normally distributed (i.e., the errors are normally distributed) with unknown mean and variance independent of the reading. Three readings were taken for which

$$\sum_{i=1}^{3} x_i = 30.04 \text{ volt} \quad \text{and} \quad \sum_{i=1}^{3} x_i^2 = 300.876$$

Would you accept the hypotheses H_0: $\sigma^2 = .01$ versus H_1: $\sigma^2 > .01$ at the 5% level of significance?

28. In testing a random number generator, the frequencies of last digits in a sample of 100 numbers were found to be:

Last digit	0	1	2	3	4	5	6	7	8	9
Frequency	12	7	9	11	9	12	8	9	11	12

Would you accept the hypothesis at the 5% level of significance that the last digits were uniformly distributed?

29. Stellar radiation tests are made to determine the source. It is known that a source of a particular kind of energy would have a waveform variance of σ_0^2. If the waveform is normally distributed, determine the optimum test for H_0: $\sigma^2 = \sigma_0^2$ versus H_1: $\sigma^2 \neq \sigma_0^2$, using a sample $(X(t_1), X(t_2), \ldots, X(t_n))$, assuming independence.

30. A manufacturer releases lots for sale only when the percentage defective is less than 10%. From large lots he draws 20 samples and releases the whole lot when he obtains two or fewer defectives. What is the probability of releasing the lot when the percentage defectives is actually 10%. (*Hint*: Test the hypotheses H_0: $\theta = .10$ versus H_1: $\theta < .10$, where θ is the probability that an item is defective.)

31. The contents of an average box of a certain brand of soap is required by law to be not less than 1.5 kg. Twenty boxes are tested and found to have an average of 1.48 kg. Is the law satisfied (at the 5% level of significance) if the variance is known to be $\sigma = .1$. What would be the probability of a lot, with actual average of 1.46 kg, being accepted?

32. A purchasing agent for a company tests certain measuring instruments by measuring a known quantity. If, in a test of 10 instruments, the sample variance was 24, determine whether or not the hypothesis that the actual variance was 19 should be accepted at the 5% level.

CHAPTER 11

Queueing Theory

On a typical day most people encounter several examples of queues. From lining up to get on public transportation to joining the line at the cafeteria or bookstore it is difficult to avoid queues. The subject of queueing theory studies the dynamic and steady-state behavior of such systems. It is a powerful and useful tool applicable to a variety of situations. One of the major applications of this theory has been in the design and analysis of telephone switching centers from which much of the terminology now used is derived.

Typically the behavior of a queue varies with time. At some point in time the system is in some "state," that is, there is a certain number of customers being served and waiting and more arrive and leave as time progresses. In most practical applications there is randomness either in the arrival or service mechanism, or both. Thus the behavior of the queue must be described in probabilistic terms. We are interested in answering such questions as "what is the probability there are n people in the queue at time t?" or "what is the probability that the server is busy at time t?." For even the simplest queueing models it turns out that the transient or time-dependent behavior is very difficult to analyze. In this chapter we will consider only the steady-state behavior of some simple queues; that is, we assume that as time tends to infinity, the probabilities of interest have settled down to an essentially constant value.

The first problem, considered in the next section, is to clarify what is meant by a queue and this is done by considering the various possibilities that arise in practice. One of the simplest assumptions that can be made in queueing theory is to assume that the arrivals form a Poisson process. These are defined and analyzed in Section 11.2. The single server queue is defined in Section 11.3 and analyzed as a birth–death process that is a direct generalization of a Poisson process. Although the single server queue with Poisson arrivals and exponential service times is one of the simplest queues, it is far from trivial to analyze. Certain modifications of it are dealt with in the final section.

11.1 *Description of Queueing Models*

Our task in this section is to describe the various types of queues that can arise. In a general queueing situation customers arrive, in some fashion, at a point to receive a service and then depart, usually from further consideration. The term "customer" here can mean anything from a T.V. set arriving at an inspection station to a car arriving at a traffic light waiting for a green light. There are three distinct parts of any queue: (i) the arrival mechanism, (ii) the queueing mechanism, and (iii) the service mechanism. There are a multitude of assumptions on these three mechanisms that can be made to build a model to approximate a given situation.

In the first instance the arrival pattern and subsequent behavior of the customers has to be modeled. Often the customers arrive in a completely random manner and a common assumption is to assume that the number of customers arriving in any time interval of length t is Poissonly distributed with parameter λt. Here λ is the average number of arrivals per unit time. From Chapter 9 such an assumption is equivalent to assuming that, regardless of what occurred prior to time t, the probability of one arrival in the small time interval $(t, t + \Delta t)$ is $\lambda \Delta t$ and the probability of more than one arrival in this interval is essentially zero. On the other hand, the customers may arrive in a fairly regular manner. For example, items on an assembly line usually pass by the workers at a constant rate. We may, however, wish to determine the effect that a small random offset from this regular spacing has on plant efficiency. Perhaps the time that each worker spends on an item is not quite constant and so the arrival at the next worker is regular with a small random offset. Between the completely random and regular arrival patterns are, of course, a variety of others.

The next part of the queueing model is usually referred to as the queueing discipline, that is, when customers arrive for service, how do they queue up, how does the server choose them for service, and how long does he take to serve them. If there is a single server, then typically the customers arrive and queue in one line for service. If the server is busy, the customer may join the queue (called a "blocked customer held" strategy) or may turn away, never to return (called a "blocked customer cleared" strategy). For example, a telephone caller may call a switchboard to find the operator busy. In frustration he decides not to call again (blocked customer cleared). A hungry man in a supermarket with one checkout counter operable will probably join the queue and wait until served (blocked customer held). There are many variations. The queue may have a length limit, such as patients arriving at a doctor's office. Whether or not a customer joins a queue may depend on how long the queue is, the longer the queue the less likely he is to join it, a situation called queueing with discouragement. If there is more than one server, the number of possible strategies grows. A customer in a bank or supermarket may join the queue of the shortest length, complicating the situation mathematically.

Finally, we have to consider the service mechanism and there are two aspects to this. First, how is the server to choose the next customer to be serviced? Usually the policy is "first come first served." In a quality control situation, however, the components to be tested may be placed in a box and the next one to be serviced chosen at random. In an inventory or storage situation the last item in may be the first item out (last come first served). In some situations each customer is assigned priority and customers chosen for service according to some priority scheme. For example, in a university computing center, student jobs are assigned priorities based on job length, and the central processor chooses the next according to priority. If the job is so important as to interrupt the processor, it would be an example of pre emptive scheduling. After choosing the next customer for service there is the second part of the service mechanism, namely, what are the characteristics of the service time? As an example, the length of time it takes a checkout person at the supermarket to total up the purchases is likely to vary with the number of purchases, which one can safely assume to be random. Typically, the service time is modeled as a random variable. In other instances the service time may be a constant. In addition the servers may not be available full time due to coffee breaks, and so forth. In a line for a bus the server (the bus) is only available when the bus arrives at a stop and then it serves (lets on board) as many customers as it is able.

This discussion points out the variety that occurs in queueing situations. In a practical situation it is important to arrive at a mathematical model that approximates the given situation. Often one is faced with the problem of choosing a less accurate model that can be analyzed easily over a more accurate but less tractable model. Once a model for the particular queue under investigation has been arrived at, what kind of information can we expect to extract from it? Usually we seek such quantities as the average queue length, average time for a customer to receive service, or perhaps some quantities on the activity of the servers. In the next section we consider the Poisson process and its structure. We will assume in later sections that the arrival process is Poisson and hence its importance for our work.

11.2   *The Poisson Process*

In certain applications the observations of interest are the points in time when an event occurs. For example, we can imagine a function of time recording when the incoming calls to a telephone exchange occur. Each time a call comes in the function increases by one and it is the times when the calls occur that is of interest. Similar functions recording the emission of radioactive particles, people entering a store, cars crossing a bridge, and so forth, can be formulated. In such cases it is convenient to consider a process $X(t)$ whose value at time t is the number of events that occurred in the interval $(0, t)$, $X(0) = 0$. At any time t, the value of $X(t)$ is an integer valued random variable. Such a function of time is an example of a random process. We will not consider random processes in general in this text since it would take a considerable treatment to do justice to the subject. We think of $X(t)$ as a counting process; each time an event occurs the process is incremented by one and the events occur at random times. Under certain assumptions the random variable $X(t)$, for any time t, will be a Poisson random variable. These assumptions, which we consider shortly, arise from "natural" physical considerations and the importance of them lies in the fact that many physical processes actually satisfy them. A random counting process $X(t)$, where $X(t)$ is a Poisson random variable, is called a Poisson process. The following four assumptions, as will be shown, define a Poisson process.

(i) $X(0) = 0$, that is, we begin observing events at time $t = 0$.
(ii) For any time instants $0 < t_1 < t_2 < t_3 < t_4$ the random variables $X(t_2) - X(t_1)$ and $X(t_4) - X(t_3)$ are independent, that is, the process has independent increments.
(iii) The probability distribution of the random variable $X(t+s) - X(t)$, $t > 0$, $s > 0$, is dependent only on s, the length of the interval, and not t.
(iv) As $\Delta t \to 0$, the probability of one event in the interval $(t, t + \Delta t)$ is approximately $\lambda \, \Delta t$, that is, proportional to the length of the interval. The probability of more than one event in the interval is essentially zero and hence the probability of no event in the interval is $1 - \lambda \, \Delta t$. (Mathematically we should express this by saying the probability of one event in the interval $(t, t + \Delta t)$ is $\lambda \, \Delta t + o(\Delta t)$, where $o(\Delta t)$ is a quantity, dependent only on Δt, such that $\lim_{\Delta t \to 0} o(\Delta t)/\Delta t \to 0$. We omit further mention of this correct mathematical approach as it will not affect the results.)

Denote by $P_k(t)$ the probability that $X(t) = k$. At the time $t + \Delta t$, where Δt is very small, there are two ways that $X(t)$ could have the value of k: either (a) $X(t) = k$ and no event occurs in the interval $(t, t + \Delta t)$, or (b) $X(t) = k - 1$ and an event occurs in the interval $(t, t + \Delta t)$. The probability that the event described in (b) will occur is, by assumption (iv), $\lambda \, \Delta t$ and the probability of the event described in (a) is $1 - \lambda \, \Delta t$. Since the random variables $X(t + \Delta t) - X(t)$ and $X(t)$ are independent, we can write

$$P_k(t + \Delta t) = P_k(t)(1 - \lambda \, \Delta t) + P_{k-1}(t)\lambda \, \Delta t \qquad (11.1)$$

Rearranging terms yields

$$\frac{P_k(t + \Delta t) - P_k(t)}{\Delta t} = -\lambda P_k(t) + \lambda P_{k-1}(t), \quad k = 1, 2, 3, \ldots$$

and letting $\Delta t \to 0$

$$P'_k(t) = -\lambda P_k(t) + \lambda P_{k-1}(t), \quad k = 1, 2, 3, \ldots \tag{11.2}$$

For $k = 0$ Eq. (11.1) becomes

$$P_0(t + \Delta t) = P_0(t)(1 - \lambda \, \Delta t)$$

or

$$\frac{P_0(t + \Delta t) - P_0(t)}{\Delta t} = -\lambda P_0(t)$$

and, as $\Delta t \to 0$ this equation becomes

$$P'_0(t) = -\lambda P_0(t)$$

The solution to this equation is

$$P_0(t) = A \, e^{-\lambda t}$$

and, since it is assumed that $X(0) = 0$, we have $P_0(0) = 1$ and consequently

$$P_0(t) = e^{-\lambda t}$$

For $k = 1$, Eq. (11.2) is

$$P'_1(t) = -\lambda P_1(t) + \lambda \, e^{-\lambda t}$$

which, using the boundary condition $P_1(0) = 0$, yields

$$P_1(t) = \lambda t \, e^{-\lambda t}$$

Solving Eq. (11.2) recursively in this manner gives the result

$$P_k(t) = \frac{(\lambda t)^k}{k!} e^{-\lambda t}, \quad k = 0, 1, 2, \ldots \tag{11.3}$$

that is, the random variable $X(t)$ is a Poisson random variable with parameter λt. The parameter λ can be interpreted as the mean rate of occurrence of events.

We conclude the section by considering some properties of the random process $X(t)$. Suppose we observe the process at time t and the next event occurs at time $t + \tau$, where τ is a random variable. To find the pdf of τ we note that

$$\text{Prob}[\tau > x] = \text{Prob}[\text{no event occurs in } (t, t + x)]$$
$$= e^{-\lambda x}$$

Consequently, the cdf of τ, $F(x)$, is

$$F(x) = \text{Prob}[\tau \leq x] = 1 - \text{Prob}[\tau > x]$$
$$= 1 - e^{-\lambda x}$$

and the corresponding pdf is

$$f(x) = \lambda e^{-\lambda x}, \quad x \geq 0 \tag{11.4}$$

A little thought will reveal that this is also the pdf of the time interval between two successive events of a Poisson process.

Suppose we are told that a single event of a Poisson process has occurred somewhere in the interval $(0, t)$. Conditioned on this information, what is the pdf of its actual occurrence time in this interval? The probability of exactly one event occurring in $(0, t)$ is

$$P_1(t) = \lambda t e^{-\lambda t}$$

On the other hand, the probability of no events in the interval $(0, s)$, one event in $(s, s + ds)$, and no event in $(s + ds, t)$ is

$$e^{-\lambda s} \cdot \lambda \, ds \cdot e^{-\lambda(t-s)}$$

since the occurrence of events in disjoint intervals are independent. Thus, the conditional pdf of the occurrence of the event in $(0, t)$, given that a single event occurred in this interval, which we denote by $g(s)$, is

$$g(s) \, ds = \frac{e^{-\lambda s} \cdot \lambda \, ds \cdot e^{-\lambda(t-s)}}{\lambda t e^{-\lambda t}}$$

or

$$g(s) = 1/t, \quad 0 \leq s \leq t$$

In other words, given that an event occurred in $(0, t)$, its actual time of occurrence is uniformly distributed over the interval.

The Poisson process is an example of a more general class of processes, called Markov processes. These have played an important role in applications.

Definition The random process $X(t)$ is called a Markov process if given the values of the process for all times t less than or equal to t_0, the values of the process at any time in the future $s > t_0$ depends only on the value of the process at time t_0.

An approximate description of a Markov process is one whose future development depends only on its present value and not on its past behavior. Fortunately such processes occur in practice and generally lead to a much simpler analysis than for other processes. In the queueing analysis of the next section we will assume that the arrivals form a Poisson process. Assumption (ii) in the formulation of the Poisson process implies that it is also a Markov process.

EXAMPLE 11.1

The arrivals of customers at a store is assumed to be a Poisson process with parameter λ. If the store decides to close its doors after the nth customer has arrived, what is the pdf of the length of time T that its doors remain open?

It has been shown that the length of time between successive events (increases) in a Poisson process with parameter λ is a random variable with pdf given in Eq. (11.4); that is, an exponential pdf. From assumption (ii) in the definition of a Poisson process the "interarrival times" are independent random variables. If T_i is the time from the $(i-1)$st arrival to the ith arrival ($T_1 =$ time from door opening to the first arrival), then

$$T = T_1 + T_2 + \cdots + T_n$$

where the T_i are independent random variables each with the pdf of Eq. (11.4). From Table 5.1 the mgf of such a random variable is

$$m_{T_i}(t) = \frac{\lambda}{(\lambda - t)} = \frac{1}{1 - t/\lambda}$$

and so the mgf of T is

$$m_T(t) = \prod_{i=1}^{n} m_{T_i}(t) = \frac{1}{(1 - t/\lambda)^n}$$

Again from Table 5.1 it can be seen that this is the mgf of a gamma pdf with parameters $\alpha = n$ and $\gamma = \lambda$, that is, the pdf

$$f_T(t) = \frac{(\lambda t)^{n-1} e^{-\lambda t}}{(n-1)!} \lambda$$

Another way of obtaining this result is to observe that the cdf of the random variable T is

$$F_T(t) = P(T \le t)$$

$$= P[n \text{ or more customers have arrived in } (0, t)]$$

$$= \sum_{k=n}^{\infty} \frac{(\lambda t)^k}{k!} e^{-\lambda t}$$

and differentiating this expression with respect to t gives

$$f_T(t) = \frac{d}{dt} F_T(t)$$

$$= \sum_{k=n}^{\infty} \left(\lambda k \frac{(\lambda t)^{k-1}}{k!} - \frac{\lambda (\lambda t)^k}{k!} \right) e^{-\lambda t}$$

$$= \frac{\lambda (\lambda t)^{n-1}}{(n-1)!} e^{-\lambda t}$$

as before.

EXAMPLE 11.2

The number of particles emitted from a radioactive source is a Poisson process with parameter λ. If n particles were observed in the interval $(0, t)$, what is the probability that k of these particles were observed in the interval $(0, \tau)$, $0 < \tau < t$?

The conditional probability

$P[k$ particles in $(0, \tau)|n$ particles in $(0, t)]$

$$= \frac{P[k \text{ in } (0, \tau) \text{ and } n \text{ in } (0, t)]}{P[n \text{ in } (0, t)]} \qquad (11.5)$$

is sought. The probability $P[k$ in $(0, \tau)$ and n in $(0, t)]$ is equal to $P[k$ in $(0, \tau)$ and $(n-k)$ in $(\tau, t)]$. Since occurrences in nonoverlapping intervals in a Poisson process are, by assumption, independent, we have

$$P[k \text{ in } (0, \tau) \text{ and } n \text{ in } (0, t)] = P[k \text{ in } (0, \tau) \text{ and } (n-k) \text{ in } (\tau, t)]$$

$$= e^{-\lambda\tau}\frac{(\lambda\tau)^k}{k!} \cdot e^{-\lambda(t-\tau)}\frac{[\lambda(t-\tau)]^{n-k}}{(n-k)!}$$

and substituting this into Eq. (11.5)

$$P[k \text{ in } (0, \tau)|n \text{ in } (0, t)] = \frac{e^{-\lambda\tau}\dfrac{(\lambda\tau)^k}{k!} \cdot e^{-\lambda(t-\tau)}\dfrac{[\lambda(t-\tau)]^{n-k}}{(n-k)!}}{e^{-\lambda t}\dfrac{(\lambda t)^n}{n!}}$$

$$= \frac{n!}{k!(n-k)!}\frac{\tau^k(t-\tau)^{n-k}}{t^n}$$

$$= \binom{n}{k}\left(\frac{\tau}{t}\right)^k\left(1-\frac{\tau}{t}\right)^{n-k}$$

from which it is seen that this is a binomial probability with parameters n and $\theta = \tau/t$.

EXAMPLE 11.3

In a Poisson process with parameter λ, what is the probability of observing an even number of events in the interval $(0, t)$?

The probability of observing an even number of events in $(0, t)$ is, by definition,

$$P[\text{even number in } (0, t)] = \sum_{\substack{k=0 \\ k=\text{even}}}^{\infty} e^{-\lambda t}\frac{(\lambda t)^k}{k!}$$

By noticing that $(-\lambda t)^k = +(\lambda t)^k$ when k is even and $-(\lambda t)^k$ when k is odd, then

$$\sum_{\substack{k=0 \\ k=\text{even}}}^{\infty} \frac{(\lambda t)^k}{k!} = \frac{1}{2}\sum_{k=0}^{\infty}\left(\frac{(\lambda t)^k}{k!} + \frac{(-\lambda t)^k}{k!}\right)$$

$$= \tfrac{1}{2}(e^{\lambda t} + e^{-\lambda t})$$

and so

$$P[\text{even number in } (0, t)] = e^{-\lambda t}\tfrac{1}{2}(e^{\lambda t} + e^{-\lambda t})$$

$$= \tfrac{1}{2}(1 + e^{-2\lambda t})$$

EXAMPLE 11.4

A traffic recorder counts the number of cars on a stretch of road. If $X(t)$ is the number of north bound cars during the interval $(0, t)$ and $Y(t)$ the number of south bound and they are independent Poisson processes with parameters λ and η, respectively, what is the probability that k of the n cars counted during this interval were north bound?

The counter records $Z(t) = X(t) + Y(t)$, the number of cars going in either direction. We are given that $Z(t) = n$ and require the conditional probability

$$P[X(t) = k \mid Z(t) = n] = \frac{P[X(t) = k \text{ and } Z(t) = n]}{P[Z(t) = n]}$$

If $X(t) = k$ and $Z(t) = n$, then $Y(t) = n - k$. Recall from Section 8.1 that the sum of two independent Poisson random variables is also a Poisson random variable with parameter the sum of the two component parameters. Consequently,

$$P[X(t) = k \mid Z(t) = n] = \frac{P[X(t) = k \text{ and } Y(t) = n - k]}{P[Z(t) = n]}$$

and since $X(t)$ and $Y(t)$ are, by assumption, independent random variables

$$P[X(t) = k \mid Z(t) = n] = \frac{P[X(t) = k]P[Y(t) = n - k]}{P[Z(t) = n]}$$

$$= \frac{\dfrac{e^{-\lambda t}(\lambda t)^k}{k!} \cdot \dfrac{e^{-\eta t}(\eta t)^{n-k}}{(n-k)!}}{\dfrac{e^{-(\lambda+\eta)t}((\lambda+\eta)t)^n}{n!}}$$

$$= \frac{n!}{k!(n-k)!} \cdot \frac{(\lambda t)^k (\eta t)^{n-k}}{((\lambda+\eta)t)^n}$$

$$= \binom{n}{k}\left(\frac{\lambda}{\lambda+\eta}\right)^k \left(1 - \frac{\lambda}{\lambda+\eta}\right)^{n-k}$$

that is, the conditional distribution is binomial with parameters n and $\theta = \lambda/(\lambda + \eta)$.

EXAMPLE 11.5

Let X_1, X_2, and X_3 be independent Poisson random variables with parameters λ_1, λ_2, and λ_3, respectively. Find the conditional distribution $P(X_1 = j, X_2 = k, X_3 = n - j - k \mid X_1 + X_2 + X_3 = n)$.

The random variable $X_1 + X_2 + X_3$ has a Poisson distribution with parameter $\lambda_1 + \lambda_2 + \lambda_3$ and

$$P(X_1 = j, X_2 = k, X_3 = n - j - k \mid X_1 + X_2 + X_3 = n)$$

$$= \frac{P(X_1 = j)P(X_2 = k)P(X_3 = n - j - k)}{P(X_1 + X_2 + X_3 = n)}$$

$$= \frac{\dfrac{e^{-\lambda_1}\lambda_1^j}{j!} \cdot \dfrac{e^{-\lambda_2}\lambda_2^k}{k!} \cdot \dfrac{e^{-\lambda_3}\lambda_3^{n-j-k}}{(n-j-k)!}}{\dfrac{e^{-(\lambda_1 + \lambda_2 + \lambda_3)}(\lambda_1 + \lambda_2 + \lambda_3)^n}{n!}}$$

$$= \frac{n!}{j!\,k!\,(n-j-k)!}\left(\frac{\lambda_1}{\lambda_1 + \lambda_2 + \lambda_3}\right)^j\left(\frac{\lambda_2}{\lambda_1 + \lambda_2 + \lambda_3}\right)^k\left(\frac{\lambda_3}{\lambda_1 + \lambda_2 + \lambda_3}\right)^{n-j-k},$$

$$j \geq 0,\ k \geq 0,\ j + k \leq n$$

that is, the conditional distribution is multinomial with parameters n, $\theta_1 = \lambda_1/(\lambda_1 + \lambda_2 + \lambda_3)$, $\theta_2 = \lambda_2/(\lambda_1 + \lambda_2 + \lambda_3)$, and $\theta_3 = \lambda_3/(\lambda_1 + \lambda_2 + \lambda_3)$.

11.3 Birth–Death Processes and the Single Server Queue

The Poisson process is important in practice for a variety of reasons, not the least of which being that it models to a high degree of accuracy certain naturally occurring processes. For example, if a traffic count is done on cars passing a point on a lightly traveled road it will likely be well approximated by a Poisson process. As the road becomes more congested, however, a greater regularity will probably appear in the results, and the Poisson assumption will no longer be valid.

Fortunately, it turns out to be easier to analyze models of queueing situations where the arrivals process is assumed to be Poisson than for the non-Poisson case. Thus even when it is known the arrivals process is not strictly a Poisson process, it is sometimes assumed to be and this lack of accuracy taken care of when interpreting results.

In this section we analyze a simple but interesting queueing situation. We will first assume that customers arrive at some point to obtain a service, according to a Poisson process with parameter λ (customers per unit of time). Next, we assume that the single server takes a random length of time to service each customer and that these times are independent random variables for each customer which have an exponential pdf

$$f(t) = \mu e^{-\mu t}, \quad t \geq 0$$

On the average μ customers are capable of being serviced per unit of time or a customer spends $(1/\mu)$ units of time, on the average, being serviced.

This distribution has a particularly interesting property. During the servicing of a customer, the probability that the server will complete service in the interval $(t, t+\Delta t)$ given that service is still in progress at time t is

$$P(t \leq T \leq t+\Delta t \mid T \geq t) = \frac{P(t \leq T \leq t+\Delta t)}{P(T \geq t)}$$

$$= \frac{\mu e^{-\mu t} \Delta t}{e^{-\mu t}} = \mu \Delta t$$

and we conclude that the probability of completion of service in the next Δt seconds is a constant, independent of how long service has been going on! The server essentially has no memory and, from the fact that the arrivals process is Markov and service times are independent, it follows that the service process is a Markov process.

This single server queue with Poisson input and exponential service time is often denoted as the $M/M/1$ queue, the M standing for Markov. More generally, a $G_1/G_2/s$ would be a queue with s servers with arrivals controlled by a distribution, represented by G_1, and the service time for each customer controlled by distribution G_2.

To begin our analysis of the $M/M/1$ queue, we analyze a slightly more general situation that will be useful in the next section as well. We will consider the queue and server as a system and say that the system is in state S_n if there are n people in the queue including the one being served (if any). From state S_n the system is only capable of making transitions to S_{n+1} or S_{n-1}, that is, either a customer completes his service and leaves the queue or, while the present customer is still being serviced, another customer joins the queue.

If $X(t)$ is the process that takes on the value n when the queue is in state S_n (it describes how the queue behaves as a function of time), then there will be certain similarities between $X(t)$ and a Poisson process. A Poisson process is increasing as time increases, that is, if the process presently has the value n, then its next value can only be $n+1$, and the probability that it increases in the interval $(t, t+\Delta t)$, $\lambda \Delta t$, is constant regardless of the state of the process at time t and how long it has been in that state. The assumptions below will define the queueing process $X(t)$. Essentially it differs from a Poisson process in that $X(t)$ can either increase or decrease from state S_n, $n \geq 1$ and that the probability of an increase or decrease may depend on the state that the process is presently in. For example, if the queue at a hot dog stand is long, a potential customer is more likely to decide not to join it than if the queue were short. The resulting queueing process will by definition be Markov but not Poisson, and is referred to as a birth–death process. It is a generalization of the Poisson process. From these considerations, it follows that the process will have the following properties:

(i) If the system is in state S_n, it can only make transitions to S_{n-1} or S_{n+1}, $n \geq 1$ (from S_0, the next state can only be S_1).

(ii) If the system is in state S_n at time t, the probability of a transition to S_{n+1} in the short interval of time, Δt, is $\alpha_n \Delta t$. The parameter α_n is referred to as the birth parameter and, for the present, we will assume it depends on the state S_n of the system.

(iii) If the system is in state S_n at time t, the probability of transition to state S_{n-1} in the arbitrarily short time interval $(t, t+\Delta t)$ is $\beta_n \Delta t$. We refer to β_n as the death parameter.

The probability of more than one transition in the interval $(t, t+\Delta t)$ is essentially zero. These properties are slightly more general than we require for this section but will be useful for more general situations considered in the next section.

From properties (i) to (iii) it follows that at time $t+\Delta t$ the system can be in state S_n in three distinct ways:

(a) By being in state S_n at time t, and no transition occurring in the interval $(t, t+\Delta t)$. This happens with probability $(1-\alpha_n \Delta t) \cdot (1-\beta_n \Delta t) \cong 1-\alpha_n \Delta t - \beta_n \Delta t$, as second-order effects $(\alpha_n \beta_n \Delta t^2)$ are negligible compared with first-order effects.

(b) By being in state S_{n-1} at time t and a transition to S_n occurring in the interval $(t, t+\Delta t)$. This happens with probability $\alpha_{n-1} \Delta t$.

(c) By being in state S_{n+1} at time t and a transition to S_n occurring in the time interval $(t, t+\Delta t)$. This happens with probability $\beta_{n+1} \Delta t$.

If $P_i(t)$ denotes the probability that the system is in state i at time t, then statements (a)–(c) enable us to write the following equations:

$$P_n(t+\Delta t) = P_n(t)(1-\alpha_n \Delta t - \beta_n \Delta t) + \alpha_{n-1} \Delta t\, P_{n-1}(t)$$

$$+ \beta_{n+1} \Delta t\, P_{n+1}(t), \quad n \geq 1$$

$$P_0(t+\Delta t) = P_0(t)(1-\alpha_0 \Delta t - \beta_0 \Delta t) + \beta_1 \Delta t P_1(t)$$

where the fact that the "birth" and "death" processes are Markov has been used. By rearranging terms and dividing by Δt we obtain

$$\frac{P_n(t+\Delta t) - P_n(t)}{\Delta t} = -(\alpha_n + \beta_n) P_n(t) + \alpha_{n-1} P_{n-1}(t) + \beta_{n+1} P_{n+1}(t), \quad n \geq 1$$

and

$$\frac{P_0(t+\Delta t) - P_0(t)}{\Delta t} = -(\alpha_0 + \beta_0) P_0(t) + \beta_1 P_1(t)$$

which, on letting $\Delta t \to 0$, becomes

$$P'_n(t) = -(\alpha_n + \beta_n) P_n(t) + \alpha_{n-1} P_{n-1}(t) + \beta_{n+1} P_{n+1}(t), \quad n \geq 1 \qquad (11.6)$$

$$P'_0(t) = -(\alpha_0 + \beta_0) P_0(t) + \beta_1 P_1(t)$$

This equation is the fundamental recursion relationship for the type of birth–death process considered here. Its solution, as a function of time, however, is reasonably complicated and we consider only the steady-state. The parameters α_n and β_n are referred to as birth and death parameters, respectively.

In the steady state, achieved as $t \to \infty$, $P_n(t)$ tends to some limiting value P_n (if the model converges, i.e., the queue size does not tend to infinity with probability 1). Allowing t to go to infinity in (11.6) yields the equation

$$(\alpha_n + \beta_n)P_n = \alpha_{n-1}P_{n-1} + \beta_{n+1}P_{n+1}, \quad n \geq 1 \tag{11.7}$$

and, for the special case $n = 0$,

$$\alpha_0 P_0 = \beta_1 P_1 \tag{11.8}$$

These equations are easily solved recursively to yield

$$P_1 = \frac{\alpha_0}{\beta_1} P_0$$

$$P_2 = \frac{\alpha_0 \alpha_1}{\beta_1 \beta_2} P_0$$

$$\vdots \tag{11.9}$$

$$P_n = \frac{\alpha_0 \alpha_1 \cdots \alpha_{n-1}}{\beta_1 \beta_2 \cdots \beta_n} P_0$$

$$\vdots$$

where P_0 is the, as yet undetermined, probability that the system is in state S_0.

Since, for a stable model in the steady state, the system must be in some state, the sum of all the probabilities in Eq. (11.9) is unity, that is,

$$P_0 \left(1 + \frac{\alpha_0}{\beta_1} + \frac{\alpha_0 \alpha_1}{\beta_1 \beta_2} + \cdots \right) = 1 \tag{11.10}$$

and from this equation we can determine the value of P_0 provided the summation in parentheses converges to a finite value. Fortunately, this summation does, in fact, converge for many cases of interest.

The Eqs. (11.8), (11.9), and (11.10) are a general steady-state solution to the $M/M/1$ queue where the queue and server parameters are dependent on the state of the system.

For any queueing situation we define the ratio of the mean service time of a single customer to the mean interarrival time of customers as the traffic intensity and denote it by ρ. As a dimensionless quantity it is measured in Erlangs, in recognition of the pioneering work of A. K. Erlang early in this century in congestion and queueing theory. In the single server queue we have $\rho = \lambda/\mu$ and it measures the ratio of the average amount of work required by the customers to the average amount of work the servers are able to provide. In this case it is often referred to as the utilization factor of the system.

We return now to the situation mentioned at the beginning of the section, namely, that of the single server $M/M/1$ queue with Poisson arrivals with parameter λ (the mean arrival rate) and exponential service times with parameter μ (the mean service rate). The customers arrive and if the server is busy form a single queue, being served on a first come first served basis. With these assumptions the process $X(t)$ describing the state of the system at time t is a birth–death process with state independent parameters $\alpha_n = \lambda$ and $\beta_n = \mu$. In this case the ratio $\rho = \lambda/\mu$ is sometimes referred to as the utilization factor. For the remainder of the section we will derive some properties of this queue, beginning with P_n, the probability that in the steady state there are n persons in the system (including the one being served).

From Eqs. (11.9) we have

$$P_1 = \frac{\lambda}{\mu} P_0 = \rho P_0$$

$$P_2 = \left(\frac{\lambda}{\mu}\right)^2 P_0 = \rho^2 P_0$$

$$\vdots$$

$$P_n = \left(\frac{\lambda}{\mu}\right)^n P_0 = \rho^n P_0$$

(11.11)

To find P_0 we equate

$$P_0(1 + \rho + \rho^2 + \cdots) = P_0\left(\frac{1}{1-\rho}\right) = 1$$

implying that $P_0 = 1 - \rho$. This gives a meaningful probability only if $0 < \rho = \lambda/\mu < 1$, which means essentially that the server, on the average, must process the customers faster than their average arrival rate, otherwise the queue length will tend to infinity.

The probability distribution

$$P_n = (1-\rho)\rho^n, \quad n = 0, 1, 2, \ldots$$

(11.12)

is the geometric distribution encountered frequently in earlier chapters. Its mean (average number of customers in the system) $E(N)$ is $E(N) = \rho/(1-\rho) = \lambda/(\mu - \lambda)$, a quantity usually denoted by L, and its variance is $\lambda\mu/(\mu - \lambda)^2$.

The mean number of customers in the queue $E(N_q)$ can be determined by noting that if the system is in state n, the number of customers is $(n-1)$ and consequently

$$E(N_q) = \sum_{n=1}^{\infty} (n-1)(1-\rho)\rho^n$$

$$= (1-\rho)\left(\sum_{n=1}^{\infty} n\rho^n - \sum_{n=1}^{\infty} \rho^n\right)$$

$$= \frac{\rho^2}{(1-\rho)} = \frac{\lambda^2}{\mu(\mu - \lambda)}$$

(11.13)

The mean time that a customer must wait in the queue T_q for service is a little more complicated to calculate. If there are n customers in the system when a customer arrives, then this new customer will have to wait $T_q = T_1 + T_2 + \cdots + T_n$, where the set $\{T_i\}$ is a set of independent and identically distributed random variables, each with an exponential pdf with parameter μ. From Chapter 8 the random variable T_q, given that there are n customers in the system, has the Erlang pdf (i.e., a gamma pdf with parameters n and μ)

$$p(t) = \frac{\mu^n t^{n-1}}{(n-1)!} e^{-\mu t}, \quad t > 0, \ n \geq 1$$

Consequently, the probability that T_q is greater than τ is

$$P(T_q > \tau) = \sum_{n=1}^{\infty} (1-\rho)\rho^n \int_{\tau}^{\infty} \frac{\mu^n t^{n-1}}{(n-1)!} e^{-\mu t} \, dt$$

$$= (1-\rho)\rho\mu \int_{\tau}^{\infty} \sum_{n=1}^{\infty} \frac{(\rho\mu t)^{n-1}}{(n-1)!} e^{-\mu t} \, dt$$

$$= (1-\rho)\rho\mu \int_{\tau}^{\infty} e^{-\mu(1-\rho)t} \, dt$$

$$= (1-\rho)\rho\mu \frac{1}{\mu(1-\rho)} e^{-\mu(1-\rho)\tau}$$

$$= \rho e^{-\mu(1-\rho)\tau}$$

The corresponding pdf for T_q is

$$g(\tau) = \frac{d}{d\tau}[1 - P(T_q > \tau)] = \rho\mu(1-\rho) e^{-\mu(1-\rho)\tau}, \quad \tau > 0$$

To complete the derivation we note that there is a finite probability that $\tau = 0$, namely, $P_0 = (1-\rho)$, the probability there are not customers in the system when the new customer arrives. Consequently,

$$E(T_q) = \int_0^{\infty} \tau g(\tau) \, d\tau$$

$$= \int_0^{\infty} \tau \cdot \rho\mu(1-\rho) e^{-\mu(1-\rho)\tau} \, d\tau$$

$$= \frac{\rho}{\mu(1-\rho)} = \frac{\lambda}{\mu(\mu-\lambda)}, \quad \mu > \lambda \tag{11.14}$$

To complete the analysis we compute the mean wait $E(T)$, which we will denote by W, that a customer has from the moment he arrives until he clears the queue. This is simply $E(T_q)$ plus the average service time per customer, $1/\mu$, and

$$W = E(T) = \frac{\lambda}{\mu(\mu-\lambda)} + \frac{1}{\mu} = \frac{1}{\mu-\lambda} \tag{11.15}$$

Notice now that

$$L = \frac{\lambda}{\mu - \lambda} = \lambda \cdot \frac{1}{\mu - \lambda} = \lambda \cdot W$$

and this relationship, referred to as Little's law, actually holds for much more general situations.

EXAMPLE 11.6

To avoid congestion a librarian would like to have fewer than five customers lined up 99% of the time. If the arrivals process is Poisson with parameter $\lambda = 1.5$ customers per minute, how fast should the service rate be?

The probability of five or more customers being in the system is

$$P = \sum_{n=5}^{\infty} (1 - \rho)\rho^n = \rho^5$$

and for this probability to be less than .01 we require

$$(\lambda/\mu)^5 \leq .01$$

or, equivalently,

$$\mu^5 \geq \frac{(\frac{3}{2})^5}{.01} = 759.375$$

which implies that $\mu > 9.12$. In order to meet the requirement, the checkout system must be capable of serving at least 9.12 customers per minute on the average.

EXAMPLE 11.7

In a university computer batch processing facility 70 jobs an hour are submitted on the average. If the average turnaround time (time to submission to time of getting job back) is to be less than 10 minutes, what should the service rate be?

We will assume here that the length of time to service a job is independent of times to service other jobs and exponentially distributed with parameter μ. It is reasonable to assume that the traffic pattern is a Poisson process. The model then reduces to a single line, single server queue for which the expected waiting time (in queue and in service) is given by Eq. (11.15) as

$$W = E(T) = \frac{1}{\mu - \lambda} = \frac{1}{\mu - 70/60}$$

We want this quantity to be less than 10 minutes, implying that $1 < 10(\mu - 7/6)$ or, equivalently, $\mu > 76/60$, and the computer must be able to handle about 1.267 jobs per minute, on the average.

EXAMPLE 11.8

A trucking firm observes that on a certain route six weighing stations are encountered by their trucks. Their drivers are paid $12.00 per hour on the road. How much does the firm pay because of the delays encountered at the stations if 25 trucks an hour arrive at the station on the average and the service is exponential with parameter $\mu = 30$ trucks per hour?

The average wait to clear service is from Eq. (11.15)

$$W = \frac{1}{\mu - \lambda} = \frac{1}{30 - 25} = .2 \text{ hours}$$

The total time lost during the trip is $(.2) \times 6 = 1.2$ hours, which costs the firm $12 \times 1.2 = \$14.40$.

EXAMPLE 11.9

A new toll booth is to be installed on a turnpike. Measurements have indicated that the traffic during 10:00 to 11:00 A.M. can be modeled as a Poisson process with parameter 7.5 cars per minute. Experience has shown that the server process is approximated by an exponential distribution with parameter $\mu = 5$ per minute. (a) If it is desired to have the average waiting time that a car spends at the toll gate during this interval less than 1.5 minutes, how many toll booths should be considered? (b) If automatic toll gates are installed, the service parameter increases to $\mu = 10$ per minute. What is the expected length of a line (including the one being served) if only two booths are installed?

The problem might appear to fall into the category of single line, multiple server with an infinite population. However, since each car is free to choose its server (rather than forming a single line and taking the next free server, or choosing one from among the free servers) and since each car presumably chooses the shortest line, it is perhaps more reasonable to model the situation as a single server, single queue with queue parameter $7.5/s$, where s is the number of toll booths and service parameter $\mu = 5$. We proceed with this model keeping in mind that we are dealing with an approximation to the physical situation.

(a) For the single queue, single server case, the expected waiting time for an arriving car to complete service is

$$W = \frac{1}{\mu - \lambda'}, \quad \text{where } \mu = 5, \quad \lambda' = \frac{7.5}{s}$$

$$= \frac{1}{5 - 7.5/s}$$

which is to be less than 1.5. A little calculation reveals that s must be two or more.

(b) Again assuming a single queue, single server discipline the expected length of the line is

$$L = E(N) = \frac{\lambda'}{\mu - \lambda'} = \frac{7.5/2}{10 - 7.5/2} = .6$$

So far only steady-state solutions have been considered for the $M/M/1$ queue. It is possible to derive expressions similar to those obtained here for an $M/G/1$ queue where the serving times do not follow the exponential distribution. In certain cases it is also possible to derive time-dependent solutions for the state probabilities by returning to the basic equations. The next example will illustrate.

EXAMPLE 11.10

A sales clerk in a store begins work at 9:00 A.M. and serves an average of 20 people in his eight-hour day. If he spends an average of 15 minutes with each customer, what is the probability he is busy serving a customer at 10:00 A.M.?

In the absence of other information we assume that the arrivals process is Poisson with parameter $\lambda = 2.5$ (customers per hour) and that the service times are exponentially distributed with parameter $\mu = 4$ (customers per hour). To obtain the required probabilities we return to the basic equations.

The probability of a transition from state 0 (no one being served) to state 1 (server busy) in the interval $(t, t + \Delta t)$ is $\lambda \Delta t$, and the probability of no transition is $(1 - \lambda \Delta t)$. Similarly, the probability of a transition from state 1 to state 0 is $\mu \Delta t$ and the probability of remaining in state 1 is $(1 - \mu \Delta t)$. Consequently, the state equations can be written as

$$P_0(t + \Delta t) = (1 - \lambda \Delta t)P_0(t) + \mu \Delta t P_1(t)$$
$$P_1(t + \Delta t) = \lambda \Delta t P_0(t) + (1 - \mu \Delta t)P_1(t)$$

and, rearranging,

$$P_0'(t) = -\lambda P_0(t) + \mu P_1(t)$$
$$P_1'(t) = \lambda P_0(t) - \mu P_1(t)$$

This pair of simultaneous differential equations is readily solved by standard techniques to yield

$$P_0(t) = \frac{\mu}{\lambda + \mu}\{1 - \exp[-(\mu + \lambda)t]\} + P_0(0)\exp[-(\mu + \lambda)t]$$

$$P_1(t) = \frac{\lambda}{\lambda + \mu}\{1 - \exp[-(\mu + \lambda)t]\} + P_1(0)\exp[-(\mu + \lambda)t]$$

where $P_0(t) + P_1(t) = 1$ for all t and $P_0(0)$ and $P_1(0)$ are the initial state probabilities. The probability that the server is busy at time t is just $P_1(t)$, and the probability that the clerk is busy at 10:00 A.M. is

$$P_1(10) = \frac{2.5}{6.5}(1 - e^{-6.5}) = \frac{2.5}{6.5} \times .9985$$

where we have assumed that he is initially not busy (i.e., $P_1(0) = 0$). Notice that by 10:00 A.M. the transient effects of the start at 9:00 A.M. have very nearly disappeared and the probabilities have essentially reached their steady-state values. This is one of the few cases where a time-dependent solution can be obtained. The steady-state solutions are obtained by letting $t \to \infty$ and setting $P_i'(t) = 0$, $i = 0, 1$. If P_0 and P_1 are these steady-state probabilities, then

$$P_0 = \frac{\mu}{\lambda + \mu} \quad \text{and} \quad P_1 = \frac{\lambda}{\lambda + \mu}$$

11.4 *Modifications of the Single Server Queue*

In this section we consider the application of the birth–death process and its steady-state probabilities to queueing situations different from the $M/M/1$ queue. In most cases it is simply a matter of interpreting the birth and death parameters correctly.

11.4.1 *THE M/M/s QUEUE*

Many queues encountered in practice have more than one server, say s, and it is interesting to develop formulas for this case. The queue has a Poisson arrivals process with parameter λ and each of the s servers has an exponential service time each with parameter μ. If all servers are busy, arrivals form a single queue to be served on a first come first served basis.

   If the system is in state S_n, the probability of an arrival in the next small interval of time Δt is $\lambda \, \Delta t$, independent of n. The probability that a customer departs from service depends on how many of the servers are presently busy. If $n < s$, then n servers are busy and the probability of a departure in the next Δt seconds is $n\mu \, \Delta t$ and $\beta_n = n\mu$. (Strictly spreaking we should apply Eq. (11.6) here and note that second-order and higher effects can be neglected.) If $n \geq s$, then $\beta_n = s\mu$. To summarize,

$$\alpha_n = \lambda \qquad \beta_n = \begin{cases} n\mu, & 0 \leq n \leq s \\ s\mu, & n \geq s \end{cases}$$

Notice how the effect of multiple servers can be taken care of by choosing the birth–death parameters correctly.

   Equation (11.10) with these parameters becomes

$$P_0 \left\{ 1 + \frac{\lambda}{\mu} + \frac{\lambda^2}{2!\mu^2} + \cdots + \frac{\lambda^{s-1}}{(s-1)!\mu^{s-1}} + \frac{\lambda^{s-1}}{(s-1)!\mu^{s-1}} \left[\sum_{k=1}^{\infty} \left(\frac{\lambda}{s\mu} \right)^k \right] \right\} = 1$$

or

$$P_0 = \left[\left(\sum_{k=0}^{s-1} \frac{\rho^k}{k!} \right) + \frac{\rho^s}{(s-1)!\,(s-\rho)} \right]^{-1} \tag{11.16}$$

The remaining probabilities are then found as

$$P_n = \begin{cases} \dfrac{1}{n!}\rho^n P_0, & n < s \\[2ex] \dfrac{1}{s!\,s^{n-s}}\rho^n P_0, & n \geq s \end{cases} \tag{11.17}$$

The expected number of people in the system can be calculated as

$$L = E(N) = \sum_{n=1}^{\infty} nP_n$$

$$= P_0 \left[\sum_{n=1}^{s} n \cdot \frac{1}{n!}\rho^n + \frac{s^s}{s!} \sum_{n=s+1}^{\infty} n\left(\frac{\rho}{s}\right)^n \right]$$

$$= P_0 \left[\rho\left(\sum_{n=0}^{s-1} \frac{\rho^n}{n!} \right) + \frac{s^s}{s!}\left(\frac{\rho}{s}\right)^{s+1} \frac{s}{1-\rho/s} + \frac{s^s}{s!}\left(\frac{\rho}{s}\right)^{s+1} \frac{1}{(1-\rho/s)^2} \right]$$

$$= \rho + P_0 \left(\frac{\rho^{s+1}}{(s-1)!\,(s-\rho)^2} \right)$$

Expressions for the other parameters can be calculated similarly but we simply report them as

$$E(N_q) = P_0 \left(\frac{\rho^{s+1}}{(s-1)!\,(s-\rho)^2} \right)$$

$$W = E(T) = \frac{P_0}{\mu} \left(\frac{\rho^s}{(s-1)!\,(s-\rho)^2} \right) + \frac{1}{\mu} \tag{11.18}$$

$$E(T_q) = \frac{P_0}{\mu} \left(\frac{\rho^s}{(s-1)!\,(s-\rho)^2} \right)$$

Notice with this system that for stability in the queue it is only necessary that $\rho < s$.

EXAMPLE 11.11

Work Example 11.9(a), this time assuming a model for the road traffic and toll booths to be an $M/M/s$ queue.

For the $M/M/s$ queue the expected total waiting time is, by Eq. (11.18),

$$W = \frac{P_0}{5} \left(\frac{(\frac{3}{2})^s}{(s-1)!\,(s-\frac{3}{2})^2} \right) + \frac{1}{5}$$

and we require the smallest s for which this quantity is less than 1.5, noting that

P_0 as given by Eq. (11.16) is a function of s. When $s = 1$ the queue is unstable in that the size tends to infinity in the steady state. When $s = 2$, $P_0 = .142$ and $E(T) = .456$ and already it is less than 1.5. For the model assumed in Example 11.9, $E(T) = .8$, which is considerably different from the value obtained here, although both required at least two toll booths. To decide which of the two models was the more valid, a careful analysis of the behavior of the motorists as they approached the toll booths would be required.

EXAMPLE 11.12

A bank manager has arranged it so that customers entering the bank join a single queue and go to the next available teller when they reach the head of the line. During the slack periods he assigns only two tellers and observes that on the average each can service .3 customers per minute. If, on the average, 15 customers enter the bank per hour, what is the probability that a customer entering the bank will have to wait for service?

The probability that a customer will have to wait for service is the probability that there are two or more in the system. Letting $\rho = .25/.30$ and using Eqs. (11.16) and (11.17), this can be written as

$$P_0 = \left[1 + \rho + \left(\frac{1}{2}\rho^2 \frac{1}{1-\rho/2}\right)\right]^{-1}$$

$$= \frac{42}{102}$$

and

$$P_1 = \rho P_0 = \frac{5}{6} \cdot \frac{42}{102} = \frac{35}{102}$$

and the probability that an arriving customer will have to wait for service is $1 - P_0 - P_1 = 25/102$.

Before leaving the $M/M/s$ queue consider the following situation. Suppose a construction firm operates a fleet of very expensive hauling trucks and its own garage. Any time a truck breaks down a repairman from the maintenance shop is found to begin work on it immediately. There is never any waiting. We assume in the model that there are an infinite number of trucks and an infinite number of repairmen. Since the probability of requiring many repairmen decreases quickly for any reasonable situation, the approximation may, in fact, be quite reasonable. This queue is referred to as the $M/M/\infty$ queue and its parameters may be determined by letting s tend to infinity in the Eqs. (11.16) and (11.17) for the $M/M/s$ queue. Thus

$$P_0 = e^{-\rho}$$

and

$$P_n = \frac{\rho^n}{n!} e^{-\rho}$$

and N has a Poisson distribution with parameter ρ. Since there is never any queue wait, $T = T_q$ and

$$W = E(T) = E(T_q) = \frac{1}{\mu}$$

11.4.2 THE M/M/1 QUEUE WITH LIMITED QUEUE SIZE

In some queueing situations the queue is, of necessity, limited by some number K, which includes the one being served. For example, in a doctor's office patients arriving when the office is full may be asked to make an appointment for some other time and are removed from further consideration. This is sometimes referred to as a "blocked customer cleared" strategy (the previous case where every arriving customer was allowed to join the line being referred to as a "blocked customer held" strategy).

This queueing discipline can be modeled by choosing the state-dependent birth and death parameters as

$$\alpha_n = \begin{cases} \lambda, & n < K \\ 0, & n \geq K \end{cases} \qquad \beta_n = \mu, \quad 0 < n \leq K$$

Substituting these variables into Eq. (11.9) gives

$$P_n = \left(\frac{\lambda}{\mu}\right)^n P_0, \quad n = 0, 1, \ldots, K \tag{11.19}$$

where

$$P_0 = \left[\sum_{i=0}^{K} \left(\frac{\lambda}{\mu}\right)^i\right]^{-1} = \frac{1-\rho}{1-\rho^{K+1}} \tag{11.20}$$

The expected number of customers in the system is

$$L = E(N) = \sum_{n=0}^{K} nP_n$$

$$= \left(\frac{1-\rho}{1-\rho^{K+1}}\right) \sum_{n=1}^{K} n\rho^n$$

$$= \rho \frac{1-(K+1)\rho^K + K\rho^{K+1}}{(1-\rho)(1-\rho^{K+1})} \tag{11.21}$$

The expected number of customers in the queue is

$$E(N_q) = \sum_{n=1}^{K} (n-1)P_n$$

$$= \sum_{n=1}^{K} nP_n - \sum_{n=1}^{K} P_n$$

$$= E(N) - (1-P_0)$$

348 *Applied Probability*

The mean waiting time in the queue can be calculated as for the unrestricted $M/M/1$ queue. A customer arriving with the queue in state S_n has a wait T_q that is the sum of n independent exponentially distributed random variables T_i, each with parameter μ. The expected value of this sum is n/μ and so

$$E(T_q) = \sum_{n=0}^{K} \frac{n}{\mu} P_n = \frac{1}{\mu} E(N)$$

Similarly the expected wait to complete service is

$$E(T) = \sum_{n=0}^{K} \frac{(n+1)}{\mu} P_n = \frac{1}{\mu} [E(N)+1] \tag{11.22}$$

Notice that because of the finite queue length allowed there is no instability possible and ρ may assume any positive value. Notice also that as $K \to \infty$ the queueing strategy approaches that of an unrestricted $M/M/1$ queue and P_0 in Eq. (11.20) tends to $(1-\rho)$ and P_n to $(1-\rho)\rho^n$ as before.

EXAMPLE 11.13

A gas station has one diesel fuel pump for trucks only and has room for three trucks only (one at the pump and two lined up waiting for service). Trucks that arrive to find the line full leave to find another gas station. What is the average time, from entering to leaving the station, that a truck takes if the average number of arrivals is 5 per hour and each truck takes an average of 6 minutes to service? What percentage of the traffic is being turned away?

We assume that the arrivals process is Poisson and the service times are exponential with parameters $\lambda = 5$, $\mu = 10$, and $\rho = \frac{1}{2}$. From Eq. (11.21) the average number of customers in the system is, with $K = 3$,

$$L = E(N) = \left(\frac{1}{2}\right) \frac{[1 - 4(\frac{1}{2})^3 + 3(\frac{1}{2})^4]}{(\frac{1}{2})[1 - (\frac{1}{2})^4]} = .733$$

and from Eq. (11.22) the average length of time a customer spends in the system is

$$W = E(T) = \frac{1}{10}(.733+1) = .173 \text{ hours}$$

or 10.38 minutes. Contrast this figure with the average service time of 6 minutes if there are no customers in the system when a truck arrives.

The percentage of traffic being turned away is $100 \times$ (probability system is full). From Eq. (11.20)

$$P_0 = \frac{(\frac{1}{2})}{1 - (\frac{1}{2})^4} = \frac{8}{15}$$

and using Eq. (11.19), the probability the system is full $= P_3 = P_0 \cdot (\frac{1}{2})^3 = \frac{1}{15}$ and 6.66% of the traffic is being turned away.

EXAMPLE 11.14

A small repair shop receives on the average 2 machines per day to be fixed. The manager however decides not to accept them for repair if he already has two in the shop (including the one he may be working on). It is bad for business to turn too many away. How fast should he work if he wants to turn fewer than 20% away?

The probability of turning a customer away is P_2 where, since $L = 2$, Eqs. (11.19) and (11.20) give

$$P_2 = \rho^2 P_0 \quad \text{and} \quad P_0 = \frac{1-\rho}{1-\rho^3} = \frac{1}{1+\rho+\rho^2}$$

In order to have $P_2 \leq .2$ it is necessary that

$$\frac{\rho^2}{1+\rho+\rho^2} \leq .2$$

or, by rearranging,

$$0 \leq .2(1+\rho+\rho^2-5\rho^2)$$

or

$$4\rho^2 - \rho - 1 < 0$$

Solving for the roots of this quadratic gives

$$\rho_{1,2} = \frac{1 \pm \sqrt{1+17}}{8} = \frac{5.123}{8}, \frac{-3.123}{8}$$

and we require

$$\rho < 5.123/8$$

or

$$\mu > \frac{8}{5.123} \times \lambda = 3.10$$

and to meet his objective he should be able to repair at least 3.10 machines per day, on the average.

11.4.3 M/M/s QUEUE WITH LIMITED QUEUE SIZE

For this case we will only derive the state probabilities. If a maximum of K customers are allowed in the system, then the birth–death parameters are given by

$$\alpha_n = \begin{cases} \lambda, & 0 \leq n < K \\ 0, & n \geq K \end{cases} \qquad \beta_n = \begin{cases} 0, & n = 0 \\ n\mu, & 0 < n \leq s \\ s\mu, & s < n \leq K \end{cases}$$

and substituting these values into Eqs. (11.9) gives the following results:

$$
P_n = \begin{cases} \dfrac{\lambda^n}{n!\mu^n}P_0, & 0 < n \le s \\[3mm] \dfrac{\lambda^n}{s!s^{n-s}\mu^n}P_0, & s < n \le K \end{cases}
\tag{11.23}
$$

and

$$
P_0 = \left(\sum_{n=0}^{s} \frac{1}{n!}\rho^n + \sum_{n=s+1}^{K} \frac{\rho^n}{s!s^{n-s}} \right)^{-1}
\tag{11.24}
$$

A special case of this formula has historical significance. Certain multiple server queues have no facility for holding customers. The telephone switching center, for example, will simply give a busy signal if all lines (servers) are busy. In such a case $K = s$ and

$$
P_n = \begin{cases} \dfrac{\rho^n}{n!}P_0, & 0 < n \le s \\[3mm] 0, & n > s \end{cases}
$$

where

$$
P_0 = \left(\sum_{n=0}^{s} \frac{\rho^n}{n!} \right)^{-1}
$$

The probability that the system is full, P_s, then gives the proportion of customers turned away:

$$
P_s = \frac{\rho^s/s'}{\sum_{n=0}^{s} \rho^n/n!}
$$

and this is often referred to as the *Erlang* loss formula. Although derived here for exponential service times, its use does not require this assumption.

EXAMPLE 11.15

How many phones should be installed in an airport lounge if it is desired to have fewer than 10% of the customers turned away, assuming that the length of the average phone call is 3 minutes and, on the average, 40 people per hour seek to use the phones?

Under the circumstances it is not unreasonable to assume that the arrivals process is Poisson and the lengths of calls are independent and exponentially distributed. The model then has s telephones (s to be determined) and we assume that a customer who arrives to find all phones in use leaves and does not return. The probability of this happening is (by the Erlang loss formula), with $\rho = \mu/\lambda = \frac{1}{2}$

$$P_s = \frac{\dfrac{1}{s!}\left(\dfrac{1}{2}\right)^s}{\displaystyle\sum_{i=0}^{s}\dfrac{1}{i!}\left(\dfrac{1}{2}\right)^i}$$

We require the smallest value of s for which $P_s \leq .10$. It is easily calculated that for $s = 1$, $P_1 = .333$, and for $s = 2$, $P_s = .0769$ and two phones are sufficient.

EXAMPLE 11.16

A barber has two chairs for waiting customers (i.e., not including the barber's chair) who arrive on the average at the rate of four per hour. He is able to cut hair at the rate of two customers per hour and is unhappy at having to turn so many people away. He is thinking of hiring a second barber (still having only two waiting chairs). How many more people would pass through the shop in an eight-hour day if he hired the second barber? If each customer paid $3.00 for a haircut and he paid the second barber a flat rate of $30 per day, would he be better off?

An average of 32 customers arrive at the shop during the day. When he is the only barber there ($K = 3$) he turns away, on the average, $32 \times P_3$ customers per day where, from Eqs. (11.23) and (11.24) (which are identical to Eqs. (11.19) and (11.20) since $s = 1$ in this case)

$$P_3 = \rho^3 P_0, \quad \rho = 2$$

where

$$P_0 = (1 + \rho + \rho^2 + \rho^3)^{-1}$$

$$= \frac{1}{15}$$

and on the average he turns away $32 \times \frac{8}{15} = 17.06$ customers per day. With two barbers however ($K = 4$), each with parameter μ,

$$P_0 = \left[1 + \rho + \frac{\rho^2}{2} + \frac{1}{2}\left(\frac{\rho^3}{2} + \frac{\rho^4}{4}\right)\right]^{-1}$$

$$= \frac{1}{9}$$

and

$$P_3 = \frac{\rho^4}{2!\,2^2}P_9 = 2P_0 = \frac{2}{9}$$

and now on the average the shop turns away $32 \times \frac{2}{9} = 7.11$ customers per day. The amount of extra income the shop takes in per day is $(17.06 - 7.11) \times 3 = $29.85 out of which the second barber's wage must come. In the absence of other considerations he is better off by himself. We assumed here, of course, Poisson arrivals and exponential service times.

In the unrestricted $M/M/1$ queue, the queue length may grow to accommodate all customers. In the case where the system can hold only K customers, arrivals when the system is full are cleared from further consideration. In some cases, however, there is no restriction on the queue size, but the longer the queue size the less likely a new arrival is to join it. To cope with such a case, assuming a single server, the only problem is to model mathematically the probability that a new arrival joins the queue when there are n people in the system. One possibility is to model this $M/M/1$ queue with "discouragement" by taking the birth parameter

$$\alpha_n = \frac{\lambda}{n+1}, \quad n \geq 0$$

indicating that a new arrival has a probability of joining the queue inversely proportional to the queue length. The death parameter is

$$\beta_n = \mu, \quad n > 0$$

independent of how many are in the queue. With these assumptions the state probabilities of Eqs. (11.9) can be written as

$$P_0 = \left(1 + \rho + \frac{1}{2!}\rho^2 + \cdots\right)^{-1} = e^{-\rho}, \quad \rho = \frac{\lambda}{\mu}$$

and

$$P_n = \frac{\rho^n}{n!} e^{-\rho}, \quad n = 0, 1, 2, \ldots$$

and the state probabilities have a Poisson distribution. In this case the model we chose for "discouragement" led to convenient expressions for the probabilities, but, needless to say, this will not always be so.

Using the birth–death equation it is possible, by choosing the parameters appropriately, to model many other situations and it is instructive to do so. For example how should the parameters be chosen for an $M/M/s$ queue with a limit of K on the system, $K > s$, such that longer queue lengths discourage new arrivals?

We have, in this chapter, dealt with only a special case, namely, that of Poisson arrivals and exponential service times, whether or not these assumptions were explicitly stated. In some cases it will be important to make other assumptions but these will generally be more difficult to deal with. For example, it may be that the service time is a constant for each arriving customer. It is not our purpose to deal with such cases here and the reader is referred to more advanced treatises.

Because of the complexity of queueing problems and their analyses, the simulation of such problems has become an important tool in understanding their behavior. Many simulation languages and computer software packages are available for this purpose, indicating the value such techniques have on system design.

Problems

1. For a Poisson distribution with parameter λ show that

$$E(X^{k+1}) = \lambda E(X^k) + \lambda \frac{d}{d\lambda} E(X^k)$$

2. Find the probability density function for the time interval between $(k+1)$ arrivals of a Poisson process with parameter λ.

3. Show that a pure birth process with parameters $\alpha_n = \lambda$ is a Poisson process by showing that the Poisson distribution satisfies Eq. (11.6).

4. Customers arriving at an information desk do so according to a Poisson process at an average rate of 40 per hour. It takes each server an average of 4 minutes to serve a customer, the actual time being an exponentially distributed random variable. How many servers should there be if the average wait in the queue is to be less than 10 minutes?

5. A single server queue has Poisson arrivals with parameter $\lambda = 10$ per hour. What should the (exponential) service parameter be in order to have the probability that a new arrival has no wait of less than .5?

6. A ticket seller at a baseball game takes an exponentially distributed random time, with mean time 1 minute, to serve customers. If the arrivals process is Poisson with parameter λ, how small must λ be in order to make the average wait in the queue less than 1.5 minutes?

7. A fire hall has two engines. If fires occur on an average of three a week and take an average of 6 hours to quell, what is the probability that there are no fire engines to answer a call?

8. Trucks arriving at a two pump gas station leave if both pumps are in use. If the trucks arrive at random and, on the average, at a rate of 10 per hour and fill up time at the pumps is an exponentially distributed random variable with mean $1/\mu = 6$ minutes, what is the fraction of customers that are turned away?

9. Find the probability that an arriving customer has to wait for service in the $M/M/s$ queue.

10. In a bacteria culture, each organism will generate a successor with probability $\lambda \, \Delta t$ in any small interval of length Δt and there are no deaths. Thus, if there are k members of the population, the probability that there will be a successor in $(t, t+\Delta t)$ is $k\lambda \, \Delta t$. Show that the probability that there will be n members in the population at time t is given by

$$P_n(t) = \binom{n-1}{n-k} e^{-k\lambda t}(1 - e^{-\lambda t})^{n-k}$$

if at time $t = 0$ the population has size k.

11. A cafeteria checkout has two cashiers working. If each is capable of clearing 20 customers per hour and customers arrive at a rate of 30 per hour, what fraction of the time are both cashiers idle?

12. A single server queueing system has Poisson arrivals with parameter $\lambda = 10$ per hour and exponential serving times with parameter $\mu = 12$ per hour.

(i) What is the average number of customers in the system?

(ii) If the number of customers in the system cannot be greater than four, what is the averaged number of customers in the system?

(iii) If a second server is employed and there is no restriction on the number in the system, what is the average number in the system?

13. In a busy department store the number of people passing by a stand who are attracted to stay and watch the clerk demonstrate the product is proportional to the size of the crowd there. If this situation is modeled by choosing the parameters $\alpha_n = (n+1)\lambda$ and $\beta_n = n\mu$, what is the average size of the crowd watching?

14. Consider a birth–death system with parameters $\alpha_k = \alpha^k\lambda$, $k \geq 0$, $0 \leq \alpha < 1$ and $\beta_k = \beta$, $k \geq 1$. Find the ratio of the probability that there are k in the system, in the steady state, to P_0.

15. At an instant in time a customer places a request for service in a queueing system with s servers. All the servers are busy and, in addition, $N-s$ customers are waiting for service. If no more customers are allowed in the system and the service times are independent and exponentially distributed with identical parameter μ, find the expected length of time that the customer spends waiting for service (in the queue). Find the expected length of time until all customers in the system have completed service.

16. Consider a pure death process with an initial ($t = 0$) population of N. Solve the time-dependent probability, $P_n(t)$, the probability that there are n in the population at time t.

17. Consider the $M/M/s$ queue and suppose that at a given instant all servers are busy but there are no customers waiting. Find the probability that the next customer who arrives will have to wait for service.

18. In a computer batch processing system, the number of jobs submitted is a Poisson process with parameter $\lambda = 10$ per hour. The cost to the company of a delay in receiving the results is $5 per minute. The cost of computing is proportional to the speed of the computer in that the computer can process jobs with an exponential distribution of times with an average of μ per hour and at a cost of $20 per job. How fast a computer should the company buy?

19. A computer network transmits packets of information from node to node. The number of messages received at a node forms a Poisson process with parameter λ. Each node must process a packet to decide whether it must be forwarded to another node or whether it is for the local computer and it does this with exponentially distributed times with parameter μ. If a node has a buffer of size $K-1$ to hold packets while a current one is being processed, and if a packet arrives at a node with a full buffer it is destroyed, determine expressions for the following quantities:

(i) the probability that the packet gets destroyed;

(ii) the average delay at a node for a packet to be transmitted to another node;

(iii) the average number of packets already in the buffer that an arriving packet sees.

20. A retail washing machine store has the following inventory control policy. It likes to keep six washing machines in stock. When it sells a machine it immediately places an order for one. The length of time it takes to process the order and for the machine to arrive is exponentially distributed with mean 1 day. If the sales form a Poisson process with parameter two machines per day, what is the probability that the store will have no machines in stock?

21. A soda bottle cleaning machine that cleans μ cases per hour on the average costs $C\mu$ dollars per hour to operate. The cases arrive at the machine at the rate of λ cases per hour on the average. If a cost of D dollars per case per hour is assigned to

the wait for the cases to be cleaned, and if the arrivals process is Poisson and service times exponential, what should μ be to minimize cost?

22. An unloading dock has room for two ships. Ships arrive at the rate of two per hour and when the dock is full have to anchor in the bay at a charge of $50 per hour. The cost of unloading a ship is $200 \times m$, where m is the number of men assigned to unload the ship. If the time to unload the ship is an exponentially distributed random variable with a mean time of $4/m$ hours, what is the optimal choice of m to minimize overall cost?

23. The length of time that a customer spends in a store is an exponentially distributed random variable with a mean of 45 minutes. If customers arrive at the store according to a Poisson process with parameter $\lambda = 12$ per hour, find an expression for the probability that there are k customers in the store.

24. A steel rolling plant has a great many machines that break down at random on the average of one per hour. The plant has two repairmen who are able to fix machines at the rate of one an hour each, on the average.
   (i) What is the probability that both servers are idle?
   (ii) What is the expected number of machines awaiting service?
   (iii) What is the probability that a machine that breaks down will have to wait for service?

25. A load of N faulty T.V. sets are delivered to a repair facility where two repairmen are to fix them. If the time that each man takes to fix a set is exponentially distributed with parameter μ, find the differential equations which describe the probability $P_n(t)$ that at time t there are still n to be fixed, including those presently being worked on.

26. During March an accountant receives requests for tax return assistance according to a Poisson process with parameter $\lambda = 3$ a day. The time he takes with each form is exponentially distributed with parameter $\mu = 4$ per day. When there is only one return in the system he works on it himself. When there is more than one, a friend helps and they are able to process, on the average, 6 per day.
   (i) What is the probability of having more than one return in the system?
   (ii) What is the probability the accountant is idle?

27. If the incoming traffic to a parking lot with N spaces is Poissonly distributed with parameter λ, if there are empty spaces available, find the differential equation for $P_n(t)$, the probability that there are n spaces occupied. For the period under consideration assume that no cars leave.

28. Find the steady state probabilities for the number of cars in a parking lot with N parking places with random arrivals with parameter λ and random departures with parameter μ.

29. Solve the birth–death process time-dependent probabilities for the parameters

$$\alpha_k = \begin{cases} \lambda, & k=0 \\ 0, & k \neq 0 \end{cases} \qquad \beta_k = \begin{cases} \mu, & k=1 \\ 0, & k \neq 1 \end{cases}$$

This corresponds to the blocked customers cleared case for the simple $M/M/1$ queue.

CHAPTER 12

Reliability Theory

Examples of attempts to make systems comprised of unreliable components, reliable, are relatively abundant in everyday experience. The spare tire in the family car is an obvious example, as is the maintaining of spares of any kind. Other examples arise in aerospace applications where computers are designed to operate even if several components fail. The system designer is usually faced with the ever-present compromise between performance, cost, complexity, and reliability. The purpose of this chapter is to introduce a few of the basic notions and principles of the subject. In the next section, some definitions pertaining to reliability are discussed and common failure laws introduced. Series and parallel connections of components are analyzed and the final section discusses some techniques used to estimate lifetime parameters

12.1   *Failure Laws and Reliability*

One of the first problems that a system designer faced with designing a reliable system from unreliable components is concerned with is how he or she is to define what is meant by the term "reliable." This is usually very much context dependent. For example, if a communication system is being designed, the reliability of it may vary considerably depending on application. A common design standard for switching centers in North America is that they should have a "down time" (be inoperable) of less than one hour in 40 years. Similarly, many military communication systems have a negligible probability of being inoperable, sometimes achieved by having backup equipment or alternative channels. On the other hand, an oil company wanting to communicate with a remote oil exploration location might be willing to accept a relatively low probability of being able to communicate, say .7, since the communications may not be of such a nature as to require instantaneous action. Thus the very definition of reliability will depend on the problem and the means of achieving a more reliable system may also be severely restricted by the problem.

A general, all-purpose, definition of reliability is the probability of a device performing its function adequately for the period of time intended under the given operating conditions. In the next two sections, the meaning of reliability for our purposes will be reflected in two functions, the reliability function $R(t)$ and the failure rate function $r(t)$.

Of crucial importance in most applications is the probability that a component is still able to perform its function at time t. We will denote this probability by $R(t)$ and call it the reliability function. Before discussing its properties it is well to discuss the manner in which things fail, and perhaps considering light bulbs is as illuminating as any other component.

Light bulbs are designed to last a certain number of hours by their filament composition and construction, bulb shape, and gas content. When a new bulb is first turned on it may fail relatively quickly due to some flaws of construction. This is sometimes referred to as the "burn-in" or "infant mortality" period. If the component passes this period, one assumes that it has no obvious faults. Another example would be a T.V. set. In this case the burn-in period is often performed at the factory to remove small bugs and fine tune the set under operating conditions before delivery to the retailer. After the initial burn-in period the component settles down to its useful operating life. Failures in this period are due to chance mechanisms such as excessive heat or cold, vibration, or stress that the component is subjected to or some inherent weakness in the component that causes failure prematurely. In a T.V. set a resistor may be closer to a hot element causing faster deterioration than expected. As the component nears its design life in age, more and more failures occur because of factors built into it. A valve may fail because the material was only designed to withstand so many operations. A T.V. tube after several years degrades in performance until it is unusable. A car, after a few hundred thousand miles with the same engine will simply wear out. This is not true of all components. For example, an electrical fuse of good material may be essentially as good as new until its rated amperage is exceeded. A steel beam may be as good as new until it is deformed beyond its limit to regain shape. However, the behavior described here is quite typical of many components.

To begin the analysis we ask ourselves the question, what is the probability that it is still operating at time t? As a function of time let us define the function $r(t)$ as the conditional pdf such that $r(t)\,\Delta t$ is the probability that a given component will fail in the time period $(t, t + \Delta t)$ given that it is still operating at time t. Recalling that $R(t)$ is the (unconditional) probability that the component is still operating at time t, we can write

$$R(t+\Delta t) = [1 - r(t)\,\Delta t]R(t)$$

and on rearranging terms we have

$$\frac{1}{R(t)} \cdot \frac{R(t+\Delta t) - R(t)}{\Delta t} = -r(t) \qquad (12.1)$$

which as $\Delta t \to 0$ becomes

$$\frac{R'(t)}{R(t)} = -r(t) = \frac{d}{dt}\ln[R(t)]$$

Integrating both sides with the initial condition $R(0)=1$

$$R(t) = \exp\left(-\int_0^t r(s)\,ds\right) \qquad (12.2)$$

which is a useful relationship between the functions $r(t)$ and $R(t)$.

The function $r(t)$ is called the failure rate function or hazard function and, as a conditional density function, $r(t)\,\Delta t$ is the fraction of those components that have survived until time t, which fail in the interval $(t, t+\Delta t)$. A typical failure rate function is shown in Figure 12.1.

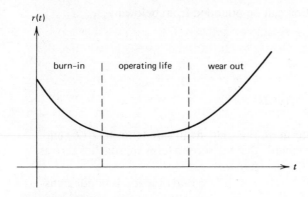

FIGURE 12.1

The time to failure for a component T is a random variable and as such has a probability density function $f(t)$ and cumulative density function $F(t)$. From the definition

$$R(t) = 1 - F(t) \qquad (12.3)$$

and, from Eq. (12.1), since $R(t+\Delta t) - R(t) = -[F(t+\Delta t) - F(t)]$,

$$r(t) = f(t)/R(t)$$

The mean time to failure of the component is given by

$$E(T) = \int_0^\infty tf(t)\,dt \qquad (12.4)$$

There are other ways of expressing this quantity. For example, since $f(t) = -R'(t)$ from Eq. (12.3)

$$E(T) = -\int_0^\infty tR'(t)\,dt$$

and integrating by parts yields

$$E(T) = -tR(t)\Big|_0^\infty + \int_0^\infty R(t)\,dt$$

At $t = 0$, $tR(t)$ clearly vanishes. If the pdf of T has a mean (i.e., the integral in Eq. (12.4) exists and is finite), then it is a matter of showing that $R(t)$ goes to zero "at a faster rate" than t tends to ∞, as t increases. The following informal argument will suffice for our purposes. We can write $E(T)$ as

$$E(T) = \int_0^\infty sf(s)\,ds$$

$$= \int_0^t sf(s)\,ds + \int_t^\infty sf(s)\,ds$$

The second term in this equation can be bounded from below by

$$t\int_t^\infty f(s)\,ds$$

and so

$$E(T) \geq \int_0^t sf(s)\,ds + t\int_t^\infty f(s)\,ds$$

As t tends to infinity the first term on the right-hand side tends to $E(T)$ and, since we assumed it finite, we conclude that the second term approaches zero as required, and, as claimed, $E(T) = \int_0^\infty R(t)\,dt$.

Typically, for real components, the failure rate function is nondecreasing beyond the burn-in region. This simply means that as time passes ever increasing fractions of components reach the end of their useful life and fail. Components that have an increasing failure rate almost from the burn-in period undergo a steady aging process throughout their useful life. Before discussing failure laws consider the following example.

EXAMPLE 12.1

To a first approximation ball bearings are made of such hard steel that in a particular application the failure rate function is a constant, $r(t) = .10$ per hour. What is the probability that a particular bearing will last 10 hours?

From Eq. (12.3) the reliability function corresponding to the failure rate function $r(t) = .10$ is

$$R(t) = \exp\left(-\int_0^t .10 \, ds\right) = \exp(-.10t), \quad t \geq 0$$

and, since $R(t) = 1 - F(t)$,

$$f(t) = -\frac{d}{dt}R(t) = (.10)\exp(-.10t)$$

The probability that a particular component will last 10 hours is

$$P(T \geq 10) = \int_{10}^{\infty} (0.1)\exp(-0.1t) \, dt$$

$$= -\exp(-0.1t)\Big|_{10}^{\infty}$$

$$= \exp(-1) = 0.368$$

In an assembly line situation, components are usually chosen according to some strategy and tested to determine quality. Such data leads to an understanding of the failure mechanism and to a model of it. Probabilistically, the model leads to a failure rate function or failure law. More commonly, perhaps, the data obtained is used to decide which one of a class of failure laws best approximates the data and the parameters in the law chosen appropriately. In the remainder of the section we describe failure laws often used in practice.

12.1.1 EXPONENTIAL DISTRIBUTION

In this instance the failure rate law is given by the exponential law

$$f(t) = \mu e^{-\mu t}, \quad t \geq 0$$

The mean time to failure is

$$E(T) = \int_0^{\infty} t\mu e^{-\mu t} \, dt = 1/\mu$$

and the failure rate function and reliability function are, respectively,

$$r(t) = \mu \qquad R(t) = P(T > t) = e^{-\mu t}$$

The fact that the failure rate function is constant means simply that at any two times t_1 and t_2 the ratios $n(t_1)/N(t_1)$ and $n(t_2)/N(t_2)$ of the items that failed during $(t_1, t_1 + \Delta t)$ and $(t_2, t_2 + \Delta t)$, $n(t_1)$ and $n(t_2)$, respectively, to the numbers of items that survived until t_1, and t_2, $N(t_1)$ and $N(t_2)$, respectively, is constant independent of time. The implication is that the components do not age, that is, if a component survives until time t, it is as good as new. Many solid-state devices actually have such a failure law. These functions are shown in Figure 12.2a for various values of μ.

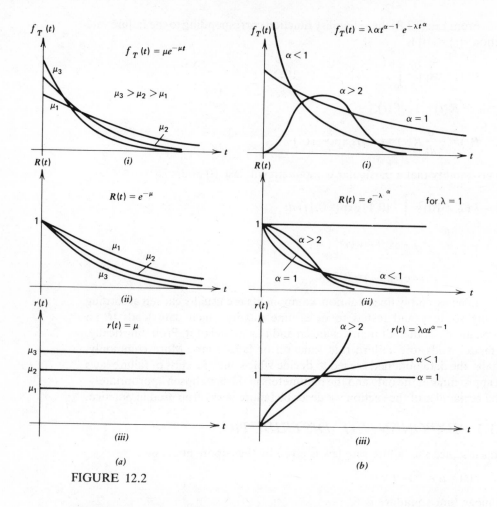

FIGURE 12.2

This law has the following interesting property. If a component has survived until t, the probability that it fails in the interval $(t+s, t+s+ds)$ is

$$P(t+s \leq T \leq t+s+ds \mid T \geq t) = \frac{\mu\, e^{-\mu(t+s)}\, ds}{e^{-\mu t}}$$

$$= \mu\, e^{-\mu s}\, ds$$

that is, the failure law of those components surviving until time t is exactly the same as the original components, justifying our characterization as a "good as new until failure" law.

12.1.2 GAMMA FAILURE LAW

The component with a gamma failure law has a time to failure with pdf

$$f(t) = \lambda^{\alpha} t^{\alpha-1}\, e^{-\lambda t} / \Gamma(\alpha), \quad \lambda, \alpha > 0, t \geq 0$$

The mean time to failure for this law is

$$E(T) = (\alpha - 1)/\lambda$$

There is no simple expression for $r(t)$ in this case.

12.1.3 THE WEIBULL FAILURE LAW

The failure law is given by the density function

$$f(t) = \lambda \alpha t^{\alpha-1} e^{-\lambda t^\alpha}, \quad \lambda, \alpha > 0, \ t \geq 0$$

The reliability function is simply

$$R(t) = \exp(-\lambda t^\alpha)$$

and the failure rate function is

$$r(t) = \lambda \alpha t^{\alpha-1}, \quad \lambda, \alpha > 0, \ t \geq 0$$

The mean time to failure is calculated as

$$E(T) = \frac{\Gamma(1+1/\alpha)}{\lambda^{1/\alpha}} \quad \text{and} \quad V(T) = \frac{1}{\lambda^{2/\alpha}} \left[\Gamma\left(1+\frac{2}{\alpha}\right) - \Gamma^2\left(1+\frac{1}{\alpha}\right) \right]$$

For $\alpha = 1$ this is simply the exponential distribution. For $\alpha = 2$ it is referred to as the Rayleigh distribution. The polynomial nature of the failure rate function makes this law particularly useful if, for example, it is desired to approximate failure rate data. These functions are shown for various values of α in Figure 12.2b.

12.1.4 TRUNCATED NORMAL DISTRIBUTION

The normal distribution is positive for all values of t, not just positive ones. However, if the mean and variance of the distribution are such that the probability the random variable is negative is negligible, then it can be used as a failure law. The failure law is

$$f(t) = \frac{1}{K\sqrt{2\pi}\sigma} \exp\left(-\frac{(t-\mu)^2}{2\sigma^2}\right), \quad 0 \leq t < \infty, \ \sigma > 0$$

where

$$K = \int_0^\infty \frac{1}{\sqrt{2\pi}\sigma} \exp\left(-\frac{(t-\mu)^2}{2\sigma^2}\right) dt$$

and this constant K is often assumed to be exactly unity since it will be very close to unity if the use of $f(t)$ as a failure law is valid. The failure rate and reliability functions can be expressed only in terms of the error function and we omit them.

Other failure functions than those mentioned here can arise in practice, but those given here are the most important. We conclude this section with examples that use the concepts introduced.

EXAMPLE 12.2

A diamond drill bit has a failure rate function $r(t) = 2 \times 10^{-6} t$ per hour. What is the probability that it will still be operational after 1000 hours of service?

With a linearly increasing failure rate function, the corresponding failure law is the Weibull law with parameter $\alpha = 2$. The reliability function for this law is just

$$R(t) = \exp(-\lambda t^2)$$

and the probability that the drill bit is still operational at $t = 1000$ is

$$R(1000) = \exp(-\lambda \, 10^6)$$

where $2\lambda = 2 \times 10^{-6}$ and

$$R(1000) = \exp(-1) = 0.368$$

EXAMPLE 12.3

Consider the following functions: (i) At^2; (ii) Bt^{-3}; (iii) e^{at}, $a > 1$. Which of these functions can be used as a failure rate function, and for those that can, find the corresponding failure law.

From the previous example, the failure law $f(t)$ and the failure rate function $r(t)$ are related by the equation

$$f(t) = r(t) \exp\left(-\int_0^t r(s) \, ds\right) \tag{12.5}$$

To check whether the proposed failure rate function is valid, we simply check that its corresponding pdf is valid.

(i) $r(t) = At^2$. In this case the integral is

$$\int_0^t r(s) \, ds = A \int_0^t s^2 \, ds = At^3/3$$

and the corresponding failure law is

$$f(t) = At^2 \exp(-At^3/3), \quad t \geq 0$$

But this is just the Weibull failure law with $\alpha = 3$ and $\lambda = A/3$ and is, of course, a valid pdf and hence $r(t) = At^2$ is a valid failure rate function.

(ii) $r(t) = Bt^{-3}$. The exponent in Eq. (12.4) is

$$\int_0^t \frac{B}{s^3} \, ds = -\frac{B}{2s^2}\bigg|_0^t$$

which is infinity, implying that $r(t) = Bt^{-3}$ is not a valid failure rate function.

(iii) $r(t) = e^{at}$. The exponent in Eq. (12.5) is

$$\int_0^t e^{as}\, ds = \frac{1}{a} e^{as}\big|_0^t = \frac{1}{a}(e^{at} - 1)$$

and for $a > 1$ this leads to a function $f(t)$ with infinite area, hence $r(t)$ is not a valid failure rate function.

EXAMPLE 12.4

What is the expected number of failures among 100 items, after 1000 hours, if the items fail independently and each has the failure rate function (i) $r(t) = 10^{-4}$, (ii) $r(t) = 10^{-4}t$.

(i) From Eq. (12.3) the reliability of a single item is

$$R(t) = P(T > t) = \exp[-G(t)]$$

where

$$G(t) = \int_0^t 10^{-4}\, ds = 10^{-4}t$$

Thus each item has a probability

$$R(1000) = \exp(-10^{-4} \cdot 1000) = \exp\left(-\frac{1}{10}\right)$$

of working after 1000 hours. The number of items still working after 1000 hours is a binomial random variable with parameters $n = 100$ and $\theta = R(1000)$ and the expected number of such items is simply $n\theta = 100\,\exp(-\frac{1}{10})$.

(ii) For this failure rate function

$$G(t) = \int_0^t 10^{-4}s\, ds = 10^{-4}t^2/2$$

and each item has a probability of

$$R(1000) = \exp[-10^{-4}(10^6)/2] = \exp(-50)$$

of surviving the 1000 hours. The expected number of survivors is then simply $100\,\exp(-50)$.

12.2 Series Connections

Most systems in practice are comprised of numerous components, each with their own failure characteristics connected together in some fashion. For example, in a T.V. receiver there are numerous transistors, transformers, resistors, and so forth and, of course, a picture tube. Generally speaking, if any one of these components fails, the whole system fails. This is not true in general as many systems permit some failures without the overall system becoming inoperable. In this section we consider a simple series connection of components, each of whose reliability is known, and determine the overall reliability of the system.

Consider the following situation as shown in Figure 12.3 with n systems $S_1, S_2, \ldots, S_n$ connected in series. The implication here is that all n systems must be operating for the overall system to work. It is also assumed that each system fails independently of the others, and that the failure law for system S_i is $f_i(t)$ with corresponding reliability and failure rate functions $R_i(t)$ and $r_i(t)$, respectively, $i = 1, 2, \ldots, n$.

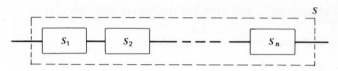

FIGURE 12.3

Since the function $R(t)$ has the interpretation of being the probability that the system is still operating at time t, this is just the probability that each of the n systems is operating at time t. Since these events are independent, the probabilities multiply and we have

$$R(t) = R_1(t)R_2(t) \cdots R_n(t) \tag{12.6}$$

Another way of expressing this result is to note that from Eq. (12.2)

$$R_i(t) = \exp\left(-\int_0^t r_i(s)\, ds\right)$$

and substituting into Eq. (12.6) gives

$$R(t) = \prod_{i=1}^n \exp\left(-\int_0^t r_i(s)\, ds\right)$$

$$= \exp\left[-\int_0^t \left(\sum_{i=1}^n r_i(s)\right) ds\right]$$

The series connection is then equivalent to a single system with reliability function $r(t) = \sum_{i=1}^n r_i(t)$. The mean time to failure for this system is

$$E(T) = \int_0^\infty R(t)\, dt$$

$$= \int_0^\infty \exp\left(-\int_0^t \sum_{i=1}^n r_i(s)\, ds\right) dt$$

In the special case when each component system has a constant failure rate function, $r_i(t) = c_i$, this can be evaluated as

$$E(T) = 1/\left(\sum_{i=1}^n c_i\right)$$

EXAMPLE 12.5

One conveyor belt system has ten bearings, all of which must work in order for the system to work. Each bearing has a constant failure rate function $r(t) = 10^{-2}$ per hour. A proposed new system has only five bearings, each with a Rayleigh distributed failure law and failure rate function $r(t) = c \times 10^{-2}t$. If both systems have the same mean time to failure, which system is preferable if it is important that the system operates continuously for eight hours (after which maintenance can check for repairs)?

The mean time to failure of the first system is

$$E(T) = 1/(10 \times 10^{-2}) = 10 \text{ hours}$$

and the reliability function is

$$R(t) = \exp(-.1t)$$

The failure rate function of the proposed system is

$$r(t) = 5 \times c \times 10^{-2}t$$

and the reliability function is

$$R(t) = \exp(-\lambda t^2)$$

The expected time to failure in this case is

$$E(T) = \Gamma(\tfrac{3}{2})/\sqrt{\lambda} = .866/\sqrt{\lambda}$$

and, since this must be identical with the original, $.866/\sqrt{\lambda} = 10$ or $\lambda = .750 \times 10^{-2}$. It must be then that $2\lambda = 5 \times c \times 10^{-2}$ or $c = .30$, although this value is not required for the problem.

To evaluate the two systems we compare the probability that each system will survive until time $t = 8$. For the first system this is

$$R_1(8) = \exp(-0.8) = .449$$

while for the proposed system it is

$$R_2(8) = \exp(-.75 \times 10^{-2} \times 64) = \exp(-.48) = .619$$

and the proposed system has a greater reliability. Notice, however, that at $t = 20$ hours

$$R_1(20) = \exp(-2) = .135$$

while, for the second system,

$$R_2(20) = \exp(-3) = .050$$

and, for this time interval, the original system has the greater reliability.

EXAMPLE 12.6 (Alternative Approach)

The system S has n components S_i connected in series. If S_i has an exponential failure law with parameter λ_i, $i = 1, \ldots, n$, find the mean time to failure of S.

There are at least two ways of considering this problem and one has already been discussed. We give an alternative approach in this example.

If T_i is the time to failure of the ith system and T the time to failure of the overall system then $T = \min(T_1, \ldots, T_n)$. From Example 7.20 the pdf of T can be written as

$$f(t) = \sum_{i=1}^{n} f_i(t) \prod_{\substack{j=1 \\ j \neq i}}^{n} \left(\int_{t}^{\infty} f_j(s)\, ds \right)$$

$$= \sum_{i=1}^{n} \lambda_i \exp\left[-\left(\sum_{k=1}^{n} \lambda_k \right) t \right]$$

and the mean of T is

$$E(T) = \int_{0}^{\infty} t f(t)\, dt$$

$$= \sum_{i=1}^{n} \lambda_i \int_{0}^{\infty} t \exp\left[-\left(\sum_{k=1}^{n} \lambda_k \right) t \right] dt$$

$$= \left(\sum_{i=1}^{n} \lambda_i \right) \cdot \frac{1}{\left(\sum_{i=1}^{n} \lambda_i \right)^2}$$

$$= \frac{1}{\left(\sum_{i=1}^{n} \lambda_i \right)}$$

EXAMPLE 12.7

In a series connection of two systems, failure does not occur independently but rather has a joint pdf of

$$f_{T_1 T_2}(t_1, t_2) = \begin{cases} \frac{1}{50}, & (t_1, t_2) \in R \\ 0, & \text{elsewhere} \end{cases}$$

where R is the region shown in Figure 12.4. Find the reliability function for the overall system.

The time to failure of the overall series connection is $T = \min(T_1, T_2)$ and, from Section 7.3, the cdf $F(t)$ of T is

$$F(t) = F_{T_1}(t) + F_{T_2}(t) - F_{T_1 T_2}(t, t)$$

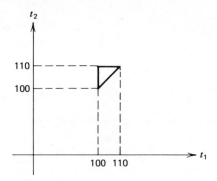

FIGURE 12.4

The three required functions are found as:

$$F_{T_1}(t) = \frac{1}{50}[(t-100) \cdot (110-t) + (t-100)^2/2]$$

$$= \frac{1}{50}(-t^2/2 + 110t - 6000), \quad 100 \le t \le 110$$

$$F_{T_2}(t) = F_{T_1 T_2}(t, t)$$

$$= \frac{1}{50}\left(\frac{(t-100)^2}{2}\right)$$

$$= \frac{1}{50}\left(\frac{t^2}{2} - 100t + 5000\right), \quad 100 \le t \le 110$$

Since $F_{T_2}(t) = F_{T_1 T_2}(t, t)$, it follows that $F(t) = F_{T_1}(t)$ and

$$R(t) = 1 - F(t)$$

$$= \begin{cases} 1 + \dfrac{t^2}{100} - \dfrac{11}{5}t + 120, & 100 \le t \le 110 \\ 1, & t \le 100 \\ 0, & t \ge 110 \end{cases}$$

12.3 Parallel Connections

Intuitively, the overall reliability of a series configuration as shown in Figure 12.3 is less than the reliability of any of its components. In fact, if $T_1, \ldots, T_n$ are the failure times of the n components, then T, the time to failure of the overall system, is just $T = \min(T_1, T_2, \ldots, T_n)$ as discussed in Example 12.6. In a similar manner it is clear that if only one system S_i is required for a parallel connection of n systems to operate, then the overall reliability is greater than or equal to that for a particular system. In fact, if T is the time to failure for S and T_i the time to failure for S_i, then $T = \max(T_1, T_2, \ldots, T_n)$.

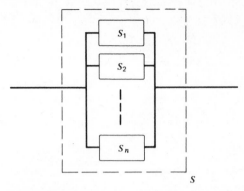

FIGURE 12.5

We will derive the reliability function for S. Since the system is not operable only if all the systems have failed, the probability that the system is still operating at time t is one minus the probability that all systems have failed by time t. The probability that the system S_i has failed by time t is $(1 - R_i(t))$, where $R_i(t)$ is its reliability. The overall reliability of the n systems is then

$$R(t) = 1 - [1 - R_1(t)][1 - R_2(t)] \cdots [1 - R_n(t)]$$

If each system has the same reliability $R_1(t)$, then

$$R(t) = 1 - [1 - R_1(t)]^n$$

$$= \sum_{j=1}^{n} \binom{n}{j} R_1(t)^j [1 - R_1(t)]^{n-j}$$

This is just one form of the binomial theorem. Clearly if we require k of the systems to operate before S will operate, then the reliability of S becomes

$$R(t) = \sum_{j=k}^{n} \binom{n}{j} R_1(t)^j [1 - R_1(t)]^{n-j}$$

In practice, of course, systems are usually some combination of parallel and series systems. Generally, techniques to deal with these systems are straightforward generalizations of the methods for series and parallel connections. For example, the reliability of the system, shown in Figure 12.6, can be

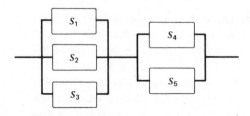

FIGURE 12.6

found by replacing systems S_1, S_2, and S_3 with a single system S_1' with reliability

$$1-[1-R_1(t)][1-R_2(t)][1-R_3(t)] = R_1'(t)$$

and, similarly, S_4 and S_5 can be replaced with a system S_4' with reliability

$$1-[1-R_4(t)][1-R_5(t)] = R_4'(t)$$

The reliability of the overall system, represented by the series connection of S_1' and S_4', is

$$R(t) = \{1-[1-R_1(t)][1-R_2(t)][1-R_3(t)]\}\{1-[1-R_4(t)][1-R_5(t)]\}$$

Sometimes, however, a system is composed in such a way that it is not easily decomposed by the above methods. Consider the system shown in Figure 12.7. When drawn in this manner, the line from S_1 to S_4 indicates that there is

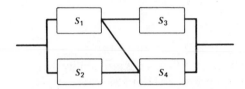

FIGURE 12.7

no path from S_2 to S_3. To analyze this system we assume for convenience that each system has the same reliability, say $R_1(t)$. Let A_i be the event that system S_i is operable. The overall system will operate if one of the following conditions is met:

(i) S_1 and S_3 operate = event $A_1 \cap A_3$
(ii) S_1 and S_4 operate = event $A_1 \cap A_4$
(iii) S_2 and S_4 operate = event $A_2 \cap A_4$

The probability that the overall system will operate is then

$$P[(A_1 \cap A_3) \cup (A_1 \cap A_4) \cup (A_2 \cap A_4)]$$
$$= P(A_1 \cap A_3) + P(A_1 \cap A_4) + P(A_2 \cap A_4)$$
$$- P(A_1 \cap A_3 \cap A_4) - P(A_1 \cap A_2 \cap A_3 \cap A_4)$$
$$- P(A_1 \cap A_2 \cap A_4) + P(A_1 \cap A_2 \cap A_3 \cap A_4)$$

Since the events A_i are assumed independent, the probabilities of these intersections multiply, and, noting that $P(A_i) = R_1(t)$, we have

$$R(t) = 3R_1^2(t) - 2R_1^3(t)$$

for the overall system reliability. The reader should compare this answer with that for the system shown in Figure 12.8 and clearly understand the difference and why a different approach was required for the first system. (This second system is easily reduced to a series combination of two parallel systems.)

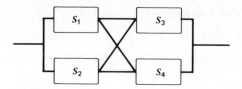

FIGURE 12.8

In a series connection all components must work for the overall system to work. In a parallel connection only one must be operable for the overall system to work. There exist systems with n components requiring at least k, $1 \le k \le n$ of the components to operate in order for the whole system to operate and a simple example of this situation is given next.

EXAMPLE 12.8

A cable consists of n strands of wire. In order for the cable to pass its rated strain, at least r of the strands must be intact. What is the probability that the cable is operable at time t if: (i) $r(t) = a = \text{constant}$; (ii) $r(t) = bt$, where $r(t)$ is the failure rate function of a single strand.

(i) The reliability function for a single strand is

$$R(t) = \exp\left(-\int_0^t r(s)\,ds\right) = \exp(-at)$$

which is the probability that a single strand is intact at time t. Since the number of strands intact is a binomial random variable, the probability that the cable is operable at time t is

$$\sum_{k=r}^{n} \binom{n}{k}(e^{-at})^k (1-e^{-at})^{n-k}$$

(ii) The reliability function for a single strand is

$$R(t) = \exp\left(-\int_0^t bs\,ds\right) = \exp\left(-\frac{bt^2}{2}\right)$$

and the probability that the cable is operable at time t is

$$\sum_{k=r}^{n} \binom{n}{k} R(t)^k [1-R(t)]^{n-k}$$

EXAMPLE 12.9

Part of a computer input-output channel is configured as shown in Figure 12.9, partly for reliability but mainly to achieve greater throughput of input and output to and from peripheral devices. If each system has reliability function $R_1(t)$, what is the reliability of the overall system?

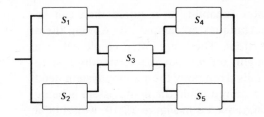

FIGURE 12.9

By straightforward computation, letting A_i be the event that system i is operating, the probability that the overall system is operating is

$$P = P[(A_1 \cap A_2) \cup (A_1 \cap A_3 \cap A_4) \cup (A_1 \cap A_3 \cap A_5)$$
$$\cup (A_2 \cap A_3 \cap A_4) \cup (A_2 \cap A_3 \cap A_5) \cup (A_2 \cap A_5)]$$

and using Eq. (1.10) this can be written as the sum of probabilities of the intersections of various sets. The final result is

$$P = 2R_1^2(t) + 2R_1^3(t) - 5R_1^4(t) + 2R_1^5(t)$$

This is rather a tedious way to achieve the result. A simpler way is to condition the events on whether or not system 3 is working. If system 3 is working, then the probability that the system is working is the probability that either 1 or 2 (or both) are working and either 4 or 5 (or both) are working, that is,

$$P[(A_1 \cup A_2) \cap (A_4 \cup A_5)] = P(A_1 \cup A_2)P(A_4 \cup A_5)$$
$$= [P(A_1) + P(A_2) - P(A_1 \cap A_2)][P(A_4)$$
$$+ P(A_5) - P(A_4 \cap A_5)]$$
$$= [2R_1(t) - R_1^2(t)]^2$$

If S_3 is not working, then the system works if 1 and 4 work or 2 and 5 work (or all 4 work), that is,

$$P[(A_1 \cap A_4) \cup (A_2 \cap A_5)] = P(A_1 \cap A_4) + P(A_2 \cap A_5)$$
$$- P(A_1 \cap A_2 \cap A_4 \cap A_5)$$
$$= 2R_1^2(t) - R_1^4(t)$$

The overall probability of the system working is

$$P = P[(A_1 \cup A_2) \cap (A_4 \cup A_5)]P(A_3)$$
$$+ P[(A_1 \cap A_4) \cup (A_2 \cap A_5)][1 - P(A_3)]$$
$$= [2R_1(t) - R_1^2(t)]^2 R_1(t) + [2R_1^2(t) - R_1^4(t)][1 - R_1(t)]$$
$$= 2R_1^2(t) + 2R_1^3(t) - 5R_1^4(t) + 2R_1^5(t)$$

as before.

In some systems, where greater reliability is required than can be afforded by a single system, other techniques are employed. Among them would be the notion of a standby system. These are often employed in a situation where the original system is highly reliable but a cheap standby system is made available in case the original fails. Hopefully the standby system need not be overly reliable since it is only required while the main system is being repaired. An example of the situation, depicted in Figure 12.10, would be a hospital power system. An auxiliary generating system, the standby system, is required because of the crucial nature of its use in a hospital environment. Hopefully, the main power supply seldom goes down and then only for short periods. The auxiliary supply is only required for the short periods while the main system is down. The effect of incorporating an unreliable system in this manner can well be to produce a highly reliable system.

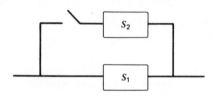

FIGURE 12.10

Assume that the main system S_1 has a failure law $f_1(t)$ and let T_1 be the random variable indicating the time to failure. Similarly, let the failure law of system S_2 be $f_2(t)$ and the corresponding random variable T_2. From the assumptions, T_1 and T_2 are independent random variables. When S_1 fails, S_2 switches in immediately and consequently the time T to failure of the overall system is $T = T_1 + T_2$ assuming no repair. The failure law of the overall system, $f(t)$, is given by (see Section 7.2)

$$f(t) = f_1(t) * f_2(t) = \int_0^t f_1(s) f_2(t-s) \, ds$$

The reliability of the overall system is then

$$R(t) = \int_t^\infty f(s) \, ds$$

The special case where each system has a constant and identical failure rate function is treated in the next example.

EXAMPLE 12.10

In the n-fold standby system shown in Figure 12.11, system S_i, which has an exponential failure law with parameter λ, is switched in when system S_{i-1} fails, $i = 2, 3, \ldots, n$. Find the overall reliability of the system.

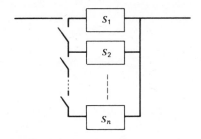

FIGURE 12.11

From the assumptions, T_i, the time to failure of S_i, is exponentially distributed with parameter λ and the time to failure of the overall system is $T = T_1 + T_2 + \cdots + T_n$, where the T_i are assumed independent random variables. We have already seen that the sum of n independent exponential random variables with parameter λ has a gamma density function

$$f(t) = \frac{\lambda^n t^{n-1}}{(n-1)!} e^{-\lambda t}, \quad t > 0$$

The reliability function for the n-fold standby system is, then,

$$R(t) = \int_t^\infty f(s)\, ds$$

$$= \int_t^\infty \frac{\lambda^n s^{n-1}}{(n-1)!} e^{-\lambda s}\, ds$$

$$= e^{-\lambda t} \left(1 + \lambda t + \frac{(\lambda t)^2}{2!} + \cdots + \frac{(\lambda t)^{n-1}}{(n-1)!} \right)$$

EXAMPLE 12.11

A hospital standby power generating unit S_2 shown in Figure 12.12 has a time to failure T_1, which is uniformly distributed in the interval (90, 100). The main system has an exponential failure law with parameter $\lambda = 1000$ hours. Find the pdf of the time to failure of the overall system.

The pdf of T_1 is

$$f_1(t) = \frac{1}{10}, \quad 90 \leq t \leq 100$$

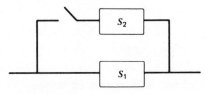

FIGURE 12.12

and the pdf of T_2 is

$$f_2(t) = 1000\, e^{-1000t}, \quad t \geq 0$$

Since the time to failure for the overall system is $T = T_1 + T_2$, from Chapter 7 the pdf of T is

$$f(t) = \int_R f_1(s) f_2(t - s)\, ds$$

where R is the region of integration shown shaded in Figure 12.13. Substituting

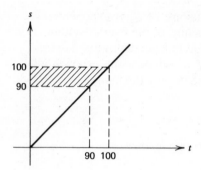

FIGURE 12.13

the pdf's gives

(i) $0 \leq t \leq 90$,

$$f(t) = \int_{90}^{100} \frac{1}{10} \cdot 1000\, e^{-1000(t-s)}\, ds$$

$$= 100\, e^{-1000t} \left(\frac{1}{1000} e^{1000s} \right) \Big|_{90}^{100}$$

$$= \frac{1}{10} e^{-1000t} (e^{10^5} - e^{9 \times 10^4})$$

(ii) $90 \leq t \leq 100$,

$$f(t) = \int_{t}^{100} \frac{1}{10} \cdot 1000\, e^{-1000(t-s)}\, ds\,.$$

$$= \frac{1}{10} \cdot e^{-1000t} (e^{10^5} - e^{103t})$$

EXAMPLE 12.12

Each of the systems shown in Figure 12.14 has an exponential failure law with identical parameter λ. Find the reliability of the overall system.

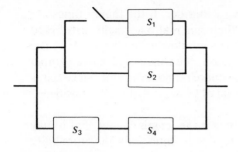

FIGURE 12.14

The reliability function for each component is

$$R_1(t) = \int_t^\infty \lambda\, e^{-\lambda s}\, ds = e^{-\lambda t}$$

The reliability of the lower branch of the system is simply $R_1^2(t) = e^{-2\lambda t}$. From Example 12.11 the reliability of the upper branch is $e^{-\lambda t}(1+\lambda t)$. The reliability of two systems in parallel with these reliabilities is

$$R(t) = 1 - (1 - e^{-2\lambda t})[1 - e^{-\lambda t}(1 + \lambda t)]$$

EXAMPLE 12.13

The systems S_1 and S_2 are connected in parallel, each having an exponential failure law with parameter λ. Determine the failure law of the overall system.

If T_i is the time to failure of system S_i, $i = 1, 2$, then the time T to failure of the overall system is $T = \max(T_1, T_2)$. From Example 7.18 the pdf of T is

$$f(t) = 2\lambda\, e^{-\lambda t}(1 - e^{-\lambda t})$$

12.4 *Life Testing and Estimation*

An important aspect of reliability is the estimation, from experimental data, of parameters such as expected lifetime and the reliability function. One of the following three strategies is commonly used:

(1) A sample of n items is placed on a test and the set of failure times observed.
(2) A sample of n items is placed on a test and the sum of the operating times to the rth failure observed. With this procedure the items that fail may or may not be replaced by good items. In either case it is referred to as a failure truncated test.
(3) A sample of n items is placed on a test for a fixed period of time and the number of failures during this time observed. Again the items that fail may or may not be replaced, and the test is called a time truncated test.

The first strategy is relatively easy to consider. If the failure times are assumed to be normally distributed, then their sum is also normally distributed and confidence limits on the mean or variance follow as in Chapter 9. If the failure times have an exponential distribution, then their sum has a gamma distribution and so is proportional to a chi-square random variable. Since this is a straightforward application of the material of Chapter 9, it is not considered any further here.

One disadvantage of the first strategy is the length of time that it may take to complete the test. For relatively long-lived components, the probability of having at least one in the test that will last much longer than the average can be quite high. This problem is overcome in the second and third strategies.

Only the second strategy will be analyzed here, the third being similar. Consider first the case where items that fail are replaced and at any given time during the test there are n items on test. The test is terminated after the rth failure has been observed. If the time taken to observe the rth failure is T_r, then the total operating time of all items is $T = nT_r$. To find the pdf of T_r we argue as follows: Let X_i be the time to failure of the ith item, $i = 1, 2, \ldots, n$ and let $T^{(1)}$ be the time to the first failure among the n items, that is,

$$T^{(1)} = \min(X_1, X_2, \ldots, X_n)$$

We will assume for the remainder of this section that each item has the identical failure law that is exponential with parameter λ, that is, $f(x) = \lambda e^{-\lambda x}$, $x \geq 0$. From Section 7.3, the pdf of $T^{(1)}$ is

$$f_1(t) = n \cdot (\lambda e^{-\lambda t})(e^{-\lambda t})^{n-1} = n\lambda e^{-n\lambda t}, \quad t \geq 0 \tag{12.7}$$

that is, $T^{(1)}$ is an exponentially distributed random variable with parameter $n\lambda$. Items with an exponential failure law are as good as new until failure, as discussed earlier. Thus, after the first failure, and the failed item replaced, there are n items as good as new. The time to the next failure (after the first failure) $T^{(2)}$ has the same pdf as $T^{(1)}$ and is independent to T_1. It follows that the time to the rth failure is

$$T_r = T^{(1)} + T^{(2)} + \cdots + T^{(r)} \tag{12.8}$$

where the $T^{(i)}$ are independent random variables, each with the pdf of Eq. (12.7). In Section 8.1 it was shown that such a sum has the gamma pdf

$$f_T(t) = \frac{(n\lambda)^r t^{r-1} e^{-n\lambda t}}{\Gamma(r)}, \quad t > 0 \tag{12.9}$$

Differentiating Eq. (12.9) with respect to λ and setting the result to zero shows that it is maximized for $\lambda = r/nt$ and so the maximum likelihood estimator of λ is r/nT_r or, equivalently, the maximum likelihood estimator of the mean time to failure (MTTF) is

$$\left(\frac{\hat{1}}{\lambda}\right) = \frac{nT_r}{r} = \frac{T_0}{r} \tag{12.10}$$

where $T_0 = nT_r$, is the total operating time of the n units, and T_r is given by Eq. (12.8).

Rather than use T_r directly we define $Y = 2\lambda T_0 = 2n\lambda T_r$ and note that the pdf of Y is

$$f_Y(y) = \frac{(n\lambda)^r (y/2n\lambda)^{r-1} e^{-y/2}}{\Gamma(r)} \cdot \frac{1}{2n\lambda}$$

$$= \frac{y^{r-1} e^{-y/2}}{2^r \Gamma(r)}$$

where the Jacobian of the transformation is $1/2n\lambda$. Thus Y is a χ^2 random variable with $2r$ degrees of freedom. From this useful fact we can immediately state that

$$P(\chi^2_{\alpha/2;2r} \le 2n\lambda T_r \le \chi^2_{1-\alpha/2;2r}) = 1 - \alpha$$

and so the $100(1-\alpha)$ percent confidence interval for the MTTF $1/\lambda$ is

$$\left(\frac{2nT_r}{\chi^2_{1-\alpha/2;2r}}, \frac{2nT_r}{\chi^2_{\alpha/2;2r}} \right)$$

The one-sided (lower) $100(1-\alpha)$ percent confidence limit is $2nT_r/\chi^2_{1-\alpha;2r}$.

Another quantity of interest is the reliability. Specifically, from the data, we are interested in estimating the parameter t_η such that

$$R(t_\eta) = P(T > t_\eta) = \eta$$

where now T is the lifetime of an individual component. If λ were known, we would have

$$e^{-\lambda t_\eta} = \eta$$

which we could write as

$$t_\eta = \left(\frac{1}{\lambda} \right) \ln\left(\frac{1}{\eta} \right)$$

From the data, the maximum likelihood estimator of $1/\lambda$ is nT_r/r and so

$$\hat{t}_\eta = \left(\frac{nT_r}{r} \right) \ln\left(\frac{1}{\eta} \right) = \left(\frac{\hat{1}}{\lambda} \right) \ln\left(\frac{1}{\eta} \right)$$

is the maximum likelihood estimate of t_η. The same result is obtained by replacing λ in Eq. (12.9) by $(1/t_\eta) \ln(1/\eta)$ and differentiating with respect to t_η. The $100(1-\alpha)$ percent confidence interval for t_η is then

$$\left[\frac{2nT_r}{\chi^2_{1-\alpha/2;2r}} \ln\left(\frac{1}{\eta} \right), \frac{2nT_r}{\chi^2_{\alpha/2;2r}} \ln\left(\frac{1}{\eta} \right) \right]$$

while the one-sided interval is

$$\frac{2nT_r}{\chi^2_{1-\alpha;2r}} \ln\left(\frac{1}{\eta}\right)$$

Other parameters can be estimated in a similar manner.

Consider now the case where the items that fail are not replaced. It will turn out that the confidence intervals for the same parameters will be exactly the same as before. Let $T_{(i)}$ be the time to the ith failure. (Note: This is not the time to failure of the ith item, but the time when the ith failure is observed.) The total operating time of the n units is

$$T_0 = \sum_{i=1}^{r} T_{(i)} + (n-r)T_{(r)} \tag{12.11}$$

The r-variate pdf of the random variables $T_{(1)}, T_{(2)}, \ldots, T_{(r)}$ is calculated using the techniques of Section 7.3. The probability there is a failure in the interval $(t_1, t_1 + dt)$ and all other failures are after t_1 is

$$(n\lambda \, e^{-\lambda t_1} \, dt)(e^{-\lambda t_1})^{n-1}$$

where the factor n arises since any of the n items could have failed in the interval. Similarly, the probability of the first failure in $(t_1, t_1 + dt)$ and the second in $(t_2, t_2 + dt)$ is

$$(n \cdot \lambda \, e^{-\lambda t_1} \, dt)[(n-1)\lambda \, e^{-\lambda t_2} \, dt](e^{-\lambda t_2})^{n-2}, \quad 0 < t_1 < t_2$$

Clearly the r-variate pdf of $T_{(1)}, T_{(2)}, \ldots, T_{(r)}$ is

$$f(t_1, t_2, \ldots, t_r) = (n\lambda \, e^{-\lambda t_1})[(n-1)\lambda \, e^{-\lambda t_2}] \cdots [(n-r+1)\lambda \, e^{-\lambda t_r}]$$
$$\times (e^{-\lambda t_r})^{n-r}$$
$$= \frac{n!}{(n-r)!} \lambda^r e^{-\lambda\left(\sum_{i=1}^{r} t_i + (n-r)t_r\right)}, \quad 0 < t_1 < t_2 < \cdots < t_r \tag{12.12}$$

We find the pdf of $Y = 2\lambda T_0$ by first finding its moment generating function:

$$m_Y(t) = E(e^{tY})$$

$$= \int \cdots \int e^{ty} f(t_1, t_2, \ldots, t_r) \, dt_1 \cdots dt_r$$

$$m_Y(t) = \int \cdots \int e^{2\lambda t[t_1 + t_2 + \cdots + t_r + (n-r)t_r]} \frac{n!}{(n-r)!} \lambda^r$$
$$\times e^{-\lambda[t_1 + t_2 + \cdots + t_r + (n-r)t_r]} \, dt_1 \cdots dt_r$$

$$= \int \cdots \int \frac{n!}{(n-r)!} \lambda^r e^{-\lambda(1-2t)[t_1 + t_2 + \cdots + t_r + (n-r)t_r]} \, dt_1 \cdots dt_r$$

$$= \frac{1}{(1-2t)^r} \int \cdots \int \frac{n!}{(n-r)!} [\lambda(1-2t)]^r$$
$$\times e^{-\lambda(1-2t)[t_1 + t_2 + \cdots + t_r + (n-r)t_r]} \, dt_1 \cdots dt_r$$

$$= \frac{1}{(1-2t)^r} \tag{12.13}$$

since the last integrand is simply the pdf of Eq. (12.12) with λ replaced by $\lambda(1-2t)$. But the mgf of Eq. (12.13) is that of a chi-square random variable with $2r$ degrees of freedom. Thus if we define T_0 by Eq. (12.10) for the case with replacement and by Eq. (12.11) for the case without replacement, then the equations for the maximum likelihood estimates and confidence intervals on the parameters are identical.

EXAMPLE 12.14

One hundred light bulbs are placed on a life test. The tenth failure (failed items replaced) occurs at 1457 hours. Find the maximum likelihood estimate of the mean time to failure and 95% two-sided confidence limits for it.
The maximum likelihood estimate of the MTTF is

$$\left(\frac{\hat{1}}{\lambda}\right) = \frac{nT_r}{r} = \frac{100 \times 1457}{10} = 14{,}570 \text{ hours}$$

and the 95% two-sided confidence limits for it are

$$\left(\frac{2nT_r}{\chi^2_{1-\alpha/2;2r}}, \frac{2nT_r}{\chi^2_{\alpha/2;2r}}\right) = \left(\frac{2 \cdot 100 \cdot 1457}{\chi^2_{.975;20}}, \frac{2 \cdot 100 \cdot 1457}{\chi^2_{.025;20}}\right)$$

$$= \left(\frac{291{,}400}{34.1696}, \frac{291{,}400}{9.5908}\right)$$

$$= (8528, 30{,}383)$$

EXAMPLE 12.15

Forty T.V. tubes are placed on a life test (failed items replaced) and the time to observe the fifth failure is 6421 hours. Find the maximum likelihood estimate of that time $t_{.8}$ for which the probability that a tube last at least this long is at least .8, and the one- and two-sided 95% confidence limits for this quantity.
The maximum likelihood estimate of $t_{.8}$ is

$$\hat{t}_{.8} = \frac{40T_r}{r} \ln\left(\frac{1}{.8}\right) = \frac{40 \times 6421}{5} \times .22314 = 11{,}462.24$$

The two-sided confidence limits for $t_{.8}$ are

$$\left[\frac{2nT_r}{\chi^2_{.975;10}} \ln\left(\frac{1}{.8}\right), \frac{2nT_r}{\chi^2_{.025;10}} \ln\left(\frac{1}{\eta}\right)\right]$$

$$= \left(\frac{513{,}680}{20.4831} \times .22314, \frac{513{,}680}{3.24697} \times .22314\right)$$

$$= (5595.96, 35{,}301.39)$$

and the one-sided confidence limit is

$$\frac{2nT_r}{\chi^2_{.95;10}} \ln\left(\frac{1}{.8}\right) = \frac{513,680}{18.307} \times .22314 = 6239.66$$

Thus we can say that we are 95% confident that the probability a tube will survive 6239 hours is at least .8.

EXAMPLE 12.16

Ten items are placed on a life test. The first failure occurs at 216 hours and the second at 272 hours at which time the test is terminated. Find the maximum likelihood estimate of the MTTF and the 95% one-sided confidence limit for it.

Using Eq. (12.9) the total operating time of the 10 units is

$$T_0 = 216 + 272 + 8 \times 272$$

The maximum likelihood estimate of the MTTF is

$$\left(\frac{\hat{1}}{\lambda}\right) = \frac{T_0}{2} = \frac{2664}{2} = 1332 \text{ hours}$$

The 95% one-sided confidence interval for the MTTF is

$$\frac{2T_0}{\chi^2_{.95;4}} = \frac{5328}{9.48773} = 561.57 \text{ hours}$$

Problems

1. If a pressure transducer has a constant failure rate function $r(t) = C\%$ per hour with an expected lifetime of 20 hours, what is the probability that it will still be operating at 25 hours?

2. For every hour beyond its expected lifetime a steel component operates, a profit of $10 is realized with no penalty for a failure before its expected life. What is the expected profit if it has a constant failure rate function $r(t) = .05$ per hour.

3. A certain component never fails before 1000 hours. After that its failure rate function is linear,

    $$r(t) = .005(t - 1000), \quad t \geq 1000$$

    Obtain an expression for (a) the reliability of the component; (b) the expected lifetime of the component.

4. A washing machine is sold with a one year warranty. The mean time to failure is 1.5 years and the failure rate function is a constant $r(t) = \frac{2}{3}$ per year. If the average repair cost to a machine is $20 (and machines that are repaired do not fail before the end of the warranty), what is the expected cost of the warranty per machine?

5. If n systems are connected in a standby configuration, what should n be if each component has a constant reliability function $r(t) = .002\%$ per hour and the mean time to failure of the overall system is to be at least 3000 hours.

6. Due to an aging process the time to failure of the standby system in a parallel connection is not independent of the failure of the main system. If the joint density

function of the failure times T_1 (of the main system) and T_2 is

$$f(t_1, t_2) = e^{-t_2}, \quad 0 < t_1 < t_2 < \infty$$

(where T_2 is time to failure of the standby system from the time origin), find the reliability of the overall system.

7. Data on the reliability of a press machine are inconclusive. Based on failure rate data it appears that the two functions $r_1(t) = 4 \times 10^{-4}t$ and $r_2(t) = .03$ per hour match the data equally well. Which would give the more pessimistic estimate of reliability at 100 hours? At 200 hours?

8. Given a parallel connection of two components, one of which has a failure rate function $r_1(t) = a$ and one of which has a failure rate function $r_2(t) = b$. What must c be if the mean time to failure of this system is to be the same as a parallel connection of two components in parallel, each with a constant failure rate function, $r(t) = c$.

9. If in each system in a parallel connection of n components, the ith system has a constant failure rate function $r_i(t) = c_i$, find the mean time to failure of the system for $n = 2, 3$.

10. Three components in a space flight system with redundancy each have an exponential failure law with parameters μ_1, μ_2, and μ_3, respectively. If at least two of the components must operate in order for the system to work, what is the mean time to failure?

11. Two power supplies operate in parallel. When both are working the failure rate function for each is constant, $r(t) = \mu$. When one of them fails the failure rate function of the remaining one becomes $r'(t) = \eta$, a constant. Find the reliability function of this system.

12. Each system in Figure 12.15 has a reliability function of $R(t)$. When the system is initially turned on each switch has a probability P of working, what is the reliability of the overall system if it is only necessary to have a working path from input to output in order for it to work?

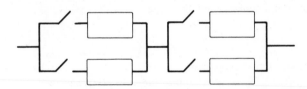

FIGURE 12.15

13. The three systems shown in Figure 12.16 each has a constant failure rate $r_i(t) = c_i$, $i = 1, 2, 3$. Find the reliability function of the overall system.

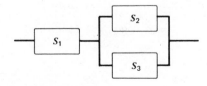

FIGURE 12.16

14. The water runoff each year into a dam reservoir is approximately a normally distributed random variable with mean $\mu = 5 \times 10^7$ gallons and variance $\sigma = 8 \times 10^6$ gallons^2. If the runoffs are independent from year to year, what is the reliability of the system if the dam will break if more than 7.5×10^7 gallons are contained in the reservoir?

15. Find the mean time to failure of n components, each with an exponential failure law with parameter μ, connected in parallel.

16. Consider the system shown in Figure 12.17. Each system has an exponential failure rate function with parameter μ. The systems fail independently except that if either S_1 or S_3 fails, then the other fails also. Find the reliability of the overall system.

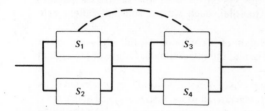

FIGURE 12.17

17. The failure data on the lifetime of an experimental battery indicate that it has a mean time to failure of 3 hours and a variance of approximately 1 hour. Compare the reliability functions obtained by assuming (i) the failure law is a Weibull law with $\alpha = 3$, and (ii) the failure law is a normal pdf with $\mu = 3$, $\sigma^2 = 1$.

18. The solar panel of a deep space satellite has to be able to cope with three different environments in which its constant failure rate function takes different values λ_1, λ_2, and λ_3, respectively. If it encounters environment i (and stays in that environment) with probability p_i, find the reliability of the satellite.

19. The failure rate function of a component is given by

$$r(t) = \begin{cases} a, & 0 < t < 100 \text{ hours} \\ b, & 100 < t \end{cases}$$

Find the expected time to failure of the component.

20. If each system shown in Figure 12.18 has a reliability function $R(t)$, can you determine which system has the greater reliability?

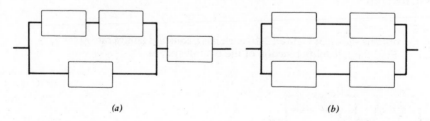

(a)   (b)

FIGURE 12.18

21. Components from an assembly line have a failure law

$$f(t) = (16 \times 10^{-4}) t\, e^{-.04t}, \quad t \geq 0, \ t \text{ in hours}$$

The components undergo a 10-hour test in the factory and those that fail are discarded. What fraction of those that are marketed last a further 10 hours.

22. A certain electrical system contains n components in parallel. A component may fail in one of two ways: either by an open circuit failure that occurs $100p\%$ of the time or by a short circuit failure. The system fails if there is at least one short circuit or if all the components are open circuited. If the probability that a component has open circuited by time t is $pF(t)$ and the probability that it has short circuited by time t is $(1-p)F(t)$, show that the reliability of the system is

$$R(t) = [1 - (1-p)F(t)]^n - [pF(t)]^n$$

23. Suppose, in Problem 22, that in order for the system to operate it is necessary that at least k of the components must work (the k components can be neither short nor open circuited). Show that the reliability is

$$R(t) = \sum_{j=k}^{n} \binom{n}{j} [1 - F(t)]^j F(t)^{n-j}$$

24. Determine the expression given in Section 12.1 for the variance of T if it has a Weibull density function with parameters α and λ.

25. Twenty electronic tubes are placed on a life test (failed items are replaced). If the fourth failure occurs at 2415 hours, estimate the mean time to failure of the tubes and find the 95% lower confidence limit for it.

26. For the data in Problem 25 estimate the time t for which we can be 99% confident that at least 90% of the tubes will last at least this long.

27. Two hundred components are placed on a life test and the first four failure times are 471, 492, 571, and 584 hours, respectively. Estimate the MTTF and find the 90% two-sided confidence interval for it.

CHAPTER 13

Quality Control and Acceptance Sampling

Both the manufacturer and buyer of goods have certain standards that a product must meet. The manufacturer knows that when his production process is working correctly the product will usually meet specifications. He would like to institute a procedure that tells him quickly when something has gone wrong. Usually he does this by sampling the end product and if there are not an excessive number of unacceptable items he assumes the process is working. The process of sampling and inspection is in itself an expensive operation and the question arises as to how much is enough to guarantee to a certain level of confidence that the process is working correctly. The subject of quality control deals with this question and is introduced in the next section.

The consumer is equally anxious not to be paying for substandard merchandise. If he or she buys large lots from a manufacturer, a complete inspection is usually not feasible. The strategy will be to choose a certain number of the items from the lot for inspection. Based on the results of the inspection the whole lot is either accepted or rejected or some intermediate course of action taken. The inspection procedure is often agreed to by both manufacturer and consumer so as not to unfairly penalize either. Such procedures are referred to as acceptance sampling and the last two sections consider certain aspects of this problem.

13.1 *Control Charts*

In any manufacturing process an inherent variability in the product is inevitable. This may be due to variations in the materials, slight fluctuations of conditions, or by the human element involved in some processes. Most products are manufactured to specification and a buyer will be reluctant to accept substandard merchandise. Usually the "natural" variations arising in the process do not exceed the specification limits (otherwise a more refined and probably more expensive process would be used). However, at times nonrandom variations may arise and these may have serious consequences for the product. These variations could arise from a slight machine malfunction, a change in operating conditions for the machine, or from some other controllable aspect of the process. The point is that some variations are due to natural random occurrences and some due to controllable malfunctions in the manufacturing environments. The small random variations in the product generally arise from many different sources, each source contributing a small fraction of the total variation. As such they are not worth tracking down and eliminating. Other variations however may be due to more serious malfunctions in the process. These are generally few in number and are very definitely worth investigating and correcting. Such variations are said to have an assignable or removable cause. If only the small natural random variations are present in the product, the process is said to be in control. Measurements taken from items produced while the process is in control will then behave like samples from a fixed population. In all other cases the process is said to be out of control.

It is very important then to be able to detect quickly and efficiently whether the process is in or out of control. The obvious method of accomplishing this is to periodically choose n items produced, measure some characteristics, and on the basis of these measurements decide whether or not the process is in control. In this guise, the problem has many elements in common with estimation and decision theory. In each case we are trying to make a decision on, or estimate, the state of a process based on randomly perturbed measurements.

The solution to the problem, proposed in the pioneering work of W. A. Shewhart, is the notion of a control chart that is widely used in such situations. Certain aspects of these control charts will be discussed in the remainder of this section. The subject is only briefly introduced here and many of the broader aspects of quality control are not included. The underlying principles, philosophy, and methodology however are similar and this section should be a useful introduction for further reading.

For a control chart, a sample of size n is chosen every so often and the characteristic of interest measured. Some function of the sample (usually its mean, variance, or range) is plotted as a function of the sample number (or time period). Limits are set, usually a central line and an upper and lower limit. A point falling outside of these limits may then be a cause for taking some action.

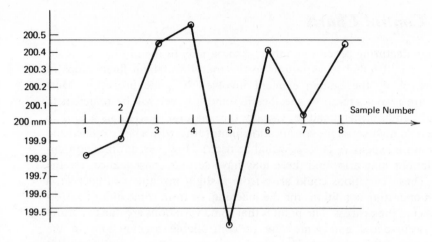

FIGURE 13.1

A typical situation is shown in Figure 13.1. While this description is simple, there are many factors that have to be decided such as how often should we sample? How many items should be taken per sample? What action should be taken when a point falls outside of the control limits on the chart? What should the control limits be? The answers to some of these problems will be dependent on the process and on experience with the process. Others may be approached mathematically.

Before analyzing the situation mathematically we discuss the situation a little further. The important notion recognized by Shewhart is the essentially statistical nature of the problem. It is to be expected that products will contain minor variations, but it is desirable that not too large a fraction of them be unacceptable. It is the job of quality control techniques to ensure this.

Suppose samples have been taken over a long period of time. The question of interest then is, do these readings represent the workings of a process in control? If the actual mean and variance of the process were unknown, they would have to be estimated from the data and in this case we are checking to see whether the data are consistent or whether, somewhere along the line, significant changes occurred in the process. If the mean and variance of the population (as opposed to the samples) were known, then the question becomes "are these data consistent with the population?"

In using control charts there are two kinds of errors that one can make. We could decide the process is out of control when in fact it is in control. Using hypothesis testing terminology of Chapter 10 this is a Type I error, or a false alarm, and its probability, $P_I = \alpha$, is called the level of significance. We could also decide that the process is in control when it is out of control, a Type II error occurring with probability $P_{II} = \beta$. The original approach of Shewhart did not consider this type of error. The consequences of these errors can be quite severe. If a production line is shut down while looking for an assignable cause of error when in fact the error is due to chance, the cost could be substantial. If a decision is reached that the process is in control when in fact it is not, the effect might be to have a large fraction of unacceptable items produced leading to a loss of material, time, and money. In some processes the assignable causes, such as deterioration due to a gradual wearing of a machine, can be predicted to some degree. Sometimes these causes occur cyclically or at random and this can make their detection with the control chart constructed with samples taken at regular intervals, difficult. In deciding whether or not the process is in control, the costs of the kinds of errors have to be kept in mind.

To begin the mathematical discussion consider a simple case where the measurement of interest on the manufactured item is a normally distributed random variable with mean μ and standard deviation σ, assumed to be known for the moment. The literature of quality control tends to have its own notation. We will use the notation of this text whenever there is a conflict. From this normal population a random sample consisting of n items is taken, the measurements being the random variables X_i, $i = 1, 2, \ldots, n$. The sample mean is

$$\bar{X} = \frac{1}{n} \sum_{i=1}^{n} X_i$$

and this random variable is also normally distributed with mean μ and standard deviation $\sigma/\sqrt{n}$. The probability that $|\bar{X} - \mu|$ is less than $k\sigma/\sqrt{n}$ is, for $k = 1, 2, 3$,

$$P(|\bar{X} - \mu| < \sigma/\sqrt{n}) = .6826$$
$$P(|\bar{X} - \mu| < 2\sigma/\sqrt{n}) = .9545$$
$$P(|\bar{X} - \mu| < 3\sigma/\sqrt{n}) = .9973$$

Experience has shown that 3σ control limits frequently yield very satisfactory performance for control charts and these are used in the U.S. and many other countries. Other nations use slightly different limits, and the results differ only in detail and not in philosophy. We use the 3σ rule throughout our discussion. As mentioned before, if μ and σ are assumed known, then, as each sample is taken, we determine whether or not the sample mean falls in the range $\mu \pm 3\sigma/\sqrt{n}$ and this is really a hypothesis test to see whether that sample is likely to have come from such a population. A plot of the sample means $\bar{X}$, with the control limits $\mu \pm 3\sigma/\sqrt{n}$, is referred to as an $\bar{X}$ chart in quality control literature. These upper and lower limits are usually expressed as $\mu \pm A\sigma$, where $A = 3/\sqrt{n}$.

Now suppose that we are only interested in one kind of machine malfunction and that it manifests itself in the readings by a shift in the mean, that is, if the machine is out of control then the readings will have a mean of $\mu + \Delta\sigma$ and a variance of $\sigma/\sqrt{n}$. With such assumptions we are testing (as in Chapter 10) which of the two hypotheses is in effect H_0: mean $= \mu$ or H_1: mean $= \mu + \Delta\sigma$. If our test is of the form—Assume the test is in control if, on a particular reading, $\mu - k\sigma/\sqrt{n} \le \bar{X} \le \mu + k\sigma/\sqrt{n}$—then the probability of a Type I error for the simple hypotheses is

$$\alpha = P_I = P(|\bar{X} - \mu| > k\sigma/\sqrt{n}) = 1 - \Phi(k) - \Phi(-k)$$

and the probability of a Type II error is

$$\beta = P_{II} = \Phi(k - \Delta\sqrt{n}) - \Phi(-k - \Delta\sqrt{n})$$

and note that P_{II} is dependent on n, the sample size at each reading. Two short examples of these procedures will illustrate.

EXAMPLE 13.1

After an automatic lathe operation the diameter of the machined piece is found to be a normally distributed random variable with mean 20 cm and standard deviation .30 mm. Each half hour, four samples of the shop output are taken and measured and the means recorded.

Reading Number	1	2	3	4	5	6	7	8
Reading (mm)	199.82	199.91	200.43	200.52	199.46	200.39	200.04	200.41

If we decide that the process is out of control when a reading exceeds a 3σ level, what are the probabilities of a Type I error and Type II error if the only possibility of machine maladjustment results in a shift of the product mean to 20.5 cm? What should the control limits be if the probability of a Type II error is to be less than .10?

A convenient way to represent the data of the examples is to plot the reading versus the sample number along with the control level. Anytime a sample mean falls outside of the control level, a corrective action is initiated. Such plots are referred to in industry as control charts and find wide application for the control of processes. The control limits for this case are $(200 \pm 3\sigma/\sqrt{n}) = (200 \pm 3 \times .30/\sqrt{4}) = (200 \pm .45)$ mm, as shown in Figure 13.1.

The 3σ control limits imply that when a sample value lies outside of the interval $(200 \pm .45)$ mm, we will assume the process is out of control. With a probability P_I this will happen when it is actually in control. Notice that both the fourth and fifth sample means lie outside the control limits. The probability of this occurring on a particular sample is

$$P_I = P(|\bar{X} - 200| > .45) = .0027$$

To calculate the probability of a Type II error we note that if the process were out of control, then the sample mean would be a normally distributed random variable with mean 200.5 and standard deviation $.30/\sqrt{4} = .15$ mm.

390 *Applied Probability*

The probability of such a variable lying in the range $(200 \pm .45)$ mm is

$$P_{\mathrm{II}} = \int_{199.55}^{200.45} \frac{\exp\left(-\dfrac{(y-200.5)^2}{2(.15)^2}\right) dy}{\sqrt{2\pi}(.15)}$$

$$= \int_{-.95}^{-.05} \frac{\exp\left(-\dfrac{z^2}{2}\right) dz}{\sqrt{2\pi}} = .309$$

In order for P_{II} to be less than .10, assume the levels are $(200 \pm X)$ mm, in which case

$$P_{\mathrm{II}} = .10 = \int_{200-X}^{200+X} \frac{\exp\left(-\dfrac{(y-200.5)^2}{2(.15)^2}\right) dy}{\sqrt{2\pi}(.15)}$$

$$= \int_{-X-.5}^{X-.5} \frac{\exp\left(-\dfrac{z^2}{2}\right) dz}{\sqrt{2\pi}}$$

and using the tables for the standard normal cdf it is found that the control limits should be approximately $(200 \pm .15)$ mm to achieve this P_{II}. The probability density functions and the errors are shown in Figure 13.2.

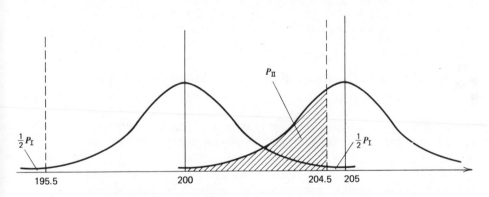

FIGURE 13.2

EXAMPLE 13.2

The weight of a measured volume of first grade cement is a normally distributed random variable with mean 42 grams and standard deviation .5 grams. Inferior cement is characterized as having a mean of 41 grams and the same variance. How many items per sample should be taken and what should the control limits be if P_{I} should be less than .05 and P_{II} should be less than .02?

Assume the control limits are $(42 \pm C)$ grams and that n items per sample are taken. The two types of error have probabilities:

$$P_{\mathrm{I}} = \int_{|y-42|>C} \frac{\exp\left(-\dfrac{(y-42)^2}{2(.5)^2/n}\right) dy}{\sqrt{2\pi}(.5/\sqrt{n})}$$

$$= \int_{|z|>C/(.5/\sqrt{n})} \frac{\exp\left(-\dfrac{z^2}{2}\right) dz}{\sqrt{2\pi}}$$

$$= 2\Phi\left(-\frac{C}{(.5/\sqrt{n})}\right) = .05 \tag{13.1}$$

$$P_{\mathrm{II}} = \int_{42-C}^{42+C} \frac{\exp\left(-\dfrac{(y-41)^2}{2(.5)^2/n}\right) dy}{\sqrt{2\pi}(.5/\sqrt{n})}$$

$$= \Phi\left(\frac{1+C}{.5/\sqrt{n}}\right) - \Phi\left(\frac{1-C}{.5/\sqrt{n}}\right) = .02 \tag{13.2}$$

Using the tables we find Eq. (13.1) implies that

$$\frac{C}{.5/\sqrt{n}} = 1.96 \quad \text{or} \quad C = \frac{.98}{\sqrt{n}}$$

Using this relation in Eq. (13.2) gives

$$\Phi(2\sqrt{n}+1.96) - \Phi(2\sqrt{n}-1.96) = .02$$

From the table of the standard normal cdf the smallest value of y such that

$$\Phi(y+1.96) - \Phi(y-1.96) \le .02$$

is approximately $y = 4.03$ implying that the number of items per sample, n, should be such that $2\sqrt{n} > 4.03$ and five samples will be required to meet the conditions.

Shewhart (1931) argued that the sample size n should be either 4 or 5. Larger sample sizes would improve the accuracy of the estimate but also increase the cost. A sample of size 4 or 5 is economical and has proved effective in practice for detecting medium to large process shifts.

As an analytical detail note that even though the underlying population is not normal, the average of n observations when $n \ge 4$ is approximately normally distributed. Thus many of the techniques used in quality control are relatively insensitive to the assumption of normality.

In addition to keeping a control chart for the mean, it is of interest to monitor the amount of variability in the process. As a measure of variability the

392 *Applied Probability*

sample standard deviation S or, more commonly, the sample range

$$R = \max_i X_i - \min_i X_i$$

can be used. The sample range is particularly useful when the test is supervised by relatively untrained personnel, and gives good results. The results for S are slightly better and if the process is monitored by computer, no more difficult to handle.

Consider first the control chart for S, where

$$S = \left(\frac{1}{n-1} \sum_{i=1}^{n} (X_i - \bar{X})^2\right)^{1/2}$$

The control chart for this variable will have a center line at $E(S)$ and upper and lower limits of $E(S) + 3\sqrt{V(S)}$ and $E(S) - 3\sqrt{V(S)}$, respectively. From Chapter 7 it is known that the random variable

$$\frac{(n-1)}{\sigma^2} S^2 = Y$$

has the χ^2 distribution with $(n-1)$ degrees of freedom, that is

$$f_Y(y) = \frac{y^{((n-3)/2)} e^{-y/2}}{2^{((n-1)/2)} \Gamma((n-1)/2)}, \quad 0 \le y < \infty$$

and from this it is easily calculated that

$$E(S) = \sigma \frac{\Gamma(n/2)}{\Gamma((n-1)/2)} \sqrt{\frac{2}{n-1}} \quad \text{and} \quad V(S) = \sigma^2 - E^2(S) \qquad (13.3)$$

In quality control literature these quantities are usually expressed as

$$E(S) = c_4 \sigma \quad \text{and} \quad V(S) = c_5^2 \sigma^2 = (1 - c_4^2) \sigma^2 \qquad (13.4)$$

where

$$c_4 = \frac{\Gamma(n/2)}{\Gamma((n-1)/2)} \sqrt{\frac{2}{n-1}} \quad \text{and} \quad c_5 = (1 - c_4^2)^{1/2}$$

and are functions of n, the size of each sample. The control limits for the sample standard deviation S, when the population variance is assumed known, are then

$$c_4 \sigma \pm 3 c_5 \sigma = (B_5 \sigma, B_6 \sigma) \qquad (13.5)$$

where $B_5 = c_4 - 3 c_5$ and $B_6 = c_4 + 3 c_5$. We refer to a chart for the sample standard deviations and these control limits as an S chart.

The sample range is easy to use in practice but the mathematics behind its use is slightly more difficult. From Eq. (7.19) it is known that the pdf of the sample range R is

$$f_R(r) = n(n-1) \int_{-\infty}^{\infty} f(y) f(y-r) \left(\int_{y-r}^{y} f(x)\, dx\right)^{n-2} dy$$

To evaluate the mean and variance of this pdf is a formidable task requiring triple integration. When $f \sim N(0, 1)$, tables for these quantities were constructed first by Tippett and later by Pearson and Hartley (1941–42). In quality control notation,

$$E(R) = d_2\sigma \quad \text{and} \quad \sigma(R) = V^{1/2}(R) = d_3\sigma \qquad (13.6)$$

and values of the constants d_2 and d_3 are given in Table 13.3 as functions of n, taken from the tables of Pearson and Hartley (1958). The control limits on the control chart for the range are then

$$d_2\sigma \pm 3d_3\sigma = (D_1\sigma, D_2\sigma) \qquad (13.7)$$

where $D_1 = d_2 - 3d_3$ and $D_2 = d_2 + 3d_3$, and the notation for all these constants is standard in quality control literature. The corresponding chart is called an R chart.

In summary of the situation, when it is assumed that the population is normal with known mean μ and standard deviation σ, the central line and lower and upper control limits are given in Table 13.2a and the constants involved are tabulated in Table 13.3 for sample sizes between 2 and 10. Of course the lower control limits for S and R cannot be negative.

The following example is artificial and is intended to give some insight into control charts. In practice, only one of the control charts for variability is used (either S or R) in conjunction with that for the mean. Note that a shift in the mean of a process would not be noticed on any variability chart but a shift on the amount of variability in the process is usually detectable on the $\bar{X}$ chart.

EXAMPLE 13.3

The data in this example contain computer generated normally distributed random variables. The intent of the example is to illustrate the detectability of shifts in the process mean or variance using the $\bar{X}$ and R charts.

In Table 13.1a the first five columns contain 25 samples of five measurements each drawn from a standard normal population $N(0, 1)$. The sixth and seventh columns contain the sample mean and range. Table 13.1b contains measurements from a normal population with mean 0 and variance 2 and Table 13.1c contains measurements from a normal population with mean 1 and variance 1. The two control charts are shown in Figures 13.3a and 13.3b, respectively.

Only the control limits for the $N(0, 1)$ data are shown on the charts and the aim is to observe the detectability of process shifts with these limits. For the $\bar{X}$ chart the limits are $\mu \pm (3/\sqrt{n})\sigma = \mu \pm A\sigma = \pm 1.342$. For the R chart the limits are $(d_2\sigma \pm 3d_3\sigma) = (D_1\sigma, D_2\sigma) = (D_1, D_2)$ and, from Table 13.3, these limits are (0, 4.918).

On the $\bar{X}$ chart the $N(0, 1)$ data (the first 25 samples) lie clearly within the control limits as they do on the R chart. The next 25 samples ($N(0, 2)$ data) also lie within the control limits indicating that for these particular samples the

TABLE 13.1a

i	1	2	3	4	5	$\bar{X}_i$	R_i
1	0.261	0.849	1.61	0.486	1.032	0.7432	1.871
2	1.254	0.169	0.995	1.082	1.19	0.938	1.085
3	2.461	−0.329	−1.027	−0.686	1.993	0.4824	3.488
4	0.552	−0.976	−0.4	−1.879	−1.235	−0.7876	2.431
5	−0.42	0.703	−0.394	−0.172	0.405	0.0244	1.123
6	−0.805	−0.244	1.755	0.104	0.06	0.174	2.56
7	−1.125	−1.969	−0.005	0.073	0.766	−0.452	2.735
8	1	0.441	−0.181	1.232	0.782	0.6548	1.413
9	−0.73	0.777	−2.069	−0.326	−1.464	−0.7624	2.846
10	−1.118	−0.013	−0.688	0.082	−1.589	−0.6652	1.671
11	0.527	0.725	0.68	−0.868	0.997	0.4122	1.865
12	0.42	−0.547	0.101	−0.358	0.957	0.1146	1.504
13	0.629	0.042	0.349	−0.907	1.9	0.4026	2.807
14	0.072	−0.065	−0.594	−1.132	−0.723	−0.4884	1.204
15	−1.492	−0.292	0.364	−0.305	3.136	0.2822	4.628
16	−1.645	−0.487	0.756	0.707	−0.405	−0.2148	2.401
17	−0.718	−0.636	0.037	0.443	−0.01	−0.1768	1.161
18	0.983	−0.834	−1.329	1.507	−1.943	−0.3232	3.45
19	1.857	0.167	−0.192	0.147	1.586	0.713	2.049
20	−1.187	−2.401	0.669	−1.016	−0.02	−0.791	3.07
21	−0.144	0.913	0.382	1.712	0.625	0.6976	1.856
22	−0.355	−1.482	0.574	−0.84	1.818	−0.057	3.3
23	1.977	1.457	0.269	−0.447	−0.574	0.5364	2.551
24	−0.606	0.274	0.879	−1.134	−0.783	−0.274	2.013
25	0.715	−0.698	−0.212	−0.103	0.144	−0.0308	1.413
					Averages	0.04609	2.26

TABLE 13.1b

1	2	3	4	5	$\bar{X}_i$	R_i
−2.191	−2.373	0.341	0.395	−1.127	−0.991	2.768
−1.274	0.036	0.618	−0.771	0.821	−0.114	2.095
−2.369	−0.347	−1.283	1.559	4.067	0.3254	6.436
−2.076	−1.199	−0.277	1.451	0.105	−0.3992	3.527
−1.659	1.245	0.441	0.503	0.067	0.1194	2.904
0.866	−1.401	−1.165	2.945	0.643	0.3776	4.346
−0.078	1.999	−1.839	−1.265	1.361	0.0356	3.838
0.621	−0.648	0.088	0.48	0.963	0.3008	1.611
0.898	2.559	−0.982	−0.779	0.564	0.452	3.541
−0.536	−1.694	−0.424	3.153	−0.046	0.0906	4.847
1.039	−2.222	−0.509	0.086	0.936	−0.134	3.261
0.134	−0.073	0.322	−1.118	1.594	0.1718	2.712
−0.086	−1.264	1.143	−2.356	0.689	−0.3748	3.499
−2.073	0.512	0.499	1.062	−0.298	−0.0596	3.135
0.752	1.679	−0.48	−0.72	0.01	0.2482	2.399

TABLE 13.1*b*—*cont.*

1	2	3	4	5	$\bar{X}_i$	R_i
−0.797	4.15	0.299	−1.098	−1.095	0.2918	5.248
1.53	0.786	0.93	0.056	0.833	0.827	1.474
0.087	−1.314	−0.059	0.986	0.769	0.0938	2.3
−0.882	1.397	1.384	2.155	−1.932	0.4244	4.087
0.489	−1.523	0.172	−1.927	−1.583	−0.8744	2.416
−3.3	3.432	−0.37	2.708	0.453	0.5846	6.732
−0.342	−2.498	−0.783	−1.135	0.609	−0.8298	3.107
−1.31	1.71	−1.158	1.409	0.781	0.2864	3.02
0.527	0.049	−0.458	−0.234	−0.364	−0.096	0.985
1.556	0.018	0.003	−0.109	−2.753	−0.8794	2.771
				Averages	−0.004912	3.322

TABLE 13.1*c*

1	2	3	4	5	$\bar{X}_i$	R_i
1.169	1.091	1.77	0.13	0.208	0.8736	1.64
1.397	1.115	0.874	1.122	1.622	1.226	0.748
0.973	0.561	0.962	0.737	−0.024	0.6418	0.997
0.801	−0.416	0.138	2.01	1.647	0.836	2.426
1.627	2.837	1.284	1.847	1.535	1.826	1.553
−0.381	0.93	−0.426	1.326	0.486	0.387	1.752
2.403	0.733	0.074	1.614	1.122	1.189	2.329
0.557	3.328	−0.209	0.068	1.978	1.144	3.537
1.912	0.615	3.081	1.096	0.995	1.54	2.466
2.048	0.215	1.107	−0.934	1.896	0.8664	2.982
0.987	−2.386	0.141	2.067	−0.49	0.0638	4.453
2.537	0.533	−0.233	−0.114	1.436	0.8318	2.77
2.249	2.175	0.582	1.824	1.189	1.604	1.667
0.803	0.639	1.575	1.559	−0.152	0.8848	1.727
1.876	0.666	2.503	−1.14	2.004	1.182	3.643
0.222	0.503	0.622	2.619	−0.35	0.7232	2.969
−0.306	1.296	1.992	1.366	0.41	0.9516	2.298
−0.385	1.261	1.378	2.604	−0.273	0.917	2.989
1.414	−0.004	0.897	−0.529	−0.012	0.3532	1.943
2.091	−0.06	0.908	0.457	2.187	1.117	2.247
1.147	2.265	1.256	2.556	1.265	1.698	1.409
1.622	0.565	1.637	0.22	1.978	1.204	1.758
0.633	2.879	0.923	0.904	0.66	1.2	2.246
0.746	1.501	2.144	1.619	0.475	1.297	1.669
1.617	−0.706	1.424	−0.549	0.728	0.5028	2.323
				Averages	1.002	2.262

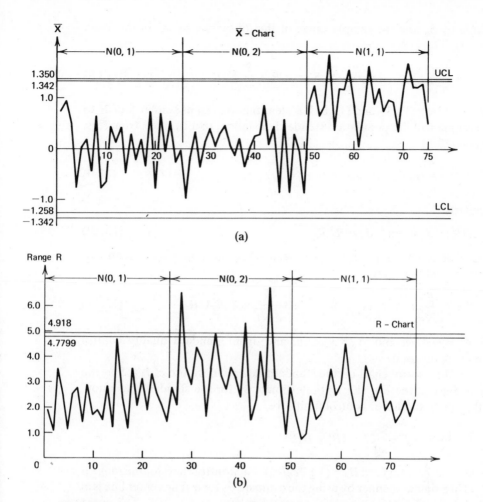

FIGURE 13.3

shift in process variance was not enough to be detected on the $\bar{X}$ chart. For these samples on the R chart, however, there is a noticeable increase in activity and three points lie outside the control limits, clearly indicating the shift in process variance. For the final 25 samples ($N(1, 1)$ data) four points lie above the upper control limit on the $\bar{X}$ chart and all points on the R chart lie within the limits, as expected. Thus, with the use of both the $\bar{X}$ chart and the R charts the chances of detecting significant process shifts are good.

In the case when the mean and variance of the population (which we still assume to be normal) are not known, then much of the foregoing has to be modified, usually by replacing μ and σ by estimates. To begin the control chart at least 20 to 25 subgroups should be obtained, each subgroup containing four or five items. Suppose m samples of n items each have been taken and denote the sample mean of the ith sample by $\bar{X}_i$, the sample standard deviation of the *i*th sample by S_i, and the sample range of the *i*th sample by R_i. In this case, denote

$$\bar{X} = \frac{1}{m}\sum_{i=1}^{m} \bar{X}_i, \qquad \bar{S} = \frac{1}{m}\sum_{i=1}^{m} S_i, \qquad \bar{R} = \frac{1}{m}\sum_{i=1}^{m} R_i \qquad (13.8)$$

From Eqs. (13.4), (13.5), and (13.6) it is clear that we can use either $\bar{S}$ or $\bar{R}$ to gain an estimate of the population standard deviation; that is, since, from Eqs. (13.4) and (13.8)

$$E(\bar{S}) = \frac{1}{m}(mc_4\sigma) = c_4\sigma \quad \text{or} \quad \hat{\sigma}_1 = \bar{S}/c_4 \qquad (13.9)$$

and similarly

$$E(\bar{R}) = d_2\sigma \quad \text{or} \quad \hat{\sigma}_2 = \bar{R}/d_2 \qquad (13.10)$$

and the control lines for the sample mean when the population mean and variance are unknown are

$$\bar{X} \pm 3\frac{\bar{S}}{c_4\sqrt{n}} = \bar{X} \pm A_3\bar{S} \quad \text{or} \quad \bar{X} \pm 3\frac{\bar{R}}{d_2\sqrt{n}} = \bar{X} \pm A_2\bar{R} \qquad (13.11)$$

where $A_2 = 3/d_2\sqrt{n}$ and $A_3 = 3/c_4\sqrt{n}$, and again standard quality control notation is being used.

For the control chart on the standard deviation we use $c_4\hat{\sigma}_1 = \bar{S}$ as the center line using (see Eq. (13.5)) the estimate $\hat{\sigma}_1$ for the standard deviation. Thus the upper and lower control limits are

$$\bar{S} \pm 3c_5\hat{\sigma}_1 = \bar{S} \pm \frac{3c_5}{c_4}\bar{S} = (B_3\bar{S}, B_4\bar{S})$$

where $B_3 = (1 - 3c_5/c_4)$ and $B_4 = (1 + 3c_5/c_4)$. The control chart for the range is obtained in a similar manner by using the estimate $\hat{\sigma}_2$ for σ. The center line is at $\bar{R} = d_2\hat{\sigma}_2$ (using Eq. (13.6)) and the control limits are

$$\bar{R} \pm 3d_3\hat{\sigma}_2 = \bar{R} \pm \frac{3d_3}{d_2}\bar{R} = (D_3\bar{R}, D_4\bar{R}) \qquad (13.12)$$

where

$$D_3 = 1 - \frac{3d_3}{d_2} \quad \text{and} \quad D_4 = 1 + \frac{3d_3}{d_2}$$

Part b of Table 13.2 summarizes the control limits on the charts when the population mean and variance are unknown and includes estimates using the biased estimate

$$\Sigma = \left(\frac{1}{n}\sum_{i=1}^{n}(X_i - \bar{X})^2\right)^{1/2}$$

of the standard deviation, which can be used instead of *S*. Table 13.3 tabulates

TABLE 13.2 Equations for the Center and Control Lines

(a) μ and σ known:

$\bar{X}$: $\quad \mu \pm A\sigma$ $\qquad\qquad$ R: $\quad d_2\sigma \pm 3d_3\sigma = (D_1\sigma, D_2\sigma)$

S: $\quad c_4\sigma \pm 3c_5\sigma = (B_5\sigma, B_6\sigma)$ $\qquad$ Σ: $\quad c_2\sigma \pm 3c_3\sigma = (B_1\sigma, B_2\sigma)$

(b) μ and σ unknown:

$\bar{X}$: $\quad \bar{\bar{X}} \pm A_1\bar{\Sigma}$ $\qquad$ R: $\quad \bar{R} \pm 3\dfrac{d_3}{d_2}\bar{R} = (D_3\bar{R}, D_4\bar{R})$

$\qquad \bar{\bar{X}} \pm A_2\bar{R}$

$\qquad \bar{\bar{X}} \pm A_3\bar{S}$ $\qquad$ S: $\quad \bar{S} \pm 3\dfrac{c_5}{c_4}\bar{S} = (B_3\bar{S}, B_4\bar{S})$

(c) Relationships between the constants:

$A = 3/\sqrt{n}$ $\quad A_1 = 3/(c_2\sqrt{n})$ $\quad A_2 = 3/(d_2\sqrt{n})$ $\quad A_3 = 3/(c_4\sqrt{n})$

$c_2 = \dfrac{\Gamma(n/2)}{\Gamma((n-1)/2)}\sqrt{\dfrac{2}{n}}$ $\quad c_3 = \left(1 - \dfrac{1}{n} - c_2^2\right)^{1/2}$ $\quad c_4 = \sqrt{\dfrac{n}{n-1}}c_2$ $\quad c_5 = (1 - c_4^2)^{1/2}$

$B_1 = c_2 - 3c_3$ $\quad B_2 = c_2 + 3c_3$ $\quad B_3 = 1 - \dfrac{3c_5}{c_4}$ $\quad B_4 = 1 + \dfrac{3c_5}{c_4}$ $\quad B_5 = c_4 - 3c_5$ $\quad B_6 = c_4 + 3c_5$

$D_1 = d_2 - 3d_3$ $\quad D_2 = d_2 + 3d_3$ $\quad D_3 = 1 - \dfrac{3d_3}{d_2}$ $\quad D_4 = 1 + \dfrac{3d_3}{d_2}$

$S = \left(\dfrac{1}{n-1}\sum\limits_{i=1}^{n}(X_i - \bar{X})^2\right)^{1/2}$ $\qquad \Sigma = \left(\dfrac{1}{n}\sum\limits_{i=1}^{n}(X_1 - \bar{X})^2\right)^{1/2}$

the various constants, which are all standard in quality control literature, for sample sizes between 2 and 10.

In starting up a control chart the estimates $\bar{\bar{X}}$, $\bar{S}$, and $\bar{R}$ will only be good if they were obtained from measurements while the process is in control. If, for example, some of the samples used to derive these estimates lay outside the limits provided by them, then some action, such as deleting these points, should be taken and new estimates derived. It is also advisable to update the estimates and the control limits every 25 samples or so taking precautions so far as is possible to ensure that only measurements taken while the process was in control are used. The following example carries on from Example 13.3.

EXAMPLE 13.4

We investigate the data of Example 13.3 as if the true population measurements were unknown. For the $\bar{X}$ chart the average of the sample averages is given as .04609. From Eqs. (13.11) and Table 13.3 the control limits for the $\bar{X}$ chart are

$$\bar{\bar{X}} \pm A_2\bar{R} = .046 \pm A_2(2.26) = (-1.258, 1.350)$$

TABLE 13.3

n	2	3	4	5	6	7	8	9	10
$d_2{}^a$	1.1284	1.6926	2.0588	2.3259	2.5344	2.7044	2.8472	2.9700	3.0775
$d_3{}^a$	.8525	.8884	.8798	.8641	.8480	.8332	.8198	.8078	.7971
c_2	.5642	.7236	.7979	.8408	.8686	.8882	.9027	.9139	.9227
c_3	.4263	.3782	.3367	.3052	.2808	.2612	.2452	.2318	.2203
c_4	.7979	.8862	.9213	.9400	.9515	.9594	.9650	.9693	.9727
c_5	.6028	.4632	.3888	.3412	.3076	.2821	.2622	.2458	.2323
A	2.1213	1.7321	1.5000	1.3416	1.2247	1.1339	1.0607	1.0000	.9487
A_1	3.7599	2.3936	1.8800	1.5958	1.4100	1.2766	1.1750	1.0942	1.0281
A_2	1.8800	1.0233	.7286	.5768	.4832	.4193	.3725	.3367	.3083
A_3	2.6587	1.9544	1.6281	1.4273	1.2871	1.1819	1.0991	1.0317	.9754
B_1	.0000	.0000	.0000	.0000	.0263	.1046	.1670	.2186	.2617
B_2	1.8429	1.8583	1.8081	1.7563	1.7109	1.6718	1.6384	1.6091	1.5838
B_3	.0000	.0000	.0000	.0000	.0303	.1178	.1850	.2392	.2836
B_4	3.2665	2.5681	2.2661	2.0889	1.9697	1.8822	1.8150	1.7608	1.7164
B_5	.0000	.0000	.0000	.0000	.0288	.1130	.1785	.2319	.2759
B_6	2.6063	2.2760	2.0878	1.9636	1.8742	1.8058	1.7515	1.7068	1.6695
D_1	.0000	.0000	.0000	.0000	.0000	.2048	.3878	.5466	.6862
D_2	3.6859	4.3578	4.6982	4.9182	5.0784	5.2040	5.3066	5.3934	5.4688
D_3	.0000	.0000	.0000	.0000	.0000	.0757	.1362	.1840	.2230
D_4	3.2665	2.5746	2.2820	2.1145	2.0038	1.9243	1.8638	1.8160	1.7770

^{a} These rows from Biometrika Tables for Statisticians, vol. 1 (2nd edition), edited by E. S. Pearson and H. O. Hartley, Cambridge University Press, London, 1958, table 20, with permission of the Biometrika Trustees.

where $A_2 = .577$ (for $n = 5$) and these limits are very close to those obtained assuming μ and σ known.

The limits for the R chart are given by Eq. (13.12) and Table 13.3 as

$$(D_3\bar{R}, D_4\bar{R}) = (2.260D_3, 2.260D_4) = (0, 4.7799)$$

and this is to be compared with the limits $(0, 4.918)$ found when μ and σ were assumed known. These control limits are plotted in Figures 13.3a and 13.3b, respectively, and for the data given there is little difference in the conclusions reached in Example 13.3 regarding whether or not the process is in control.

For some final observations on the data of Example 13.3 let us attempt to estimate the population standard deviation from the sample range and standard deviation. From Eq. (13.10) the required estimate is

$$\hat{\sigma}_2 = \bar{R}/d_2$$

From Table 13.3 the required constant is $d_2 = 2.3259$.

For the three populations we have the following table:

DATA FROM	POPULATION VARIANCE	$\hat{\sigma}_2$
Table 13.1a	1.000	.972
Table 13.1b	1.414	1.428
Table 13.1c	1.000	.973

The agreement between the estimates and true values appears to be quite reasonable in all cases.

In some instances it is not necessary to take actual measurements to determine the acceptability of a product. A simple decision can be made on a "yes-no" basis whether or not it is acceptable either by special measuring devices or a simple visual inspection of one or more product attributes. In these circumstances a p chart can be used. The parameter p refers to the parameter of the binomial distribution denoted by θ in this book. We will use θ for this parameter in the analyses but refer to the chart as a p chart to agree with standard quality control notation. The p chart is based on observing the fraction of defectives in a sample. When the fraction of defectives exceeds certain limits the process is halted. While this approach is simpler and usually cheaper to operate, it suffers in that the information on the process is much less precise than before.

With each sample we suppose that N items are drawn for inspection. If at the ith sample D_i items are found to be defective, we record D_i/N, the estimate of the probability an item is defective. When the process is in control there will be some probability θ, assumed known for the moment, due to chance mechanisms, that an item will be defective. The sample average of the *i*th sample

$$\bar{X}_i = D_i/N,$$

is binomially distributed with mean θ and variance $(\theta(1-\theta)/N)$. The control limits are then $\theta \pm 3[\theta(1-\theta)/N]^{1/2}$. The false alarm probability is

$$P_{\mathrm{I}} = P(|\bar{X}_i - \theta| \geq 3[\theta(1-\theta)/N]^{1/2})$$

When the probability θ is not known exactly the cumulative estimate $(\sum_1^k X_i)/k$ over k previous samples can be used and the expressions remain unchanged when this estimate is substituted for θ. In many cases the values of N and θ involved are such that the expression for P_{I} can be approximated by a normal distribution. Sometimes when the size of the samples does not vary, the actual number of defective items is plotted and the control limits adjusted accordingly, but this is entirely equivalent to the p chart. Maintaining and revising the chart is done in a manner similar to that for the $\bar{X}$, R, and S charts.

EXAMPLE 13.5

A cereal manufacturer makes boxes of cereal that contain nominally 20 grams. The actual contents are normally distributed with mean 20.25 grams and standard deviation .1 gram. Because of legal implications boxes containing less than 20 grams are not sold. For a particular run the readings on the number of defectives in samples of size 200 are:

Sample Number	1	2	3	4	5	6
Number of Defectives	4	8	7	12	17	14

If it is known that $\theta = .05$, plot the control chart for these readings. What should the control limits be if it is desired to have $P_{\mathrm{I}} = .01$?

Assuming that three control limits are to be used, these limits are $.05 \pm 3(.05 \times .95/200)^{12} = (.004, .096)$.

The probability that a sample exceeds the control limits $(.05 \pm x)$ is

$$1 - P(.05 - x < D_i/N < .05 + x) = P_{\mathrm{I}} = .01$$

Although $\theta = .05$ is small, we use the normal approximation to the binomial distribution to obtain

$$1 - P\left(.05 - x < \frac{D_i}{N} < .05 + x\right)$$

$$= 1 - P\left(-\frac{xN}{\sqrt{N\theta(1-\theta)}} < \frac{D_i - N}{\sqrt{N\theta(1-\theta)}} < \frac{+xN}{\sqrt{N\theta(1-\theta)}}\right)$$

$$\cong 1 - \Phi\left(\frac{xN}{\sqrt{N\theta(1-\theta)}}\right) + \Phi\left(\frac{-xN}{\sqrt{N\theta(1-\theta)}}\right)$$

$$= 2\Phi\left(\frac{-xN}{\sqrt{N\theta(1-\theta)}}\right) = .01$$

and from tables

$$\frac{xN}{\sqrt{N\theta(1-\theta)}} = 2.575$$

or equivalently

$$x = \sqrt{\frac{.05 \times .95}{200}} \times 2.575 = .039$$

and the limits for $P_1 = .01$ are $(.011, .089)$. The p chart is shown in Figure 13.4.

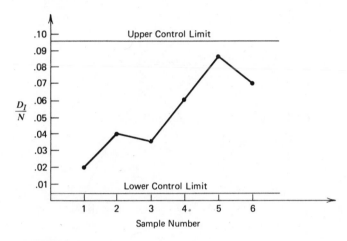

FIGURE 13.4

In some cases, rather than judging an item defective or not, it is the number of defects per item that interests us. A typical example is the number of defective rivets on an airplane fuselage or wing. The common assumption here is that the number of defects follows a Poisson law with parameter λ, its mean and variance, and it is desired to set control limits on the number of defectives. The 3σ control limits would simply be $(\lambda \pm 3\sqrt{\lambda})$, but recall that the Poisson distribution is not symmetric about λ. When λ is large (>10) and X is a Poisson random variable with parameter λ, then $(X-\lambda)/\sqrt{\lambda}$ is approximately a normal random variable and this fact can be used in calculating error probabilities. The resulting plot of sample number versus sample mean with the control levels is referred to as a c chart, as in quality control literature the Poisson parameter is usually denoted by c rather than λ. As with the binomial distribution we use λ for this parameter but refer to the resulting chart as a c chart. Often only the upper control level is of interest.

EXAMPLE 13.6

Each 100 meter length of carpet is inspected for defects as it is manufactured. The results of 10 such inspections are

Sample Number	1	2	3	4	5	6	7	8	9	10
Number of Defects	41	71	65	32	47	39	57	63	52	29

Is the process in control (use 3σ control limits)? What is the probability of a Type I error for such a process?

The sample mean of the observations is 49.6 defects per carpet roll. The control limits are $(49.6 \pm 3\sqrt{49.6}) = (28.5, 70.7)$ and from the observations we would declare the process out of control with the second sample. The c chart is shown in Figure 13.5.

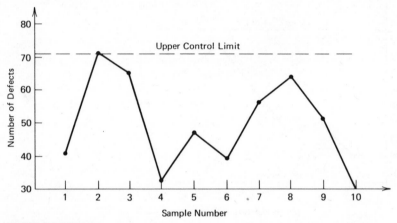

FIGURE 13.5

The probability of a Type I error, assuming that only the upper control limit is of interest to us and using the normal approximation to the Poisson distribution, is

$$P_I = P(\text{Number of defects exceeds 70.7, given a mean of 49.6})$$

$$= P(D > 70.7)$$

$$= P\left(\frac{D - 49.6}{\sqrt{49.6}} > \frac{70.7 - 49.6}{\sqrt{49.6}}\right)$$

$$= P\left(\frac{D - 49.6}{49.6} > 3\right)$$

$$\cong \int_3^\infty \frac{\exp(-x^2/2)\,dx}{\sqrt{2\pi}}$$

$$= .0015$$

When θ is unknown it is estimated from the past data as the average number of defects per item. In this case we are really testing the hypothesis that there is an unknown but constant mean and variance to the process versus the hypothesis that it is varying in some uncontrolled manner or has varied from the norm and is stable at some undesirable level. Of course, in order for the Poisson assumption to hold, certain conditions would have to be met. In particular, the defects on the item or sample would have to be independent of one another, and the defects must occur at random and not clustered over the item.

Various modifications of the Shewhart control charts are possible. One weakness of these charts is their inability to use all the information available from the plotted points. To partially remedy this, the idea of a cumulative sum control chart (cusum chart) has been introduced and its use is briefly outlined here. In this chart the deviation from the mean of each sample is found, say,

$$Y_k = \bar{X}_k - \mu$$

The cusum chart is then a plot of $\sum_{k=1}^{r} Y_k$ versus r, $r = 1, 2, 3, \ldots$. To determine control, a "V-mask" is constructed according to the performance desired from the chart. If any points on the chart lie under the mask when it is properly positioned on the chart, the process is assumed to be out of control. Although more difficult to implement than an ordinary control chart, experience has shown it to be more effective in detecting moderate shifts in the process mean. It can also be used for parameters other than the mean.

There are many other considerations of control charts not touched upon here. It is an interesting and very useful application of probabilistic and statistical methods that, if correctly applied, can result in substantial savings in most manufacturing processes.

13.2   Acceptance Sampling by Attributes—Single Sampling Plans

A consumer who regularly receives large shipments of items would like to be assured that he is not accepting a shipment with an excessively large number of defectives. Presumably the producer has already taken steps to ensure that the quality of the product remains inside certain limits, but now it is a consumer problem. Usually complete inspection of the lot is costly and, because of errors in the inspection process, may not be as effective as it implies.

Typically a consumer devises a sampling plan whereby he or she chooses a certain number of samples from each lot, inspects them, and accepts or rejects the whole lot on the basis of the inspection. If the consumer performs some measurement on the samples and bases the acceptance or rejection on the actual values, the sampling plan is said to be by variables. If each article inspected, however, is judged to be good or bad (nondefective/defective, acceptable/not acceptable) and the lot is accepted or rejected based on the number of defectives found in the sample, the plan is said to be by attribute and it is this situation that will be considered in this section. A sampling plan of this nature is an effective and economic solution to the consumer's problem. Using such a plan the consumer may occasionally accept a lot with a higher than acceptable proportion of defective items in it, but in the long run will be guaranteed of a certain level of quality. Often the sampling plan used is agreed upon by both producer and consumer. Rejected lots may either be returned to the producer for rectification or else some credit arrangement agreed upon.

Suppose that each lot contains N items from which n are drawn for inspection. The sampling plan will be to accept the lot if the number of defective items found in the sample D is less than or equal to some predetermined number c, called the acceptance number. If the actual number of defectives in the lot was known to be M, then the probability of obtaining k defectives in a sample of n items is given by

$$P(D=k)=\frac{\binom{M}{k}\binom{N-M}{n-k}}{\binom{N}{n}}, \quad k=0,1,\ldots,M \tag{13.13}$$

the hypergeometric distribution, and the probability of accepting the lot is simply

$$P(A)=\sum_{k=0}^{c}\frac{\binom{M}{k}\binom{N-M}{n-k}}{\binom{N}{n}}$$

The actual number of defectives in the lot is, of course, unknown. What may be known, however, is that this manufacturing process has in the past produced items that on the average are $100\theta\%$ defective. Choosing N items from such a process, the probability of obtaining D defectives is given by the binomial distribution

$$P(D)=\binom{N}{D}\theta^{D}(1-\theta)^{N-D}, \quad D=0,1,2,\ldots,N \tag{13.14}$$

For a lot of size N with exactly D defectives in it, the probability that a sample of size n from this lot will have d defectives in it is

$$P(d \text{ in } n)=\sum_{D}P(d \text{ in } n|D \text{ in } N)P(D \text{ in } N) \tag{13.15}$$

The conditional probability $P(d \text{ in } n|D \text{ in } N)$ is just the hypergeometric probability of Eq. (13.13), that is,

$$P(d \text{ in } n \mid D \text{ in } N) = \frac{\binom{D}{d}\binom{N-D}{n-d}}{\binom{N}{n}}$$

Using Eq. (13.14) and the above in Eq. (13.15) yields

$$P(d \text{ in } n) = \sum_{D=0}^{N} \frac{\binom{D}{d}\binom{N-D}{n-d}}{\binom{N}{n}} \binom{N}{D} \theta^D (1-\theta)^{N-D}$$

$$= \sum_{D=d}^{N-n+d} \frac{D!}{d!(D-d)!} \frac{(N-D)!}{(n-d)!(N-D-n+d)!}$$

$$\times \frac{n!(N-n)!N!}{N!D!(N-D)!} \theta^D (1-\theta)^{N-D}$$

$$= \binom{n}{d} \theta^d (1-\theta)^{n-d} \sum_{D=d}^{N-n+d} \binom{N-n}{D-d} \theta^{D-d} (1-\theta)^{N-n-D+d}$$

$$= \binom{n}{d} \theta^d (1-\theta)^{n-d}$$

since the summation in the equation second to last is easily shown to be unity, the sum of a binomial distribution with parameters $N - n$ and θ. Consequently, the probability of obtaining d defectives in a sample of size n from such a population is a binomial probability as opposed to the hypergeometric probability found in the previous situation. The probability of acceptance in this case is

$$P(A, \theta) = \sum_{k=0}^{c} \binom{n}{k} \theta^k (1-\theta)^{n-k}$$

If $n\theta$ is moderate, then we can approximate this quantity with the Poisson distribution

$$P(A, \theta) = \sum_{k=0}^{c} e^{-\lambda} \frac{\lambda^k}{k!}$$

where $\lambda = n\theta$. The problem of choosing the acceptance/rejection number c will be discussed shortly.

A plot of the probability of acceptance of the lot versus the fraction defective is called the operating characteristic or OC curve of the sampling plan. It is the single most important characteristic of the plan. Of course the probability of acceptance of the lot, $P(A)$, is really a conditional probability conditioned on the number of defectives in the lot. In the case where we assume there are D defectives in the lot and the number of defectives in the sample is given by Eq. (13.13), the resulting OC curve is called a Type A OC curve. It is really a continuous curve plotting $P(A)$ versus $\theta = D/N$, $D = 0, 1, \ldots, N$.

When it is assumed that the lot comes from a process making $100\theta\%$ defective items, then the number of defectives in the lot and in the sample has a binomial distribution and the resulting (continuous) OC curve of $P(A, \theta)$ versus θ is called a Type B OC curve. Our concern will be mainly with Type B OC curves. For such curves the size of the lot is immaterial, the sampling plan depending only on the sample size n and the acceptance number c.

Suppose a lot arrives at a consumer's loading dock. In a typical sampling plan a random sample of size n is chosen. If c or fewer of these items are found defective, the entire lot is accepted. If $c + 1$ or more items are found defective the entire lot is rejected. Such a scheme is referred to as a single sampling plan. To ensure that the sample is "random," a table of random numbers is often used. If the lot size is N, then with the parameters N, n, and c either Type A or B OC curves can be found. Figure 13.6 shows typical curves. If a consumer was willing to accept lots with $100p_1\%$ items (or fewer) defective but not lots with a higher percentage defective, then his ideal OC curve would be as shown in that figure. Of course such ideal curves are impossible to achieve. In practice sampling plans are designed to give both consumer and producer a certain level of protection, the producer from having too many "good" lots rejected by the consumer, and the consumer from accepting too many "bad" lots. The following two examples examine some elementary aspects of OC curves.

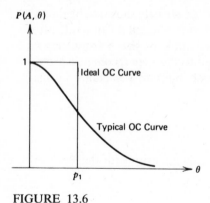

FIGURE 13.6

EXAMPLE 13.7

It has been suggested that the OC curves, for single sampling plans that are "scaled" versions of one another, are similar. Disprove this by comparing the Type B OC curves for the three sampling plans with parameters $n = 5$, $c = 1$, $n = 10$, $c = 2$, and $n = 15$, $c = 3$.

The three curves are shown in Figure 13.7. It is easily seen that the notion that the same protection is always obtained by taking the same fraction

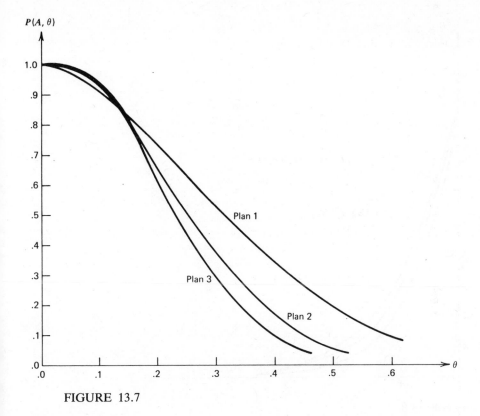

$P(A, \theta)$

Plan 1

Plan 3

Plan 2

FIGURE 13.7

of the lot for the sample is dispelled. For example, for $\theta = .30$, the probability of accepting a lot with $n = 5$, $c = 1$ plan is $.528$ while for the $n = 15$, $c = 3$ plan it is $.297$, showing considerably more discrimination.

EXAMPLE 13.8

Show the difference between Type A and Type B OC curves by finding the OC curves for the sampling plans (i) $N = 20$, $n = 10$, $c = 0$, (ii) $N = 50$, $n = 10$, $c = 0$, and (iii) $N = 100$, $n = 10$, $c = 0$.

The Type B curve, which is found by using the binomial probabilities is independent of N, is shown in Figure 13.8. For Type A curves when $c = 0$, the probability of acceptance is

$$P(A) = \frac{\binom{D}{0}\binom{N-D}{n}}{\binom{N}{n}}, \quad D = 0, 1, \ldots, N$$

and this is plotted against $p = D/N$. As a general rule, when $N > 10n$ the two curves are essentially the same for all values of p and its equivalent value of

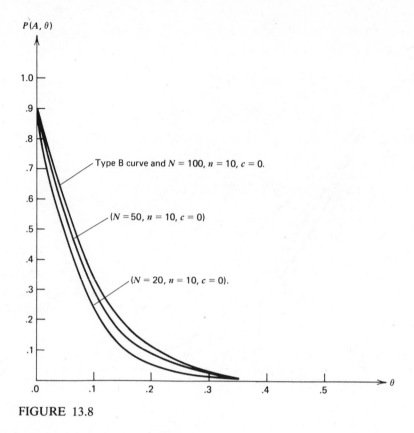

$P(A, \theta)$

FIGURE 13.8

D/N. The Type A OC curve for plan (iii) is essentially identical to the Type B curve in this case.

For a given sample size, making the acceptance number c lower makes the probability of acceptance lower. There is an intuitive feeling that making $c = 0$ gives the best results of all since no sample defectives are allowed for lot acceptance. However, the performance of the sampling plan is contained in the OC curve and these indicate that sampling plans with $c > 0$ are superior. Their drawback is that they require a larger sample from the lot (and hence a more expensive inspection process). The OC curves resulting, however, have steeper slopes. Increasing the size of the sample and keeping c/n fixed forces the OC curve towards the ideal.

We have mentioned that there are two types of errors possible in accepting/rejecting a lot. Suppose the producer and consumer have agreed that any lot with a fraction of $\theta \leq p_1$ defective items is acceptable while a lot with a fraction of $\theta \geq p_2$ defective items is unacceptable. A lot that is judged to have a fraction defective θ, $p_1 < \theta < p_2$, is referred to as an indifferent lot and these lots, as well as those rejected, are subject to a special arrangement (such as being returned to producer to undergo 100% inspection with all defective items replaced with good items, or else being sold to the consumer, as is, at a cut-rate).

In such a situation there is a possibility that even though the fraction defective in the lot is less than p_1 it will be rejected. This is referred to as a Type I error and occurs with probability P_I (or α). It is also referred to as the producer's risk since it is the probability that the consumer will not accept a "good" lot. There is also the possibility that a lot with fraction defective greater than p_2 will be accepted by the consumer. This is referred to as a Type II error, with probability P_{II} (or β). It is also called the consumer's risk since it is the probability the consumer will accept an unacceptable lot. These quantities are shown graphically for a typical OC curve in Figure 13.9. The parameters p_1 and p_2 are called the acceptable quality level (AQL) and lot tolerance percent defective (LTPD), respectively. The parameter p_2 is also referred to as the rejectable quality level (RQL).

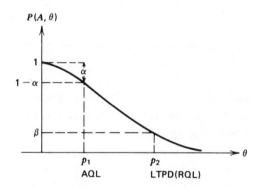

FIGURE 13.9

The AQL is a percent defective that the sampling plan will accept most of the time if the process percent defective is at most AQL. If p_a is the percent defective for which $P(A, p_a) = a$, then AQL is often, but not always, taken as $p_{.95}$, that is, for any sampling plan, if the process average is less than $p_{.95}$, then 95% of the lots on the average will be accepted. Similarly, the LTPD is often, but not always, taken as $p_{.10}$, and often identified with RQL. The point $p_{.50}$ is usually called the indifference point and processes with this percent defective have a 50–50 chance of being accepted.

The situation described here is quite analogous to an hypothesis testing situation. We could test the hypotheses $H_0: \theta = p_1$ versus $H_1: \theta = p_2$ or consider the composite hypothesis $H_0: \theta < p_1$ versus $H_1: \theta > p_2$. It is also an estimation problem where based on a sample of size N we try to estimate the percent defective in the lot.

Suppose now that the producer and consumer agree that a lot which is not acceptable is to be rectified, that is, it is returned to the producer who inspects all items in the lot and replaces unacceptable ones with known acceptable items. Thus a lot that was originally rejected now contains only acceptable items. Such a procedure would, of course, not be used if the testing was destructive. The consumer might be interested in the average number of defective items in the lots he accepts (including the rectified lots). If the average fraction defective in the producer's lots is θ, then the fraction of lots initially rejected is $1 - P(A, \theta)$ and these, after rectification, contain no defectives. An acceptable lot would have an average fraction defective of θ. If, in the samples of size n for such acceptable lots, the defectives found were replaced into the lot, then the average fraction defective over all lots accepted is $\theta P(A, \theta)$ and this quantity, as a function of θ, is referred to as the average outgoing quality (AOQ) curve:

$$AOQ(\theta) = \theta P(A, \theta)$$

If in the accepted lots the defectives found in the sample are replaced with good items, then, on the average, the fraction defective in the lot is $\theta((N-n)/N)$ and, in this case

$$AOQ(\theta) = P(A, \theta)\theta(1 - n/N)$$

When n/N is small these two expressions are equivalent. Sometimes the defectives found in the sample are removed but not replaced—we do not consider that option here.

A typical AOQ curve is shown in Figure 13.10. It is simply the product of the OC curve with the straight line of slope 1. Notice $AOQ(0) = 0 = AOQ(1)$ since $P(A, 1) = 0$ and the AOQ curve always lies underneath the OC curve.

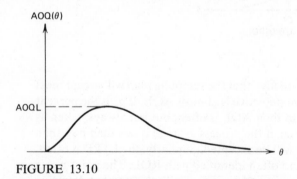

FIGURE 13.10

The peak of the curve appears between p_1 and p_2. For $\theta < p_1$ almost all lots are accepted and contain a very low fraction of defective items. For $\theta > p_2$ most lots will be rejected and all defective items in such lots will be replaced with good items, enhancing the outgoing quality. It is only between p_1 and p_2 that the two factors combine to produce the worst outgoing quality. The maximum of this curve is called the average outgoing quality limit (AOQL) and is the worst possible fraction defective among lots accepted by the consumer. It is possible that different sampling plans will lead to the same AOQL.

The quantities $P(A, \theta)$ and $AOQ(\theta)$ are primarily concerned with protection. Two other quantities, the average sample number (ASN) and the average total inspection (ATI), are primarily concerned with the cost of the sampling plan. The $ASN(\theta)$ is the average number of pieces needed to be inspected from a lot drawn from a process with $100\theta\%$ defective, in order to reach a decision about accepting or rejecting a lot. For a single sampling plan we always inspect n items and so $ASN(\theta) = n$ regardless of θ, and as a plot against θ it is a horizontal line. It might be argued that, in a single sampling plan, once $c + 1$ defective items are found we could stop inspecting and reject the lot. Although this would reduce the ASN, it is seldom done in practice as companies prefer to keep complete quality control charts, such information being useful in dealing with the producer and often leading to quality improvement. The ASN for other sampling plans tends to be more complicated.

The $ATI(\theta)$ curve is the average number of items inspected per lot, as a function of the process fraction defective θ. For a single sampling plan, such as those considered so far, for every lot accepted n items are inspected while for every lot rejected (we will assume) all N items are inspected to replace all defectives. Thus we have, for such a plan,

$$ATI(\theta) = nP(A, \theta) + N[1 - P(A, \theta)]$$

Since $P(A, \theta)$ is a function of θ, so also is $ATI(\theta)$ and for $\theta = 0$ it is n while for larger values of θ, where most lots are rejected, it approaches N.

EXAMPLE 13.9

Sketch the OC, AOQ, ASN, and ATI curves for the sampling plan $N = 1000$, $n = 100$, $c = 2$. Estimate the AOQL for the plan from these curves (use Type B calculations).

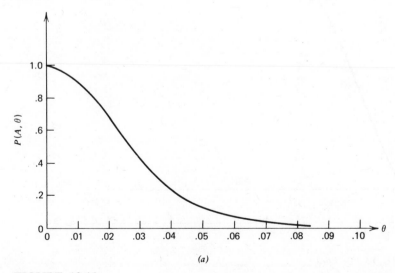

(a)

FIGURE 13.11

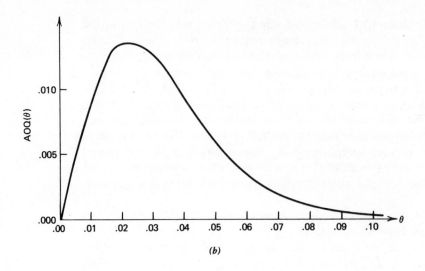

(b)

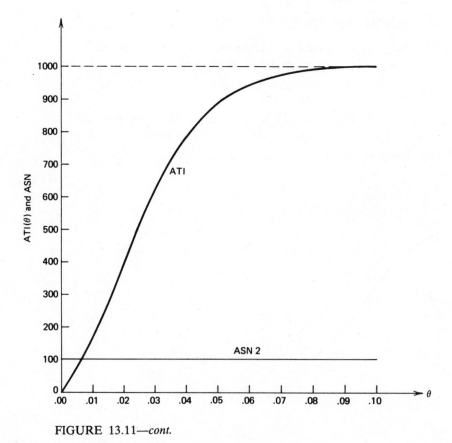

FIGURE 13.11—cont.

414 *Applied Probability*

The curves are shown in Figures 13.11a, 13.11b, and 13.11c. From Figure 13.11a, we obtain the values $p'_{95} = .004$, $p'_{50} = .027$, and $p'_{10} = .052$. From Figure 13.11b the value for AOQL is estimated as .0135. The approach of the ATI curve to N, the lot size, with increasing θ, is shown in Figure 13.11c.

The problem of finding a suitable sampling plan either single as considered in this section or double to be considered in the next section, is generally an approximate iterative procedure to find that OC curve that comes closest to the desired performance points $(p_1, 1-\alpha)$ and (p_2, β), where $p_1 = $ AQL, $p_2 = $ RQL, $\alpha = $ producer's risk, and $\beta = $ consumer's risk. We consider only one such procedure here for single sampling plans. Many other procedures exist but tend to be too involved to include here.

To begin the procedure for the single sampling plan, assume $c = 0$. Find the largest sample size, n_l, such that $P(D \le 0, p_1) \ge 1-\alpha$, where D is the number of defectives. The Poisson approximation is often used in such calculations, using the parameter $\lambda = n_l p_1$. Using any $n < n_l$ will also satisfy the required inequality. Now find the smallest n, n_s, such that $P(D \le 0, p_2) \le \beta$, where $P(D \le 0, p_2)$ is calculated using p_2 (or $\lambda' = n_s p_2$). Again any $n > n_s$ will also satisfy this inequality. Now, if $n_s \le n_l$, then we can use a single sampling plan with parameters $n = n_s$ and $c = 0$. If $n_s > n_l$, then we increment c by 1 and repeat the procedure until eventually a c is found for which $n_s \le n_l$ and, for this value, the single sampling plan with the parameters n_s and c meets the requirements.

EXAMPLE 13.10

Devise a single sampling plan for the parameters AQL $= p_1 = .02$, RQL $= p_2 = .10$, $\alpha = .05$, and $\beta = .05$.

As suggested in the procedure we use the Poisson approximation throughout. For $c = 0$, the largest n for which $P(d \le 0, .02) \ge .95$, $\lambda = n \times .02$, occurs for $\lambda = .05$ or $n_l = 2$, where the value is obtained from the table of the Poisson distribution. Similarly, the smallest value of n for which $P(d \le 0, .10) \le .05$ occurs for $\lambda = 3.0$ or $n_s = 30$. Since $n_s > n_l$, we increment c by 1 and try again. The following table is constructed in this manner:

c	n_s	n_l
0	30	2
1	48	15
2	63	40
3	78	65
4	92	95

We conclude that the single sampling plan with $n = 92$ and $c = 4$ satisfies the requirement. The tabulation of the OC curve for this plan is:

p	.01	.02	.03	.04	.05	.06
$P(A, p)$	.9976	.9633	.8568	.6924	.5104	.3471

p	.07	.08	.09	.10	.11	.12
$P(A, p)$	.2204	.1319	.0751	.0408	.0213	.0108

The procedure is approximate in that using the Poisson distribution is an approximation and this distribution is tabulated for only certain λ and sometimes interpolation has to be used. It is, nonetheless, typical of the kind of computations required.

A procedure developed by Peach and Littauer (1946) is similar in nature but a little more convenient in that it allows standard tables to be set up for it.

The problem of devising sampling plans is so important that various sets of tables exist. The Dodge–Romig Sampling Inspection Tables, for example, contain four sets of tables as well as the OC curves for all the plans. All of the plans are constructed assuming that all rejected lots are rectified and that the sampling plan that minimizes inspection (ATI) is given. It also assumes that the process percent defective is known. The first set of tables are indexed by LTPD for a consumer's risk of .10. The second set gives the equivalent information for double sampling plans (discussed in the next section). The third and fourth sets of tables give the single and double sampling plans indexed by the AOQL for a consumer's risk of .10.

To assist producers and consumers many sampling standards have been devised. These are usually in the form of extensive rules and tables to be used to ensure a given performance. One of the important sampling standards, known as MIL-STD-105D in the United States and ABC-STD-105D elsewhere (for America, Britain, and Canada, the international participants in its development). For this standard the AQL to be maintained and the lot size are assumed known. It introduces the notions of reduced, normal, and tightened inspection and specific rules for moving from one to the other as the lot quality varies. Single and multiple sampling plans are used. This standard is constantly under review by international committees and modifications appear from time to time.

13.3 Other Acceptance Sampling Plans

Under certain circumstances savings can be realized by a modification of the single sampling plan. If rejection or acceptance is obvious before the complete sample has been inspected, then certain short-cuts can be effected. The double sampling scheme takes advantage of this possibility and is described by the following strategy: An initial sample of n_1 is chosen from the lot. If the number of defectives d_1 is less than or equal to some preassigned number c_1, the lot is accepted. If d_1 is greater than c_2, the lot is rejected. If $c_1 < d_1 \leq c_2$, then a further n_2 samples are taken and inspected. If d_2 is the number of defectives found among these further n_2 items, then the lot is accepted if $d_1 + d_2 \leq c_2$ and rejected otherwise. This scheme has the intuitive appeal to only use "extra" samples when there is some doubt and, if designed correctly, can be very effective. Of course this plan can be extended to more than two samples but the complexity increases although the average number of samples inspected per lot decreases.

It is straightforward to calculate an expression for the OC curve for the double sample plan although more complicated than the equivalent single sample plan expression. A typical curve is shown in Figure 13.12. Similarly, expressions for the AOQ, ASN, and ATI curves can be found.

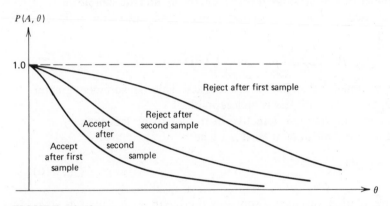

FIGURE 13.12

Other variations of the single sample plan, such as multiple sampling plans (more than two samples), sequential sampling plans, and continuous sampling plans are also used. These will not be discussed here. Another type of sampling plan, sampling by variables, will be only briefly discussed.

When each item of the N samples drawn from a lot is inspected by performing measurements on it, and the lot is accepted or rejected based on the actual values of these measurements (rather than a simple acceptable/nonacceptable type of measurement), the plan is referred to as one by variables. Such plans, which have both advantages and disadvantages over the simpler plans by attributes, are gaining wider popularity in industrial applications as more is learned on how to use them effectively. Before considering a simple analysis we mention some of the characteristics of such plans.

Performing actual measurements on items is usually more time consuming and expensive than a simple attribute inspection. For example, to determine whether a coffee can contains more than a minimum allowable weight is a simpler mechanical operation than to actually measure and record the weight of each can in the lot. However, sampling by variables generally supplies more useful information about the product, often leading to an equivalent acceptance decision with fewer samples than a decision by attributes. The actual measurements can also be used to give an understanding of the product process and how to correct its malfunctions and improve its quality.

Typically we assume that the measurements are normally distributed. This is not only convenient mathematically but it is also satisfied in many cases. In any particular situation the applicability of the assumption should be carefully checked. For the remainder of the section we consider only the simple situation where it is assumed that the acceptance of the lot will be based on the sample mean and that the variance of the population is known. This is referred to as a sampling plan by variables with known variability as opposed to the case where the population variance is unknown and must be estimated from the sample.

We suppose that N items are drawn from the lot and the sample mean of some characteristic is found. The sampling plan is to satisfy the following parameters:

$$P_I = \alpha, \quad P_{II} = \beta, \quad AQL = p_1 \quad \text{and} \quad LTPD = p_2$$

In addition it is assumed that an item with a characteristic measurement greater than x_0 is acceptable and with less is unacceptable.

In order for the fraction defective in the lot to be less than p_1, it is necessary that it have a mean of at least μ_1, where

$$p_1 = \Phi\left(\frac{x_0 - \mu_1}{\sigma}\right) \tag{13.16}$$

and σ is the population variance. From this the mean μ_1 is found as $\mu_1 = x_0 - \sigma z_{p_1}$. If the true mean of the population was less than μ_2, where

$$p_2 = \Phi\left(\frac{x_0 - \mu_2}{\sigma}\right) \tag{13.17}$$

then the fraction defective would exceed p_2 and $\mu_2 = x_0 - \sigma z_{p_2}$. We require that a lot that has mean μ_1 (i.e., an acceptable lot) be rejected less than $100\alpha\%$ of the time and a lot that has μ_2 be accepted less than $100\beta\%$ of the time.

To achieve these quantities we must choose the number of samples N and constant C so that

(i) $P(\bar{X} < C \mid \text{mean} = \mu_1) = \alpha = \Phi\left(\dfrac{C - \mu_1}{\sigma/\sqrt{N}}\right)$ \hfill (13.18)

(ii) $P(\bar{X} > C \mid \text{mean} = \mu_2) = \beta = 1 - \Phi\left(\dfrac{C - \mu_2}{\sigma/\sqrt{N}}\right)$ \hfill (13.19)

Of course, the solution N to these equations for a given μ_1, μ_2, C, and α will not generally be an integer and the next largest integer must be chosen. The solution to these equations is found by equating

$$C - \mu_1 = z_\alpha \sigma / \sqrt{N} \quad \text{and} \quad C - \mu_2 = -z_\beta \sigma / \sqrt{N}$$

and subtracting these equations gives

$$\mu_2 - \mu_1 = (z_\alpha + z_\beta)\sigma / \sqrt{N}$$

or

$$N = \left(\frac{\sigma(z_\alpha + z_\beta)}{\mu_2 - \mu_1} \right)^2 = \left(\frac{z_\alpha + z_\beta}{z_{p_1} - z_{p_2}} \right)^2$$

With this value of N the value of C can be found as

$$C = \mu_1 + \sigma z_\alpha / \sqrt{N} = \mu_2 - \sigma z_\beta / \sqrt{N}$$

For the case when $\mu_1 > \mu_2$, the situation is illustrated in Figure 13.13. The

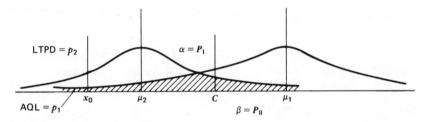

FIGURE 13.13

operating characteristic for such a sampling plan is a plot of $P(\text{acceptance} \mid \text{population mean} = \mu)$ and a sketch of this quantity is shown in Figure 13.14. These arguments are exactly the same as those used in the hypothesis testing situations of Chapter 10. An example will illustrate the use of this technique.

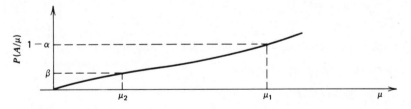

FIGURE 13.14

EXAMPLE 13.11

The breaking tension of a twine is considered to be a normal random variable with variance 25 kg^2. The twine is considered acceptable if its breaking tension exceeds 50 kg. Devise an acceptance sampling plan for the sample that has the following parameters:

$$P_I = .05, \quad P_{II} = .15, \quad AQL = .05, \quad LTPD = .10$$

The first step is to use Eqs. (13.16) and (13.17) to determine population means that would correspond to acceptable and nonacceptable lots, respectively:

$$AQL = .05 = \Phi\left(\frac{50 - \mu_1}{5}\right) \quad \text{and} \quad LTPD = .10 = \Phi\left(\frac{50 - \mu_2}{5}\right)$$

which yields the values $\mu_1 = 58.23$ and $\mu_2 = 56.45$. Using these quantities in Eqs. (13.18) and (13.19) gives

$$P_I = .05 = \Phi\left(\frac{C - 58.23}{5/\sqrt{N}}\right) \quad \text{and} \quad P_{II} = .15 = 1 - \Phi\left(\frac{C - 56.45}{5/\sqrt{N}}\right)$$

which corresponds to the equations

$$-1.645 = \frac{C - 58.23}{5/\sqrt{N}} \quad \text{and} \quad 1.033 = \frac{C - 56.45}{5/\sqrt{N}}$$

The solution of these equations is $N = 52$ and $C = 57.09$. The sampling plan to meet the requirements is then to accept the lot if the sample mean, for a sample of size 52, is greater than 57.09.

EXAMPLE 13.12

Devise a sampling scheme by attributes for the previous example.

We use the techniques of the previous section to determine the sample size N and acceptance number c to meet the specifications. Using the Poisson approximation to the hypergeometric, the probability of accepting a lot that is defective (10% defective) is

$$P_{II} = P(\text{accepting a defective lot})$$

$$= \sum_{k=0}^{c} e^{-(.10)N} \frac{(.10N)^k}{k!} = .15$$

while the probability of rejecting an acceptable lot (5% defective) is

$$P_I = P(\text{rejecting an acceptable lot})$$

$$= \sum_{k=c+1}^{\infty} e^{-(.05N)} \frac{(.05N)^k}{k!} = .05$$

Thus values for N and c satisfying

$$\sum_{k=0}^{c} e^{-p_1 N} \frac{(p_1 N)^k}{k!} \geq .95 \quad \text{and} \quad \sum_{k=0}^{c} e^{-p_2 N} \frac{(p_2 N)^k}{k!} \leq .15$$

are required. Using the same procedure for the design of single sampling acceptance plans given in the previous section, we find that for $c = 15$ and $N = 200$ the errors are

$$P_{\mathrm{I}} = .049 \qquad P_{\mathrm{II}} = .0157$$

for $p_1 = .05$ and $p_2 = .10$. Notice how many more samples are required for this case than for the sampling by variables plan of the previous example. This will in general be true for similar situations.

Problems

1. The weight of nails from a particular process is assumed to be a normally distributed random variable with mean 8.0 grams and standard deviation .1 gram. Four nails are chosen from the process each hour and their mean weight recorded:

Sample Number	1	2	3	4	5	6	7	8	9	10
Mean (gm)	8.075	7.919	8.101	8.073	7.88	7.974	7.810	8.027	7.895	7.917

   Is this process in control if 3σ limits are used? When a machine goes out of control the mean shifts to 8.2 grams. What is the probability of the Type I and II errors if 3σ control limits are used? What should the control limits be if P_{I} is to be less than .05.

2. The human error in a meter reading is assumed to be normally distributed with mean 0 and variance .04. Twenty-five samples are taken, each sample consisting of the mean of four individual readings. The twenty-five means are:

   $-.089, \quad -.075, \quad .120, \quad .84, \quad -.039, \quad .008, \quad -.009, \quad -.096, \quad .060,$
   $.114, \quad .019, \quad -.100, \quad -.041, \quad -.001, \quad .028, \quad -.123, \quad .151, \quad .008,$
   $.075, \quad .012, \quad -.045, \quad .022, \quad -.024, \quad .099, \quad -.015.$

   Construct an $\bar{X}$ chart for these means and determine whether or not the process is in control.

3. Forty samples of weights of cereal boxes, each sample containing five boxes, are taken. If $\sum \bar{X}_i = 1223.42$ grams and $\sum S_i = 4.11$, find the 3σ limits for the $\bar{X}$ charts.

4. If in Problem 3 the process mean was 30.0 and the process variance .1, what would the control limits be?

5. In the manufacture of resistors, the actual resistance is assumed to be a normally distributed random variable. A sample containing four resistors is taken from the machine output every hour and the results of 10 such consecutive samples are:

SAMPLE NO.	X_1	X_2	X_3	X_4
1	10.037	9.954	9.794	10.157
2	10.008	10.106	9.897	10.093
3	10.055	9.836	9.835	9.935
4	9.942	10.007	10.020	10.065
5	9.914	9.974	9.969	10.040
6	9.881	10.191	10.110	10.025
7	9.598	10.016	9.861	10.136
8	10.181	10.154	10.291	9.984
9	10.210	9.971	9.953	9.963
10	10.105	10.244	10.153	10.027

Determine the $\bar{X}$ and R charts for these data, assuming that the mean and variance of the process is 10 ohms and .01 ohms^2, respectively. Is the process in control?

6. In Problem 5 find the $\bar{X}$ and R charts assuming that the mean and variance are unknown. Does this assumption make much difference to the conclusions of Problem 5?

7. In Problem 5 find the $\bar{X}$ and $\bar{S}$ charts and determine whether the process is in control.

8. From the data of Problem 5 estimate the variance using $\bar{X}$, $\bar{R}$, and $\bar{S}$. The first five samples of the data are actually drawn from a $N(10, .01)$ population while the last five are from a $N(10, .04)$ population.

The following 90 samples are used in the next five problems. The first 30 are drawn from a $N(0, .01)$ population, the second 30 from a $N(.5, .01)$ population, and the last 30 from a $N(.5, .25)$ population:

$N(0, .01)$	.109	−.034	−.120	−.094	.028	.037	.131	.054	.189	−.098
	−.071	.157	.171	−.121	−.093	.119	−.092	.037	.038	.197
	.011	−.139	.020	−.103	.171	.051	−.244	−.041	.101	−.211

$N(.5, .01)$	.547	.468	.433	.536	.429	.580	.464	.455	.308	.567
	.502	.521	.370	.584	.634	.442	.541	.474	.509	.503
	.373	.440	.474	.505	.427	.586	.517	.514	.399	.412

$N(.5, .25)$	−.014	1.417	.896	.138	.724	1.520	1.274	1.068	−.002	.585
	.342	.386	.665	1.044	.578	.308	.742	1.208	.347	.816
	1.232	.095	.186	.353	−.254	.704	.281	.514	.516	.356

9. Construct $\bar{X}$ and R charts for this data in two ways: (i) using samples of size 3; (ii) using samples of size 6. Use control limits derived assuming an $N(0, .01)$ population.

10. Estimate σ using $\bar{S}$ and $\bar{R}$: (i) using all of the data and samples of size 6; (ii) dividing the data into three parts and using samples of size 6, estimate σ for each part.

11. Construct $\bar{X}$ and $\bar{S}$ charts using samples of size 5 and assuming the mean and variance are unknown.

12. Find the unbiased sample variance for all 90 readings above and compare this with the unbiased sample variance for the first 60 readings. How do these estimates compare with those of Problem 10?

13. Estimate σ from $\bar{R}$ using only the last 30 readings and dividing these into samples of size 3. Compare with that found in Problem 10.

14. A sample of 100 resistors is taken from a production process every hour and resistors not within 5% of the nominal value are rejected. The results of 25 samples are:

Sample Number	1	2	3	4	5	6	7	8	9	10	11	12	13
No. of Defects	3	7	4	4	1	5	7	1	5	4	3	3	2

	14	15	16	17	18	19	20	21	22	23	24	25
	8	1	0	4	3	3	5	1	7	6	4	4

Find the p chart for the process and determine whether the process is in control. Assuming the process is in control and that $\theta = .04$, find the false alarm probability.

15. A manufacturing process produces 2% defective items. At regular intervals n items are taken for testing. If the process suddenly changes to producing 4%

defective items, how big should n be to detect this change at the first sample after the change with probability .90?

16. The average number of defects in 100 meters of carpeting is 4.0 when the process is in control. Find the 3σ control limits of the process and find the false alarm probability.

17. A single sampling plan contains lots of size $N=1000$ and uses samples of size $n=50$. If $c=5$ find the (i) OC curve, (ii) AOQ curve (and the AOQL), (iii) ASN curve, and (iv) ATI curve.

18. Repeat Problem 17 when $N=2000$, $n=100$, and $c=10$.

19. A double sampling plan uses lots of size $N=500$. If $n_1=n_2=35$, $c_1=3$, and $c_2=8$, find the OC curve for the plan.

20. Repeat Problem 19 when $N=1000$, $n_1=n_2=50$, $c_1=5$, and $c_2=10$.

21. For the single sampling plan $N=200$, $n=20$, $c=1$, find $p_{.10}$, $p_{.50}$, and $p_{.95}$.

22. Find a single sampling plan that meets, as well as possible, the specifications $p_1=.01$, $p_2=.05$, $\alpha=.05$, and $\beta=.05$.

23. Do Problem 22 for $p_1=.02$, $p_2=.04$, $\alpha=.05$, $\beta=.10$.

24. Show that Type B OC curves which have an acceptance number of 0 have no point of inflection while those for $c>0$ do.

25. Show that the mean and variance of the hypergeometric distribution of Eq. (13.15) are

$$E(D)=\frac{nM}{N} \qquad V(D)=\left(\frac{nM}{N}\cdot\frac{N-M}{N}\cdot\frac{N-n}{N-1}\right)$$

26. Discuss the considerations you would take into account in the construction of a triple sampling plan.

27. Sketch the OC and AOQ curves when the number of items per sample is 4 and the lot is accepted if 0 or 1 defective items are found. Find the AOQ limit.

28. A sampling plan for a certain product uses samples of size $n=100$ and $c=2$. Find the OC curve for this plan and estimate the AQL for a producer's risk of .05.

29. A single sampling plan has the parameters $n=20$, $c=1$. Find the OC curve and from it estimate the LTPD if the consumer's risk is .05.

30. A double sampling plan has the parameters $n_1=30$, $n_2=30$, $c_1=0$, and $c_2=1$. What is the probability of accepting a lot from a process with 5% defective? Estimate the LTPD of this plan if the consumer's risk is .05. Find the ASN curve for this plan.

31. If in Problem 29 the lot size is 1000 and the process has 5% defective, find the ATI curve assuming that rejected lots are rectified.

32. The diameter of an electric motor bearing is a normally distributed random variable with mean 10 cm and variance 10^{-4} cm^2. The specification calls for diameters in the range $10\pm.02$ cm. If the process is being monitored by an $\bar{X}$ chart, find the probability that a lot will be rejected by a consumer using a single sampling plan with parameters $n=40$, $c=1$.

33. The weight of steel ingots is a normally distributed random variable with mean 1000 kg and variance 100 kg^2. If ingots weighing less than 985 kg are rejected, devise a variables sampling plan with AQL$=.04$, LTPD$=.10$, $\alpha=.05$, and $\beta=.05$.

34. The percentage of impurities in a certain high grade alloy is a normally distributed random variable with mean 2.0% and standard deviation .1%. It is acceptable if it has less than 2.1% impurities. Find a variables sampling plan with AQL$=.02$, LTPD$=.04$, and $\alpha=\beta=.05$.

35. In Problem 34 find a plan by attributes satisfying the same requirements and compare the number of samples needed for the sampling by variables plan.

CHAPTER 14

Linear Models for Data

The contents of the first four sections of this chapter represent a departure from the previous material in the book. To this point we have been concerned with either a single random variable, a bivariate random variable, or a random sample consisting of n independent random variables with a common pdf. In this chapter we introduce situations where one of the variables x is controlled and hence is not a random variable. The dependent variable Y is random and it is the relationship between the random Y and the nonrandom x that is to be studied. Such a study is referred to as a regression analysis.

   As an example of this situation the hardness of steel might depend on the temperature of it at a certain stage of its processing. The temperature can usually be controlled and in this context would not be regarded as random. The outcome of an experiment would then be a sample $\{(x_1, Y_1), (x_1, Y_1), \ldots, (x_n, Y_n)\}$ of size n, say, of the random variable Y, dependent on the n values of the controlled variable x. Based on this sample, which will be assumed random in the sense that $Y_1, Y_2, \ldots, Y_n$ are independent random variables, information on the relationship between x and Y is sought.

In this chapter it will be assumed that the relationship between x and Y is linear in that $E(Y|x) = \alpha + \beta x$, which is the expected value of Y at the point x. Nonlinear relationships are considered only informally in the examples and problems. Methods to estimate the parameters α and β from the random sample are derived along with indications (confidence intervals and hypothesis tests) as to how reliable the estimates are. Section 14.4 extends the results to the multivariable linear relationship.

The final section of the chapter returns to the case where it is the relationship between two random variables X and Y that is to be studied via a random sample of joint observations. A reason for investigating the situation where both X and Y are random might be economic. For example, the tensile strength of steel reinforcing rods is a random variable that can usually only be found by destructive testing. If it is related to the hardness X, which is found by a relatively simple nondestructive test, then, depending on the relationship between X and Y, we may be able to use hardness measurements to estimate tensile strength. Our main concern in this final section will be to consider confidence intervals and hypothesis tests on the sample correlation coefficient.

The role of x as a controlled variable in the first four sections of the chapter and as an observation of a random variable X in the final section should be carefully noted. Both of these situations arise in practice.

14.1 *Least Squares and Linear Regression*

For the first three sections of this chapter we will be interested in the situation where an experiment has been performed by controlling one variable x (either implicitly or explicitly) and observing another variable Y assumed to be random. The result is a sequence of observations $\{(x_1, y_1), (x_2, y_2), \ldots, (x_n, y_n)\}$, where the y_i's are particular realizations of a random variable Y dependent in some manner on the controlled variable x. This is a familiar situation to scientists and engineers. The randomness of the observations Y can arise in a variety of ways. For example, if a plot of land is divided into sections and each section treated with a certain dosage of fertilizer, the yield of a crop would depend not only on the dosage but also on such factors as water runoff patterns, minor variations in earth consistency, minor variations in section size, variations in seed quantity planted, yield measurement inconsistencies, and so forth.

We view the experiment as setting the controlled variable x, which is either known exactly or can be measured without significant error, and observing Y. An observation consists of a pair $(x_i, Y = y_i)$ and we will assume that, for any two readings (x_i, Y_i), (x_j, Y_j), the random variables Y_i and Y_j are independent, even though x_i may equal x_j. The sequence of observations $\{(x_1, y_1), \ldots, (x_n, y_n)\}$, in which the x_i need not be distinct, can be viewed graphically in a so-called scatter diagram, an example of which is shown in Figure 14.1.

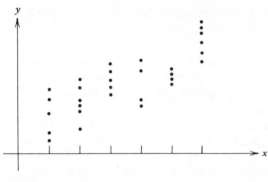

FIGURE 14.1

The problem now is to extract useful information from the data gathered. This can only be done in a statistical sense by making additional assumptions on the nature of the relationship between x and Y. We first assume that the mean of Y is a linear function of x and write this as

$$\mu(x) = E(Y|x) = \int yf(y|x)\,dy = \alpha + \beta x$$

where the pdf of y, $f(y|x)$ is not a conditional pdf since x is not a random variable. It is written in this manner only because of its intuitive relationship with the notion of conditional pdfs. We further assume that higher central moments of Y are independent of x, that is, $E\{[Y - \mu(x)]^j|x\}$ depends on j but not on x. In particular, we assume that the variance of Y, $V(Y|x)$, is the constant σ^2.

These are the simplest nontrivial assumptions that can be made and lead to interesting results. In practice, the function $\mu(x)$ may be more complicated than linear, and assuming a more general relationship will lead to considerably more complicated results. Some indications of how more general functions are dealt with are given in the examples and problems. Of course any relationship can be approximated by a linear one over a small enough range, and in this sense the results obtained here will be useful.

With the assumptions made, that $\mu(x) = \alpha + \beta x$, $V(Y|=x) = \sigma^2$, and the Y_i are independent random variables, we can model the observed random variable Y_i as

$$Y_i = \alpha + \beta x_i + E_i, \quad i = 1, \ldots, n$$

where the E_i are mutually independent random variables with mean 0 and variance σ^2. Such a model is useful to keep in mind for the analysis to follow.

From the sample $\{(x_1, Y_1), (x_2, Y_2), \ldots, (x_n, Y_n)\}$ our first task is to obtain estimates of the coefficients α and β. The coefficient α is the intercept of the regression line and β is the slope of the regression line, also called the regression coefficient. These are unknown and must be obtained from the data. For our purpose we will choose the line with slope b and intercept a that minimizes the error sum of squares defined as

$$\text{Error sum of squares } e = \sum_{i=1}^{n} (y_i - a - bx_i)^2 = \sum_{i=1}^{n} e_i^2$$

It is the sum of the squares of the vertical distances of the proposed line to the sample points and the line is to be chosen to minimize these distances. The minimization of the error sum of squares is achieved by differentiating e with respect to a and b and then checking that the solutions to these equations do indeed yield a minimum. The first partial derivatives are

$$\frac{\partial e}{\partial a} = -2 \sum_{i=1}^{n} (y_i - a - bx_i) = 0$$

$$\frac{\partial e}{\partial b} = -2 \sum_{i=1}^{n} (y_i - a - bx_i)x_i = 0 \tag{14.1}$$

Introducing the notation $\bar{y} = (1/n)\sum y_i$ and $\bar{x} = (1/n)\sum x_i$, these equations read

$$\bar{y} = a + b\bar{x}$$

$$\frac{1}{n} \sum_{i=1}^{n} x_i y_i = a\bar{x} + b\frac{1}{n} \sum_{i=1}^{n} x_i^2$$

which, on substituting $a = \bar{y} - b\bar{x}$ into the second equation, yields

$$\frac{1}{n} \sum_{i=1}^{n} x_i y_i - \bar{y}\bar{x} = b\left(\frac{1}{n} \sum_{k=1}^{n} x_i^2 - (\bar{x})^2\right)$$

or

$$b = \frac{\sum\limits_{i=1}^{n} (x_i - \bar{x})(y_i - \bar{y})}{\sum\limits_{i=1}^{n} (x_i - \bar{x})^2}, \quad a = \bar{y} - b\bar{x} \tag{14.2}$$

To check that these values minimize the error sum of squares e, we observe that if we perturbed these values, say to $a + \Delta a$ and $b + \Delta b$, the resulting error sum of squares can be written

$$e' = \sum_{1}^{n} [y_i - a - \Delta a - (b + \Delta b)x_i]^2$$

$$= \sum_{1}^{n} (y_i - a - bx_i)^2 - 2 \sum_{1}^{n} (y_i - a - bx_i)(\Delta a + \Delta b x_i)$$

$$+ \sum_{1}^{n} (\Delta a + \Delta b x_i)^2$$

The middle term of the equation vanishes by using Eqs. (14.1) and the last term as the sum of squares is positive while the first term is the supposed minimum error sum of squares. But this implies that choosing a and b different from the values of Eq. (14.2) increases the minimum error sum of squares and hence these values are indeed optimum.

It was assumed, without any knowledge of the distribution of Y (dependent on x), that the regression of Y on x was the straight line $\mu(x) = \alpha + \beta x$ and consequently the values of a and b derived in Eq. (14.2), as random variables, form the least squares estimators of a and b, that is,

$$A = \bar{Y} - B\bar{x}, \quad B = \frac{\sum_{i=1}^{n} (x_i - \bar{x})(Y_i - \bar{Y})}{\sum_{i=1}^{n} (x_i - \bar{x})^2} \quad (14.3)$$

These mean square estimators are, in fact, unbiased estimators for, using the fact that by assumption $E(Y_i) = \alpha + \beta x_i$, it follows that $E(\bar{Y}) = \alpha + \beta \bar{x}$ and

$$E(A) = \alpha + \beta \bar{x} - \beta \bar{x} = \alpha$$

$$E(B) = \frac{\sum_{i=1}^{n} (x_i - \bar{x})[E(Y_i) - E(\bar{Y})]}{\sum_{i=1}^{n} (x_i - \bar{x})^2}$$

$$= \frac{\beta \sum_{i=1}^{n} (x_i - \bar{x})^2}{\sum_{i=1}^{n} (x_i - \bar{x})^2} = \beta$$

Notice that the estimated regression line always goes through the point $(\bar{x}, \bar{y})$. The estimators A and B given in Eq. (14.3) are also linear in that they can both easily be expressed as linear combinations of the Y_i. It turns out that they are also minimum variance estimators. Estimates for a given set of observations using Eq. (14.3) will be denoted by a and b.

EXAMPLE 14.1

The yield of alcohol, in liters, of a fermentation process is related to the temperature during fermentation. The results of a sequence of experiments are as follows:

x (°C)	35	40	45	50	55	60
Y (liters)	20.2	23.1	23.2	23.6	25.8	26.3

Find the least mean squares line of Y on x. Estimate the average yield if the fermentation temperature is 48°.

428 *Applied Probability*

From the data given the following quantities are computed:

$$\bar{x} = \frac{1}{6} \sum_{i=1}^{6} x_i = 47.5, \qquad \bar{y} = \frac{1}{6} \sum_{i=1}^{6} y_i = 23.7$$

$$b = \frac{\sum_{i=1}^{6} (x_i - \bar{x})(y_i - \bar{y})}{\sum_{i=1}^{6} (x_i - \bar{x})^2} = \frac{97.50}{437.50} = .225, \quad a = \bar{y} - b\bar{x} = 13.01$$

The least mean square line is then $a + bx$. The estimate of the process yield for a process temperature of 48°C is

$$a + b.48 = 23.81$$

In order to proceed further we make some additional assumptions that will simplify the analysis. We assume that the sample $\{(x_1, Y_1), (x_2, Y_2), \ldots, (x_n, Y_n)\}$ is random in that the random variables $Y_1, Y_2, \ldots, Y_n$ are independent, and that Y_i is a normally distributed random variable with mean $\alpha + \beta x_i$ and variance σ^2 (a constant).

Under these assumptions we can derive maximum likelihood estimates for α and β. Not surprisingly they will turn out to be identical with the least squares estimates. With the assumption of normality, however, we can get more information on the characteristics of the regression line.

The likelihood function for the sample assuming a regression line with parameters α and β is

$$L(\mathbf{y}) = \left(\frac{1}{\sqrt{2\pi}\sigma}\right)^n \prod_{i=1}^{n} \exp\left(-\frac{(y_i - \alpha - \beta x_i)^2}{2\sigma^2}\right)$$

$$= \left(\frac{1}{\sqrt{2\pi}\sigma}\right)^n \exp\left(-\frac{1}{2\sigma^2} \sum_{i=1}^{n} (y_i - \alpha - \beta x_i)^2\right)$$

The value (α, β) at which this function assumes its maximum is found by differentiating with respect to α and β. As in Chapter 9, however, it is sufficient to look for the maximum of the log of $L(y)$ since the log function is monotonically increasing:

$$\frac{\partial \log L(\mathbf{y})}{\partial \alpha} = \frac{1}{\sigma^2} \sum_{i=1}^{n} (y_i - \alpha - \beta x_i) = 0$$

$$\frac{\partial \log L(\mathbf{y})}{\partial \beta} = \frac{1}{\sigma^2} \sum_{i=1}^{n} (y_i - \alpha - \beta x_i)x_i = 0$$

But these equations are precisely Eqs. (14.1) multiplied by a constant, and so the maximum likelihood estimators of α and β are precisely those of Eq. (14.3).

It can be argued that the least squares criterion and the normal pdf are "matched" in that minimizing error sum of squares and maximizing likelihood functions using normal densities often yield the same result. However, other densities often yield the same results as the next example shows.

EXAMPLE 14.2

The lifetime of a bearing is assumed to be an exponentially distributed random variable with mean $\alpha + \beta x$, where x is the percentage of a tungsten alloy used in its manufacture. If n_1 bearings with x_1 percent of the alloy are tested and fail at times $(y_1, y_2, \ldots, y_{n_1})$ and n_2 bearings with x_2 percent of the alloy are tested and fail at times $(z_1, z_2, \ldots, z_{n_2})$, find maximum likelihood estimates of the parameters α and β.

The random variables Y_i, $i = 1, \ldots, n_1$ are independent, each having the pdf

$$f_Y(y_i) = \frac{1}{(\alpha + \beta x_1)} \exp[-(\alpha + \beta x_1) y_i]$$

while the Z_i, $i = 1, \ldots, n_2$ have the pdf

$$f_Z(z_i) = \frac{1}{(\alpha + \beta x_2)} \exp[-(\alpha + \beta x_2) z_i]$$

The likelihood function is given by

$$L(\mathbf{y}, \mathbf{z}) = \frac{1}{(\alpha + \beta x_1)^{n_1} (\alpha + \beta x_2)^{n_2}}$$

$$\cdot \exp\left[-\frac{1}{(\alpha + \beta x_1)} \left(\sum_1^{n_1} y_i \right) - \frac{1}{(\alpha + \beta x_2)} \left(\sum_1^{n_2} z_i \right) \right]$$

and the maximum likelihood estimates of the parameters a and b are found by solving the simultaneous equations

$$\frac{\partial L}{\partial \alpha} = 0 = \left(\frac{-n_1}{(\alpha + \beta x_1)} - \frac{n_2}{(\alpha + \beta x_2)} + \frac{\sum y_i}{(\alpha + \beta x_1)^2} + \frac{\sum z_i}{(\alpha + \beta x_2)^2} \right) L = 0$$

and

$$\frac{\partial L}{\partial \beta} = 0 = \left(-\frac{n_1 x_1}{(\alpha + \beta x_1)} - \frac{n_2 x_2}{(\alpha + \beta x_2)} + \frac{x_1 \sum y_i}{(\alpha + \beta x_1)^2} + \frac{x_2 \sum z_i}{(\alpha + \beta x_2)^2} \right) L = 0$$

These equations are easily reduced to the set

$$(\alpha + \beta x_1) = \frac{1}{n_1} \sum_1^{n_1} y_i = \bar{y}, \qquad (\alpha + \beta x_2) = \frac{1}{n_2} \sum_1^{n_2} z_i = \bar{z}$$

and these are solved to yield

$$\alpha = \bar{y} - \beta x_1 = \frac{x_1 \bar{z} - x_2 \bar{y}}{x_1 - x_2}, \qquad \beta = \frac{\bar{y} - \bar{z}}{x_1 - x_2}$$

Thus the maximum likelihood estimates of α and β are

$$A = \frac{x_1 \bar{Z} - x_2 \bar{Y}}{x_1 - x_2} \quad \text{and} \quad B = \frac{\bar{Y} - \bar{Z}}{x_1 - x_2} \tag{14.4}$$

It can easily be checked that the regression line goes through the points $(x_1, \bar{y})$ and $(x_2, \bar{z})$ as expected.

The estimators of Eq. (14.4) are also the least squares estimators, as manipulating Eq. (14.3) for the special set of observations considered here will show.

In the following two sections, using the assumption of normality, we will find confidence limits on the parameters of the regression line and consider hypothesis tests on them.

With the assumption that the regression of Y on x is $\alpha + \beta x$ and that the variance of Y, σ^2 is independent of x, the variance of the estimators of the parameters α and β given by Eq. (14.3) can be calculated. Letting $d = \sum (x_i - \bar{x})^2$,

$$B = \frac{1}{d}\left(\sum_{i=1}^{n} (x_i - \bar{x})Y_i - \sum_{i=1}^{n} (x_i - \bar{x})\bar{Y} \right)$$

$$= \frac{1}{d}\left(\sum_{i=1}^{n} (x_i - \bar{x})Y_i - \sum_{i=1}^{n} \frac{1}{n}\left(\sum_{j=1}^{n} (x_j - \bar{x}) \right) Y_i \right)$$

$$= \frac{1}{d}\left(\sum_{i=1}^{n} (x_i - \bar{x})Y_i \right)$$

and so

$$V(B) = \frac{1}{d^2}\sigma^2 \sum_{i=1}^{n} (x_i - \bar{x})^2 = \frac{\sigma^2}{d} \tag{14.5}$$

Similarly, to find the variance of A notice that

$$A = \frac{1}{n}\sum_{i=1}^{n} Y_i - \frac{\bar{x}}{d}\sum_{i=1}^{n} (x_i - \bar{x})Y_i$$

$$= \sum_{i=1}^{n} \left(\frac{1}{n} - \frac{\bar{x}}{d}(x_i - \bar{x}) \right) Y_i$$

and so

$$V(A) = \sum_{i=1}^{n} \left[\frac{1}{n^2} - \frac{2\bar{x}}{nd}(x_i - \bar{x}) + \left(\frac{\bar{x}}{d} \right)^2 (x_i - \bar{x})^2 \right] \sigma^2$$

$$= \frac{\sigma^2}{n} + \frac{\bar{x}^2 \sigma^2}{d} \tag{14.6}$$

For a given value of x the mean of Y has the estimator $A + Bx$. The expected value of this statistic is $a + bx$ while the variance is calculated as

$$V(A + Bx) = V(A + B(x - \bar{x}) + B\bar{x})$$

$$= V(B(x - \bar{x}) + \bar{y})$$

$$= \frac{\sigma^2}{n} + \frac{\sigma^2(x - \bar{x})^2}{d} \qquad (14.7)$$

In these equations we have assumed that the variance of Y is known. In cases where it is not it has to be estimated and the equations assume added complexity.

EXAMPLE 14.3

Show that in least squares analysis the data can be translated without changing the slope of the line. Find the translation that gives a zero intercept.

Let us translate all measured data by $x' = x + r$ and $y' = y + s$. From Eq. (14.2) the coefficient b' for the transformed data is

$$b' = \frac{\sum (x_i' - x')(y_i' - \bar{y}')}{\sum (x_i' - \bar{x}')^2}$$

and, since $\bar{x}' = \bar{x} + r$, $\bar{y}' = \bar{y} + s$, it readily follows that $b' = b$.

The intercept for the transformed data is

$$a' = \bar{y}' - b\bar{x}' = \bar{y} + s - b(\bar{x} + r)$$

and if we choose $r = -\bar{x}$ and $s = -\bar{y}$, then $a' = 0$, that is, the intercept of the transformed data is zero.

EXAMPLE 14.4

The hardness of steel, in standardized units, varies with the percentage of tungsten added. The results of y readings are:

% tungsten (x)	0.20	0.40	0.60	0.80	1.00	1.20
Hardness (y)	.13	.24	.51	.72	1.05	1.34

Find the best estimates for regression curves of the form $\alpha + \beta x$ and $\alpha + \beta x^2$ and compare the resulting error sum of squares.

For the regression line $\alpha + \beta x$, the least squares estimates of the parameters α and β are given by Eq. (14.3):

$$B = \frac{\sum (x_i - \bar{x})(Y_i - \bar{Y})}{\sum (x_i - \bar{x})^2}, \quad b = 1.24 \qquad A = \bar{Y} - B\bar{x}, \quad a = -.20$$

and the resulting error sum of squares is .019.

To determine the least squares estimates for the parameters α and β in the assumed regression curve, we proceed as in the derivation for Eqs. (14.3). If

$$e = \sum_1^n (y_i - \alpha - \beta x_i^2)^2$$

then

$$\frac{\partial e}{\partial \alpha} = -2 \sum_{i=1}^n (y_i - \alpha - \beta x_i^2) = 0$$

and

$$\frac{\partial e}{\partial \beta} = -2 \sum_{i=1}^n (y_i - \alpha - \beta x_i^2) x_i^2 = 0$$

and the solutions to these equations give the estimators

$$B = \frac{\sum Y_i x_i^2 - \bar{Y} \sum x_i^2}{\sum x_i^3 - \bar{x} \sum x_i^2} \quad \text{and} \quad A = \bar{Y} - B\bar{x}$$

For the data given, $b = 1.27$, $a = .22$ and the resulting error sum of squares is .213, which is considerably more than for the linear model. In this case, with the information given, linear dependence of the hardness on the percentage of tungsten would seem a more appropriate model.

In determining expressions for the variance of the estimators A and B it was assumed that the variance of Y, σ^2, was known although knowledge of σ^2 was not required for the computational form of the estimators themselves. In the next two sections confidence intervals and hypothesis tests on the regression parameters and equation will be considered. For these situations when it is not assumed that σ^2 is known, it is necessary to have an estimate of it. To complete this section it will be shown that

$$S_y^2 = \frac{1}{(n-2)} \sum_{i=1}^n (Y_i - A - Bx_i)^2 \tag{14.8}$$

is an unbiased estimator for σ^2, the true variance of Y. To verify this, calculate the expected value:

$$E(S_y^2) = \frac{1}{n-2} \sum_{i=1}^n E(Y_i - A - Bx_i)^2$$

$$= \frac{1}{n-2} \sum_{i=1}^n E[Y_i - \bar{Y} - B(x_i - \bar{x})]^2$$

using the fact that $\bar{Y} = A + B\bar{x}$. Expanding the square yields

$$E(S_y^2) = \frac{1}{n-2} \left[\sum_{i=1}^n E[(Y_i - \bar{Y})^2] - 2E\left(B \sum_{i=1}^n (Y_i - \bar{Y})(x_i - \bar{x}) \right) \right.$$

$$\left. + E(B^2) \sum_{i=1}^n (x_i - \bar{x})^2 \right] \tag{14.9}$$

and notice that

$$\sum_{i=1}^{n} (Y_i - \bar{Y})(x_i - \bar{x}) = \sum_{i=1}^{n} (x_i - \bar{x})Y_i$$

$$= B \sum_{i=1}^{n} (x_i - \bar{x})^2$$

using Eq. (14.3). Equation (14.8) can be written as

$$E(S_y^2) = \frac{1}{n-2}\left(\sum_{i=1}^{n} E[(Y_i - \bar{Y})^2] - E(B^2) \sum_{i=1}^{n} (x_i - \bar{x})^2\right)$$

$$= \frac{1}{n-2}\left(\sum_{i=1}^{n} E[(Y_i - \bar{Y})^2] - \sigma^2 - \beta^2 \sum_{i=1}^{n} (x_i - \bar{x})^2\right)$$

where Eq. (14.5) was used. The first summation on the right-hand side is calculated as

$$\sum_{i=1}^{n} E[(Y_i - \bar{Y})^2] = (n-1)\sigma^2 + \beta^2 \sum_{i=1}^{n} (x_i - \bar{x})^2$$

and so

$$E(S_y^2) = \frac{1}{n-2}\left((n-1)\sigma^2 + \beta^2 \sum_{i=1}^{n} (x_i - \bar{x})^2 - \sigma^2 - \beta^2 \sum_{i=1}^{n} (x_i - \bar{x})^2\right)$$

$$= \sigma^2$$

and the estimate is unbiased.

In these computations notice that

$$(a + bx_i - \bar{y}) = (a + bx_i - a - b\bar{x})$$

$$= b(x_i - \bar{x})$$

and, by using Eqs. (14.1),

$$\sum (y_i - a - bx_i)(a + bx_i - \bar{y}) = 0$$

Since

$$y_i - \bar{y} = (y_i - a - bx_i) + (a + bx_i - \bar{y})$$

we can write

$$\sum (y_i - \bar{y})^2 = \sum (y_i - a - bx_i)^2 + \sum (a + bx_i - \bar{y})^2$$

$$= \sum (y_i - a - bx_i)^2 + b^2 d$$

and

$$S_y^2 = \frac{1}{(n-2)}\left(\sum_i (Y_i - \bar{Y})^2 - B \sum_i (x_i - \bar{x})(Y_i - \bar{Y})\right)$$

which is often computationally more convenient since the quantities involved often have to be calculated anyway.

EXAMPLE 14.5

The tensile strength of a synthetic fiber is related to the temperature at a certain stage of its manufacture. The results of an experiment are as follows:

Temperature (°C) x	120	130	140	150	160	170	180	190
Tensile strength Y (kg/cm^2)		62.32	66.52	72.14	76.44	83.00	87.04	91.55	96.92

(i) Find the regression line of Y on x.
(ii) What value of Y would you estimate for $x = 172°C$.
(iii) Find $V(B)$. How should x_i be chosen to minimize $V(B)$, assuming that the temperature must be in the range 120 to 190°C.
(iv) Estimate the variance of Y.

(i) From Eqs. (14.3) the least squares estimates of the regression line parameters are

$$b = \frac{\sum\limits_{1}^{8} (x_i - \bar{x})(y_i - \bar{y})}{\sum\limits_{1}^{8} (x_i - \bar{x})^2}$$

$$= \frac{2096.55}{4200} = .499, \quad (\bar{x} = 155, \bar{y} = 79.49)$$

and

$$a = \bar{y} - b\bar{x} = 1.99$$

(ii) For $x = 172°C$ the estimate of the tensile strength is $a + bx = 87.99$ kg/cm^2.

(iii) The variance of the estimator B is by Eq. (14.5)

$$V(B) = \frac{\sigma^2}{\sum\limits_{1}^{8} (x_i - \bar{x})^2} = \frac{\sigma^2}{4200}$$

for a given value of σ^2, the variance of Y. Since σ^2 is fixed, $V(B)$ is minimized by maximizing the denominator, and a little thought on the matter will show that this is accomplished by concentrating the values of x at each end of the scale while keeping $\bar{x}$ at the center. For this case we should take four readings at $x = 120$ and four at $x = 190$.

(iv) From Eq. (14.8) the estimator for the variance of Y is

$$s_y^2 = \frac{1}{6} \sum_1^8 (y_i - a - bx_i)^2$$

$$= \frac{1}{6} \left(\sum_{i=1}^8 (y_i - \bar{y})^2 - b \sum_{i=1}^8 (x_i - \bar{x})(y_i - \bar{y}) \right)$$

and so s_y^2 is .313.

14.2 Confidence Intervals for the Parameters

Suppose initially that the variance of the Y population, assumed to be normal with mean $\alpha + \beta x$ for a particular x value and constant variance σ^2, is known. Since A and B can be expressed as linear combinations of the Y_i, they also are normally distributed. Their means are α and β, respectively, and their variances are given by Eqs. (14.6) and (14.5), respectively. It should be mentioned that the parameter α will play a dual role in this section and the next being both a parameter of the regression line and of the level of significance. Since no confusion can arise in the use of these parameters the notation will not be changed.

At the $100(1-\alpha)\%$ level of significance it follows immediately that the confidence limits for α are

$$A \pm z_{1-\alpha/2} \sigma_A \tag{14.10}$$

where

$$\sigma_A = \sigma \left(\frac{1}{n} + \frac{\bar{x}^2}{d} \right)^{1/2}$$

and $z_{1-\alpha/2}$ is defined for the normal random variable Z as in Chapter 5

$$P(Z < z_{1-\alpha/2}) = 1 - \alpha/2$$

Similarly the $100(1-\alpha)\%$ level of significance confidence limits on β are

$$B \pm z_{1-\alpha/2} \sigma_B \tag{14.11}$$

where

$$\sigma_B = \frac{\sigma}{\left(\sum_{i=1}^n (x_i - \bar{x})^2 \right)^{1/2}}$$

Confidence limits for the mean of Y at a prescribed point x given by $A + Bx$ are established in a similar manner. The mean of $A + Bx$ is $\alpha + \beta x$ while

$$V(A + Bx) = V(A) + x^2 V(B) + 2x \, \text{Cov}(A, B)$$

which is easily shown to be

$$V(A + Bx) = \sigma^2\left(\frac{1}{n} + \frac{(x - \bar{x})^2}{d}\right)$$

The $100(1 - \alpha)\%$ confidence interval for y at the point x is then

$$A + Bx \pm z_{1-\alpha/2}\sigma\left(\frac{1}{n} + \frac{(x - \bar{x})^2}{d}\right)^{1/2}$$

In practice the more important situation is when the variance of Y is unknown and the estimate of it must be used, and this changes the statistics and hence the intervals. To determine these statistics notice that $[(n-2)/\sigma^2]S_y^2$, where S_y^2 is given by Eq. (14.8), is a χ^2 random variable with $(n-2)$ degrees of freedom. Furthermore, from the results of Chapter 7, since

$$\frac{A - \alpha}{\sqrt{V(A)}} \quad \text{and} \quad \frac{B - \beta}{\sqrt{V(B)}}$$

are standard normal random variables independent of S_y^2, we conclude that the ratios

$$\frac{A - \alpha}{\sqrt{V(A) \cdot \dfrac{(n-2)}{\sigma^2} \dfrac{S_y^2}{(n-2)}}} \quad \text{and} \quad \frac{B - \beta}{\sqrt{V(B) \cdot \dfrac{(n-2)}{\sigma^2} \dfrac{S_y^2}{(n-2)}}}$$

both have a t-distribution with $(n-2)$ degrees of freedom. The $100(1-\alpha)\%$ confidence intervals for both α and β follow immediately as

for α: $\quad \left[A \pm t_{1-\alpha/2;n-2} \cdot \left(\frac{V(A)S_y^2}{\sigma^2}\right)^{1/2}\right]$

$$= \left[A \pm t_{1-\alpha/2;n-2}S_y\left(\frac{\frac{1}{n}\sum_{i=1}^{n}x_i^2}{\sum_{i=1}^{n}(x_i - \bar{x})^2}\right)^{1/2}\right] \qquad (14.12)$$

for β: $\quad \left[B \pm t_{1-\alpha/2;n-2} \cdot \left(\frac{V(B)\sigma_y^2}{\sigma^2}\right)^{1/2}\right]$

$$= \left[B \pm t_{1-\alpha/2;n-2}S_y\bigg/\left(\sum_{i=1}^{n}(x_i - \bar{x})^2\right)^{1/2}\right] \qquad (14.13)$$

where, it will be recalled, if T is a random variable with a t-distribution with n degrees of freedom, then $t_{\eta;n}$ is defined by the relation

$$P(T < t_{\eta;n}) = \eta.$$

To derive confidence limits for the point estimate $A + Bx$ of the mean of Y for a given x, in the case when the variance is unknown, notice that

$$\frac{(A+Bx)-(\alpha+\beta x)}{\sqrt{V(Y|x)}}$$

is a standard normal random variable where $V(Y|x)$ is given by Eq. (14.7). The ratio

$$\frac{(A+Bx)-(\alpha+\beta x)}{\sqrt{V(Y|x)}} = \frac{(A+Bx)-(\alpha+\beta x)}{S_y\left(\dfrac{1}{n}+\dfrac{(x-\bar{x})^2}{\displaystyle\sum_{i=1}^{n}(x_i-\bar{x})^2}\right)^{1/2}}$$

has a t-distribution with $(n-2)$ degrees of freedom. Consequently, the $100(1-\alpha)\%$ confidence limits for $(\alpha+\beta x)$ are

$$A + Bx \pm t_{1-\alpha/2;n-2}S_y\left|\dfrac{1}{n}+\dfrac{(x-\bar{x})^2}{\displaystyle\sum_{i=1}^{n}(x_i-\bar{x})^2}\right|^{1/2} \qquad (14.14)$$

These limits can be plotted for a particular level of significance α as functions of x and it is easily checked that the limits will be narrowest when $x = \bar{x}$ and the second term inside the brackets vanishes. All other terms are independent of x. A sketch of the situation is shown in Figure 14.2.

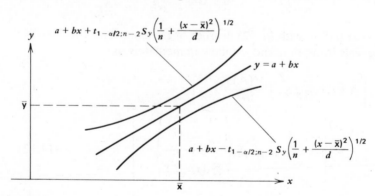

FIGURE 14.2

EXAMPLE 14.6

Determine the 95% confidence intervals for the parameters α and β of Example 14.1 assuming σ is known and equal to .8.

From Eq. (14.6) using $z_{.975} = 1.96$,

$$\sigma_A = \sqrt{V(A)} = (0.8)\left(\frac{1}{6}+\frac{(47.5)^2}{437.50}\right)^{1/2} = .661$$

where $d = \sum_{i=1}(x_i - \bar{x})^2 = 437.50$. Consequently, the confidence interval for α is

$$a \pm z_{1-\alpha/2}\sigma_A = 13.01 \pm 1.96 \times .661 = (11.71, 14.31)$$

Calculated in the same manner, the confidence interval for β is

$$b \pm z_{1-\alpha/2}\sigma_B = .225 \pm 1.96 \times \frac{.8}{(437.50)^{1/2}} = (.150, .300)$$

EXAMPLE 14.7

Repeat Example 14.4 for the case when σ is not assumed to be known. For the 95% confidence intervals of the parameters, note that $t_{.975;10} = 2.228$. Since the variance is assumed unknown, Eq. (14.8)

$$s_y^2 = \frac{1}{(n-2)} \sum_{i=1}^{n} (y_i - a - bx_i)^2$$

is used to estimate it and $s_y^2 = .120$. Using Eq. (14.12) the 95% confidence interval for α is

$$-2.846 \pm 2.228 \times 3.256 = (-10.116, 4.424)$$

The confidence interval for β is

$$.116 \pm 2.228 \times \frac{.346}{(496.27)^{1/2}} = (.081, .151)$$

EXAMPLE 14.8

Consider the following data:

x:	0	1	2	3	4	5	6	7	8	9
y:	.16	.09	.08	.23	.60	.39	.55	.75	.81	.85

Assuming that $\sigma^2 = .01$:

(i) Find the regression line of y on x.
(ii) Find a 95% confidence interval for a and b.
(iii) Find a 90% confidence interval for the expected value of y at $x = 2.5$.

(i) The regression parameters are

$$b = \frac{\sum\limits_{0}^{9}(x_i - \bar{x})(y_i - \bar{y})}{\sum\limits_{0}^{9}(x_i - \bar{x})^2} = \frac{7.875}{82.50} = .093$$

and

$$a = .032.$$

(ii) Since $\sigma^2 = .01$, the 95% confidence limits on α are

$$a \pm z_{.975}\sigma\left(.1 + \frac{(4.5)^2}{82.50}\right)^{1/2} = (-.082, .147)$$

and the confidence limits on β are

$$b \pm z_{.975}(\sigma/\sqrt{82.50}) = (.071, .115)$$

(iii) The 90% confidence limits on the estimate of Y at $x = 2.5$ are

$$a + bx \pm z_{.95}\sigma\left(\frac{1}{10} + \frac{4}{82.50}\right)^{1/2} = (.212, .318)$$

EXAMPLE 14.9

Repeat Example 14.8 assuming that the variance of Y is unknown. The estimate of the variance is

$$s_y^2 = \frac{1}{8}\sum_1^{10}(y_i - a - bx_i)^2 = .01258$$

The regression line is not dependent on the variance of the Y population. The 95% confidence intervals for a and b are, respectively,

$$a \pm t_{.975;8}s_y\left(\frac{\sum_1^{10}x_i^2}{d}\right)^{1/2} = (-.448, .513)$$

$$b \pm t_{.975;8}s_y\left(\frac{1}{d}\right)^{1/2} = (.0645, .1215)$$

The 90% confidence interval for the estimate of Y at $x = 2.5$ is

$$a + bx \pm t_{.975;8}s_y\left(\frac{1}{10} + \frac{(x - \bar{x})^2}{d}\right)^{1/2} = (.165, .364)$$

The data for Example 14.8 were generated by using the model $Y = (0.1)x + E_i$, where E_i were chosen as independent normal random variances with mean 0 and variance .01, and the above intervals reflect this model well.

The above confidence intervals are for the mean value of the random variable Y for a particular value of the controlled variable x. In some applications a probabilistic statement about the actual value of Y is far more useful. From the properties of the estimators already discussed it can be shown that with probability $1 - \alpha$ the random variable Y, corresponding to a value x of the controlled variable, will lie in the interval

$$A + Bx \pm t_{\alpha/2;n-2}S_y\left(1 + \frac{1}{n} + \frac{(x - \bar{x})^2}{\sum(x_i - \bar{x})^2}\right)^{1/2}$$

and this is referred to as the $100(1-\alpha)\%$ prediction interval for Y (for a given x). Notice that it is wider than the confidence interval for $E(Y|x)$ for the same value of x but is narrowest at the same point, namely, $x = \bar{x}$. If more than one prediction interval is required the estimators A, B, and S_y must be used on different data since otherwise the intervals obtained would not be independent. Other techniques, beyond the scope of this presentation, exist for determining multiple prediction intervals from the same data.

A summary of the confidence interval formulae of this section is given in Table 14.1 for the cases where σ is unknown.

TABLE 14.1 *Confidence Intervals, σ Unknown*

PARAMETER	ESTIMATOR	$100(1-\alpha)\%$ CONFIDENCE INTERVAL
α	$A = \bar{Y} - B\bar{x}$	$A \pm t_{\alpha/2;n-2} S_y \left(\dfrac{1}{n} + \dfrac{\bar{x}^2}{d} \right)^{1/2}$
β	$B = \dfrac{\sum (x_i - \bar{x})(Y_i - \bar{Y})}{d}$	$B \pm t_{\alpha/2;n-2} S_y \left(\dfrac{1}{d} \right)^{1/2}$
$E(Y\|x) = \alpha + \beta x$	$A + Bx$	$A + Bx \pm t_{\alpha/2;n-2} S_y \left(\dfrac{1}{n} + \dfrac{(x - \bar{x})^2}{d} \right)^{1/2}$

14.3 *Hypothesis Testing on the Parameters*

In some cases it may be more relevant to consider a hypothesis testing situation rather than confidence limits, although the two techniques for gaining information on the random variable Y are not unrelated. In this section two tests on the parameters a and b of the regression line are derived.

To test the hypotheses H_0: $\alpha = \alpha_0$ versus H_1: $\alpha \neq \alpha_0$ at the level of significance α, recall from the previous section that the random variable

$$\frac{A - \alpha_0}{\sqrt{V(A)\dfrac{S_y^2}{\sigma^2}}} = \frac{A - \alpha_0}{S_y \left(\dfrac{\sum\limits_{i=1}^{n} x_i^2}{n \sum\limits_{i=1}^{n} (x_i - \bar{x})^2} \right)^{1/2}} = T \qquad (14.15)$$

has a t-distribution with $(n-2)$ degrees of freedom. The test will be of the form: Reject H_0 if $|A - \alpha_0| > c$ for some constant c chosen so that the probability of this statistic exceeding c when $\alpha = \alpha_0$ is less than or equal to α. Since the variable $t_{1-\alpha/2;n-2}$ is such that

$$P(|T| > t_{1-\alpha/2;n-2}) = \alpha$$

and consequently

$$P\left[|A-\alpha_0|>t_{1-\alpha/2;n-2}\cdot S_y\left(\frac{\sum\limits_{i=1}^{n}x_i^2}{n\sum\limits_{i=1}^{n}(x_i-\bar{x})^2}\right)^{1/2}\right]=\alpha$$

then H_0 is rejected if

$$|A-\alpha_0|>t_{1-\alpha/2;n-2}S_y\left(\frac{\sum\limits_{i=1}^{n}x_i^2}{n\sum\limits_{i=1}^{n}(x_i-\bar{x})^2}\right)^{1/2}$$

To test the hypotheses H_0: $\alpha=\alpha_0$ versus H_1: $\alpha>\alpha_0$ the test is of the form: Reject H_0 if $T>c$ where c is chosen so that

$$P(T>c)=\alpha$$

that is, $c=t_{1-\alpha;n-2}$ and T is given by Eq. (14.15). The test is then to reject H_0 if

$$(A-\alpha_0)>t_{1-\alpha;n-2}S_y\left(\frac{\sum\limits_{i=1}^{n}x_i^2}{n\sum\limits_{i=1}^{n}(x_i-\bar{x})^2}\right)^{1/2}$$

The tests on the parameter b follow in precisely the same manner and H_0: $\beta=\beta_0$ versus H_1: $\beta\neq\beta_0$ the test is to reject H_0 if

$$|B-\beta_0|>t_{1-\alpha/2;n-2}S_y\left(\frac{1}{\sum\limits_{i=1}^{n}(x_i-\bar{x})^2}\right)^{1/2} \tag{14.16}$$

EXAMPLE 14.10

The following data were collected:

x	10	15	20	25	30	35
y	182	164	139	117	97	74

(a) Find the least squares regression line of y on x.
(b) Test the hypothesis that $b=-5$ at the 5% level of significance.
(c) Find the 90% confidence interval for the mean of y when $x=22$.

By direct calculation $a=226.93$ and $b=-4.36$ and $s_y^2=2.533$. To test the hypotheses H_0: $\beta=-5$ versus H_1: $\beta\neq-5$ Eq. (14.15) is used and H_0 is rejected if

$$|b+5| > t_{.975;r} s_y \left(\frac{1}{\sum (x_i - \bar{x})^2} \right)^{1/2} = 2.776 \times \left(\frac{2.533}{437.5} \right)^{1/2}$$

$$= .211$$

Since $|b+5| = |-4.36+5| = .64 > .211$, the hypothesis that $\beta = -5$ is rejected at the 5% level of significance.

The confidence interval for the mean of y when $x = 22$ is given by Eq. (14.13) and by calculation, $a + b22 = 131.01$ and

$$s_y \left(\frac{1}{6} + \frac{(22 - \bar{x})^2}{\sum (x_i - \bar{x})^2} \right)^{1/2} = 0.651$$

From tables, $t_{.95;4} = 2.132$ the 90% confidence interval for $\alpha + 22 \cdot \beta$ is

$$(131.01 \pm 2.132 \times 0.651) = (129.62, 132.40)$$

EXAMPLE 14.11

For the data of Example 14.8 test (i) the hypothesis that $\alpha = 0$, (ii) the hypothesis that $\beta = .1$, both at the 1% level of significance and assuming that σ^2 is unknown.

The least squares estimates for α and β are, from Example 14.8, .032 and .093, respectively. To test the hypothesis $H_0: \alpha = 0$ versus $H_1: \alpha \neq 0$ we use the statistic

$$t = \frac{a}{s_y (\sum x_i^2 / nd)^{1/2}} = \frac{.032}{.112 \left(\frac{285}{10 \times 82.50} \right)^{1/2}}$$

$$= .486$$

where, from Example 14.9, $s_y = .112$. Since $|t| < t_{.995;8} = 3.355$ we accept the hypothesis that $\alpha = 0$ at the 1% level of significance.

To test the hypothesis $H_0: \beta = .1$ versus $H_1: \beta \neq .1$ we use the statistic

$$t' = \frac{|b - .1|}{s_y \left(\frac{1}{\sum (x_i - \bar{x})^2} \right)^{1/2}} = \frac{.007}{.112 \left(\frac{1}{82.50} \right)^{1/2}}$$

$$= .567$$

Since $|t'| < t_{.995;8} = 3.355$ we accept the hypothesis that $\beta = .1$ at the 1% level of significance. These results are again consistent with the model used to generate Example 14.8.

A particularly important hypothesis to test is $\beta = 0$ in the sense that if one cannot reject the hypothesis that the slope of the regression line is zero, then there is no justification to suppose that Y is linearly related to x at all. The test of Eq. (14.16) could be used directly to test this hypothesis but the following discussion is instructive.

In showing that the estimator of the variance of Y, S_y^2, of Eq. (14.8) is unbiased, it was observed that

$$\sum (y_i - \bar{y})^2 = \sum (y_i - \hat{y}_i)^2 + \sum (\hat{y}_i - \bar{y})^2$$

where $\hat{y}_i = a + bx_i$. The first term on the right-hand side of this equation is $(n-2)s_y^2$ while the second term is $b^2 \sum (x_i - \bar{x})^2$. The summation $\sum (y_i - \bar{y})^2$ is called the sum of squares about the mean, $\sum (y_i - \hat{y}_i)^2$ the sum of squares about the regression, and $\sum (\hat{y}_i - \bar{y})^2$ the sum of squares due to regression. For a given set of observations the sum of squares about the regression is small if the sum of squares due to regression is close to the sum of squares about the mean. Consequently, if the ratio

$$\frac{\sum (\hat{y}_i - \bar{y})^2}{\sum (y_i - \hat{y}_i)^2}$$

is close to unity, then the sum of squares about the regression is small compared to the sum of squares about the mean. In such a case we conclude that most of the variation about the mean can be explained by the regression and the linear model is adequate.

Consider now the sum of squares about the mean $\sum (Y_i - \bar{Y})^2$ as a random variable. Since $\sum (Y_i - \bar{Y}) = 0$, the distribution of this statistic has only $(n-1)$ degrees of freedom. Similarly, $\sum (\hat{Y}_i - \bar{Y})^2 = B^2 \sum (x_i - \bar{x})^2$ can be computed from the single statistic B and has one degree of freedom. Consequently, the sum of squares about regression has $(n-2)$ degrees of freedom, by subtraction. We have already observed that $(n-2)S_y^2$ has a chi-square distribution with $(n-2)$ degrees of freedom. If $\beta = 0$, it can be shown that $\sum (\hat{Y}_i - \bar{Y})^2$ has a chi-square distribution with one degree of freedom.

If X and Y are two independent random variables with chi-square distributions with n_1 and n_2 degrees of freedom, respectively, then their ratio

$$\frac{(X/n_1)}{(Y/n_2)}$$

has an F-distribution with n_1 and n_2 degrees of freedom. This distribution is tabulated in many books but has not been included here because of the limited use we have for it. From the above discussion it follows that if $\beta = 0$, then the ratio

$$F = \frac{\sum (\hat{Y}_i - \bar{Y})^2}{\dfrac{1}{(n-2)} \sum (Y_i - \hat{Y}_i)^2} = \frac{B^2}{(S_y^2 / \sum (x_i - \bar{x})^2)}$$

has an F-distribution with 1 and $(n-2)$ degrees of freedom. This fact can be used to test $\beta = 0$ by comparing F with the $100(1-\alpha)$ percentage point of the F-distribution with 1 and $(n-2)$ degrees of freedom.

Notice however that the use of the F-distribution can be avoided since it was observed earlier in the section that

$$\frac{B}{S_y/[\sum (x_i - \bar{x})^2]^{1/2}} = \sqrt{F}$$

has a t-distribution with $(n-2)$ degrees of freedom and hence the F test for $\beta = 0$ is exactly the same as that given by Eq. (14.16).

14.4 *Multiple Linear Regression*

The previous sections considered the case where the random variable Y depended on the one controlled variable x. In many cases, however, it can depend on several variables. For example, the yield from a plot of land would depend on the mean temperature of the growing period, the amount of fertilizer applied, the chemical composition of the soil, and so forth. The quality of steel produced would depend on the carbon content, the temperature profile of its treatment, and other parameters.

In this section we consider that Y is normally distributed with mean $\mu(x_1, x_2) = \alpha + \beta_1 x_1 + \beta_2 x_2$ and that the mean of Y, σ^2 does not depend on x_1 and x_2. The extension to more than two controlled variables is straightforward but cumbersome and is best dealt with by using matrix terminology.

To find the least squares estimate of the three parameters α, β_1, β_2 we proceed exactly as in the case of the single variable and differentiate the error sum of squares

$$e = \sum_{i=1}^{n} (y_i - a - b_1 x_{1i} - b_2 x_{2i})^2$$

with respect to a, b_1, and b_2 and setting the results to zero. The results are

$$\frac{\partial}{\partial a}: \quad \sum_{i=1}^{n} (y_i - a - b_1 x_{1i} - b_2 x_{2i}) = 0$$

$$\frac{\partial}{\partial b_1}: \quad \sum_{i=1}^{n} (y_i - a - b_1 x_{1i} - b_2 x_{2i}) x_{1i} = 0 \qquad (14.17)$$

$$\frac{\partial}{\partial b_2}: \quad \sum_{i=1}^{n} (y_i - a - b_1 x_{1i} - b_2 x_{2i}) x_{2i} = 0$$

where the sample $\{(x_{11}, x_{21}, y_1), (x_{12}, x_{22}, y_2), \ldots, (x_{1n}, x_{2n}, y_n)\}$ is of size n. Letting

$$\bar{x}_1 = \frac{1}{n} \sum_{i=1}^{n} x_{1i}, \qquad \bar{x}_2 = \frac{1}{n} \sum_{i=1}^{n} x_{2i}, \qquad \bar{y} = \frac{1}{n} \sum_{i=1}^{n} y_i$$

and

$$s_1^2 = \sum (x_{1i} - \bar{x}_1)^2 \qquad s_2^2 = \sum (x_{2i} - \bar{x}_2)^2$$

$$r_{12} = \sum (x_{1i} - \bar{x}_1)(x_{2i} - \bar{x}_2)$$

the solution of Eqs. (14.17) can be found by simple substitution methods to yield the estimators

$$B_1 = \frac{1}{c}[s_2^2 \sum Y_i(x_{1i} - \bar{x}_1) - r_{12} \sum Y_i(x_{2i} - \bar{x}_2)]$$

$$\quad (14.18)$$

$$B_2 = \frac{1}{c}[s_1^2 \sum Y_i(x_{2i} - \bar{x}_2) - r_{12} \sum Y_i(x_{1i} - \bar{x}_1)]$$

$$A = \bar{Y} - B_1\bar{x}_1 - B_2\bar{x}_2$$

where $c = s_1^2 s_2^2 - r_{12}^2$. The random variables A, B_1, and B_2 are easily seen to be linear sums of the random variables Y_i and as such are normal random variables with means α, β_1, and β_2 and their variances can be calculated as

$$V(B_1) = \frac{\sigma^2 s_2^2}{c} \qquad V(B_2) = \frac{\sigma^2 s_1^2}{c} \qquad (14.19)$$

and

$$V(A) = \frac{\sigma^2}{n} + \frac{\sigma^2}{c}(\bar{x}_1^2 s_2^2 + \bar{x}_2^2 s_1^2 - 2\bar{x}_1\bar{x}_2 r_{12}) \qquad (14.20)$$

As in the case of the single variable, for a given pair of controlled variable values (x_1, x_2) the estimator for $E(Y|x_1, x_2)$ at this point is

$$A + B_1 x_1 + B_2 x_2$$

which has mean $\alpha + \beta_1 x_1 + \beta_2 x_2$ and variance

$$\frac{\sigma^2}{n} + \frac{\sigma^2}{c}[(x_1\bar{x}_1)^2 s_2^2 + (x_2 - \bar{x}_2)^2 s_1^2 - 2r_{12}(x_1 - \bar{x}_1)(x_2 - \bar{x}_2)]$$

With these assumptions, confidence intervals for the parameters α, β_1, and β_2 and $E(Y|x_1, x_2)$ can be found and hypothesis tests performed in much the same manner as in the previous two sections. Of course, if the variance of the random variable Y is unknown, it must be estimated, which leads to more complicated, but straightforward, computations.

EXAMPLE 14.12

The amount of sediment deposit at the mouth of a river channel is a function of the precipitation (measured at a particular point upriver) and pollution in the water, which affects riverbed plant life (and hence cohesion of the river bed). The following data were collected over a period:

PRECIPITATION (cm)	POLLUTION (Normalized Count)	SEDIMENT (cm)
1.64	11	0.20
4.81	74	3.14
2.24	31	1.21
2.41	40	1.42
0.98	22	0.34
1.78	51	1.62
1.04	36	1.14
1.12	17	0.85
2.45	29	1.70
2.16	20	1.16

Find the regression of the sediment on pollution and precipitation. Find an estimate of the sediment when the precipitation is 1.5 cm and pollution 30. The parameters required in the computation are

$$\bar{x}_1 = \frac{1}{10} \sum_{i=1}^{10} x_{1i} = 2.063 \qquad \bar{x}_2 = \frac{1}{10} \sum_{i=1}^{10} x_{2i} = 33.1 \qquad \bar{y} = \frac{1}{10} \sum_{i=1}^{10} y_i = 1.278$$

$$s_1^2 = \sum_{i=1}^{10} (x_{1i} - \bar{x}_1)^2 = 11.225 \qquad s_2^2 = \sum_{i=1}^{10} (x_{2i} - \bar{x}_2)^2 = 3112.9$$

$$r_{12} = \sum_{i=1}^{10} (x_{1i} - \bar{x}_1)(x_{2i} - \bar{x}_2) = 140.037$$

The estimates of the regression coefficients are

$b_1 = 0.335$

$b_2 = 0.025$

$a = 1.407$

and the estimate of sediment when precipitation is 1.5 cm and pollution is 30 is

$$y = 1.407 + 0.335 \times 1.5 + 0.025 \times 30$$

$$= 2.660 \text{ cm}$$

EXAMPLE 14.13

Consider the following data:

x_1	4.0	4.2	4.4	4.6	4.8	5.0
x_2	17.1	16.5	18.2	17.9	19.0	18.9
y	7.1	8.2	8.1	9.8	11.6	13.0

Find the regression equation of y on x_1 and x_2. Find the 95% confidence interval for the regression coefficient β_1. (Assume the variance of Y is 2.)

From the data and Eqs. (14.18) to (14.20)

$$\bar{x}_1 = 4.5 \qquad \bar{x}_2 = 17.93 \qquad s_1^2 = 0.7 \qquad s_2^2 = 4.893$$
$$c = .801 \qquad r_{12} = 1.620$$

and, using these quantities, the regression coefficients are calculated as

$$b_1 = 6.881 \qquad b_2 = -0.418 \qquad \text{and} \quad a = -13.841$$

Since B_1 is a normally distributed random variable with mean β_1 and variance, using Eq. (14.19), $\sigma^2 s_2^2 / c = 12.218$, the 95% confidence intervals are readily found as:

$$(b_1 \pm \sqrt{12.218} \times 1.96) = (6.881 \pm \sqrt{12.218} \times 1.96)$$
$$= (6.881 \pm 6.770)$$
$$= (.111, 13.651)$$

14.5   Correlation Analysis

In the previous sections of this chapter, values of the random variable Y were observed at controllable values of the variable x and the relationship between the two variables considered. In this section the situation where X and Y are both random variables is considered and some understanding of the relation between them is sought. A random sample for this case is a sequence of observations $\{(x_1, y_1), (x_2, y_2), \ldots, (x_n, y_n)\}$ of the bivariate random variable. It will be assumed that (X, Y) are jointly normally distributed with means μ_x, μ_y, variances σ_x^2, σ_y^2, and correlation coefficient ρ_{xy}.

Before proceeding with the analysis, some facts on the correlation coefficient, discussed in Chapter 6, are recalled. The value of the correlation coefficient of any two random variables (not necessarily normal) gives an indication of the linearity of the relationship between them in the sense that values of ρ_{xy} close to $+1$ or -1 indicate a linear relationship.

When two random variables are independent, then they are uncorrelated and $\rho_{xy} = 0$. The converse to this statement holds when X and Y are normally distributed, as will be assumed for the remainder of the section. In this case

$$E(Y \mid X = x) = \mu_y + \rho_{xy} \frac{\sigma_y}{\sigma_x} (x - \mu_x)$$

that is, the regression curve of Y on X is linear. The joint pdf of X and Y is

$$f(x, y) = \frac{1}{2\pi\sigma_x\sigma_y\sqrt{1-\rho_{xy}^2}} \exp\left\{ -\frac{1}{2(1-\rho_{xy}^2)}\left[\left(\frac{x-\mu_x}{\sigma_x}\right)^2 \right.\right.$$
$$\left.\left. -2\rho_{xy}\left(\frac{x-\mu_x}{\sigma_x}\right) \times \left(\frac{y-\mu_y}{\sigma_y}\right) + \left(\frac{y-\mu_y}{\sigma_y}\right)^2 \right] \right\}$$

and the likelihood function for a random sample of size n is

$$L(x, y) = \prod_{i=1}^{n} f(x_i, y_i)$$

or, since the natural log of this quantity will be more useful,

$$\ln L(x, y) = \sum_{i=1}^{n} \ln f(x_i, y_i)$$

$$= -n \ln(2\pi\sigma_x\sigma_y\sqrt{1-\rho_{xy}^2})$$

$$-\frac{1}{2(1-\rho_{xy}^2)}\left\{ \sum_{i=1}^{n} \left[\left(\frac{x_i-\mu_x}{\sigma_x}\right)^2 - 2\rho_{xy}\left(\frac{x_i-\mu_x}{\sigma_x}\right)\left(\frac{y_i-\mu_y}{\sigma_y}\right) \right. \right.$$

$$\left. \left. +\left(\frac{y_i-\mu_y}{\sigma_y}\right)^2 \right] \right\} \tag{14.21}$$

The maximum likelihood estimates of the parameters μ_x, μ_y, σ_x, σ_y, and ρ_{xy} denoted by m_x, m_y, s_x, s_y, and r_{xy}, respectively, can be found by differentiating Eq. (14.21) with respect to each parameter and equating the result to zero. The equations are long but with some manipulation it can be shown that

$$m_x = \frac{1}{n}\sum_{i=1}^{n} x_i \qquad m_y = \frac{1}{n}\sum_{i=1}^{n} y_i$$

$$s_x = \left(\frac{1}{n}\sum_{i=1}^{n}(x_i-m_x)^2\right)^{1/2} \qquad s_y = \left(\frac{1}{n}\sum_{i=1}^{n}(y_i-m_y)^2\right)^{1/2}$$

$$r_{xy} = \frac{\sum_{i=1}^{n}(x_i-m_x)(y_i-m_y)}{\left(\sum_{i=1}^{n}(x_i-m_x)^2 \sum_{i=1}^{n}(y_i-m_y)^2\right)^{1/2}} \tag{14.22}$$

This last quantity is also referred to as the sample correlation coefficient. It can be shown that r_{xy} is a biased estimate of ρ_{xy} unless $\rho_{xy} = 0$.

The sample correlation coefficient is a complicated statistic whose distribution function is very difficult to compute, even for the restrictive case considered here when the variables are assumed normal. When $\rho_{xy} = 0$ (and X and Y are independent), it can be shown that the statistic

$$T = r_{xy}\left(\frac{n-2}{1-r_{xy}^2}\right)^{1/2} \tag{14.23}$$

has a t-distribution with $(n-2)$ degrees of freedom. The test for the hypothesis H_0: $\rho_{xy} = 0$ versus H_1: $\rho_{xy} \neq 0$ with level of significance α is to reject the hypothesis H_0 when $|T| > t_{1-\alpha/2;n-2}$. To test the hypothesis H_0: $\rho_{xy} = 0$ versus H_1: $\rho_{xy} > 0$, and reject H_0 if $T > t_{1-\alpha;n-2}$.

EXAMPLE 14.14

The mean temperature X during a month, and average weight Y of tomatoes picked at maturity during that month, is assumed to be a bivariate normal random variable. Readings at various locations and at various times of the year were

X MEAN TEMP. (°C)	Y MEAN WEIGHT (grams)	X	Y	X	Y
21.6	120	27.1	177	28.1	197
28.7	184	28.2	221	20.4	140
24.1	98	26.4	184	17.6	52
24.5	127	19.4	82	22.2	131
19.6	43	23.2	117	31.6	245
22.4	72	21.6	104	27.2	208
27.1	191	27.4	196	26.4	176
23.2	112	28.2	211	25.4	145
29.4	243	24.3	131	29.2	206
22.4	90	24.2	145	19.2	71

Test the hypothesis H_0: $\rho_{xy} = 0$ versus H_0: $\rho_{xy} \neq 0$ at the 5% level of significance.

To calculate the statistic T of Eq. (14.23) we calculate the parameters

$$m_x = \frac{1}{30} \sum_{i=1}^{30} x_i = 24.68$$

$$m_y = \frac{1}{30} \sum_{i=1}^{30} y_i = 147.3$$

$$s_x = \frac{1}{30} \sum_{i=1}^{30} (x_i - m_x)^2 = 12.264$$

$$s_y = \frac{1}{30} \sum_{i=1}^{30} (y_i - m_y)^2 = 3,123.21$$

and, from Eq. (14.22)

$$r_{xy} = .931$$

The statistic T is

$$T = r_{xy} \left(\frac{n-2}{1-r_{xy}^2} \right)^{1/2}$$

$$= .931 \left(\frac{28}{1-.931^2} \right)^{1/2} = 13.448$$

From the table of the t-distribution the threshold for rejecting H_0 is $t_{.975;28} = 2.048$ and H_0 is rejected if $|T| > 2.048$. Hence for the data presented H_0 is rejected at the 5% level of significance.

Because of the difficulty in obtaining the distribution of r_{xy}, approximations are often used to establish confidence limits. It was shown by Fisher that the random variable

$$Z = \frac{1}{2}\ln\left(\frac{1+r_{xy}}{1-r_{xy}}\right)$$

has a distribution that is asymptotically normal with mean

$$E(Z) = \frac{1}{2}\ln\left(\frac{1+\rho}{1-\rho}\right)$$

and variance

$$V(Z) = \frac{1}{n-3}$$

The approximate $100(1-\alpha)\%$ confidence interval for the statistic Z is then

$$\frac{1}{2}\ln\left(\frac{1+r_{xy}}{1-r_{xy}}\right) \pm \frac{z_{1-\alpha/2}}{\sqrt{n-3}}$$

The corresponding approximate $100(1-\alpha)\%$ confidence interval for r_{xy} is

$$\left\{\tanh\left[\frac{1}{2}\ln\left(\frac{1+r_{xy}}{1-r_{xy}}\right) - \frac{z_{1-\alpha/2}}{\sqrt{n-3}}\right], \tanh\left[\frac{1}{2}\ln\left(\frac{1+r_{xy}}{1-r_{xy}}\right) + \frac{z_{1-\alpha/2}}{\sqrt{n-3}}\right]\right\}$$

the approximation to the true confidence interval being good for large values of n. The confidence intervals for various sample sizes and for $\alpha = .05$ and $\alpha = .01$ are shown in Figures 14.3 and 14.4, respectively.

EXAMPLE 14.15

For the data of Example 14.14, find the 95% confidence limits for the sample correlation coefficient.

For $\alpha = .05$, $z_{1-\alpha/2} = 1.96$, and the 95% confidence limits for the sample correlation coefficient are

$$\left\{\tanh\left[\frac{1}{2}\ln\left(\frac{1+.931}{1-.931}\right) - \frac{1.96}{\sqrt{27}}\right], \tanh\left[\frac{1}{2}\ln\left(\frac{1+.931}{1-.931}\right) + \frac{1.96}{\sqrt{27}}\right]\right\}$$

or $(0.858, 0.967)$. This confidence interval can also be obtained graphically from Figure 14.3 for $n = 30$, and $\rho_{xy} = .931$.

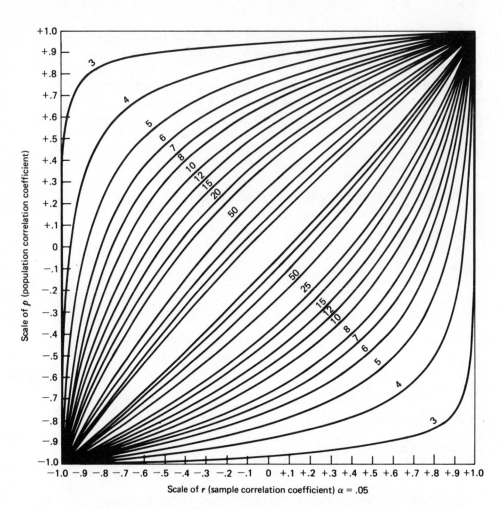

Problems

1. The yield Y of ten experimental acres, in response to various dosages of a new fertilizer x, both variables reported in normalized units, is:

Dosage	1	2	3	4	5	6	7	8	9	10
Yield	4.1	5.0	6.3	7.7	8.5	9.2	10.0	11.9	14.2	15.6

Find the regression line for these data.

2. Under what conditions on the sample will the slope of the regression line be zero?

3. The efficiency of workers on an automobile assembly line measured as the average number of defective operations performed per day (over a 10 day period) versus

452   *Applied Probability*

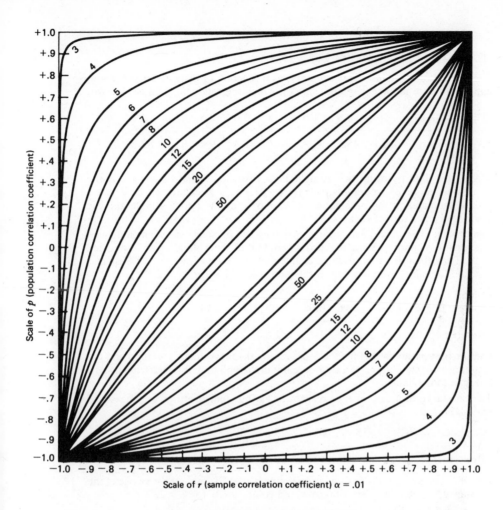

Scale of r (sample correlation coefficient) $\alpha = .01$

FIGURE 14.4   From "Biometrika Tables for Statisticians", Vol. 1, 2nd edition. Edited by E. S. Pearson and H. O. Hartley, Cambridge University Press, London, 1958, Table 15, with permission of the Biometrika Trustees.

the noise level (in decibels) in the plant controlled by a variety of means was

Noise level (db)	4.3	5.1	6.2	7.3	8.0
Average number of defective operations	72.1	81.2	98.6	112.6	131.4

Find the regression line for this data. How should the constants a_1, a_2, b_1, and b_2 be chosen so that the regression line for the transformed variables

$$x' = a_1 x + b_1$$

$$y' = a_2 y + b_2$$

is a line of slope 1 through the origin in the (x', y') plane.

4. Show that the mean square error of the regression line is zero iff

$$[\sum (x_i - \bar{x})(y_i - \bar{y})]^2 = \sum (x_i - \bar{x})^2 \sum (y_i - \bar{y})^2$$

5. The density of algae in a river tributary, as a function of distance from the sea, appears to be a nonlinear relationship on which the following data are known:

Distance (km)	0.1	0.5	1.0	1.5	2.0	2.5	3.0	3.5	4.0
Algae density $(cm^3)^{-1}$	31.0	45.2	54.9	66.1	82.4	79.4	85.1	82.3	81.4

Model this data by using two straight lines over appropriate intervals. If the regression curve is a parabola, $\mu(x) = \alpha + \beta x + \gamma x^2$, calculate the least mean square estimates of the parameters α, β, and γ for the above sample and the associated error sum of squares.

6. In a driving experiment, five groups, each with ten persons, are given controlled amounts of alcohol to bring the alcohol content in their blood up to a specified level. They then drive an obstacle course and the average number of obstacles touched for each group is noted. If the data recorded were

Blood alcohol content (mg/cm^3)	0.06	0.08	0.10	0.12	0.14
Number of obstacles	2.3	4.2	6.2	9.2	13.9

find the best parabola that fits these data using the least squares method and the resulting error sum of squares. Compare this error with that obtained from using a linear fit to the data.

7. The resistance of a carbon-type resistor at various temperatures was measured with the results:

Temperature (°C)	15	20	25	30	35
Resistance (ohm)	49.90	50.11	50.45	50.84	51.06

Find the regression of the resistance on temperature. Estimate the variance of the resistance. What value of resistance would you estimate for $T = 22°C$?

8. The random variable Y is measured against an independent variable x:

x	0.0	0.1	0.2	0.3	0.4	0.5
y	100.2	111.6	119.4	128.2	140.6	148.7

Find the least squares line for these data and estimate the variance of Y. Find the point estimate of Y for $x = 0.25$.

9. Suppose that the regression of the mean of the random variable Y on x is linear, $E(Y|x) = \beta x$, and the variance of Y, σ^2 is independent of x. Derive an unbiased estimate for σ^2.

10. Show that the estimators of Eq. (14.3) can be expressed as linear sums of the Y_i.

11. The actual weight of steel ingots is a normally distributed random variable depending on its pouring temperature:

Pouring temperature (°C)	1370	1375	1380	1385	1390
Ingot weight (kg)	10.10	10.14	10.22	10.28	10.31

Find the regression line of weight on the temperature. Test the hypothesis that $\alpha = -5.00$ at the 5% level of significance.

12. In Problem 11, test the hypothesis that $\beta = .01$ at the 5% level of significance.

13. For six successive weeks the number of absentee workers in a plant per week is monitored as a function of average outdoor temperature at 12 noon for the week, with the following results:

Average temperature (°C)	22.1	25.6	19.4	27.6	22.4	23.2
Number of absentee workers	30	45	28	48	34	35

Find the regression coefficients for these data and the 95% confidence limits for α assuming normality and that the variance of y is equal to 2.0.

14. The weight of people of a given height is assumed to be a normal random variable with variance σ^2 independent of height. The weight recorded for each height is actually the mean weight of 10 people of that height.

Height (m)	1.40	1.50	1.60	1.70	1.80	1.90
Weight (kg)	57.6	60.2	68.4	73.9	83.4	89.4

Find the regression line of y on x. Determine the 90% confidence limits for the parameter α and for the mean as a function of x.

15. For Problem 14 test the hypothesis that $\beta = 60$ at a 10% level of significance.

16. For the data of Problem 7 find the 95% confidence limits for $E(R \mid T = 22°C)$.

17. The stopping distance of a car going 100 km/hr on a particular dry surface for a range of temperatures (0 to 30°C) is a normal random variable. The following were the results of an experiment:

T = Temperature (°C)	0	5	10	15	20	25	30
D = Distance (m)		20.24	20.26	19.92	19.54	19.21	18.96	18.46

Find the least squares regression line of D on T and the 95% confidence interval for β. Find the 95% confidence interval for $E(D \mid x = 15)$.

18. In a nationwide study it is desired to determine the effect of air pollution on a certain kind of lung disease. The pollution figures were given on a normalized scale from 0 to 10 and the incidence of lung disease as a percentage of the population having the disease, within a locality. In five localities the figures obtained were:

Pollution figure (normalized)	2.6	5.1	4.6	8.2	6.3
Incidence of lung disease (%)	.14	.62	.52	1.40	.97

Use a 5% level of significance on the regression coefficient to test whether $\beta = 0$ versus $\beta \neq 0$.

19. In a study to determine the relationship between the weight of a baby and the weight of its mother, it was found that, in a sample of size one hundred, the results were (x = weight of mother, y = weight of baby):

$\sum x_i = 5422$ kg, $\sum y_i = 398$ kg, $\sum x_i^2 = 294{,}650$ kg^2, $\sum y_i^2 = 1772$ kg^2,

$\sum x_i y_i = 21{,}872$ kg^2

At the 5% level of significance, would you accept the hypothesis that the regression coefficient β is 0 against its being positive?

20. For the data of Problem 7 test the hypothesis that the regression coefficient $\beta = .05$ ohm/degree at the 5% level of significance.

21. For a given value of x the random variable Y is assumed to be normally distributed with variance independent of x. A random sample of size 10 was taken with results

x	0	1	2	3	4	5	6	7	8	9	10
y	4.2	4.6	5.4	5.6	5.9	6.6	7.2	7.4	8.0	8.1	8.9

Determine the least squares line for these data. Test the hypothesis that $\alpha = 4.0$ at the 5% level of significance. Test the hypothesis that $\beta = 0.50$ at the 5% level of significance.

22. For Problem 17 test the hypothesis that $\beta = -.06$ m/deg at the 5% level of significance.

23. Find 95% confidence limits for the estimate of y at $x = 5.0$ in Problem 21.

24. The amount of runoff from rainstorms depends on both the precipitation and length of time the storm lasts. The following data were collected from six storms:

Time (min)	4.1	26.2	27.1	10.2	62.4	41.0
Precipitation (cm)	1.2	4.2	5.1	3.0	8.3	7.0
Runoff (cm)	0.5	1.1	1.3	0.9	1.7	1.4

Find estimates of the parameters α, β_1, and β_2 and the associated error sum of squares.

25. In a certain problem the sums for a sample of size 10 were computed as

$$\sum x_i = 51.6, \quad \sum y_i = 85.4, \quad \sum x_i y_i = 417.6, \quad \sum x_i^2 = 349.7, \quad \sum y_i^2 = 756.2$$

Compute the sample correlation coefficient and find 95% confidence limits for it.

26. Verify Eqs. (14.18).

27. Show that Eqs. (14.22) are a solution of Eqs. (14.21).

28. Verify Eqs. (14.19) and (14.20).

29. Assuming that both the height and weight of Problem 14 are normal random variables, calculate the sample correlation coefficient. Would you accept the hypothesis that $\rho = 0$ at the 5% level of significance.

30. Suppose $E(Y|x) = \beta x$ and $V(Y|x) = \sigma^2$. Show that the least squares estimator of β is

$$B = \sum x_i Y_i / \sum x_i^2$$

and find its mean and variance. Show that

$$S_y^2 = \frac{1}{n-1} \sum (Y_i - Bx_i)^2$$

is an unbiased estimator for σ^2.

31. Test the hypothesis $H_0: \rho = 0$ versus $H_1: \rho \neq 0$ when the sample is of size 30 and the sample correlation coefficient $r_{xy} = 0.3$ at the 5% level of significance.

CHAPTER 15

Information Theory and Coding

Most people are well familiar with examples of communication systems such as radio, T.V., telephones, telegraph, and so forth. The analysis of communication systems and their performance in the presence of noise is an important aspect of their design and involves considerable knowledge of probability. This is particularly true of digital communication systems that have gained prominence over the last several years. The purpose of this chapter is to introduce some fundamental notions underlying the science of communications. We will examine only a few idealized situations and not prove the main theorems, the principal aim being to demonstrate the applicability of probability to these problems.

The purpose of a communication system is to transmit information. There are two properties of a system that we will be interested in, namely, the quantity of information being transmitted (in a given time interval) and the accuracy with which it is transmitted. Clearly a measure of information is required to answer the first question. The problem of defining such a measure was tackled by R. V. Hartley in the late 1920s and by C. E. Shannon in the late 1940s and has led to the development of what is now called information theory. We will discuss in this chapter some of the main ideas of information theory including two theorems that are often called the fundamental theorems of communication. They have had a profound impact on the study of communications and we will attempt to give an insight into their meaning here.

15.1 Uncertainty, Information, and Entropy

In this chapter we will only be considering discrete memoryless sources that can be defined as follows. Each unit of time one element from the set $X = \{x_1, x_2, \ldots, x_K\}$ is chosen by the source for transmission and the choices are independent from one time instant to the next. At any particular instant of time the element x_i is chosen with probability p_i and we call the set $(X, p) = \{(x_1, p_1), \ldots, (x_K, p_K)\}$, $\sum_1^K p_i = 1$, the source ensemble. When the probability distribution of the ensemble is understood, it will also be denoted by X. Thus a source sequence is of the form $x_{i_1}, x_{i_2}, x_{i_3}, \ldots$. Such a source is termed a discrete memoryless source (DMS) or sometimes zero-memory source. The term memoryless or zero memory comes from the fact that the symbol produced by the source at any instant of time is independent from all previous choices.

We would like to have some notion as to how much information is produced by such a source. The idea of information is closely linked to that of uncertainty. For example, if p_i, the probability that x_i is produced by the source at any given time is very low, then it is more surprising when it appears than when a symbol of higher probability appears.

Definition If x_i is a symbol with probability p_i, we say that x_i has

$$I(x_i) = \log(1/p_i) = -\log(p_i)$$

units of information associated with its occurrence.

Another way of stating this is by saying that we have gained $I(x_i)$ units of information when we have observed x_i to occur. When we use logs to the base 2 the units of information will be bits and we will use the base 2 exclusively in this chapter. A change of base affects the formula only by a multiplicative constant since

$$\log_a x = \left(\frac{1}{\log_b a}\right) \log_b x$$

For convenience logs to the base 2 are tabulated in Table 15.1 and $p \log_2 p$ in Table 15.2 at the end of this chapter. This way of defining information may appear a little strange. It can be shown, however, that if we set down what properties we would like an information measure to possess, then under a certain set of reasonable and desirable properties the log function results. We will not engage in any philosophical discussion of this definition but simply note that if p_i is small, then $I(x_i)$ is large, and if p_i is large, $I(x_i)$ is small, agreeing with our previous comments.

An important quantity in information theory is the entropy of an ensemble.

Definition The entropy of the ensemble $(X, p) = \{(x_1, p_1), \dots, (x_K, p_K)\}$ is given by the expression

$$H(X) = -\sum_{i=1}^{K} p_i \log p_i \qquad (15.1)$$

which is just the average information associated with its elements.

The name entropy derives from thermodynamics where similar quantities are encountered.

EXAMPLE 15.1

An experiment consists of tossing a biased coin for which the probability of a head is p. What is the average amount of information gained (or uncertainty removed) by knowing the outcome of the experiment?

The ensemble in this case is $(X, p) = \{(H, p), (T, 1-p)\}$ and, by Eq. (15.1),

$$H(X) = -p \log p - (1-p) \log(1-p)$$

This particular function, as a function of p, occurs in several places in information theory. It is readily established that it attains its maximum when $p = \frac{1}{2}$, as can be seen by differentiating $H(X)$ with respect to p and setting the result equal to zero. This maximum value is

$$H(X) = -\tfrac{1}{2} \log \tfrac{1}{2} - \tfrac{1}{2} \log \tfrac{1}{2}$$

$$= -\log \tfrac{1}{2} = \log 2 = 1$$

recalling that all logs are to the base 2 unless specified otherwise. As $p \to 0$ the behavior of the function $p \log p$ can be examined by applying l'Hopital's rule, that is,

$$\lim_{p \to 0} p \log p = \lim_{p \to 0} \frac{\log p}{\dfrac{1}{p}}$$

$$= \lim_{p \to 0} \frac{\dfrac{d}{dp}(\log p)}{\dfrac{d}{dp}\left(\dfrac{1}{p}\right)} = \lim_{p \to 0} -\frac{\left(\dfrac{1}{p}\right)}{\left(\dfrac{1}{p}\right)^2}$$

$$= \lim_{p \to 0} p = 0$$

The general shape of the curve is shown in Figure 15.1.

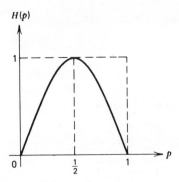

$H(p)$

1

0    $\frac{1}{2}$ 1 p

FIGURE 15.1

EXAMPLE 15.2

A fair coin is tossed until a head is obtained and the number of tosses required is noted. What is the entropy associated with this experiment?

In this experiment the ensemble is

$$(X, p) = \left\{ \left(1, \frac{1}{2}\right), \left(2, \frac{1}{2^2}\right), \left(3, \frac{1}{2^3}\right), \cdots \right\}$$

and hence the associated entropy is given by

$$
\begin{aligned}
H(X) &= -\sum_{i=1}^{\infty} p_i \log p_i \\
&= -\sum_{i=1}^{\infty} \frac{1}{2^i} \log\left(\frac{1}{2^i}\right) \\
&= \sum_{i=1}^{\infty} \frac{1}{2^i} \log(2^i) \\
&= \sum_{i=1}^{\infty} \frac{i}{2^i} = 2
\end{aligned}
$$

since this last summation is essentially the mean of a geometric distribution with parameter $\frac{1}{2}$.

This example merely illustrates that the notion of entropy is not restricted to the case where the ensemble is finite. In fact, entropy can be defined for continuous random variables and pdf's but we are only concerned with the discrete case.

Suppose that we are given two ensembles $(X, p) = \{(x_1, p_1), \ldots, (x_K, p_K)\}$ and $(Y, q) = \{(y_1, q_1), \ldots, (y_L, q_L)\}$ and a joint (Cartesian product) ensemble $(X \times Y) = \{[(x_i, y_j), p_{ij}], 1 \le i \le K, 1 \le j \le L\}$. We can think of X and Y as random variables and $X \times Y$ as a bivariate random variable with joint distribution p_{ij}. Note that we do not necessarily take X and Y to be independent random variables. The entropies of the three ensembles are, respectively,

$$H(X) = -\sum_{i=1}^{K} p_i \log p_i \qquad H(Y) = -\sum_{i=1}^{L} q_i \log q_i$$

$$H(X \times Y) = H(X, Y) = -\sum_{i=1}^{K} \sum_{j=1}^{L} p_{ij} \log p_{ij}$$

If X and Y are independent ensembles, then $p_{ij} = p_i q_j$ and

$$H(X, Y) = -\sum_{i=1}^{K} \sum_{j=1}^{L} p_i q_j \log(p_i q_j)$$

$$= -\sum_{i=1}^{K} p_i \sum_{j=1}^{L} q_j (\log p_i + \log q_j)$$

$$= -\sum_{i=1}^{K} p_i \log p_i \left(\sum_{j=1}^{L} q_j \right) - \left(\sum_{i=1}^{K} p_i \right) \sum_{j=1}^{L} q_j \log q_j$$

$$= -\sum_{i=1}^{K} p_i \log p_i - \sum_{j=1}^{L} q_j \log q_j$$

$$= H(X) + H(Y) \tag{15.2}$$

We will require three other functions in Section 15.3 and it is convenient to introduce them here. The average amount of information gained (or uncertainty removed) about the X ensemble by observing the outcome of the Y ensemble is

$$H(X|Y) = -\sum_{i=1}^{K} \sum_{j=1}^{L} p_{ij} \log \left(\frac{p_{ij}}{q_j} \right)$$

and, by symmetry,

$$H(Y|X) = -\sum_{i=1}^{K} \sum_{j=1}^{L} p_{ij} \log \left(\frac{p_{ij}}{p_i} \right)$$

We call the difference between $H(X)$ and $H(X|Y)$ the mutual information between the X and Y ensembles and denote it by $I(X, Y)$:

$$I(X, Y) = H(X) - H(X|Y)$$

This quantity can be written conveniently as

$$I(X, Y) = -\sum_{i=1}^{K} \sum_{j=1}^{L} p_{ij} \log p_i + \sum_{i=1}^{K} \sum_{j=1}^{L} p_{ij} \log \left(\frac{p_{ij}}{q_j} \right)$$

$$= \sum_{i=1}^{K} \sum_{j=1}^{L} p_{ij} \log \left(\frac{p_{ij}}{p_i q_j} \right) \tag{15.3}$$

From the symmetry in this equation it is apparent that

$$I(X, Y) = H(Y) - H(Y|X)$$

Intuitively, if our interpretation of the function $I(X, Y)$ is correct, then if X and Y are independent ensembles, we should have $I(X, Y) = 0$. One interpretation of this equation is that if X and Y are independent, then knowledge of the outcome in one ensemble yields no information about the outcome in the other ensemble. From the definition in Eq. (15.3) of mutual information, if X and Y are independent, then $p_{ij} = p_i q_j$ and

$$I(X, Y) = \sum_{i=1}^{K} \sum_{j=1}^{L} p_i q_j \log\left(\frac{p_i q_j}{p_i q_j}\right)$$

$$= \sum_{i=1}^{K} \sum_{j=1}^{L} p_i q_j \log(1) = 0$$

The functions defined here can have interesting interpretations depending on the context in which they are used. Their properties have been well investigated and only their elementary properties were given here. It can be shown, for example, that $I(X, Y) \geq 0$. The following identity is useful to show the relationships between these various functions:

$$H(X, Y) = H(X|Y) + H(Y|X) + I(X, Y) \tag{15.4}$$

This equation, although formidable in appearance, is quickly tamed by noting that

$$\frac{1}{p_{ij}} = \frac{p_{ij}}{p_i q_j} \cdot \frac{p_i}{p_{ij}} \cdot \frac{q_j}{p_{ij}}$$

and so

$$H(X, Y) = \sum_{i=1}^{K} \sum_{j=1}^{L} p_{ij} \log\left(\frac{1}{p_{ij}}\right)$$

$$= \sum_{i=1}^{K} \sum_{j=1}^{L} p_{ij} \log\left(\frac{p_{ij}}{p_i q_j} \cdot \frac{p_i}{p_{ij}} \cdot \frac{q_j}{p_{ij}}\right)$$

$$= \sum_{i=1}^{K} \sum_{j=1}^{L} p_{ij} \left[\log\left(\frac{p_{ij}}{p_i q_j}\right) + \log\left(\frac{p_i}{p_{ij}}\right) + \log\left(\frac{q_j}{p_{ij}}\right)\right]$$

$$= I(X, Y) + H(Y|X) + H(X|Y)$$

as was required to show.

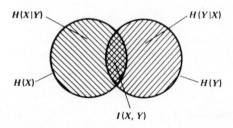

FIGURE 15.2

A convenient diagram, similar to a Venn diagram, which assists in remembering the relationships between these functions is shown in Figure 15.2. The "union" between the two "sets" represents $H(X, Y)$.

A few examples of how these notions are used, follow.

EXAMPLE 15.3

Two experiments of the following nature are performed. Two shots are fired at one target, the target having a probability of $\frac{1}{2}$ of being hit on each shot. For the second experiment four shots are fired at another target, this target having a probability of $\frac{1}{4}$ of being hit on each shot. Which experiment has less uncertainty associated with it?

On the first experiment either 0, 1, or 2 shots hit the target with probabilities $\frac{1}{4}$, $\frac{1}{2}$, and $\frac{1}{4}$, respectively. The entropy $H_1(X)$ associated with this ensemble is then

$$H_1(X) = -\tfrac{1}{4} \log_2 \tfrac{1}{4} - \tfrac{1}{2} \log_2 \tfrac{1}{2} - \tfrac{1}{4} \log_2 \tfrac{1}{4}$$

$$= \tfrac{1}{2} + \tfrac{1}{2} + \tfrac{1}{2} = 1.5$$

The possible outcomes for the second experiment are 0, 1, 2, 3, or 4 shots with probabilities $(\tfrac{3}{4})^4$, $4 \cdot \tfrac{1}{4}(\tfrac{3}{4})^3$, $6(\tfrac{1}{4})^2(\tfrac{3}{4})^2$, $4 \cdot (\tfrac{1}{4})^3(\tfrac{3}{4})$, and $(\tfrac{1}{4})^4$, respectively. The entropy $H_2(X)$ associated with this experiment is

$$H_2(X) = -(\tfrac{3}{4})^4 \log_2(\tfrac{3}{4})^4 - (\tfrac{3}{4})^3 \log_2(\tfrac{3}{4})^3 - 6(\tfrac{1}{4})^2(\tfrac{3}{4})^2 \log_2[6(\tfrac{1}{4})^2(\tfrac{3}{4})^2]$$

$$- 3(\tfrac{1}{4})^3 \log_2[3(\tfrac{1}{4})^3] - (\tfrac{1}{4})^4 \log_2(\tfrac{1}{4})^4$$

$$= 1.762$$

Since $H_1(X) < H_2(X)$, experiment 1 has less uncertainty associated with it.

EXAMPLE 15.4

Two urns contain eight balls each. Urn one contains four white, three red, and one blue. While urn two contains two white, three red, and three blue. One ball is drawn from each urn. Which drawing has more uncertainty associated with it?

Drawing from urn one the probability of getting a white, red, or blue ball is $\frac{4}{8}$, $\frac{3}{8}$, and $\frac{1}{8}$, respectively, and the associated entropy is

$$H_1(X) = -(\tfrac{4}{8}) \log(\tfrac{4}{8}) - (\tfrac{3}{8}) \log(\tfrac{3}{8}) - (\tfrac{1}{8}) \log(\tfrac{1}{8})$$

$$= 1.406$$

The probabilities for the second urn drawing are $\frac{2}{8}$, $\frac{3}{8}$, and $\frac{3}{8}$, respectively, and the associated entropy is

$$H_2(X) = -(\tfrac{2}{8}) \log(\tfrac{2}{8}) - (\tfrac{3}{8}) \log(\tfrac{3}{8}) - (\tfrac{3}{8}) \log(\tfrac{3}{8})$$

$$= 1.561$$

and since $H_1(X) < H_2(X)$, there is more uncertainty associated with the second experiment.

EXAMPLE 15.5 (Raisbeck, 1964)*

Among nine given coins one is known to be counterfeit, identified only by its having a different weight than the remaining eight. Devise a strategy whereby the counterfeit coin can be identified in three weighings, and whether it is lighter or heavier than the remaining eight coins.

Putting an equal number of coins on each side of the balance results in one of three conditions; either the balance tips left, balances, or tips right. Thus the most information that can be obtained from a weighing is $\log_2(3) = 1.585$ bits. Consequently, the amount of information that could be obtained from three weighings is 4.755 bits.

Now imagine that each coin has a number placed on it, from 1 to 9. The outcome of the experiment will be of the form (i, L), indicating, for example, that the ith coin is the counterfeit one and it is lighter (L) than the others. There are 18 possible such outcomes, $\{(i, L), (i, H), i = 1, 2, \ldots, 9\}$ and we assume they are equally likely. The amount of information needed to specify one of the 18 possibilities is

$\log_2(18) = 4.170$ bits

and, since $4.170 < 4.755$, we conclude that it is not impossible to identify the coin in three weighings.

In order to accomplish this, however, we should maximize the information obtained from each weighing. Let p_l, p_b, and p_r be the probabilities that the balance tips left, balances, and tips right, respectively. The amount of information gained from each weighing then is

$$H(X) = -p_l \log_2 p_l - p_b \log_2 p_b - p_r \log_2 p_r$$

and it has been commented on (see also Problem 15.5) that this quantity is maximized when $p_l = p_b = p_r = \frac{1}{3}$.

Suppose that n coins are placed in the left pan and n in the right pan, $n \leq 4$. The following probabilities are easily calculated.

p_l = probability balance tips left
  = probability that coin is heavier and is among the n placed on left pan + probability that coin is lighter and is among the n placed on right pan.

$$= \frac{1}{2} \cdot \frac{n}{9} + \frac{1}{2} \cdot \frac{n}{9} = \frac{n}{9} = p_r$$

$$p_b = \frac{(9 - 2n)}{9}$$

* Adapted from *Information Theory; An Introduction for Scientists and Engineers*, by G. Raisbeck, MIT Press, Cambridge, Massachusetts, 1964, with permission of G. Raisbeck and the MIT Press.

In order to maximize the information obtained from the first weighing, each probability must equal $\frac{1}{3}$ and three coins must be placed in each pan. Let us call those in the left pan coins 1, 2, and 3, those in the right pan 4, 5, and 6, and the remaining ones 7, 8, and 9. There are three cases to consider:

(i) Scales balance. In this case the counterfeit coin is among 7, 8, and 9 and coins 1 to 6 are known good. The same calculation applied to the second weighing requires a $\frac{1}{3}$ probability of a balance, which requires that one of the three coins—say number 9—be *omitted* from the second weighing. Repeating the tactic of the first weighing, one could weigh 7 against 8. If this balances, 9 is the bad coin and one more weighing will tell whether it is lighter or heavier. If 7 does not balance 8, then one more weighing of either against 9 will tell whether 7 or 8 is bad and whether it is heavier or lighter.

(ii) If the scales do not initially balance (say they tilt to the left), then 7, 8, and 9 are good coins. To produce a probability of $\frac{1}{3}$ of a balance on the next weighing, it is necessary to remove two coins entirely—say 1 and 4. To produce a probability of $\frac{1}{3}$ that the balance tips to the right on the next weighing, it suffices to exchange 2 and 5 and leave 3 and 6 alone. If scales now balance, then it is known that either 1 or 4 is counterfeit and one more weighing determines which one and whether it is heavier or lighter. If scales now tip right, then either 2 or 5 is the counterfeit coin while if they still tip left 3 or 6 is the counterfeit coin. In either case one more weighing will find it.

(iii) If scales initially tip right, the case can be treated as in (i).

EXAMPLE 15.6

A joint ensemble $X \times Y$ is generated in the following manner: The X ensemble is $(x_1, .4), (x_2, .6)$ and the Y ensemble is found as follows. If x_1 occurs, then with probability .9 y_1 occurs and with probability .1 y_2 occurs. If x_2 occurs, then with probability .2 y_1 occurs and with probability .8 y_2 occurs. Calculate the quantities $H(X), H(Y), H(X, Y), H(X|Y), H(Y|X), I(X, Y)$ and verify Eq. (15.4). Verify also

$$H(X, Y) = H(X|Y) + H(Y) = H(Y|X) + H(X)$$

The generation of the probabilities for the joint ensemble $X \times Y$ can be viewed as in the diagram of Figure 15.3. The probability of y_1, q_1 is calculated as

$$q_1 = .4 \times .9 + .6 \times .2 = .48$$

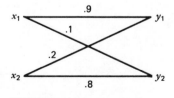

FIGURE 15.3

Similarly, the probability for y_2, q_2 is

$$q_2 = .4 \times .1 + .6 \times .8 = .52$$

The Y ensemble is then $\{(y_1, .48), (y_2, .52)\}$ and the corresponding entropies for the X and Y ensembles are

$$H(X) = -(.4) \log_2(.4) - (.6) \log_2(.6) = .971$$

and

$$H(Y) = -(.48) \log_2(.48) - (.52) \log_2(.52) = .999$$

The joint ensemble is

$$((x_1, y_1), .36), ((x_1, y_2), .04), ((x_2, y_1), .12), ((x_2, y_2), .48)$$

and its entropy is

$$H(X, Y) = -(.36) \log_2(.36) - (.04) \log_2(.04)$$
$$- (.12) \log_2(.12) - (.48) \log_2(.48)$$
$$= 1.592$$

The remaining probabilities and entropies are calculated in a like manner and

$$H(X \mid Y) = -\sum_{i=1}^{2} \sum_{j=1}^{2} p_{ij} \log_2\left(\frac{p_{ij}}{q_j}\right) = .593$$

$$H(Y \mid X) = -\sum_{i=1}^{2} \sum_{j=1}^{2} p_{ij} \log_2\left(\frac{p_{ij}}{p_i}\right) = .621$$

$$I(X, Y) = \sum_{i=1}^{2} \sum_{j=1}^{2} p_{ij} \log_2\left(\frac{p_{ij}}{p_i q_j}\right) = .378$$

Now to check that

$$H(X, Y) = H(X \mid Y) + H(Y)$$

is straightforward:

$$H(X, Y) = -\sum_{i} \sum_{j} p_{ij} \log_2(p_{ij})$$

$$= -\sum_{i} \sum_{j} p_{ij} \log_2\left(\frac{p_{ij}}{q_j} \cdot q_j\right)$$

$$= -\sum_{i} \sum_{j} p_{ij} \log\left(\frac{p_{ij}}{q_j}\right) - \sum_{j} \left(\sum_{i} p_{ij}\right) \log_2(q_j)$$

and, since $\sum_i p_{ij} = q_j$, the result follows, and similarly

$$H(X, Y) = H(Y \mid X) + H(X) \qquad (15.5)$$

Substituting values from above gives the identity $1.592 = .621 + .971$. Equation (15.4) states that

$$H(X, Y) = H(X \mid Y) + H(Y \mid X) + I(X, Y) \tag{15.6}$$

and substituting values obtained yields the identity $1.592 = .593 + .621 + .378$. It is also simple to verify the other relations derived.

The method of generating the joint ensemble (X, Y) in this last example will be used in Section 15.3 where the ensemble X will be interpreted as the set of input symbols to a transmission channel, the ensemble Y will be the set of output symbols of the channel, and the probabilities p_{ij}, the probability that symbol x_i was transmitted and y_j received. The probabilities p_{ij} account for distortion or noise on the channel. Usually, but not always, the input ensemble X and the output ensemble Y will have the same number of symbols in them (in fact, have the same symbols). This subject is discussed further in Section 15.3.

15.2 Discrete Sources and the First Coding Theorem

An interesting problem in communications and other areas is the efficient representation of data. As an example, consider the Morse code where the letters of the alphabet and numerals are each encoded into strings of marks and spaces (or dots and dashes). We will discuss the construction and properties of such codes in this section.

To define the problem more concisely consider an ensemble $(X, p) = \{(x_1, p_1), \ldots, (x_K, p_K)\}$. We want to encode the ith symbol into a sequence of binary digits 0 and 1. We could encode with alphabets larger than $\{0, 1\}$ but most of the essential ideas are contained in the binary coding problem. As an example suppose we wish to encode the ensemble representing the outcomes of a die throwing experiment, that is, $(X, p) = \{(1, \frac{1}{6}), (2, \frac{1}{6}), \ldots, (6, \frac{1}{6})\}$. One such code might be

$$
\begin{aligned}
x_1 &= 1 \to 0 \\
x_2 &= 2 \to 1 \\
x_3 &= 3 \to 00 \\
x_4 &= 4 \to 01 \\
x_5 &= 5 \to 001 \\
x_6 &= 6 \to 011
\end{aligned}
\tag{15.7}
$$

In the general ensemble (X, p), if the ith symbol x_i is encoded into a binary sequence containing l_i digits, then we define the average length of the code as

$$\bar{l} = \sum_{i=1}^{K} p_i l_i$$

The average length of the code in Eq. (15.7) is

$$\bar{l} = \tfrac{1}{6} \cdot 1 + \tfrac{1}{6} \cdot 1 + \tfrac{1}{6} \cdot 2 + \tfrac{1}{6} \cdot 2 + \tfrac{1}{6} \cdot 3 + \tfrac{1}{6} \cdot 3 = 2$$

The code of Eq. (15.7), however, turns out not to be a good code. For example, if we encode the sequence $x_3 x_4 x_2$ the result is

$$x_3 x_4 x_2 \quad 00|01|1$$

On receiving the sequence 00011, however, the decoder is in trouble. It would be unable to decide whether the transmitter sent $x_3 x_4 x_2$ or $x_3 x_6$ since we do not want to send a space between each symbol. Actually, a space could be interpreted as just another symbol x_0, say, so there is no loss in considering the problem without spaces. Clearly, one property that we would like any code that we construct to have is that it be uniquely decodeable, that is, any sequence of received bits (zeros or ones) that represents the encoding of a sequence $x_{i_1} x_{i_2} \cdots$ can be decoded uniquely into this same sequence. We will consider a slightly stronger property, namely that as soon as the binary sequence representing an ensemble symbol is received it can be decoded. We call a code with this property instantaneously decodeable. It turns out that not all codes which are uniquely decodeable are instantaneously decodeable, but every instantaneously decodeable code is uniquely decodeable.

We now state the problem. For a given ensemble (X, p) we want to construct an instantaneously uniquely decodeable code with minimum average length. The following example illustrates one procedure that is efficient in that it results in codes which are instantaneously and uniquely decodeable but whose average length is not in general minimum.

EXAMPLE 15.7 (Shannon–Fano Encoding)

Determine an instantaneously and uniquely decodeable code for the ensemble $(X, p) = \{(x_1, .3), (x_2, .15), (x_3, .15), (x_4, .1), (x_5, .1), (x_6, .1), (x_7, .05), (x_8, .05)\}$.

The Shannon–Fano encoding procedure is as follows: List all ensemble symbols in order of decreasing probability. Divide the list into two parts such that the sum of the probabilities of the ensemble symbols in each part is as close to $\tfrac{1}{2}$ as possible. As the first code symbol assign a zero to all ensemble symbols in the top part and a one to all those in the lower part. Divide each part as before and repeat the process until each part contains only a single ensemble symbol. The procedure is best illustrated by example as in Figure 15.4.

The code is instantaneously decodeable and has an average length

$$\bar{l} = .3 \times 2 + .15 \times 2 + .15 \times 3 + .1 \times 3 + .1 \times 4 + .1 \times 4 + .05 \times 4 + .05 \times 4$$

$$= 2.85$$

Every binary code can be represented as a binary tree as shown in Figure 15.5 for the code of Example 15.7. Moving from one node in the tree to a node

x_1	.3	0	0		
x_2	.15	0	1		
x_3	.15	1	0	0	
x_4	.1	1	0	1	
x_5	.1	1	1	0	0
x_6	.1	1	1	0	1
x_7	.05	1	1	1	0
x_8	.05	1	1	1	1

FIGURE 15.4

on the right adds either a zero or a one to the sequence depending on whether the upper or lower branch is taken. The end points of the tree are the code words. Such tree representations are useful in discussing codes with the prefix property. A binary sequence $\underline{u} = (u_1, u_2, \ldots, u_r)$ is called a prefix of the binary sequence $\underline{v} = (v_1, v_2, \ldots, v_s)$, $s > r$ if $\underline{v}$ is of the form $\underline{v} = (u_1, u_2, \ldots, u_r, v_{r+1}, \ldots, v_s)$, that is, if the first r bits of $\underline{v}$ is the sequence $\underline{u}$. A code for the ensemble (X, p) is called a prefix code if no code word is the prefix of another code. If a code is a prefix code, then it simply means that in tracing along the code tree (representing the code) from the origin (leftmost point) to a node representing a code word, no other code word is encountered. The code constructed in Example 15.7 is a prefix code.

It is possible to prove several properties of prefix codes that are always instantaneously and uniquely decodeable. Suppose a prefix code for the ensemble $(X, p) = \{(x_1, p_1), \ldots, (x_K, p_K)\}$ has been constructed and the code

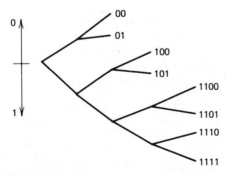

FIGURE 15.5

word for x_i has length l_i, $i = 1, 2, \ldots, K$. Then it is true that

$$\sum_{i=1}^{K} 2^{-l_i} \leq 1 \tag{15.8}$$

and this inequality is referred to as the Kraft inequality. Conversely, if the code lengths of any code for the ensemble (X, p) satisfy Eq. (15.8), then a prefix code with these lengths can be constructed. Although the proofs of these statements are not difficult, we omit them.

The Shannon–Fano encoding procedure of Example 15.7 always yields a prefix code, but is generally not optimum; that is, it is often possible to determine a prefix code with smaller average length. Before giving a procedure which always results in an optimum code (the Huffman procedure of Example 15.8), the so-called first coding theorem or noiseless coding theorem is discussed.

The first coding theorem relates these ideas to the entropy of the ensemble. Before stating it note that we are, in general, interested in sequences of ensemble symbols and their encoding. In particular, consider $X \times X \times \cdots \times X = X^n$, the n-fold Cartesian product that we can consider as the set of all n-tuples over X or all sequences containing n-symbols from X. The probability of an element in X^n is simply the product of the probability of elements in its sequence, that is, if the element is $x_{i_1} x_{i_2} \cdots x_{i_n}$, then its probability is $p_{i_1} p_{i_2} \cdots p_{i_n}$. This is a consequence of the fact that we have assumed the source produces symbols independently. The entropy of X^n is, by Eq. (15.2), $nH(X)$, that is

$$H(X^n) = H(X) + H(X) + \cdots + H(X) = nH(X) \tag{15.9}$$

It can be shown that for any discrete ensemble $(Y, q) = \{(y_1, q_1), \ldots, (y_L, q_L)\}$ an instantaneously decodeable binary code can be constructed with average length $\bar{l}$ that satisfies the inequalities

$$H(Y) \leq \bar{l} < H(Y) + 1 \tag{15.10}$$

where $H(Y)$ is the entropy of (Y, q). Thus for the (X, p) ensemble in particular we can construct an instantaneously decodeable code with average length $\bar{l}_1$ satisfying

$$H(X) \leq \bar{l}_1 < H(X) + 1$$

Consider now a new ensemble $X \times X = X^2$. By Eq. (15.9) the entropy of this ensemble (assuming independence) is $2H(X)$ and by Eq. (15.10) we can construct a new code (different from that for X) with average length $\bar{l}_2$ satisfying

$$2H(X) \leq \bar{l}_2 < 2H(X) + 1$$

or

$$H(X) \leq \frac{\bar{l}_2}{2} < H(X) + \tfrac{1}{2}$$

In general, we can construct a code for X^n with average distance $\bar{l}_n$ that satisfies

$$H(X) \le \frac{\bar{l}_n}{n} < H(X) + \frac{1}{n} \tag{15.11}$$

THEOREM 15.1   (*The First Coding Theorem*)

It is possible to construct an instantaneously decodeable code for the discrete memoryless source with ensemble (X, p) such that the average length of the code per ensemble symbol, $\bar{l}_n/n$, approaches $H(X)$, that is,

$$\lim_{n \to \infty} \frac{\bar{l}_n}{n} = H(X)$$

The theorem follows from Eq. (15.11) by allowing n to approach infinity and noting that $\bar{l}_n$ is the average length of code word assigned to a string of n ensemble symbols from X. Hence $\bar{l}_n/n$ has the meaning of the average number of code symbols per ensemble symbol. The best code, in the sense of minimum average length $\bar{l}_1$, for the ensemble X has an average length somewhere between $H(X)$ and $H(X)+1$. The best code for X^n has an average length per ensemble symbol $\bar{l}_n/n$ somewhere between $H(X)$ and $H(X)+1/n$, and $\bar{l}_n/n \le \bar{l}_1$. The price that we pay for such improved performance can be seen as follows. The number of code words required for X is K while the number of code words required for X^n is K^n and it is, in general, more difficult to construct codes for the larger number of code words. In addition, when a code word for an element $\mathbf{x} = (x_{i_1} x_{i_2} \cdots x_{i_n}) \in X^n$ is transmitted, the receiver must await until the entire code word is received until it can begin to decode, even for the first symbol x_{i_1}. Thus even though codes for X^n can be more efficient than those for X, this is achieved at the cost of complexity and delay. This type of compromise is typical of those encountered in many communication systems.

For coding over an alphabet of D symbols, say $\{0, 1, \ldots, D-1\}$, the first coding theorem becomes

$$\lim_{n \to \infty} \frac{\bar{l}_n}{n} = \frac{H(X)}{\log_2 D}$$

where, as before, $\bar{l}_n$ is the average code word length for X^n over this alphabet. Only binary alphabets will be used in this chapter although some of the problems consider extensions to other alphabets.

The next example gives an encoding method that results in instantaneously decodeable codes with the smallest possible average length, that is, an optimal coding procedure. Thus for a given ensemble (X, p) the best possible average length $\bar{l}$ is such that $H(X) \le \bar{l} < H(X)+1$, but the lower bound of $H(X)$ on $\bar{l}$ may not be attainable. The lowest possible average length is always attained by the procedure of the next example, called the Huffman encoding procedure.

EXAMPLE 15.8 (Huffman Encoding)

Use the Huffman encoding procedure to find the minimum possible average length code for the ensemble $(X, p) = \{(x_1, .4), (x_2, .2), (x_3, .2), (x_4, .1), (x_5, .1)\}$.

The first step in the Huffman encoding procedure is, as with the Shannon–Fano procedure, to list the ensemble in order of decreasing probability, as in step I of Figure 15.6. The two ensemble elements of lowest probability are assigned a zero and a one, arbitrarily. At step II these two elements are thought of as being combined into a single element, with a probability equal to the sum of the two original probabilities. This new element is then inserted into the list in step II at the appropriate place and the procedure is repeated until an ensemble with only one element is reached. The code word for one of the original ensemble symbols is found by tracing the sequence of bits assigned to it and to its successors. While being difficult to express in words, the procedure is actually quite simple and best understood by studying Figure 15.6.

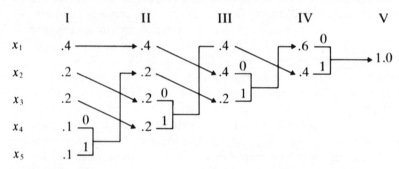

FIGURE 15.6

The code words resulting from the procedure in Figure 15.6 are

x_1	0 0
x_2	1 0
x_3	1 1
x_4	0 1 0
x_5	0 1 1

and the average length of the code is

$$\bar{l} = .4 \times 2 + .2 \times 2 + .2 \times 2 + .1 \times 3 + .1 \times 3 = 2.2$$

and no other instantaneously decodeable code will have a shorter average length.

When going from one step to the next in the Huffman encoding procedure there is some ambiguity possible. This can sometimes result in different code word lengths and quite a different code. The average code word length using this procedure is, however, always the same for the same ensemble. The next example illustrates this point.

EXAMPLE 15.9

Construct two different Huffman codes for the ensemble

$$(X, p) = \{(x_1, .55), (x_2, .15), (x_3, .15), (x_4, .10), (x_5, .05)\}$$

Using the procedure of Example 15.8 the two codes are constructed as in Figure 15.7:

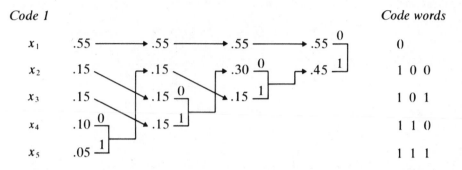

Code 1   *Code words*

	Code words
x_1 .55 → .55 → .55 → .55 0	0
x_2 .15 → .15 → .30 0 → .45 1	1 0 0
x_3 .15 → .15 0 → .15 1	1 0 1
x_4 .10 0 → .15 1	1 1 0
x_5 .05 1	1 1 1

Code 2   *Code words*

	Code words
x_1 .55 → .55 → .55 → .55 0	0
x_2 .15 → .15 → .30 0 → .45 1	1 0
x_3 .15 → .15 0 → .15 1	1 0 0
x_4 .10 0 → .15 1	1 0 1 0
x_5 .05 1	1 0 1 1

FIGURE 15.7

The average code word lengths are:

for code 1: $\bar{l}_1 = 1 \times .55 + 3 \times .15 + 3 \times .15 + 3 \times .10 + 3 \times .05 = 1.90$
for code 2: $\bar{l}_2 = 1 \times .55 + 2 \times .15 + 3 \times .15 + 4 \times .10 + 4 \times .05 = 1.90$

With a table of logarithms to the base 2 we can calculate the entropy of the ensemble as

$$H(X) = -\sum_{i=1}^{5} p_i \log p_i = 1.844$$

and we verify that $H(X) \le \bar{l} < H(X) + 1$.

Under certain conditions we can achieve the lower bound $\bar{l} = H(X)$ as the next example shows.

EXAMPLE 15.10

Find the minimum average length for an instantaneously decodeable code for the ensemble $(X, p) = \{(x_1, .25), (x_2, .25), (x_3, .125), (x_4, .125), (x_5, .125), (x_6, .0625), (x_7, .0625)\}$.

A Huffman code for this ensemble is constructed:

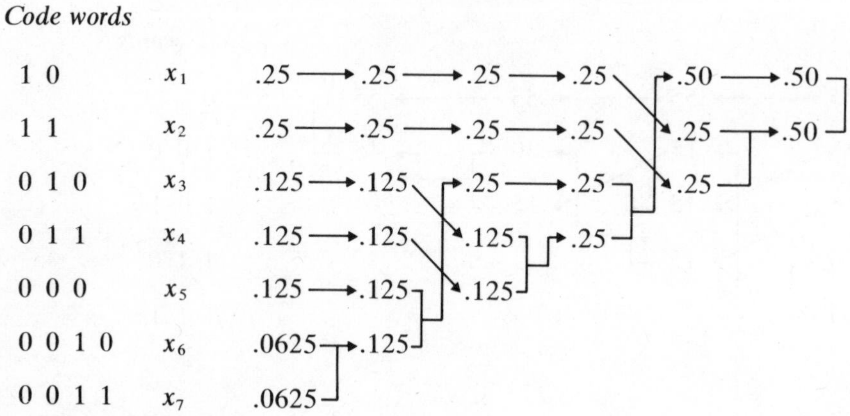

Code words

1 0	x_1	.25 → .25 → .25 → .25 → .50 → .50
1 1	x_2	.25 → .25 → .25 → .25 → .25 → .50
0 1 0	x_3	.125 → .125 → .25 → .25 → .25
0 1 1	x_4	.125 → .125 → .125 → .25
0 0 0	x_5	.125 → .125 → .125
0 0 1 0	x_6	.0625 → .125
0 0 1 1	x_7	.0625

FIGURE 15.8

The average length of the code is

$$\bar{l} = 2 \times .25 + 2 \times .25 + 3 \times .125 + 3 \times .125 + 3 \times .125 + 4 \times .0625 \\ + 4 \times .0625$$

$$= 2.625$$

and the entropy of the ensemble is

$$H(X) = -\sum_{i=1}^{7} p_i \log p_i = 2.625$$

and, in this case, $H(X) = \bar{l}$.

We might well ask under what conditions it will be true that $H(X) = \bar{l}$. Suppose (X, p) is such that the probability of x_i is 2^{-l_i}, l_i a positive integer, $i = 1, \ldots, K$. In this case

$$\sum_{i=1}^{K} 2^{-l_i} = 1$$

and this implies from Eq. (15.8) that we can construct a prefix code such that the length of the code word associated with x_i (which has probability 2^{-l_i}) is l_i.

For such a code the average length $\bar{l}$ is

$$\bar{l} = \sum_{i=1}^{K} \frac{1}{2^{l_i}} \cdot l_i$$

and the entropy is

$$H(X) = -\sum_{i=1}^{K} \left(\frac{1}{2^{l_i}}\right) \log\left(\frac{1}{2^{l_i}}\right) = \sum_{i=1}^{K} \frac{1}{2^{l_i}} \cdot l_i$$

and hence, in this case, $H(X) = \bar{l}$.

EXAMPLE 15.11

For the ensemble $(X, p) = \{(x_1, .41), (x_2, .17), (x_3, .17), (x_4, .17), (x_5, .08)\}$ construct a Shannon–Fano and a Huffman code and determine their average lengths.

The Shannon–Fano code is

x_1	.41	0	0	
x_2	.17	0	1	
x_3	.17	1	0	
x_4	.17	1	1	0
x_5	.08	1	1	1

FIGURE 15.9

and the average code word length is $\bar{l} = 2.42$. The Huffman code is

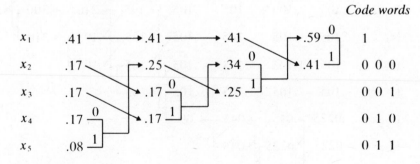

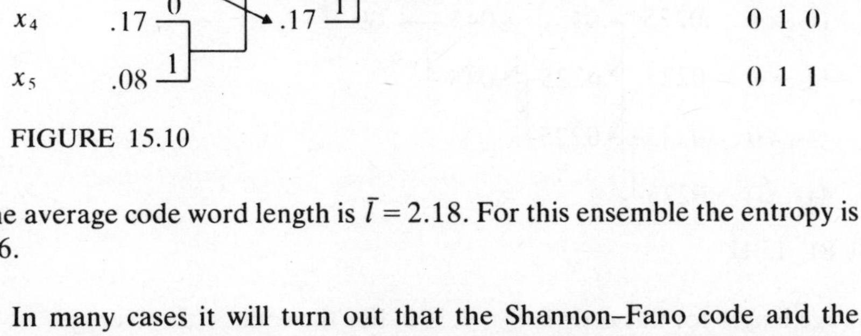

FIGURE 15.10

and the average code word length is $\bar{l} = 2.18$. For this ensemble the entropy is 2.1226.

In many cases it will turn out that the Shannon–Fano code and the optimum Huffman code have the same average length. It is an interesting exercise to try and construct codes for which this is not true, as above.

The following example demonstrates in a simple manner the meaning of the first coding theorem.

EXAMPLE 15.12

Find the minimum possible average length for an instantaneous code for the ensemble $(X, p) = \{(x_1, .7), (x_2, .15), (x_3, .15)\}$. Find the minimum possible average length for an instantaneous code for the ensemble X^2.

The Huffman code for X is:

Code words
x_1 .7 — 0
x_2 .15 — 1 0
x_3 .15 — 1 1
$\bar{l}_1 = .7 + 2\times.15 + 2\times.15 = 1.3$

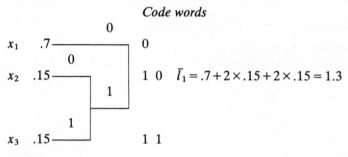

FIGURE 15.11

The Huffman code for X^2 is:

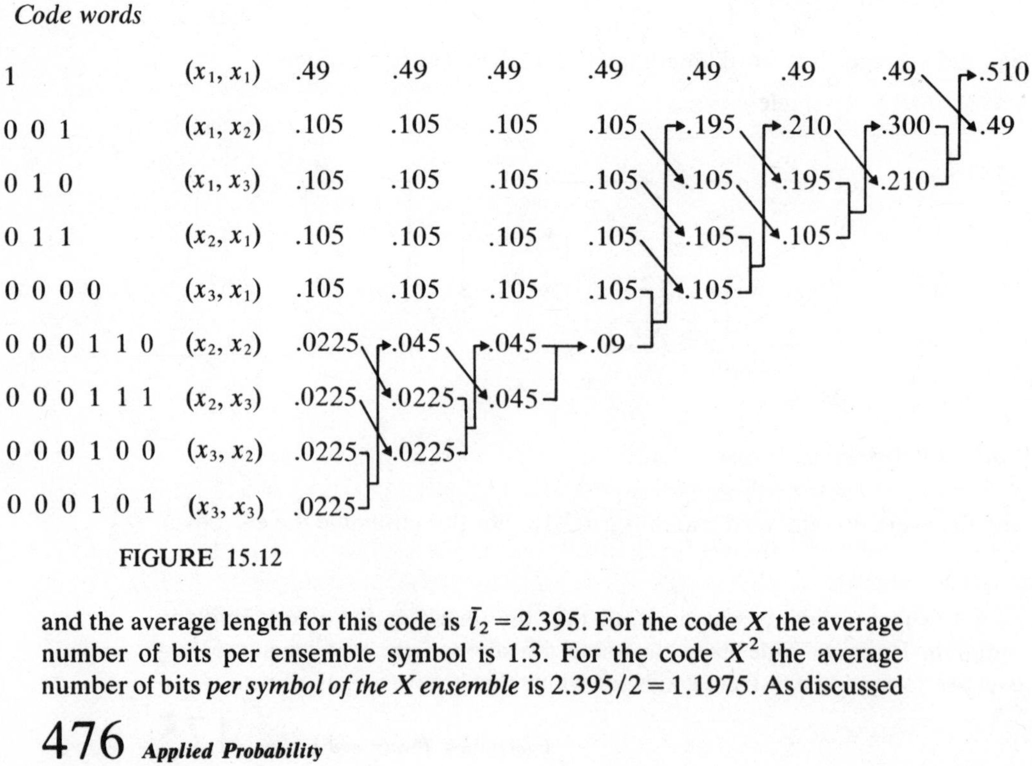

Code words
1 (x_1, x_1) .49 .49 .49 .49 .49 .49 .49 .510
0 0 1 (x_1, x_2) .105 .105 .105 .105 .195 .210 .300 .49
0 1 0 (x_1, x_3) .105 .105 .105 .105 .105 .195 .210
0 1 1 (x_2, x_1) .105 .105 .105 .105 .105 .105
0 0 0 0 (x_3, x_1) .105 .105 .105 .105 .105
0 0 0 1 1 0 (x_2, x_2) .0225 .045 .045 .09
0 0 0 1 1 1 (x_2, x_3) .0225 .0225 .045
0 0 0 1 0 0 (x_3, x_2) .0225 .0225
0 0 0 1 0 1 (x_3, x_3) .0225

FIGURE 15.12

and the average length for this code is $\bar{l}_2 = 2.395$. For the code X the average number of bits per ensemble symbol is 1.3. For the code X^2 the average number of bits *per symbol of the X ensemble* is $2.395/2 = 1.1975$. As discussed in the text this increase in efficiency is achieved at the cost of a more complex code and greater delay in decoding. To further elaborate on this point, if we wanted to encode an ensemble sequence, say $x_1 x_2 x_1 x_1$, using the code of X this would be $0|10|0|0$ and using the code for X^2 this would be $0\ 0\ 1|1$. We could, of course, construct a code for X^n and achieve perhaps a lower average code length per ensemble symbol. The absolute lower limit is $H(X) = 1.181$ and so coding X^n for $n > 2$ will simply lead to greater complexity and delay without a significant increase in efficiency.

EXAMPLE 15.13

A joint ensemble (X, Y) has probabilities represented by the matrix

	y_1	y_2	y_3	y_4
x_1	.4	.02	.02	.02
x_2	.05	.03	.06	.4

Compare the following two strategies for coding this joint ensemble over a binary alphabet: (i) the Huffman codes for the X and Y ensemble separately found and used in concatenation as a code for (X, Y); (ii) the Huffman code for (X, Y) directly is used.
  (i) The Huffman code for the X ensemble is just

x_2	.54	0
x_1	.46	1

and for the Y ensemble is

		I	II	III	IV
1	y_1	.45 ⟶ .45		.55 ⟶ 1	
0 0	y_4	.42 ⟶ .42		.45	
0 1 0	y_3	.08	.13		
0 1 1	y_2	.05			

The average length for the X code is 1 and that for the Y code is 1.68. The average length of the concatenated code (for example the element (x_2, y_3) of (X, Y) is coded as 0 010) is then 2.68.
  (ii) The Huffman code for the (X, Y) ensemble is shown in Figure 15.13 and it has average length 2.09. This is slightly better than the average length of 2.68 obtained with method (i). This was to be expected since combining codes as with method (i) leads to some obvious inefficiency. This would in general be true even if the X and Y ensembles were independent.

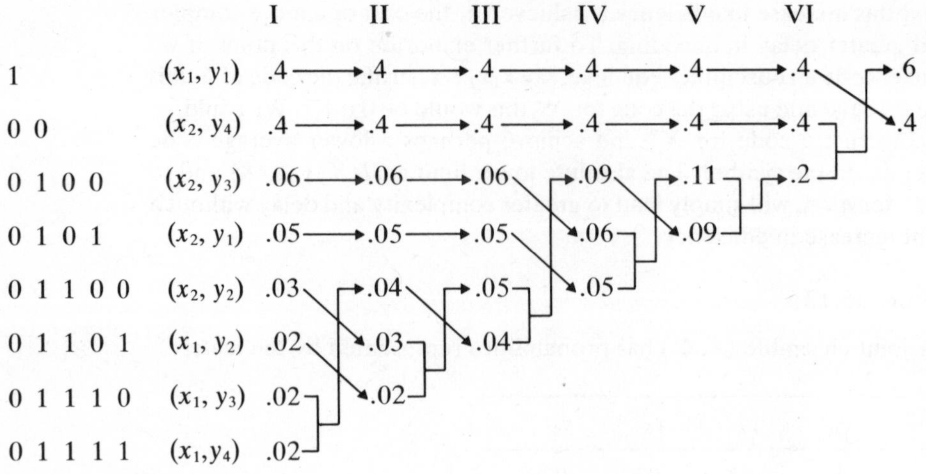

		I	II	III	IV	V	VI	
1	(x_1, y_1)	.4 →	.4 →	.4 →	.4 →	.4 →	.4 ↘	.6
0 0	(x_2, y_4)	.4 →	.4 →	.4 →	.4 →	.4 →	.4 ↘	.4
0 1 0 0	(x_2, y_3)	.06 →	.06 →	.06 ↘	.09 ↗	.11 ↗	.2	
0 1 0 1	(x_2, y_1)	.05 →	.05 →	.05 ↘	.06 ↘	.09		
0 1 1 0 0	(x_2, y_2)	.03 ↗	.04 ↗	.05 ↘	.05			
0 1 1 0 1	(x_1, y_2)	.02 ↗	.03 ↘	.04				
0 1 1 1 0	(x_1, y_3)	.02 ↘	.02					
0 1 1 1 1	(x_1, y_4)	.02						

FIGURE 15.13

One interpretation of the first coding theorem is that the entropy of an ensemble is simply the number of bits per ensemble symbol required to encode the source. There are many generalizations to the encoding. For example, we might want to encode using more than the two code symbols, such as, {0, 1, 2}. Such a problem is not really any more difficult than the binary coding problem and the interested reader is referred to Reza (1951) and Abramson (1963) for further reading. This situation is also considered in the problems. We note here only that determining the number of unused nodes at the initial stage is the only difficulty and this number is $a - 2$ minus the remainder of $m - 2$ when divided by $a - 1$, where a is the size of the alphabet and m is the size of the ensemble being encoded.

15.3 Discrete Channels and the Second Coding Theorem

In the previous section we were concerned with the problem of efficiently representing data. Here we will briefly discuss the coding of data to detect and correct errors that may occur in transmission. Thus the term code in this section has a totally different meaning from that held in the previous section.

We begin by introducing the binary symmetric channel. A simple model of a communication system would be to imagine a black box that emits either a zero or one at each instant of time, independently and with equal probability. These bits are to be transmitted over some channel or transmission medium (e.g., telephone line, atmosphere, optical fiber, etc.) for use by the receiver. We will *model* the effect of the channel by assuming that the channel either leaves a bit unchanged with probability $q = 1 - p$ or changes the bit (either from 0 to 1 or 1 to 0) with probability p. It does this independently for each bit as it traverses the channel. It is as though there is a genie in the channel who, as each bit is presented for transmission, performs some experiment. If the experiment is a success (with probability p), he changes the bit. This channel can be represented as in the diagram in Figure 15.14. For obvious reasons this is referred to as a binary symmetric channel (BSC). Whether or not this is an accurate model for the particular channel in use will depend on many factors. It is quite a good model for a digital communication system on a space channel but quite a poor model for digital transmission over a telephone line.

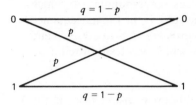

FIGURE 15.14

If we transmit a single bit through the BSC, the probability that it will be received in error is p. Suppose now that this probability of error is unacceptably high. We might adopt a different strategy in order to improve the performance. For example, suppose a "0" is presented for transmission. Rather than send a single 0 we transmit three zeros. Similarly, rather than transmit a single 1 we transmit 111. For each information bit presented we send three channel bits, shown in Figure 15.15. This will slow down the rate at which information bits

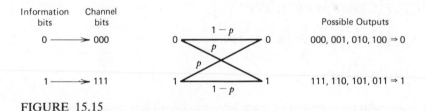

FIGURE 15.15

are being sent, but it should, intuitively at least, improve the accuracy. We analyze the situation. Regardless of which codeword (000 or 111) is transmitted, there are eight possibilities for the channel output. Suppose 110 is received. How are we to decide which of the two code words was transmitted? We can calculate that

$$P(000 \text{ sent} \mid 110 \text{ received}) = p^2(1-p)$$

and

$$P(111 \text{ sent} \mid 110 \text{ received}) = p(1-p)^2$$

For $p < \frac{1}{2}$ we have

$$\frac{p^k(1-p)^{n-k}}{p^{k+1}(1-p)^{n-k-1}} = \frac{1-p}{p} > 1 \tag{15.12}$$

and so $p^k(1-p)^{n-k}$ is a decreasing function of k for fixed n and consequently $0 < p < \frac{1}{2}$. It follows that

$$P(000|110) < P(111|110)$$

and in this case we would decide that a 1 had been transmitted. In Figure 15.15 the possible outputs are divided into two classes corresponding to which decision will be made by the receiver by using the rule: If the binary sequence $\mathbf{r} = (r_1, r_2, r_3)$ is received and the possible code words are $\mathbf{C}_1 = (a_1, a_2, a_3)$ and $\mathbf{C}_2 = (b_1, b_2, b_3)$, choose code word $\mathbf{C}_1$ as the one which was transmitted if

$$P(\mathbf{C}_1|\mathbf{r}_1) \geq P(\mathbf{C}_2|\mathbf{r})$$

Such a rule is referred to as a maximum likelihood decision rule. If the information bit 0 is transmitted, then the probability that we will correctly decode the received sequence of three bits into a 0 is the probability that it is one of the four sequences 000, 001, 010, 100 and so

$$P(\text{correct decoding}|0 \text{ transmitted}) = q^3 + \binom{3}{1}pq^2 = P(C|0)$$

Similarly, we have

$$P(\text{correct decoding}|1 \text{ transmitted}) = q^3 + \binom{3}{1}pq^2 = P(C|1)$$

and consequently the probability of correct decoding in this situation is

$$P(C) = P(C|0)P(0) + P(C|1)P(1)$$
$$= \left(q^3 + \binom{3}{1}pq^2\right)\frac{1}{2} + \left(q^3 + \binom{3}{1}pq^2\right)\frac{1}{2}$$
$$= q^3 + \binom{3}{1}pq^2$$

If $p = .01$, then, with no coding at all, each information bit is received correctly 99% of the time. If we employ this coding scheme, each information bit is received correctly with probability

$$(.99)^3 + 3(.01)(.99)^2 \approx .9997$$

or about 99.97% of the time. In the first case the probability of making an error is .01 and in this last case the probability of making an error is .0003. The improvement is due to the fact that we are transmitting three channel bits for every information bit.

We now formalize the coding situation. A source produces information bits 0 and 1 equally likely and independently. The coder divides this information stream into blocks of k bits each and note that there are $M = 2^k$ possible distinct blocks. For each k bit block an n bit code word $n > k$ is assigned (by some rule yet to be described) and we denote these n-tuples by $\mathbf{u}_i = (u_1^i, u_2^i, \ldots, u_n^i)$, $u_j^i \in \{0, 1\}$, $i = 1, 2, \ldots, 2^k$. If the n-bit sequence $\mathbf{r}$ is received, then it is decoded into the code word $\mathbf{u}_i$, where

$$P(\mathbf{u}_i | \mathbf{r}) \geq P(\mathbf{u}_j | \mathbf{r}) \quad \text{for all } j \neq i,\ j = 1, \ldots, 2^k \qquad (15.13)$$

In the case of equality, say $P(\mathbf{u}_i | \mathbf{r}) = P(\mathbf{u}_k | \mathbf{r})$, an arbitrary decision is made as to whether $\mathbf{u}_i$ or $\mathbf{u}_k$ was transmitted. This decoding rule will minimize the probability of a decoding error for the source and channel described. We say that such a code has rate $R = k/n$ and we call it an (n, k) code indicating that the code words are of length n and each code word represents k information bits. The rate of the code is just the fraction of information bits that is being transmitted. If the n-tuples $\mathbf{u}_i$ and $\mathbf{r}$ differ in exactly k places, then

$$P(\mathbf{u}_i | \mathbf{r}) = p^k (1-p)^{n-k}$$

by Eq. (15.12). The decoding rule of Eq. (15.13) then says that if $\mathbf{r}$ is received, decode it into that code word that differs from $\mathbf{r}$ in the fewest number of places. This is also called the "nearest neighbor" decoding rule. The following example illustrates the coding and decoding behavior for a particular code.

EXAMPLE 15.14   (The Hamming (7, 4) Code)

The Hamming $(7, 4)$ code is constructed as follows: The information sequence (i_1, i_2, i_3, i_4) is encoded into the code word $(i_1, i_2, i_3, i_4, i_1 + i_2 + i_4, i_1 + i_3 + i_4, i_1 + i_2 + i_3)$, where the addition is modulo 2 with no carry, that is, $1 + 0 = 0 + 1 = 1$, $0 + 0 = 1 + 1 = 0$. Find the probability that a code word will be correctly decoded if it is transmitted on a BSC with a probability $p = 0.01$ of making a bit error.

The complete code is:

	i_1	i_2	i_3	i_4	$(i_1 + i_2 + i_4)$	$(i_1 + i_3 + i_4)$	$(i_1 + i_2 + i_3)$
$\mathbf{u}_1 =$	0	0	0	0	0	0	0
$\mathbf{u}_2 =$	0	0	0	1	1	1	0
$\mathbf{u}_3 =$	0	0	1	0	0	1	1
$\mathbf{u}_4 =$	0	1	0	0	1	0	1
$\mathbf{u}_5 =$	1	0	0	0	1	1	1
$\mathbf{u}_6 =$	0	0	1	1	1	0	1
$\mathbf{u}_7 =$	0	1	0	1	0	1	1
$\mathbf{u}_8 =$	1	0	0	1	0	0	1
$\mathbf{u}_9 =$	0	1	1	0	1	1	0
$\mathbf{u}_{10} =$	1	0	1	0	1	0	0
$\mathbf{u}_{11} =$	1	1	0	0	0	1	0
$\mathbf{u}_{12} =$	0	1	1	1	0	0	0
$\mathbf{u}_{13} =$	1	0	1	1	0	1	0
$\mathbf{u}_{14} =$	1	1	0	1	1	0	0
$\mathbf{u}_{15} =$	1	1	1	0	0	0	1
$\mathbf{u}_{16} =$	1	1	1	1	1	1	1

This code has the remarkable property (which readers can check for themselves) that any binary 7-tuple is either one of these code words or differs from a code word in exactly one position. Furthermore, any two of these code words disagree in at least three places. Thus if $\mathbf{r}$ is a received word, using our nearest neighbor decoding rule, we will decode it as $\mathbf{u}_i$ if either (i) $\mathbf{r} = \mathbf{u}_i$ or (ii) $\mathbf{r}$ differs from $\mathbf{u}_i$ in exactly one position. This decoding rule uniquely decodes any received word $\mathbf{r}$. If a code word $\mathbf{u}_i$ is transmitted, then the probability of decoding correctly is the probability that either no error or exactly one error was made, that is

$$P(C\,|\,\mathbf{u}_i) = (1-p)^7 + \binom{7}{1}p(1-p)^6$$

and the overall probability of correct decoding is

$$P(C) = \sum_{i=1}^{16} P(C\,|\,\mathbf{u}_i)P(\mathbf{u}_i) = P(C\,|\,\mathbf{u}_{16})$$

since $P(C|\mathbf{u}_i)$ does not depend on i and $P(\mathbf{u}_i) = \frac{1}{16}$. For $p = .01$ we calculate that

$$P(C) = .99796$$

and this is the probability that the four information bits will be received correctly. If no coding were employed, then the probability that four information bits transmitted are correctly received is

$$P(C) = (1 - .01)^4 = .96059$$

and, again, the improvement is at the expense of decreased channel throughput.

The code listed here might appear cumbersome to implement, but actually it is not. It can be mechanized very easily as shown in Figure 15.16. At the receiving end we can detect very easily whether or not the seven bits

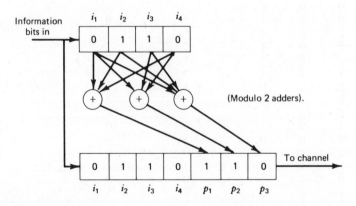

FIGURE 15.16

received are, in fact, a code word. Notice that, modulo 2, the following three equations must be true:

$$i_1 + i_2 + i_4 + p_1 = 0, \quad i_1 + i_3 + i_4 + p_2 = 0, \quad \text{and} \quad i_1 + i_2 + i_3 + p_4 = 0$$

(i.e., an even number of the summands of each equation must be 1). Thus, for the received seven bits $r_1, r_2, \ldots, r_7$ to be a code word, it is only necessary to perform these sums (modulo 2) as shown in Figure 15.17 and if any of them is a 1, the seven bits do not form a code word. Thus an error has been made in which case we can correct it (generally difficult to mechanize) or throw the word away and ask for a retransmission (a typical strategy in computer communications where such codes are often used).

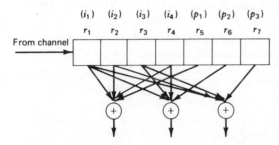

(If any output is a one, an error has been made.)

FIGURE 15.17

The code in this example is termed a parity check code. The four information bits are used to form three parity checks and the seven information plus parity bits are transmitted as a code word. It is referred to as a $(7, 4)$ code and, in general, an (n, k) parity check code has code words of length n of which k are information bits and the remaining $(n - k)$ are parity checks. In this example it was shown that if we use the Hamming $(7, 4)$ code, the probability of correct decoding on a BSC with $p = 0.01$ is .99796. It turns out, and this will be the meaning of the coding theorem, that for certain cases it is possible to decrease the probability of error on a channel by using codes of longer length, but the same throughput. For example, with a $(7, 4)$ code we achieved a probability of error of .00204 (as opposed to .03941 without decoding). If we use a $(7l, 4l)$ code, where l is a positive integer, it is possible to make the probability of error as close to zero as desired by choosing l sufficiently large as long as the rate $R = \frac{4}{7}$ of the code is less than a quantity C, the capacity of the channel, to be defined shortly. Notice in this formulation we are not changing the fraction of information bits to channel bits, that is, the channel throughput is the same. Making the code longer simply allows us to design better codes, for the same rate. Of course, the lower probability of error is achieved at the expense of more complex codes and a larger delay in decoding since the decoder cannot begin to decode until it receives all n bits of the code word.

In preparation for stating the second coding theorem, we define the capacity of a channel referred to previously. Suppose the zeros and ones presented to the channel (forget about coding for the moment) are not equally likely but $p_0 = r$ and $p_1 = s$, $r + s = 1$. We denote the input ensemble by $X = \{(0, r), (1, s)\}$. If these are transmitted on a BSC with a probability of error p, then the probability of an output 0 is

$$p_{0|0} p_0 + p_{0|1} p_1 = (1 - p)r + ps$$

where $p_{j|i}$ is the probability the output is j given the input is i. Similarly, the probability of an output 1 is

$$p_{1|0} p_0 + p_{1|1} p_1 = pr + (1 - p)s$$

Denote the output ensemble by $Y = \{(0, (1 - p)r + ps), (1, pr + (1 - p)s)\}$. The Cartesian product $X \times Y = \{[(i, j), p_{ij}], i, j = 0, 1\}$ is also found since $p_{ij} = p_{j|i} p_i$. Notice that all these quantities depend only on the parameters r, s, and p.

Now suppose that we were free to choose r (and hence $s = 1 - r$) but that the channel crossover probability p (for the BSC) is fixed. After all, we have some control over the frequency of 0's and 1's (and hence r) being presented to the channel but little control over the channel itself. We would like to choose r so that, on each use of the channel, the maximum amount of information is transferred. But the mutual information $I(X, Y)$, given in Section 15.1, has the interpretation of the amount of information that the X ensemble gives about the Y ensemble. It is therefore reasonable to define the channel capacity for the BSC as

$$C = \max_{0 \le r \le 1} I(X, Y)$$

where X and Y refer to the input and output ensembles of the BSC, respectively.

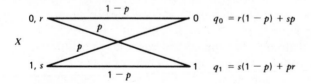

FIGURE 15.18

The calculation of capacity for the BSC is tedious but instructive. The mutual information for the BSC, depicted again in Figure 15.18, is

$$I(X, Y) = \sum_{i=0}^{1} \sum_{j=0}^{1} p_{ij} \log\left(\frac{p_{ij}}{p_i q_j}\right)$$

The joint probabilities are

$$p_{00} = r(1 - p), \quad p_{01} = rp, \quad p_{10} = sp = (1 - r)p, \quad p_{11} = (1 - r)(1 - p)$$

and the mutual information can be written as

$$I(X, Y) = [(1-p)r] \log_2\left(\frac{(1-p)r}{r[(1-p)r+(1-r)p]}\right)$$

$$+ rp \log_2\left(\frac{rp}{r[rp+(1-p)(1-r)]}\right)$$

$$+ [(1-r)p] \log_2\left(\frac{(1-r)p}{(1-r)[r(1-p)+(1-r)p]}\right)$$

$$+ (1-r)(1-p) \log_2\left(\frac{(1-r)(1-p)}{(1-r)[rp+(1-p)(1-r)]}\right)$$

and, on rearranging terms and expanding logarithms,

$$I(X, Y) = p \log p + q \log q - [r(1-p)+(1-r)p] \log[r(1-p)+(1-r)p]$$

$$- [rp+(1-r)(1-p)] \log[rp+(1-r)(1-p)]$$

We are interested in the value of r that maximizes this expression. The logs are to the base 2, but multiplying by $\log_e 2$ will convert them to natural logs and will not alter the value of r that maximizes the expression. The expression for r obtained is complicated but the maximum value that $I(X, Y)$ assumes is

$$C = \max_{0 \le r \le 1} I(X, Y) = 1 + p \log_2 p + (1-p) \log_2(1-p)$$

and this is the capacity of a BSC. For example, when $p = 0.01$, $C = .92$.

In general, the capacity of a channel is defined for an input ensemble $X = \{(x_i, p_i), i = 0, \dots, K-1\}$, and output ensemble $Y = \{(y_j, q_j), j = 0, \dots, L-1\}$, and the joint ensemble $(X, Y) = \{(x_i, y_j, p_{ij}), i = 0, \dots, K-1, j = 0, \dots, L-1\}$ as

$$C = \max_{p_1, \dots, p_K} I(X, Y)$$

that is, the input probability distribution is "matched" to the channel to maximize the amount of information transfer. The situation is shown in Figure 15.19. Generally, the problem of finding the capacity of a channel can be a

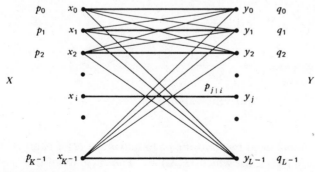

FIGURE 15.19

challenging one. The next example deals with one case that is not too difficult. Channels such as those of Figure 15.18, 15.19, or 15.20 are called discrete memoryless channels. The discrete part refers to the finite number of inputs and outputs. Memoryless refers to the assumption that each time the channel is used, symbols are treated independently.

EXAMPLE 15.15

Find the capacity of the binary erasure channel (BEC) shown in Figure 15.20.

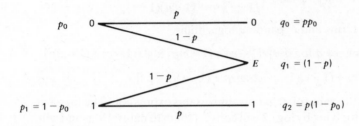

FIGURE 15.20

The mutual information for this channel is given by

$$I(X, Y) = \sum_{i=0}^{1} \sum_{j=0}^{2} p_{ij} \log\left(\frac{p_{ij}}{p_i q_j}\right)$$

and the probabilities p_{ij} are easily calculated as

$$p_{00} = pp_0 \qquad p_{01} = (1-p)p_0 \qquad p_{02} = 0$$
$$p_{10} = 0 \qquad p_{11} = (1-p)(1-p_0) \qquad p_{12} = p(1-p_0)$$

Substituting these values gives

$$I(X, Y) = (pp_0)\log_2\left(\frac{pp_0}{p_0(pp_0)}\right) + [(1-p)p_0]\log_2\left(\frac{(1-p)p_0}{(1-p)p_0}\right)$$

$$+ [(1-p)(1-p_0)]\log_2\left(\frac{(1-p)(1-p_0)}{(1-p)(1-p)}\right)$$

$$+ [p(1-p_0)]\log_2\left(\frac{p(1-p_0)}{(1-p_0)p(1-p_0)}\right)$$

$$= p[-p_0\log_2 p_0 - (1-p_0)\log_2(1-p_0)]$$

$$= pH(X)$$

Since p is fixed, $I(X, Y)$ is maximized by choosing p_0 to maximize $H(X)$ and from Figure 15.1 this value is $p_0 = \frac{1}{2}$, for which $H(X) = 1$. The capacity of the BEC is then p.

The binary erasure channel is a model for a communication system where no decision is made on a received bit unless we are sure that it is either a 0 or a 1. The erasure E simply indicates that we were not sure enough of the received bit to make a decision. In practice, of course, a "bit" is transmitted in some time interval $(0, T)$ on a channel as a voltage waveform (or electromagnetic wave), usually in some modulated form. While in the channel it undergoes various distortions and usually has noise of various types affecting it. It arrives at the receiver sometimes in poor shape and the job of the receiver is to decide, based on the received waveform, whether a "0" or "1" was sent during the interval $(0, T)$. The theory used in system design depends often on the characteristics of the channel. Figure 15.21 shows a simplified block diagram of a binary communication system model. The source emits symbols from the ensemble $X = \{(x_i, p_i), i = 0, \ldots, K - 1\}$ at a certain rate. The coder, according to some strategy, maps each symbol x_i into a string of bits that are then modulated in some manner for transmission on the channel. The combination of modulator, channel, and demodulator is the discrete channel (e.g., BSC or BEC).

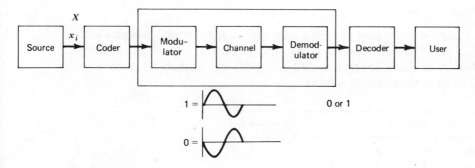

FIGURE 15.21

To tie these ideas with the notion of the capacity of a channel it should first be noted that no concept of time has yet been associated with either the source or the channel. Suppose, in fact, the source emits a certain number of symbols per second and the channel is capable of being used a certain number of times per second. Then the following theorem is perhaps the cornerstone of communication theory. It is called the noisy coding theorem or sometimes Shannon's second coding theorem.

THEOREM 15.2   (*The Second Coding Theorem*)

If an information source has entropy $H(X)$ and a channel has capacity C, then, if $H(X) \leq C$, there exists a coding scheme for which the source output can be transmitted over the channel and recovered with an arbitrarily small probability of error. If $H(X) > C$, it is not possible to transmit and recover the information with an arbitrarily small probability of error.

The proof of the theorem is not very difficult, but is not very illuminating and will be omitted. In the theorem both $H(X)$ and C are in bits per sec. Notice the theorem says that, in spite of noise on the channel, we can communicate essentially without error as long as we do not try to transmit too fast. In the late 1940s when Shannon's paper appeared, this was a revolutionary concept, it being felt that noise on a channel provided a kind of lower limit to the quality of transmission.

The price that one pays for high quality transmission is coding. In a binary communication, as shown in Figure 15.21, the source output is first converted to a string of binary digits, perhaps using a Huffman code, for example. The string of digits is then encoded with an error correcting code, such as the Hamming code discussed in Example 15.14. The coder block of Figure 15.21 generally has to perform both kinds of coding.

To summarize, using probability theory some crucial theorems on various aspects of communications can be demonstrated. For the remainder of the chapter error correcting and detecting codes will be discussed. Consider first the following example.

EXAMPLE 15.16

A code is formed by grouping information symbols into 3-tuples and adding an overall parity check. Find the probability that errors will go undetected if these code words are transmitted on a BSC with probability of error p.

The code words are as follows:

	i_1	i_2	i_3	$i_1 + i_2 + i_3$
$\mathbf{u}_1$	0	0	0	0
$\mathbf{u}_2$	0	0	1	1
$\mathbf{u}_3$	0	1	0	1
$\mathbf{u}_4$	1	0	0	1
$\mathbf{u}_5$	0	1	1	0
$\mathbf{u}_6$	1	0	1	0
$\mathbf{u}_7$	1	1	0	0
$\mathbf{u}_8$	1	1	1	1

and every code word has an even number of ones in it. In this situation we are not interested in correcting errors, we simply detect errors. If a word is received with an odd number of ones in it, then the receiver is alerted to the fact that either one or three errors have occurred and no further action is taken. If two or four errors occur, they are not detected. Thus the probability that errors will be undetected, $P(E)$, is

$$P(E) = \sum_{i=1}^{8} P(E|\mathbf{u}_i)P(\mathbf{u}_i)$$

where $P(\mathbf{u}_i)$ is the probability that $\mathbf{u}_i$ is chosen for transmission and $P(E|\mathbf{u}_i)$ is the probability of undetected errors given that $\mathbf{u}_i$ was transmitted. It is assumed that $P(\mathbf{u}_i) = \frac{1}{8}$ and, for any i, $1 \le i \le 8$,

$$P(E \mid \mathbf{u}_i) = \binom{4}{2} p^2 (1-p)^2 + \binom{4}{4} p^4 = P(E)$$

that is, the probability that either 2 or 4 errors occurred.

If $p = .01$, for example, without coding, on the average 40 out of every 4000 bits transmitted will be in error and the receiver will be unaware of them. In the same time frame that 4000 information bits were sent, 3000 could have been transmitted along with 1000 parity checks. On the average 40 errors will still be made, but the probability of not detecting an error is about $.588 \times 10^{-4}$, that is, on the average less than .25 errors in the 4000 received bits will not be detected.

A more graphical description of these error detecting and correcting codes is possible. By the distance between two binary n-tuples is meant the number of corresponding places in which they disagree. For example, the distance between the two 6-tuples

$$\mathbf{x} = 1 \ \ 0 \ \ 0 \ \ 1 \ \ 1 \ \ 0$$

$$\mathbf{y} = 1 \ \ 1 \ \ 0 \ \ 1 \ \ 0 \ \ 1$$

is three since they have different elements in the second, fifth, and sixth places. We write this as $d(\mathbf{x}, \mathbf{y}) = 3$. We can now view the set of all n-tuples as a space with a distance described on it. For $n = 3$ this is just the cube as shown in Figure 15.22. The distance between any two 3-tuples is the smallest number of edges which must be traversed in order to go from one to the other.

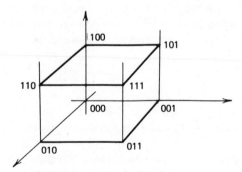

FIGURE 15.22

If an n-tuple $\mathbf{x}$ is transmitted and e errors are made, then, clearly the received n-tuple $\mathbf{r}$ is distance e from $\mathbf{x}$. It is difficult to view the n-dimensional analog of Figure 15.22 but a rough representation is attempted in Figure 15.23 for $e = 4$. Now suppose in the code, no two code words have distance less than

Information Theory and Coding 489

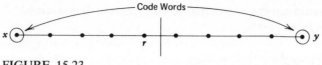

FIGURE 15.23

d. In order for one code word to be transmitted and the corresponding received word **r** end up closer to another code word **y**, more than $(d-1)/2$ errors would have to be made (assume *d* is an odd number). For $d = 9$ the situation is sketched in Figure 15.23. If four or fewer errors are made and **x** transmitted, then **r** will still be closer to **x** than to **y**.

In general, if any two words in a code are distance at least *d* apart, then we can surround each code point with a "sphere" of radius $((d-1)/2)$ (i.e., all points of distance $((d-1)/2)$ or less from the code word) and these spheres will not intersect. The nearest neighbor decoding rule then is to decode received word **r** to code word **x** if it lies in the sphere about **x**. Of course, if more than $((d-1)/2)$ errors are made, **r** may end up in the wrong sphere and a decoding error will be made.

In the above strategy we were interested in actually correcting errors made in transmission. If fewer than $((d-1)/2)$ errors are made using a code with distance *d*, then all the errors using the sphere argument can be corrected. Usually it is quite expensive to correct errors and a more economical strategy is simply to detect the errors. In this scheme we simply try to determine whether or not the received word **r** is a code word. If it is, we assume it is the correct code word and take off the parity bits leaving the information bits. If it is not, we make no attempt to correct the errors but discard the entire word and ask for a retransmission. Such a strategy is commonly employed in practice and can be quite effective. The final example illustrates the procedure, as did Example 15.16 to a lesser degree. It is mentioned in passing that for codes considered here (called linear codes) the minimum distance between any two code words is the same as the weight (number of nonzero positions) of a code word of minimum nonzero weight.

EXAMPLE 15.17

Add an overall parity check to the (7, 4) Hamming code of Example 15.14 and show that the resulting code is capable of correcting any one error but can detect any three errors.

It was stated in Example 15.15 that the (7, 4) code has distance 3 and so adding an overall parity check (to make it an (8, 4) code) will not change that. A careful examination to the code words of that code shows that they either have three or four 1's in them, with the exception of $\mathbf{u}_{16}$ which has seven. A little experimentation will quickly convince the reader that the distance between any two words with three 1's or any two with four 1's, is 4, and the distance between one with three 1's and one with four 1's is at least 3. If we add an overall parity check, the code words with three 1's will have a 1 added, those with four 1's will have a 0 added and $\mathbf{u}_{16}$ will have a 1 added. A little thought will show that the minimum distance between any two code words of the (8, 4) code is now 4. If a code word in transmission incurs four errors, it could result in a valid code word at the receiver. If one, two, or three errors are made, then the received word cannot possibly be a code word and this fact can be detected, although they cannot be corrected (only one error could be corrected).

This chapter has been an ambitious introduction to information theory and coding. Perhaps more important than the results derived and discussed is the recognition of the role that probability theory plays in the subject. Probability is indispensable to the study of advanced communication and information processing systems and the material introduced in this chapter is but one aspect of a broad and interesting area.

Problems

1. A first experiment I consists in selecting a number from 1 to 900 at random. A second experiment II consists in determining its remainder modulo 30. Find the entropy of experiment I and the conditional entropy of I given II.

2. How many bits of information are contained in the statement, "my birthday was a Friday"? How many bits in the statement, "my birthday was Friday, September 12"?

3. A rectangle is divided into 64 squares by 4 vertical and 16 horizontal strips. A point of the rectangle is chosen at random. How much uncertainty is removed by specifying the row number?

4. Two urns each contain 12 balls: urn 1 contains four red, six white, and two blue; urn 2 contains three red, six white, and three blue. If two balls are drawn from each, without replacement, which situation has the greater uncertainty?

5. Show that the maximum entropy for the ensemble $X = \{(x_1, p_1), (x_2, p_2), \ldots, (x_n, p_n)\}$ is achieved when $p_1 = \cdots = p_n = 1/n$.

6. In a particular application of Morse code the probability of a dot is .45, a dash .45, and a space is .10. In a message of 100 such symbols, assumed independent, what is the entropy of the message?

7. Using the same method as in Example 15.5, show that only three weighings are required to determine one counterfeit coin from 11 good coins and whether it is heavier or lighter. A balance, with no weights, is available.

8. Among n coins, one is known to be lighter than the others. What is the minimum number of weighings needed to find the light coin?

9. Using the same method for generating a joint ensemble as employed in Example 15.6, determine the conditions of p_1 and p_2 in the Figure 15.24, for which the X and Y ensembles are independent.

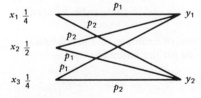

FIGURE 15.24

10. When an ensemble Y is continuous with pdf $f(x)$, the entropy is defined as

$$H(X) = -\int_{-\infty}^{\infty} f(x) \ln[f(x)]\, dx$$

where natural logs are usually used. Find the entropy when $f(x)$ is a normal pdf with mean μ and variance σ^2

11. With the definition of entropy of continuous ensembles as in the previous problem, find the entropy of the one-sided exponential pdf with parameter α.

12. Construct a binary Huffman code for the English alphabet, including a space, with the following probabilities:

Space	A	B	C	D	E	F
.186	.064	.013	.022	.032	.103	.021

G	H	I	J	K	L	M
.015	.047	.058	.001	.005	.032	.020

N	O	P	Q	R	S	T
.057	.063	.015	.061	.048	.051	.080

U	V	W	X	Y	Z
.023	.008	.018	.001	.016	.001

(For Morse code, which uses a dot, dash, and space with one dash taking the same length as three dots and a space taking the same length as a dot, the average length is 6.0 bits per letter.)

13. Construct optimal Shannon–Fano and Huffman binary codes for the ensemble $X = \{(x_1, .7), (x_2, .3)\}$. Repeat for X_2 and X_3 and compare the average word lengths obtained.

14. Which of the following three codes for the ensemble $X = \{x_1, x_2, x_3, x_4, x_5, x_6\}$ has the prefix property:

	I	II	III
x_1	0	111	1
x_2	10	1101	011
x_3	110	1100	010
x_4	1110	10	001
x_5	1011	01	000
x_6	1101	00	110

For those codes that do not, construct a sequence of 0's and 1's that can be decoded ambiguously.

15. Extend the Shannon–Fano binary encoding method of Example 15.7 to construct a ternary (three symbol alphabet 0, 1, and 2) code for the ensemble

$$X = \{(x_1, .2), (x_2, .2), (x_3, .1), (x_4, .1), (x_5, .1), (x_6, .1),$$

$$(x_7, .05), (x_8, .05), (x_9, .05), (x_{10}, .05)\}$$

492 *Applied Probability*

16. Under what condition on the probabilities of an ensemble will the average length of the Huffman code over an alphabet of D symbols be $H(X)/\log_2 D$?

17. For a prefix code with word lengths $l_1, l_2, \ldots, l_K$ over an alphabet with D letters to exist it is necessary and sufficient that $\sum D^{-l_i} \le 1$. (Note that this is a generalization of Eq. (15.8) where an alphabet with two letters, $D = 2$, was assumed.) What is the size of the smallest alphabet that can be used if we require a code with four words of length 2 and five words of length 4?

18. Extend the Huffman coding procedure of Example 15.8 to construct an optimal ternary code (an alphabet of size 3) for the ensemble of Problem 15.15.

19. To indicate that coding can be a little more complicated than as presented here, consider the set of code words 101, 00110, 10111, 11001. This set happens to be uniquely decodeable but it is not a prefix code. Construct two different sequences of code words that are identical in at least the first 11 bits.

20. Construct a Huffman code over an alphabet with four symbols for an ensemble with probabilities .3, .1, .1, .08, .08, .06, .06, .06, .06, .05, .05.

21. Construct a binary Huffman code for an ensemble X with probabilites .25, .25, .125, .125, .0625, .0625, .0625, .0625. Show how you could use this code directly to construct an optimal code for the mth extension of X. Would this procedure be valid for an optimal code of the mth extension of *any* ensemble X?

22. A company wishes to store information on its customers in the form (name, telephone number, monthly bill). If $X = \{\text{names}\}$, $Y = \{\text{telephone numbers}\}$, and $Z = \{\text{monthly bill}\}$ are the three ensembles (assumed independent) with entropies of 13.5, 12.0, and 9.2 bits, how many bits per customer would be required? If the ensembles are not independent (e.g., the customer name uniquely specifies the telephone number) and, in fact, $H(X, Y, Z) = 24.2$ bits, what is the minimum number of bits required per customer?

23. Two binary symmetric channels are placed in cascade, as shown in Figure 15.25. What is the capacity of the overall channel?

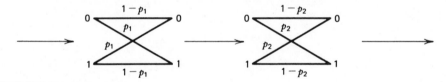

FIGURE 15.25

24. Find the capacity of the channel shown in Figure 15.26.

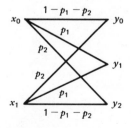

FIGURE 15.26

25. In a $(15, 11)$ code there are four parity checks determined by the following equations:

$$p_1 = i_1 + i_2 + i_4 + i_5 + i_7 + i_9 + i_{11}$$

$$p_2 = i_1 + i_3 + i_4 + i_6 + i_7 + i_{10} + i_{11}$$

$$p_3 = i_2 + i_3 + i_4 + i_8 + i_9 + i_{10} + i_{11}$$

$$p_4 = i_5 + i_6 + i_7 + i_8 + i_9 + i_{10} + i_{11}$$

Show that any two code words have distance at least 3.

26. Convince yourself that adding an overall parity check to the code of Problem 25 gives a code with distance 4.

TABLE 15.1 $-\log_2 p$

	.00	.01	.02	.03	.04	.05	.06	.07	.08	.09
.0	0	6.6439	5.6439	5.0589	4.6439	4.3219	4.0589	3.8365	3.6439	3.4739
.1	3.3219	3.1844	3.0589	2.9434	2.8365	2.737	2.6439	2.5564	2.4739	2.3959
.2	2.3219	2.2515	2.1844	2.1203	2.0589	2	1.9434	1.889	1.8365	1.7859
.3	1.737	1.6897	1.6439	1.5995	1.5564	1.5146	1.4739	1.4344	1.3959	1.3585
.4	1.3219	1.2863	1.2515	1.2176	1.1844	1.152	1.1203	1.0893	1.0589	1.0291
.5	1	0.97143	0.94342	0.91594	0.88897	0.8625	0.8365	0.81097	0.78588	0.76121
.6	0.73697	0.71312	0.68966	0.66658	0.64386	0.62149	0.59946	0.57777	0.55639	0.53533
.7	0.51457	0.49411	0.47393	0.45403	0.4344	0.41504	0.39593	0.37707	0.35845	0.34008
.8	0.32193	0.30401	0.2863	0.26882	0.25154	0.23447	0.21759	0.20091	0.18442	0.16812
.9	0.152	0.13606	0.12029	0.1047	0.089267	0.074001	0.058894	0.043943	0.029146	0.0145

TABLE 15.2 $-p \log_2 p$

	.00	.01	.02	.03	.04	.05	.06	.07	.08	.09
.0	0	0.066439	0.11288	0.15177	0.18575	0.2161	0.24353	0.26856	0.29151	0.31265
.1	0.33219	0.35029	0.36707	0.38264	0.39711	0.41054	0.42302	0.43459	0.44531	0.45523
.2	0.46439	0.47282	0.48057	0.48767	0.49413	0.5	0.50529	0.51002	0.51422	0.5179
.3	0.52109	0.52379	0.52603	0.52782	0.52917	0.5301	0.53062	0.53073	0.53045	0.5298
.4	0.52877	0.52738	0.52565	0.52356	0.52115	0.5184	0.51534	0.51196	0.50827	0.50428
.5	0.5	0.49543	0.49058	0.48545	0.48004	0.47437	0.46844	0.46225	0.45581	0.44912
.6	0.44218	0.435	0.42759	0.41994	0.41207	0.40397	0.39564	0.3871	0.37835	0.36938
.7	0.3602	0.35082	0.34123	0.33144	0.32146	0.31128	0.30091	0.29034	0.27959	0.26866
.8	0.25754	0.24625	0.23477	0.22312	0.21129	0.1993	0.18713	0.17479	0.16229	0.14963
.9	0.1368	0.12382	0.11067	0.097369	0.083911	0.070301	0.056538	0.042625	0.028563	0.014355

APPENDIX A

Review of Set Theory

In this appendix those notions from set theory used in the text are briefly reviewed. No proofs of any statements are given.

A set will be taken to mean a collection of individual entities that we refer to as elements. The element x is said to belong to the set A if x is one of its elements, and this fact is written $x \in A$. If x is not a member of the set A, we write $x \notin A$. In most cases there is some set which contains all the elements under consideration and this is referred to as the universal set, which we denote by S. In a probabilistic experiment it would be a sample space for that experiment.

There are essentially two ways of designating a set; either by listing all of its elements or by some word or mathematical statement on the conditions to be satisfied for membership in the set. For example, if the set A represents the set of possible outcomes when rolling a dice, then we would write

$$A = \{1, 2, 3, 4, 5, 6\}$$

while if the set B was the set of all possible integers, then we might write

$$B = \{x \mid x \text{ is a positive integer}\}$$

where the statement or mathematical condition on the right of the vertical line defines the membership for the set.

A set B is called a subset of A if it is true that $x \in B$ implies $x \in A$ for every $x \in B$. This condition is also described by saying that B is included in A and the notion of set inclusion is written $B \subseteq A$ (or $A \supseteq B$). If $B \subseteq A$ and there is an element $x \in A$ but $x \notin B$, then we write $B \subset A$ and say that B is a proper subset of A. Writing $B \subseteq A$ allows for the possibility that the two sets are equal, that is, contain precisely the same elements.

The empty set or the null set is designated by $\varnothing$ and is the set that contains no elements. It turns out to be a useful notion to have in proving statements about sets. If A is a set in the universal set S, then its complement, denoted by $\bar{A}$, is defined by

$$\bar{A} = \{x \in S \mid x \notin A\}$$

that is, it is the set of all those elements of S that are not in A.

Given two sets A and B, there are several ways of deriving a third set. We define the union of A and B as the set

$$A \cup B = \{x \in S \mid x \in A \text{ or } x \in B \text{ or both}\}$$

The intersection of the two sets is

$$A \cap B = \{x \in S \,|\, x \in A \text{ and } x \in B\}$$

If $A \cap B = \varnothing$, then we say that A and B are disjoint sets. The symmetric difference of the sets A and B is

$$A \triangle B = (A \cap \bar{B}) \cup (\bar{A} \cap B)$$

These operations on sets are conveniently depicted by the use of Venn diagrams. In such a diagram the set A in the universal set S is shown (shaded) in Figure A.1 as some closed area in the square S. Points inside the shaded area A

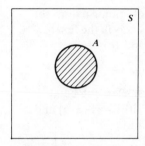

FIGURE A.1

correspond to those points of S in A. The use of these diagrams is only for convenience in visualizing the relationship between sets and is not to be taken as an actual representation of the sets. The shaded regions in Figure A.2 show the complement, union, intersection, and symmetric difference of sets.

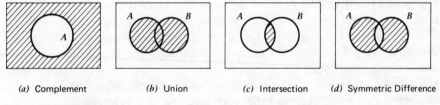

(a) Complement   (b) Union   (c) Intersection   (d) Symmetric Difference

FIGURE A.2

If $A_1, A_2, \ldots, A_n$ is a collection of n sets, then we sometimes use the shorthand notation

$$\bigcup_{i=1}^{n} A_i = A_1 \cup A_2 \cup \cdots \cup A_n$$

and

$$\bigcap_{i=1}^{n} A_i = A_1 \cap A_2 \cap \cdots \cap A_n$$

Many facts can be proved about set operations and we list a few of these here:

(i) $A \cap (B \cup C) = (A \cap B) \cup (A \cap C)$ (intersection distributes over union)
(ii) $A \cup (B \cap C) = (A \cup B) \cap (A \cup C)$ (union distributes over intersection)
(iii) $A \cup \varnothing = A$, $A \cap S = A$, $A \cap \bar{A} = \varnothing$, $\overline{(\bar{A})} = A$
(iv) $\overline{(A \cup B)} = \bar{A} \cap \bar{B}$ (more generally, $\overline{(\bigcup_i A_i)} = \bigcap_i (\bar{A}_i)$)
(v) $\overline{(A \cap B)} = \bar{A} \cup \bar{B}$ (more generally, $\overline{(\bigcap_i A_i)} = \bigcup_i \bar{A}_i$)

The last two relationships are referred to as deMorgan's laws. The proofs of all of these relationships and many more can be found in almost any reference on set theory (e.g., Halmos (1960)).

There is another particularly important method for obtaining a third set from two given sets and this construction will be used extensively in the first few chapters of the text. The Cartesian product of the sets A and B is

$$A \times B = \{(a, b) | a \in A, b \in B\}$$

It is the collection of ordered pairs of elements, the first element being a member of A and the second a member of B. For example, if $A = \{1, 2, 5\}$ and $B = \{\alpha, \beta\}$, then

$$A \times B = \{(1, \alpha), (1, \beta), (2, \alpha), (2, \beta), (5, \alpha), (5, \beta)\}$$

In general, $A \times B \neq B \times A$ as sets. If $B = A$, then $A \times B$ is sometimes written as A^2. Perhaps the most familiar example of Cartesian products is R^n defined as

$$R^n = \{(x_1, x_2, \ldots, x_n) | x_i \in R, i = 1, \ldots, n\}$$

where R is the set of all real numbers. R^n is just Euclidean n-space and for $n = 3$ it corresponds to "real" space. Elements of R^n are called n-tuples over R. Elements of A^4 would be called 4-tuples over the set A. Cartesian products are important in the study of probability when we are concerned with repetitions of an experiment. As shown in the text, if S is the set of all possible outcomes for an experiment, then $S^n = S \times S \times \cdots \times S$ (n times) represents the set of all possible outcomes for n repetitions of the experiment.

The subject of set theory is, of course, considerably more extensive than the few brief notions presented here and the interested reader is referred to Halmos (1960) for a more extensive treatment. The level of coverage in this appendix is nonetheless quite adequate for the purposes of this text.

APPENDIX B

Cumulative Distribution Function of the Binomial Distribution

$$\sum_{j=r}^{n}\binom{n}{j}\theta^{j}(1-\theta)^{n-j}*$$

n	r	$\theta=0.05$	$\theta=0.10$	$\theta=0.15$	$\theta=0.20$	$\theta=0.25$	$\theta=0.30$	$\theta=0.35$	$\theta=0.40$	$\theta=0.45$	$\theta=0.50$
1	0	1.00000	1.00000	1.00000	1.00000	1.00000	1.00000	1.00000	1.00000	1.00000	1.00000
	1	0.05000	0.10000	0.15000	0.20000	0.25000	0.30000	0.35000	0.40000	0.45000	0.50000
2	0	1.00000	1.00000	1.00000	1.00000	1.00000	1.00000	1.00000	1.00000	1.00000	1.00000
	1	0.09750	0.19000	0.27750	0.36000	0.43750	0.51000	0.57750	0.64000	0.69750	0.75000
	2	0.00250	0.01000	0.02250	0.04000	0.06250	0.09000	0.12250	0.16000	0.20250	0.25000
3	0	1.00000	1.00000	1.00000	1.00000	1.00000	1.00000	1.00000	1.00000	1.00000	1.00000
	1	0.14263	0.27100	0.38588	0.48800	0.57813	0.65700	0.72538	0.78400	0.83363	0.87500
	2	0.00725	0.02800	0.06075	0.10400	0.15625	0.21600	0.28175	0.35200	0.42525	0.50000
	3	0.00013	0.00100	0.00338	0.00800	0.01563	0.02700	0.04288	0.06400	0.09113	0.12500
4	0	1.00000	1.00000	1.00000	1.00000	1.00000	1.00000	1.00000	1.00000	1.00000	1.00000
	1	0.18549	0.34390	0.47799	0.59040	0.68359	0.75990	0.82149	0.87040	0.90849	0.93750
	2	0.01402	0.05230	0.10952	0.18080	0.26172	0.34830	0.43702	0.52480	0.60902	0.68750
	3	0.00048	0.00370	0.01198	0.02720	0.05078	0.08370	0.12648	0.17920	0.24148	0.31250
	4	0.00001	0.00010	0.00051	0.00160	0.00391	0.00810	0.01501	0.02560	0.04101	0.06250
5	0	1.00000	1.00000	1.00000	1.00000	1.00000	1.00000	1.00000	1.00000	1.00000	1.00000
	1	0.22622	0.40951	0.55629	0.67232	0.76270	0.83193	0.88397	0.92224	0.94967	0.96875
	2	0.02259	0.08146	0.16479	0.26272	0.36719	0.47178	0.57159	0.66304	0.74378	0.81250
	3	0.00116	0.00856	0.02661	0.05792	0.10352	0.16308	0.23517	0.31744	0.40687	0.50000
	4	0.00003	0.00046	0.00223	0.00672	0.01563	0.03078	0.05402	0.08704	0.13122	0.18750
	5	0.00000	0.00001	0.00008	0.00032	0.00098	0.00243	0.00525	0.01024	0.01845	0.03125
6	0	1.00000	1.00000	1.00000	1.00000	1.00000	1.00000	1.00000	1.00000	1.00000	1.00000
	1	0.26491	0.46856	0.62285	0.73786	0.82202	0.88235	0.92458	0.95334	0.97232	0.93438
	2	0.03277	0.11427	0.22352	0.34464	0.46606	0.57983	0.68092	0.76672	0.83643	0.89063
	3	0.00223	0.01585	0.04734	0.09888	0.16943	0.25569	0.35291	0.45568	0.55848	0.65625
	4	0.00009	0.00127	0.00589	0.01696	0.03760	0.07047	0.11742	0.17920	0.25526	0.34375
	5	0.00000	0.00006	0.00040	0.00160	0.00464	0.01094	0.02232	0.04096	0.06920	0.10938
	6		0.00000	0.00001	0.00006	0.00024	0.00073	0.00184	0.00410	0.00830	0.01563
7	0	1.00000	1.00000	1.00000	1.00000	1.00000	1.00000	1.00000	1.00000	1.00000	1.00000
	1	0.30166	0.52170	0.67942	0.79028	0.86652	0.91765	0.95098	0.97201	0.98478	0.99219
	2	0.04438	0.14969	0.28342	0.42328	0.55505	0.67058	0.76620	0.84137	0.89758	0.93750
	3	0.00376	0.02569	0.07377	0.14803	0.24359	0.35293	0.46772	0.58010	0.68356	0.77344
	4	0.00019	0.00273	0.01210	0.03334	0.07056	0.12604	0.19985	0.28979	0.39171	0.50000

* Abridged from *Tables for the Cumulative Binomial Probability Distribution*, Harvard University Press, Cambridge Mass., 1955, tables I to VI, with permission of the Harvard University Press.

n	r	$\theta = 0.05$	$\theta = 0.10$	$\theta = 0.15$	$\theta = 0.20$	$\theta = 0.25$	$\theta = 0.30$	$\theta = 0.35$	$\theta = 0.40$	$\theta = 0.45$	$\theta = 0.50$
	5	0.00001	0.00018	0.00122	0.00467	0.01288	0.02880	0.05561	0.09626	0.15293	0.22656
	6	0.00000	0.00001	0.00007	0.00037	0.00134	0.00379	0.00901	0.01884	0.03571	0.06250
	7		0.00000	0.00000	0.00001	0.00006	0.00022	0.00064	0.00164	0.00374	0.00781
8	0	1.00000	1.00000	1.00000	1.00000	1.00000	1.00000	1.00000	1.00000	1.00000	1.00000
	1	0.33658	0.56953	0.72751	0.83223	0.89989	0.94235	0.96814	0.98320	0.99163	0.99609
	2	0.05724	0.18690	0.34282	0.49668	0.63292	0.74470	0.83087	0.89362	0.93682	0.96484
	3	0.00579	0.03809	0.10521	0.20308	0.32146	0.44823	0.57219	0.68461	0.77987	0.85547
	4	0.00037	0.00502	0.02135	0.05628	0.11382	0.19410	0.29360	0.40591	0.52304	0.63672
	5	0.00002	0.00043	0.00285	0.01041	0.02730	0.05797	0.10609	0.17367	0.26038	0.36328
	6	0.00000	0.00002	0.00024	0.00123	0.00423	0.01129	0.02532	0.04981	0.08846	0.14453
	7		0.00000	0.00001	0.00008	0.00038	0.00129	0.00357	0.00852	0.01812	0.03516
	8			0.00000	0.00000	0.00002	0.00007	0.00023	0.00066	0.00168	0.00391
9	0	1.00000	1.00000	1.00000	1.00000	1.00000	1.00000	1.00000	1.00000	1.00000	1.00000
	1	0.36975	0.61258	0.76838	0.86578	0.92492	0.95965	0.97929	0.98992	0.99539	0.99805
	2	0.07121	0.22516	0.40052	0.56379	0.69966	0.80400	0.87891	0.92946	0.96148	0.98047
	3	0.00836	0.05297	0.14085	0.26180	0.39932	0.53717	0.66273	0.76821	0.85050	0.91016
	4	0.00064	0.00833	0.03393	0.08564	0.16573	0.27034	0.39111	0.51739	0.63862	0.74609
	5	0.00003	0.00089	0.00563	0.01958	0.04893	0.09881	0.17172	0.26657	0.37858	0.50000
	6	0.00000	0.00006	0.00063	0.00307	0.00999	0.02529	0.05359	0.09935	0.16582	0.25391
	7		0.00000	0.00005	0.00031	0.00134	0.00429	0.01118	0.02503	0.04977	0.08984
	8			0.00000	0.00002	0.00011	0.00043	0.11140	0.00380	0.00908	0.01953
	9			0.00000	0.00000	0.00002	0.00008	0.00026	0.00076	0.00195
10	0	1.00000	1.00000	1.00000	1.00000	1.00000	1.00000	1.00000	1.00000	1.00000	1.00000
	1	0.40126	0.65132	0.80313	0.89263	0.94369	0.97175	0.98654	0.99395	0.99747	0.99902
	2	0.08614	0.26390	0.45570	0.62419	0.75597	0.85069	0.91405	0.95364	0.97674	0.98926
	3	0.01150	0.07019	0.17980	0.32220	0.47441	0.61722	0.73839	0.83271	0.90044	0.94531
	4	0.00103	0.01280	0.04997	0.12087	0.22412	0.35039	0.48617	0.61772	0.73396	0.82813
	5	0.00006	0.00163	0.00987	0.03279	0.07813	0.15027	0.24850	0.36690	0.49560	0.62305
	6	0.00000	0.00015	0.00138	0.00637	0.01973	0.04735	0.09493	0.16624	0.26156	0.37695
	7		0.00001	0.00013	0.00086	0.00351	0.01059	0.02602	0.05476	0.10199	0.17188
	8		0.00000	0.00001	0.00008	0.00042	0.00159	0.00482	0.01229	0.02739	0.05469
	9			0.00000	0.00000	0.00003	0.00014	0.00054	0.00168	0.00450	0.01074
	10					0.00000	0.00001	0.00003	0.00010	0.00034	0.00098
11	0	1.00000	1.00000	1.00000	1.00000	1.00000	1.00000	1.00000	1.00000	1.00000	1.00000
	1	0.43120	0.68619	0.83266	0.91410	0.95776	0.98023	0.99125	0.99637	0.99861	0.99951
	2	0.10189	0.30264	0.50781	0.67788	0.80290	0.88701	0.93942	0.96977	0.98607	0.99414
	3	0.01524	0.08956	0.22119	0.38260	0.54480	0.68726	0.79987	0.88108	0.93478	0.96729
	4	0.00155	0.01853	0.06944	0.16114	0.28670	0.43044	0.57445	0.70372	0.80888	0.88672
	5	0.00011	0.00275	0.01589	0.05041	0.11463	0.21030	0.33169	0.46723	0.60286	0.72559
	6	0.00001	0.00030	0.00266	0.01165	0.03433	0.07822	0.14868	0.24650	0.36688	0.50000
	7	0.00000	0.00002	0.00032	0.00197	0.00756	0.02162	0.05014	0.09935	0.17380	0.27441
	8		0.00000	0.00003	0.00024	0.00119	0.00429	0.01224	0.02928	0.06096	0.11328
	9			0.00000	0.00002	0.00013	0.00058	0.00204	0.00592	0.01480	0.03271
	10				0.00000	0.00001	0.00005	0.00021	0.00073	0.00221	0.00586
	11					0.00000	0.00000	0.00001	0.00004	0.00015	0.00049
12	0	1.00000	1.00000	1.00000	1.00000	1.00000	1.00000	1.00000	1.00000	1.00000	1.00000
	1	0.45964	0.71757	0.85776	0.93128	0.96832	0.98616	0.99431	0.99782	0.99923	0.99976
	2	0.11836	0.34100	0.55654	0.72512	0.84162	0.91497	0.95756	0.98041	0.99171	0.99683
	3	0.01957	0.11087	0.26418	0.44165	0.60932	0.74718	0.84871	0.91656	0.95786	0.98071
	4	0.00224	0.02564	0.09221	0.20543	0.35122	0.50748	0.65335	0.77466	0.86553	0.92700

n	r	θ = 0.05	θ = 0.10	θ = 0.15	θ = 0.20	θ = 0.25	θ = 0.30	θ = 0.35	θ = 0.40	θ = 0.45	θ = 0.50
	5	0.00018	0.00433	0.02392	0.07256	0.15764	0.27634	0.41665	0.56182	0.69557	0.80615
	6	0.00001	0.00054	0.00464	0.01941	0.05440	0.11785	0.21274	0.33479	0.47307	0.61279
	7	0.00000	0.00005	0.00067	0.00390	0.01425	0.03860	0.08463	0.15821	0.26069	0.38721
	8		0.00000	0.00007	0.00058	0.00278	0.00949	0.02551	0.05731	0.11174	0.19385
	9			0.00001	0.00006	0.00039	0.00169	0.00561	0.01527	0.03557	0.07300
	10			0.00000	0.00000	0.00004	0.00021	0.00085	0.00281	0.00788	0.01929
	11				0.00000	0.00000	0.00002	0.00008	0.00032	0.00108	0.00317
	12						0.00000	0.00000	0.00002	0.00007	0.00024
3	0	1.00000	1.00000	1.00000	1.00000	1.00000	1.00000	1.00000	1.00000	1.00000	1.00000
	1	0.48666	0.74581	0.87909	0.94502	0.97624	0.99031	0.99630	0.99869	0.99958	0.99988
	2	0.13542	0.37866	0.60172	0.76635	0.87329	0.93633	0.97042	0.98737	0.99510	0.99829
	3	0.02451	0.13388	0.30804	0.49835	0.66740	0.79752	0.88681	0.94210	0.97309	0.98877
	4	0.00310	0.03416	0.11800	0.25268	0.41575	0.57939	0.72173	0.83142	0.90708	0.95386
	5	0.00029	0.00646	0.03416	0.09913	0.20604	0.34569	0.49950	0.64696	0.77205	0.86658
	6	0.00002	0.00092	0.00753	0.03004	0.08021	0.16540	0.28411	0.42560	0.57319	0.70947
	7	0.00000	0.00010	0.00127	0.00700	0.02429	0.06238	0.12947	0.22884	0.35626	0.50000
	8		0.00001	0.00016	0.00125	0.00565	0.01822	0.04620	0.09767	0.17877	0.29053
	9		0.00000	0.00002	0.00017	0.00099	0.00403	0.01257	0.03208	0.06985	0.13342
	10			0.00000	0.00002	0.00013	0.00065	0.00251	0.00779	0.02034	0.04614
	11				0.00000	0.00001	0.00007	0.00035	0.00132	0.00414	0.01123
	12					0.00000	0.00000	0.00003	0.00014	0.00052	0.00171
	13							0.00000	0.00001	0.00003	0.00012
4	0	1.00000	1.00000	1.00000	1.00000	1.00000	1.00000	1.00000	1.00000	1.00000	1.00000
	1	0.51233	0.77123	0.89723	0.95602	0.98218	0.99322	0.99760	0.99922	0.99977	0.99994
	2	0.15299	0.41537	0.64333	0.80209	0.89903	0.95252	0.97948	0.99190	0.99711	0.99908
	3	0.03005	0.15836	0.35209	0.55195	0.71887	0.83916	0.91607	0.96021	0.98299	0.99353
	4	0.00417	0.04413	0.14651	0.30181	0.47866	0.64483	0.77950	0.87569	0.93678	0.97131
	5	0.00043	0.00923	0.04674	0.12984	0.25847	0.41580	0.57728	0.72074	0.83281	0.91022
	6	0.00003	0.00147	0.01153	0.04385	0.11167	0.21948	0.35949	0.51415	0.66268	0.78802
	7	0.00000	0.00018	0.00221	0.01161	0.03827	0.09328	0.18359	0.30755	0.45388	0.60474
	8		0.00002	0.00033	0.00240	0.01031	0.03147	0.07534	0.15014	0.25864	0.39526
	9		0.00000	0.00004	0.00038	0.00215	0.00829	0.02434	0.05832	0.11886	0.21198
	10			0.00000	0.00005	0.00034	0.00167	0.00604	0.01751	0.04262	0.08978
	11				0.00000	0.00004	0.00025	0.00111	0.00391	0.01143	0.02869
	12					0.00000	0.00003	0.00014	0.00061	0.00215	0.00647
	13						0.00000	0.00001	0.00006	0.00025	0.00092
	14							0.00000	0.00000	0.00001	0.00006
5	0	1.00000	1.00000	1.00000	1.00000	1.00000	1.00000	1.00000	1.00000	1.00000	1.00000
	1	0.53671	0.79411	0.91265	0.96482	0.98664	0.99525	0.99844	0.99953	0.99987	0.99997
	2	0.17095	0.45096	0.68141	0.83287	0.91982	0.96473	0.98582	0.99483	0.99831	0.99951
	3	0.03620	0.18406	0.39577	0.60198	0.76391	0.87317	0.93827	0.97289	0.98935	0.99631
	4	0.00547	0.05556	0.17734	0.35184	0.53871	0.70313	0.82730	0.90950	0.95758	0.98242
	5	0.00061	0.01272	0.06171	0.16423	0.31351	0.48451	0.64806	0.78272	0.87960	0.94077
	6	0.00005	0.00225	0.01681	0.06105	0.14837	0.27838	0.43572	0.59678	0.73924	0.84912
	7	0.00000	0.00031	0.00361	0.01806	0.05662	0.13114	0.24516	0.39019	0.54784	0.69638
	8		0.00003	0.00061	0.00424	0.01730	0.05001	0.11323	0.21310	0.34650	0.50000
	9		0.00000	0.00008	0.00078	0.00419	0.01524	0.04219	0.09505	0.18176	0.30362
	10			0.00001	0.00011	0.00079	0.00365	0.01244	0.03383	0.07693	0.15088
	11			0.00000	0.00001	0.00012	0.00067	0.00283	0.00935	0.02547	0.05923
	12				0.00000	0.00001	0.00009	0.00048	0.00193	0.00633	0.01758
	13					0.00000	0.00001	0.00006	0.00028	0.00111	0.00369
	14						0.00000	0.00000	0.00003	0.00012	0.00049
	15								0.00000	0.00001	0.00003

n	r	$\theta=0.05$	$\theta=0.10$	$\theta=0.15$	$\theta=0.20$	$\theta=0.25$	$\theta=0.30$	$\theta=0.35$	$\theta=0.40$	$\theta=0.45$	$\theta=0.50$
16	0	1.00000	1.00000	1.00000	1.00000	1.00000	1.00000	1.00000	1.00000	1.00000	1.00000
	1	0.55987	0.81470	0.92575	0.97185	0.98998	0.99668	0.99898	0.99972	0.99993	0.99998
	2	0.18924	0.48527	0.71610	0.85926	0.93652	0.97389	0.99024	0.99671	0.99901	0.99974
	3	0.04294	0.21075	0.43862	0.64816	0.80289	0.90064	0.95491	0.98166	0.99338	0.99791
	4	0.00700	0.06841	0.21011	0.40187	0.59501	0.75414	0.86614	0.93485	0.97187	0.98936
	5	0.00086	0.01700	0.07905	0.20175	0.36981	0.55010	0.71079	0.83343	0.91469	0.96159
	6	0.00008	0.00330	0.02354	0.08169	0.18965	0.34022	0.51004	0.67116	0.80240	0.89494
	7	0.00001	0.00050	0.00559	0.02666	0.07956	0.17531	0.31185	0.47283	0.63397	0.77275
	8	0.00000	0.00006	0.00106	0.00700	0.02713	0.07435	0.15941	0.28394	0.43710	0.59819
	9		0.00001	0.00016	0.00148	0.00747	0.02567	0.06706	0.14227	0.25589	0.40181
	10		0.00000	0.00002	0.00025	0.00164	0.00713	0.02286	0.05832	0.12410	0.22725
	11			0.00000	0.00003	0.00029	0.00157	0.00620	0.01914	0.04862	0.10506
	12				0.00000	0.00004	0.00027	0.00130	0.00490	0.01494	0.03841
	13					0.00000	0.00003	0.00020	0.00094	0.00346	0.01064
	14						0.00000	0.00002	0.00013	0.00056	0.00209
	15							0.00000	0.00001	0.00006	0.00026
	16								0.00000	0.00000	0.00002
17	0	1.00000	1.00000	1.00000	1.00000	1.00000	1.00000	1.00000	1.00000	1.00000	1.00000
	1	0.58188	0.83323	0.93689	0.97748	0.99248	0.99767	0.99934	0.99983	0.99996	0.99999
	2	0.20777	0.51821	0.74755	0.88178	0.94989	0.98072	0.99330	0.99791	0.99943	0.99986
	3	0.05025	0.23820	0.48024	0.69038	0.83630	0.92261	0.96727	0.98768	0.99591	0.99883
	4	0.00880	0.08264	0.24439	0.45112	0.64698	0.79809	0.89721	0.95358	0.98155	0.99364
	5	0.00116	0.02214	0.09871	0.24178	0.42611	0.61131	0.76516	0.87400	0.94042	0.97548
	6	0.00012	0.00467	0.03187	0.10570	0.23469	0.40318	0.58030	0.73607	0.85293	0.92827
	7	0.00001	0.00078	0.00828	0.03766	0.10708	0.22478	0.38122	0.55216	0.70976	0.83385
	8	0.00000	0.00011	0.00174	0.01093	0.04024	0.10464	0.21276	0.35949	0.52569	0.68547
	9		0.00001	0.00030	0.00258	0.01238	0.04028	0.09938	0.19894	0.33744	0.50000
	10		0.00000	0.00004	0.00049	0.00310	0.01269	0.03833	0.09190	0.18341	0.31453
	11			0.00000	0.00008	0.00063	0.00324	0.01203	0.03481	0.08259	0.16615
	12				0.00001	0.00010	0.00066	0.00301	0.01059	0.03010	0.07173
	13				0.00000	0.00001	0.00010	0.00059	0.00252	0.00862	0.02452
	14					0.00000	0.00001	0.00009	0.00045	0.00187	0.00636
	15						0.00000	0.00001	0.00006	0.00029	0.00117
	16							0.00000	0.00000	0.00003	0.00014
	17									0.00000	0.00001
18	0	1.00000	1.00000	1.00000	1.00000	1.00000	1.00000	1.00000	1.00000	1.00000	
	1	0.60279	0.84991	0.94635	0.98199	0.99436	0.99837	0.99957	0.99990	0.99998	1.00000
	2	0.22648	0.54972	0.77595	0.90092	0.96054	0.98581	0.99541	0.99868	0.99967	0.99993
	3	0.05813	0.26620	0.52034	0.72866	0.86469	0.94005	0.97638	0.99177	0.99749	0.99934
	4	0.01087	0.09820	0.27976	0.49897	0.69431	0.83545	0.92173	0.96722	0.98802	0.99623
	5	0.00155	0.02819	0.12056	0.28365	0.48133	0.66735	0.81138	0.90583	0.95893	0.98456
	6	0.00017	0.00642	0.04190	0.13292	0.28255	0.46562	0.64500	0.79124	0.89230	0.95187
	7	0.00002	0.00117	0.01182	0.05127	0.13898	0.27830	0.45090	0.62572	0.77419	0.88106
	8	0.00000	0.00017	0.00272	0.01628	0.05695	0.14068	0.27172	0.43656	0.60852	0.75966
	9		0.00002	0.00051	0.00425	0.01935	0.05959	0.13906	0.26316	0.42215	0.59274
	10		0.00000	0.00008	0.00091	0.00542	0.02097	0.05969	0.13471	0.25272	0.40726
	11			0.00001	0.00016	0.00124	0.00607	0.02123	0.05765	0.12796	0.24034
	12			0.00000	0.00002	0.00023	0.00143	0.00617	0.02028	0.05372	0.11894
	13				0.00000	0.00003	0.00027	0.00144	0.00575	0.01829	0.04813
	14					0.00000	0.00004	0.00026	0.00128	0.00491	0.01544
	15						0.00000	0.00004	0.00021	0.00100	0.00377
	16							0.00000	0.00003	0.00014	0.00066
	17								0.00000	0.00001	0.00007
	18									0.00000	0.00000

r	θ = 0.05	θ = 0.10	θ = 0.15	θ = 0.20	θ = 0.25	θ = 0.30	θ = 0.35	θ = 0.40	θ = 0.45	θ = 0.50
0	1.00000	1.00000	1.00000	1.00000	1.00000	1.00000	1.00000	1.00000	1.00000	
1	0.62265	0.86491	0.95440	0.98559	0.99577	0.99886	0.99972	0.99994	0.99999	1.00000
2	0.24529	0.57974	0.80151	0.91713	0.96899	0.98958	0.99687	0.99917	0.99981	0.99996
3	0.06655	0.29456	0.55868	0.76311	0.88866	0.95378	0.98304	0.99454	0.99847	0.99964
4	0.01324	0.11500	0.31585	0.54491	0.73691	0.86683	0.94086	0.97704	0.99228	0.99779
5	0.00201	0.03519	0.14444	0.32671	0.53458	0.71778	0.85000	0.93039	0.97202	0.99039
6	0.00024	0.00859	0.05370	0.16306	0.33224	0.52614	0.70324	0.83708	0.92229	0.96822
7	0.00002	0.00170	0.01633	0.06760	0.17488	0.33450	0.51883	0.69193	0.82734	0.91647
8	0.00000	0.00027	0.00408	0.02328	0.07746	0.18197	0.33443	0.51222	0.68307	0.82036
9		0.00004	0.00084	0.00666	0.02875	0.08392	0.18549	0.33252	0.50602	0.67620
10		0.00000	0.00014	0.00158	0.00890	0.03255	0.08747	0.18609	0.32896	0.50000
11			0.00002	0.00031	0.00229	0.01054	0.03469	0.08847	0.18410	0.32380
12			0.00000	0.00005	0.00048	0.00282	0.01144	0.03523	0.08713	0.17964
13				0.00001	0.00008	0.00062	0.00309	0.01156	0.03423	0.08353
14				0.00000	0.00001	0.00011	0.00067	0.00307	0.01093	0.03178
15					0.00000	0.00001	0.00012	0.00064	0.00276	0.00961
16						0.00000	0.00001	0.00010	0.00053	0.00221
17							0.00000	0.00001	0.00007	0.00036
18								0.00000	0.00001	0.00004
19									0.00000	0.00000
0	1.00000	1.00000	1.00000	1.00000	1.00000	1.00000	1.00000	1.00000	1.00000	
1	0.64151	0.87842	0.96124	0.98847	0.99683	0.99920	0.99982	0.99996	0.99999	1.00000
2	0.26416	0.60825	0.82444	0.93082	0.97569	0.99236	0.99787	0.99948	0.99989	0.99998
3	0.07548	0.32307	0.59510	0.79392	0.90874	0.96452	0.98788	0.99639	0.99907	0.99980
4	0.01590	0.13295	0.35227	0.58855	0.77484	0.89291	0.95562	0.98404	0.99507	0.99871
5	0.00257	0.04317	0.17015	0.37035	0.58616	0.76249	0.88180	0.94905	0.98114	0.99409
6	0.00033	0.01125	0.06731	0.19579	0.38283	0.58363	0.75460	0.87440	0.94467	0.97931
7	0.00003	0.00239	0.02194	0.08669	0.21422	0.39199	0.58337	0.74999	0.87007	0.94234
8	0.00000	0.00042	0.00592	0.03214	0.10181	0.22773	0.39897	0.58411	0.74799	0.86841
9		0.00006	0.00133	0.00998	0.04093	0.11333	0.23762	0.40440	0.58569	0.74828
10		0.00001	0.00025	0.00259	0.01386	0.04796	0.12178	0.24466	0.40864	0.58810
11		0.00000	0.00004	0.00056	0.00394	0.01714	0.05317	0.12752	0.24929	0.41190
12			0.00000	0.00010	0.00094	0.00514	0.01958	0.05653	0.13076	0.25172
13				0.00002	0.00018	0.00128	0.00602	0.02103	0.05803	0.13159
14				0.00000	0.00003	0.00026	0.00152	0.00647	0.02141	0.05766
15					0.00000	0.00004	0.00031	0.00161	0.00643	0.02069
16						0.00001	0.00005	0.00032	0.00153	0.00591
17						0.00000	0.00001	0.00005	0.00028	0.00129
18							0.00000	0.00001	0.00004	0.00020
19								0.00000	0.00000	0.00002
20										0.00000

APPENDIX C

Cumulative Distribution Function of the Poisson Distribution

$$\left(\sum_{j=c}^{\infty} \frac{\lambda^j e^{-\lambda}}{j!} \right)^*$$

	$\lambda = .010$	$\lambda = .02$	$\lambda = .03$	$\lambda = .04$	$\lambda = .05$	$\lambda = .06$	$\lambda = .07$	$\lambda = .08$	$\lambda = .09$
0	1.0000000	1.0000000	1.0000000	1.0000000	1.0000000	1.0000000	1.0000000	1.0000000	1.0000000
1	.0099502	.0198013	.0295545	.0392106	.0487706	.0582355	.0676062	.0768837	.0860688
2	.0000497	.0001973	.0004411	.0007790	.0012091	.0017296	.0023386	.0030343	.0038150
3	.0000002	.0000013	.0000044	.0000104	.0000201	.0000344	.0000542	.0000804	.0001136
4				.0000001	.0000003	.0000005	.0000009	.0000016	.0000025

	$\lambda = .10$	$\lambda = .20$	$\lambda = .30$	$\lambda = .40$	$\lambda = .50$	$\lambda = .60$	$\lambda = .70$	$\lambda = .80$	$\lambda = .90$
0	1.0000000	1.0000000	1.0000000	1.0000000	1.000000	1.000000	1.000000	1.000000	1.000000
1	.0951626	.1812692	.2591818	.3296800	.393469	.451188	.503415	.550671	.593430
2	.0046788	.0175231	.0369363	.0615519	.090204	.121901	.155805	.191208	.227518
3	.0001547	.0011485	.0035995	.0079263	.014388	.023115	.034142	.047423	.062857
4	.0000038	.0000568	.0002658	.0007763	.001752	.003358	.005753	.009080	.013459
5		.0000023	.0000158	.0000612	.000172	.000394	.000786	.001411	.002344
6		.0000001	.0000008	.0000040	.000014	.000039	.000090	.000184	.000343
7				.0000002	.000001	.000003	.000009	.000021	.000043
8							.000001	.000002	.000005
9									

	$\lambda = 1.0$	$\lambda = 2.0$	$\lambda = 3.0$	$\lambda = 4.0$	$\lambda = 5.0$	$\lambda = 6.0$	$\lambda = 7.0$	$\lambda = 8.0$	$\lambda = 9.0$
0	1.000000	1.000000	1.000000	1.000000	1.000000	1.000000	1.000000	1.000000	1.000000
1	.632121	.864665	.950213	.981684	.993262	.997521	.999088	.999665	.999877
2	.264241	.593994	.800852	.908422	.959572	.982649	.992705	.996981	.998766
3	.080301	.323324	.576810	.761897	.875348	.938031	.970364	.986246	.993768
4	.018988	.142877	.352768	.566530	.734974	.848796	.918235	.957620	.978774
5	.003660	.052653	.184737	.371163	.559507	.714943	.827008	.900368	.945036
6	.000594	.016564	.083918	.214870	.384039	.554320	.699292	.808764	.884309
7	.000083	.004534	.033509	.110674	.237817	.393697	.550289	.686626	.793219
8	.000010	.001097	.011905	.051134	.133372	.256020	.401286	.547039	.676103
9	.000001	.000237	.003803	.021363	.068094	.152763	.270909	.407453	.544347
10		.000046	.001102	.008132	.031828	.083924	.169504	.283376	.412592
11		.000008	.000292	.002840	.013695	.042621	.098521	.184114	.294012
12		.000001	.000071	.000915	.005453	.020092	.053350	.111924	.196992
13			.000016	.000274	.002019	.008827	.027000	.063797	.124227
14			.000003	.000076	.000698	.003628	.012811	.034181	.073851

* Abridged from *Poisson's Exponential Binomial Limit*, by E. C. Molina. D. Van Nostrand Company Inc., New York, 1942, table II, with permission of the D. Van Nostrand Company, Inc.

	$\lambda=1.0$	$\lambda=2.0$	$\lambda=3.0$	$\lambda=4.0$	$\lambda=5.0$	$\lambda=6.0$	$\lambda=7.0$	$\lambda=8.0$	$\lambda=9.0$
15			.000001	.000020	.000226	.001400	.005717	.017257	.041466
16				.000005	.000069	.000509	.002407	.008231	.022036
17				.000001	.000020	.000175	.000958	.003718	.011106
18					.000005	.000057	.000362	.001594	.005320
19					.000001	.000018	.000130	.000650	.002426
20						.000005	.000044	.000253	.001056
21						.000001	.000014	.000094	.000439
22							.000005	.000033	.000175
23							.000001	.000011	.000067
24								.000004	.000025
25								.000001	.000009
26									.000003
27									.000001

	$\lambda=10.0$	$\lambda=11.0$	$\lambda=12.0$	$\lambda=13.0$	$\lambda=14.0$	$\lambda=15.0$	$\lambda=16$	$\lambda=17$	$\lambda=18$	$\lambda=19$
0	1.000000	1.000000	1.000000	1.000000	1.000000					
1	.999955	.999983	.999994	.999998	.999999	1.000000	1.000000	1.000000		
2	.999501	.999800	.999920	.999968	.999988	.999995	.999998	.999999	1.000000	1.000000
3	.997231	.998789	.999478	.999777	.999906	.999961	.999984	.999993	.999997	.999999
4	.989664	.995084	.997708	.998950	.999526	.999789	.999907	.999959	.999982	.999992
5	.970747	.984895	.992400	.996260	.998195	.999143	.999600	.999815	.999916	.999962
6	.932914	.962480	.979659	.989266	.994468	.997208	.998616	.999325	.999676	.999846
7	.869859	.921386	.954178	.974113	.985772	.992368	.995994	.997938	.998957	.999480
8	.779779	.856808	.910496	.945972	.968380	.981998	.990000	.994567	.997107	.998487
9	.667180	.768015	.844972	.900242	.937945	.962554	.978013	.987404	.992944	.996127
10	.542070	.659489	.757608	.834188	.890601	.930146	.956702	.973875	.984619	.991144
11	.416960	.540111	.652771	.748318	.824319	.881536	.922604	.950876	.969634	.981678
12	.303224	.420733	.538403	.646835	.739960	.815248	.873007	.915331	.945113	.965327
13	.208444	.311303	.424035	.536895	.641542	.732389	.806878	.864976	.908331	.939439
14	.135536	.218709	.318464	.426955	.535552	.636782	.725489	.799127	.857402	.901601
15	.083458	.145956	.227975	.324868	.429563	.534346	.632473	.719167	.791923	.850250
16	.048740	.092604	.155584	.236393	.330640	.431910	.533255	.628546	.713347	.785206
17	.027042	.055924	.101291	.164507	.244082	.335877	.434038	.532262	.624950	.707966
18	.014278	.032191	.062966	.109535	.172799	.251141	.340656	.435977	.531352	.621639
19	.007187	.017687	.037417	.069833	.117357	.180528	.257651	.345042	.437755	.530516
20	.003454	.009289	.021280	.042669	.076505	.124781	.187751	.263678	.349084	.439393
21	.001588	.004671	.011598	.025012	.047908	.082971	.131832	.194519	.269280	.352826
22	.000700	.002252	.006065	.014081	.028844	.053106	.089227	.138534	.200876	.274503
23	.000296	.001042	.003047	.007622	.016712	.032744	.058241	.095272	.144910	.206861
24	.000120	.000464	.001473	.003972	.009328	.019465	.036686	.063296	.101110	.150983
25	.000047	.000199	.000686	.001994	.005020	.011165	.022315	.040646	.068260	.106746
26	.000018	.000082	.000308	.000966	.002608	.006185	.013119	.025245	.044608	.073126
27	.000006	.000033	.000133	.000452	.001309	.003312	.007459	.015174	.028234	.048557
28	.000002	.000013	.000056	.000204	.000635	.001716	.004105	.008834	.017318	.031268
29	.000001	.000005	.000023	.000089	.000298	.000861	.002189	.004984	.010300	.019536
30		.000002	.000009	.000038	.000136	.000418	.001131	.002727	.005944	.011850
31		.000001	.000003	.000016	.000060	.000197	.000567	.001448	.003331	.006982
32			.000001	.000006	.000026	.000090	.000276	.000747	.001813	.003998
33				.000002	.000011	.000040	.000131	.000375	.000960	.002227
34				.000001	.000004	.000017	.000060	.000183	.000494	.001207
35					.000002	.000007	.000027	.000087	.000248	.000637
36					.000001	.000003	.000012	.000040	.000121	.000327
37						.000001	.000005	.000018	.000058	.000164
38							.000002	.000008	.000027	.000080
39							.000001	.000003	.000012	.000038

	λ=10.0	λ=11.0	λ=12.0	λ=13.0	λ=14.0	λ=15.0	λ=16	λ=17	λ=18	λ=19
40								.000001	.000005	.000018
41								.000001	.000002	.000008
42									.000001	.000004
43										.000002
44										.000001

	λ=20	λ=21	λ=22	λ=23	λ=24	λ=25	λ=26	λ=27	λ=28	λ=29
3	1.000000	1.000000	1.000000							
4	.999997	.999999	.999999	1.000000	1.000000					
5	.999983	.999993	.999997	.999999	.999999	1.000000	1.000000			
6	.999928	.999967	.999985	.999993	.999997	.999999	.999999	1.000000	1.000000	
7	.999745	.999876	.999941	.999972	.999987	.999994	.999997	.999999	.999999	1.000000
8	.999221	.999605	.999803	.999903	.999953	.999977	.999989	.999995	.999998	.999999
9	.997913	.998894	.999423	.999703	.999849	.999925	.999963	.999982	.999991	.999996
10	.995005	.997234	.998495	.999194	.999575	.999779	.999886	.999942	.999971	.999986
11	.989188	.993749	.996453	.998023	.998915	.999414	.999687	.999836	.999914	.999956
12	.978613	.987095	.992370	.995573	.997476	.998584	.999218	.999574	.999771	.999878
13	.960988	.975451	.984884	.990878	.994598	.996856	.998200	.998985	.999436	.999690
14	.933872	.956641	.972215	.982572	.989284	.993533	.996164	.997762	.998714	.999271
15	.895136	.928426	.952307	.968926	.980175	.987598	.992383	.995403	.997270	.998403
16	.843487	.888925	.923108	.948002	.965600	.977707	.985830	.991156	.994574	.996725
17	.778926	.837081	.882960	.917923	.943728	.962252	.975182	.983991	.989857	.993682
18	.702972	.773037	.831004	.877229	.912874	.939525	.958895	.972610	.982088	.988492
19	.618578	.698320	.767502	.825231	.871721	.907959	.935371	.955539	.970003	.980131
20	.529743	.615737	.693973	.762286	.819739	.866425	.903179	.931281	.952193	.967369
21	.440907	.529026	.613091	.689900	.757361	.814508	.861330	.898532	.927259	.948863
22	.356302	.442314	.528358	.610619	.686072	.752701	.809517	.856426	.894014	.923308
23	.279389	.359544	.443625	.527734	.608302	.682467	.748283	.804750	.851702	.889622
24	.212507	.283971	.362576	.444850	.527150	.606124	.679063	.744087	.800191	.847149
25	.156773	.217845	.288281	.365419	.445999	.526602	.604073	.675842	.740096	.795826
26	.112185	.162299	.222901	.292343	.368093	.447079	.526085	.602137	.672789	.736293
27	.077887	.117435	.167580	.227698	.296181	.370614	.448096	.525597	.600305	.669889
28	.052481	.082541	.122503	.172631	.232258	.299814	.372996	.449057	.525136	.598567
29	.034334	.056370	.087086	.127397	.177468	.236599	.303260	.375251	.449967	.524698
30	.021816	.037419	.060217	.091521	.132124	.182104	.240738	.306535	.377390	.450829
31	.013475	.024153	.040514	.064017	.095848	.136691	.186553	.244690	.309651	.379422
32	.008092	.015166	.026531	.043611	.067764	.100068	.141107	.190825	.248468	.312622
33	.004727	.009269	.016918	.028943	.046701	.071456	.104182	.145377	.194933	.252085
34	.002688	.005516	.010509	.018721	.031383	.049780	.075089	.108192	.149509	.198885
35	.001489	.003198	.006362	.011806	.020570	.033842	.052842	.078663	.112101	.153509
36	.000804	.001807	.003755	.007261	.013155	.022458	.036316	.055883	.082175	.115912
37	.000423	.000996	.002162	.004358	.008212	.014552	.024380	.038798	.058899	.085625
38	.000217	.000536	.001215	.002553	.005006	.009211	.015993	.026331	.041285	.061887
39	.000109	.000281	.000667	.001461	.002980	.005696	.010254	.017473	.028306	.043771
40	.000053	.000144	.000357	.000817	.001734	.003444	.006429	.011340	.018987	.030300
41	.000025	.000072	.000187	.000446	.000987	.002036	.003942	.007200	.012465	.020533
42	.000012	.000035	.000096	.000238	.000549	.001177	.002365	.004474	.008010	.013625
43	.000005	.000017	.000048	.000125	.000299	.000666	.001389	.002722	.005040	.008856
44	.000002	.000008	.000024	.000064	.000159	.000369	.000798	.001622	.003107	.005639
45	.000001	.000004	.000011	.000032	.000083	.000200	.000450	.000946	.001876	.003519
46		.000002	.000005	.000016	.000042	.000106	.000248	.000541	.001110	.002152
47		.000001	.000002	.000008	.000021	.000055	.000134	.000303	.000644	.001291
48			.000001	.000004	.000010	.000028	.000071	.000167	.000367	.000759
49				.000002	.000005	.000014	.000037	.000090	.000205	.000438

	$\lambda=20$	$\lambda=21$	$\lambda=22$	$\lambda=23$	$\lambda=24$	$\lambda=25$	$\lambda=26$	$\lambda=27$	$\lambda=28$	$\lambda=29$
50				.000001	.000002	.000007	.000019	.000048	.000112	.000248
51					.000001	.000003	.000009	.000025	.000060	.000138
52						.000002	.000005	.000013	.000032	.000075
53						.000001	.000002	.000006	.000016	.000040
54							.000001	.000003	.000008	.000021
55								.000001	.000004	.000011
56								.000001	.000002	.000006
57									.000001	.000003
58										.000001
59										.000001

APPENDIX D

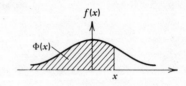

Tabulation of the Standard
*Normal Distribution**

* Abridged from *Biometrika Tables for Statisticians*, vol. 1 (2nd edition), edited by E. S. Pearson and H. O. Hartley, Cambridge University Press, London, 1958, table 1, with permission of the Biometrika Trustees.

x	Φ(x)	f(x)	x	Φ(x)	f(x)	x	Φ(x)	f(x)
.00	.50000	.39894	.50	.69146	.35207	1.00	.84134	.24197
.01	.50399	.39892	.51	.69497	.35029	1.01	.84375	.23955
.02	.50798	.39886	.52	.69847	.34849	1.02	.84614	.23713
.03	.51197	.39876	.53	.70194	.34667	1.03	.84850	.23471
.04	.51595	.39862	.54	.70540	.34482	1.04	.85083	.23230
.05	.51994	.39844	.55	.70884	.34294	1.05	.85314	.22988
.06	.52392	.39822	.56	.71226	.34105	1.06	.85543	.22747
.07	.52790	.39797	.57	.71566	.33912	1.07	.85769	.22506
.08	.53188	.39767	.58	.71904	.33718	1.08	.85993	.22265
.09	.53586	.39733	.59	.72240	.33521	1.09	.86214	.22025
.10	.53983	.39695	.60	.72575	.33322	1.10	.86433	.21785
.11	.54380	.39654	.61	.72907	.33121	1.11	.86650	.21546
.12	.54776	.39608	.62	.73237	.32918	1.12	.86864	.21307
.13	.55172	.39559	.63	.73565	.32713	1.13	.87076	.21069
.14	.55567	.39505	.64	.73891	.32506	1.14	.87286	.20831
.15	.55982	.39448	.65	.74215	.32297	1.15	.87493	.20594
.16	.56356	.39387	.66	.74537	.32086	1.16	.87698	.20357
.17	.56750	.39322	.67	.74857	.31874	1.17	.87900	.20121
.18	.57142	.39253	.68	.75175	.31659	1.18	.88100	.19886
.19	.57535	.39181	.69	.75490	.31443	1.19	.88298	.19652
.20	.57926	.39104	.70	.75804	.31225	1.20	.88493	.19419
.21	.58317	.39024	.71	.76115	.31006	1.21	.88686	.19186
.22	.58706	.38940	.72	.76424	.30785	1.22	.88877	.18954
.23	.59095	.38853	.73	.76730	.30563	1.23	.89065	.18724
.24	.59484	.38762	.74	.77035	.30339	1.24	.89251	.18494
.25	.59871	.38667	.75	.77337	.30114	1.25	.89435	.18265
.26	.60257	.38568	.76	.77637	.29887	1.26	.89617	.18037
.27	.60643	.38466	.77	.77935	.29659	1.27	.89796	.17810
.28	.61026	.38361	.78	.78230	.29431	1.28	.89973	.17585
.29	.61409	.38251	.79	.78524	.29200	1.29	.90147	.17360
.30	.61791	.38139	.80	.78814	.28969	1.30	.90320	.17137
.31	.62172	.38023	.81	.79103	.28737	1.31	.90490	.16915
.32	.62552	.37903	.82	.79389	.28504	1.32	.90658	.16694
.33	.62930	.37780	.83	.79673	.28269	1.33	.90824	.16474
.34	.63307	.37654	.84	.79955	.28034	1.34	.90988	.16256
.35	.63683	.37524	.85	.80234	.27798	1.35	.91149	.16038
.36	.64058	.37391	.86	.80511	.27562	1.36	.91308	.15822
.37	.64431	.37255	.87	.80785	.27324	1.37	.91466	.15608
.38	.64803	.37115	.88	.81057	.27086	1.38	.91621	.15395
.39	.65173	.36973	.89	.81327	.26848	1.39	.91774	.15183
.40	.65542	.36827	.90	.81594	.26609	1.40	.91924	.14973
.41	.65910	.36678	.91	.81859	.26369	1.41	.92073	.14764
.42	.66276	.36526	.92	.82121	.26129	1.42	.92220	.14556
.43	.66640	.36371	.93	.82381	.25888	1.43	.92364	.14350
.44	.67003	.36213	.94	.82639	.25647	1.44	.92507	.14146
.45	.67365	.36053	.95	.82894	.25406	1.45	.92647	.13943
.46	.67724	.35889	.96	.83147	.25164	1.46	.92786	.13742
.47	.68082	.35723	.97	.83398	.24923	1.47	.92922	.13542
.48	.68439	.35553	.98	.83646	.24681	1.48	.93056	.13344
.49	.68793	.35381	.99	.83891	.24439	1.49	.93189	.13147
.50	.69146	.35207	1.00	.84134	.24197	1.50	.93319	.12952

x	Φ(x)	f(x)	x	Φ(x)	f(x)	x	Φ(x)	f(x)
1.50	.93319	.12952	2.00	.97725	.05399	2.50	.99379	.01753
1.51	.93448	.12758	2.01	.97778	.05292	2.51	.99396	.01709
1.52	.93574	.12566	2.02	.97831	.05186	2.52	.99413	.01667
1.53	.93699	.12376	2.03	.97882	.05082	2.53	.99430	.01625
1.54	.93822	.12188	2.04	.97932	.04980	2.54	.99446	.01585
1.55	.93943	.12001	2.05	.97982	.04879	2.55	.99461	.01545
1.56	.94062	.11816	2.06	.98030	.04780	2.56	.99477	.01506
1.57	.94179	.11632	2.07	.98077	.04682	2.57	.99492	.01468
1.58	.94295	.11450	2.08	.98124	.04586	2.58	.99506	.01431
1.59	.94408	.11270	2.09	.98169	.04491	2.59	.99520	.01394
1.60	.94520	.11092	2.10	.98214	.04398	2.60	.99534	.01358
1.61	.94630	.10915	2.11	.98257	.04307	2.61	.99547	.01323
1.62	.94738	.10741	2.12	.98300	.04217	2.62	.99560	.01289
1.63	.94845	.10567	2.13	.98341	.04128	2.63	.99573	.01256
1.64	.94950	.10396	2.14	.98382	.04041	2.64	.99585	.01223
1.65	.95053	.10226	2.15	.98422	.03955	2.65	.99598	.01191
1.66	.95154	.10059	2.16	.98461	.03871	2.66	.99609	.01160
1.67	.95254	.09893	2.17	.98500	.03788	2.67	.99621	.01130
1.68	.95352	.09728	2.18	.98537	.03706	2.68	.99632	.01100
1.69	.95449	.09566	2.19	.98574	.03626	2.69	.99643	.01071
1.70	.95543	.09405	2.20	.98610	.03547	2.70	.99653	.01042
1.71	.95637	.09246	2.21	.98645	.03470	2.71	.99664	.01014
1.72	.95728	.09089	2.22	.98679	.03394	2.72	.99674	.00987
1.73	.95818	.08933	2.23	.98713	.03319	2.73	.99683	.00961
1.74	.95907	.08780	2.24	.98745	.03246	2.74	.99693	.00935
1.75	.95994	.08628	2.25	.98778	.03174	2.75	.99702	.00909
1.76	.96080	.08478	2.26	.98809	.03103	2.76	.99711	.00885
1.77	.96164	.08329	2.27	.98840	.03034	2.77	.99720	.00861
1.78	.96246	.08183	2.28	.98870	.02965	2.78	.99728	.00837
1.79	.96327	.08038	2.29	.98899	.02898	2.79	.99736	.00814
1.80	.96407	.07895	2.30	.98928	.02833	2.80	.99744	.00792
1.81	.96485	.07754	2.31	.98956	.02768	2.81	.99752	.00770
1.82	.96562	.07614	2.32	.98983	.02705	2.82	.99760	.00748
1.83	.96638	.07477	2.33	.99010	.02643	2.83	.99767	.00727
1.84	.96712	.07341	2.34	.99036	.02582	2.84	.99774	.00707
1.85	.96784	.07206	2.35	.99061	.02522	2.85	.99781	.00687
1.86	.96856	.07074	2.36	.99086	.02463	2.86	.99788	.00668
1.87	.96926	.06943	2.37	.99111	.02406	2.87	.99795	.00649
1.88	.96995	.06814	2.38	.99134	.02349	2.88	.99801	.00631
1.89	.97062	.06687	2.39	.99158	.02294	2.89	.99807	.00613
1.90	.97128	.06562	2.40	.99180	.02239	2.90	.99813	.00595
1.91	.97193	.06438	2.41	.99202	.02186	2.91	.99819	.00578
1.92	.97257	.06316	2.42	.99224	.02134	2.92	.99825	.00562
1.93	.97320	.06195	2.43	.99245	.02083	2.93	.99831	.00545
1.94	.97381	.06077	2.44	.99266	.02033	2.94	.99836	.00530
1.95	.97441	.05959	2.45	.99286	.01984	2.95	.99841	.00514
1.96	.97500	.05844	2.46	.99305	.01936	2.96	.99846	.00499
1.97	.97558	.05730	2.47	.99324	.01889	2.97	.99851	.00485
1.98	.97615	.05618	2.48	.99343	.01842	2.98	.99856	.00471
1.99	.97670	.05508	2.49	.99361	.01797	2.99	.99861	.00457
2.00	.97725	.05399	2.50	.99379	.01753	3.00	.99865	.00443

x	Φ(x)	f(x)	x	Φ(x)	f(x)	x	Φ(x)	f(x)
3.00	.99865	.00443	3.50	.99977	.00087	4.00	.99997	.00013
3.01	.99869	.00430	3.51	.99978	.00084	4.01	.99997	.00013
3.02	.99874	.00417	3.52	.99978	.00081	4.02	.99997	.00012
3.03	.99878	.00405	3.53	.99979	.00079	4.03	.99997	.00012
3.04	.99882	.00393	3.54	.99980	.00076	4.04	.99997	.00011
3.05	.99886	.00381	3.55	.99981	.00073	4.05	.99997	.00011
3.06	.99889	.00370	3.56	.99981	.00071	4.06	.99998	.00011
3.07	.99893	.00358	3.57	.99982	.00068	4.07	.99998	.00010
3.08	.99897	.00348	3.58	.99983	.00066	4.08	.99998	.00010
3.09	.99900	.00337	3.59	.99983	.00063	4.09	.99998	.00009
3.10	.99903	.00327	3.60	.99984	.00061	4.10	.99998	.00009
3.11	.99906	.00317	3.61	.99985	.00059	4.11	.99998	.00009
3.12	.99910	.00307	3.62	.99985	.00057	4.12	.99998	.00008
3.13	.99913	.00298	3.63	.99986	.00055	4.13	.99998	.00008
3.14	.99916	.00288	3.64	.99986	.00053	4.14	.99998	.00008
3.15	.99918	.00279	3.65	.99987	.00051	4.15	.99998	.00007
3.16	.99921	.00271	3.66	.99987	.00049	4.16	.99998	.00007
3.17	.99924	.00262	3.67	.99988	.00047	4.17	.99998	.00007
3.18	.99926	.00254	3.68	.99988	.00046	4.18	.99999	.00006
3.19	.99929	.00246	3.69	.99989	.00044	4.19	.99999	.00006
3.20	.99931	.00238	3.70	.99989	.00042	4.20	.99999	.00006
3.21	.99934	.00231	3.71	.99990	.00041	4.21	.99999	.00006
3.22	.99936	.00224	3.72	.99990	.00039	4.22	.99999	.00005
3.23	.99938	.00216	3.73	.99990	.00038	4.23	.99999	.00005
3.24	.99940	.00210	3.74	.99991	.00037	4.24	.99999	.00005
3.25	.99942	.00203	3.75	.99991	.00035	4.25	.99999	.00005
3.26	.99944	.00196	3.76	.99992	.00034	4.26	.99999	.00005
3.27	.99946	.00190	3.77	.99992	.00033	4.27	.99999	.00004
3.28	.99948	.00184	3.78	.99992	.00031	4.28	.99999	.00004
3.29	.99950	.00178	3.79	.99992	.00030	4.29	.99999	.00004
3.30	.99952	.00172	3.80	.99993	.00029	4.30	.99999	.00004
3.31	.99953	.00167	3.81	.99993	.00028	4.31	.99999	.00004
3.32	.99955	.00161	3.82	.99993	.00027	4.32	.99999	.00004
3.33	.99957	.00156	3.83	.99994	.00026	4.33	.99999	.00003
3.34	.99958	.00151	3.84	.99994	.00025	4.34	.99999	.00003
3.35	.99960	.00146	3.85	.99994	.00024	4.35	.99999	.00003
3.36	.99961	.00141	3.86	.99994	.00023	4.36	.99999	.00003
3.37	.99962	.00136	3.87	.99995	.00022	4.37	.99999	.00003
3.38	.99964	.00132	3.88	.99995	.00021	4.38	.99999	.00003
3.39	.99965	.00127	3.89	.99995	.00021	4.39	.99999	.00003
3.40	.99966	.00123	3.90	.99995	.00020	4.40	.99999	.00002
3.41	.99968	.00119	3.91	.99995	.00019	4.41	.99999	.00002
3.42	.99969	.00115	3.92	.99996	.00018	4.42	1.00000	.00002
3.43	.99970	.00111	3.93	.99996	.00018	4.43	1.00000	.00002
3.44	.99971	.00107	3.94	.99996	.00017	4.44	1.00000	.00002
3.45	.99972	.00104	3.95	.99996	.00016	4.45	1.00000	.00002
3.46	.99973	.00100	3.96	.99996	.00016	4.46	1.00000	.00002
3.47	.99974	.00097	3.97	.99996	.00015	4.47	1.00000	.00002
3.48	.99975	.00094	3.98	.99997	.00014	4.48	1.00000	.00002
3.49	.99976	.00090	3.99	.99997	.00014	4.49	1.00000	.00002
3.50	.99977	.00087	4.00	.99997	.00013	4.50	1.00000	.00002

APPENDIX E

Percentage Points of the Chi-Square Distribution*

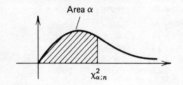

Area α

$\chi^2_{\alpha;n}$

* From *Biometrika Tables for Statisticians*, vol. 1, edited by E. S. Pearson
and H. O. Hartley, Cambridge University Press, London (1958), table
8, with permission of the Biometrika Trustees.

n	0.999	0.995	0.990	0.975	0.950	0.900	0.750	0.500	0.250	0.100	0.050	0.025	0.010	0.005
1	10.828	7.87944	6.63490	5.02389	3.84146	2.70554	1.32330	0.454937	0.1015308	0.0157908	393214.10^{-8}	982069.10^{-9}	157088.10^{-9}	392704.10^{-10}
2	13.816	10.5966	9.21034	7.37776	5.99147	4.60517	2.77259	1.38629	0.575364	0.210720	0.102587	0.0506356	0.0201007	0.0100251
3	16.266	12.8381	11.3449	9.34840	7.81473	6.25139	4.10835	2.36597	1.212534	0.584375	0.351846	0.215795	0.114832	0.0717212
4	18.467	14.8602	13.2767	11.1433	9.48773	7.77944	5.38527	3.35670	1.92255	1.063623	0.710721	0.484419	0.297110	0.206990
5	20.515	16.7496	15.0863	12.8325	11.0705	9.23635	6.62568	4.35146	2.67460	1.61031	1.145476	0.831211	0.554300	0.411740
6	22.458	18.5476	16.8119	14.4494	12.5916	10.6446	7.84080	5.34812	3.45460	2.20413	1.63539	1.237347	0.872085	0.675727
7	24.322	20.2777	18.4753	16.0128	14.0671	12.0170	9.03715	6.34581	4.25485	2.83311	2.16735	1.68987	1.239043	0.989265
8	26.125	21.9550	20.0902	17.5346	15.5073	13.3616	10.2188	7.34412	5.07064	3.48954	2.73264	2.17973	1.646482	1.344419
9	27.877	23.5893	21.6660	19.0228	16.9190	14.6837	11.3887	8.34283	5.89883	4.16816	3.32511	2.70039	2.087912	1.734926
10	29.588	25.1882	23.2093	20.4831	18.3070	15.9871	12.5489	9.34182	6.73720	4.86518	3.94030	3.24697	2.55821	2.15585
11	31.264	26.7569	24.7250	21.9200	19.6751	17.2750	13.7007	10.3410	7.58412	5.57779	4.57481	3.81575	3.05347	2.60321
12	32.909	28.2995	26.2170	23.3367	21.0261	18.5494	14.8454	11.3403	8.43842	6.30380	5.22603	4.40379	3.57056	3.07382
13	34.528	29.8194	27.6883	24.7356	22.3621	19.8119	15.9839	12.3398	9.29906	7.04150	5.89186	5.00874	4.10691	3.56503
14	36.123	31.3193	29.1413	26.1190	23.6848	21.0642	17.1170	13.3393	10.1653	7.78953	6.57063	5.62872	4.66043	4.07468
15	37.697	32.8013	30.5779	27.4884	24.9958	22.3072	18.2451	14.3389	11.0365	8.54675	7.26094	6.26214	5.22935	4.60094
16	39.252	34.2672	31.9999	28.8454	26.2962	23.5418	19.3688	15.3385	11.9122	9.31223	7.96164	6.90766	5.81221	5.14224
17	40.790	35.7185	33.4087	30.1910	27.5871	24.7690	20.4887	16.3381	12.7919	10.0852	8.67176	7.56418	6.40776	5.69724
18	42.312	37.1564	34.8053	31.5264	28.8693	25.9894	21.6049	17.3379	13.6753	10.8649	9.39046	8.23075	7.01491	6.26481
19	43.820	38.5822	36.1908	32.8523	30.1435	27.2036	22.7178	18.3376	14.5620	11.6509	10.1170	8.90655	7.63273	6.84398
20	45.315	39.9968	37.5662	34.1696	31.4104	28.4120	23.8277	19.3374	15.4518	12.4426	10.8508	9.59083	8.26040	7.43386
21	46.797	41.4010	38.9321	35.4789	32.6705	29.6151	24.9348	20.3372	16.3444	13.2396	11.5913	10.28293	8.89720	8.03366
22	48.268	42.7956	40.2894	36.7807	33.9244	30.8133	26.0393	21.3370	17.2396	14.0415	12.3380	10.9823	9.54249	8.64272
23	49.728	44.1813	41.6384	38.0757	35.1725	32.0069	27.1413	22.3369	18.1373	14.8479	13.0905	11.6885	10.19567	9.26042
24	51.179	45.5585	42.9798	39.3641	36.4151	33.1963	28.2412	23.3367	19.0372	15.6587	13.8484	12.4011	10.8564	9.88623
25	52.620	46.9278	44.3141	40.6465	37.6525	34.3816	29.3389	24.3366	19.9393	16.4734	14.6114	13.1197	11.5240	10.5197
26	54.052	48.2899	45.6417	41.9232	38.8852	35.5631	30.4345	25.3364	20.8434	17.2919	15.3791	13.8439	12.1981	11.1603
27	55.476	49.6449	46.9630	43.1944	40.1133	36.7412	31.5284	26.3363	21.7494	18.1138	16.1513	14.5733	12.8786	11.8076
28	56.892	50.9933	48.2782	44.4607	41.3372	37.9159	32.6205	27.3363	22.6572	18.9392	16.9279	15.3079	13.5648	12.4613
29	58.302	52.3356	49.5879	45.7222	42.5569	39.0875	33.7109	28.3362	23.5666	19.7677	17.7083	16.0471	14.2565	13.1211
30	59.703	53.6720	50.8922	46.9792	43.7729	40.2560	34.7998	29.3360	24.4776	20.5992	18.4926	16.7908	14.9535	13.7867
40	73.402	66.7659	63.6907	59.3417	55.7585	51.8050	45.6160	39.3354	33.6603	29.0505	26.5093	24.4331	22.1643	20.7065
50	86.661	79.4900	76.1539	71.4202	67.5048	63.1671	56.3336	49.3349	42.9421	37.6886	34.7642	32.3574	29.7067	27.9907
60	99.607	91.9517	88.3794	83.2976	79.0819	74.3970	66.9814	59.3347	52.2938	46.4589	43.1879	40.4817	37.4848	35.5346
70	112.317	104.215	100.425	95.0231	90.5312	85.5271	77.5766	69.3344	61.6983	55.3290	51.7393	48.7576	45.4418	43.2752
80	124.839	116.321	112.329	106.629	101.879	96.5782	88.1303	79.3343	71.1445	64.2778	60.3915	57.1532	53.5400	51.1720
90	137.208	128.299	124.116	118.136	113.145	107.565	98.6499	89.3342	80.6247	73.2912	69.1260	65.6466	61.7541	59.1963
100	149.449	140.169	135.807	129.561	124.342	118.498	109.141	99.3341	90.1332	82.3581	77.9295	74.2219	70.0648	67.3276

APPENDIX F

Percentage Points of the t-Distribution*

Area α $t_{\alpha;n}$

n	$\alpha = 0.6$	0.75	0.9	0.95	0.975	0.99	0.995	0.9975	0.999	0.9995
1	0.325	1.000	3.078	6.314	12.706	31.821	63.657	127.32	318.31	636.62
2	.289	0.816	1.886	2.920	4.303	6.965	9.925	14.089	22.326	31.598
3	.277	.765	1.638	2.353	3.182	4.541	5.841	7.453	10.213	12.924
4	.271	.741	1.533	2.132	2.776	3.747	4.604	5.598	7.173	8.610
5	0.267	0.727	1.476	2.015	2.571	3.365	4.032	4.773	5.893	6.869
6	.265	.718	1.440	1.943	2.447	3.143	3.707	4.317	5.208	5.959
7	.263	.711	1.415	1.895	2.365	2.998	3.499	4.029	4.785	5.408
8	.262	.706	1.397	1.860	2.306	2.896	3.355	3.833	4.501	5.041
9	.261	.703	1.383	1.833	2.262	2.821	3.250	3.690	4.297	4.781
10	0.260	0.700	1.372	1.812	2.228	2.764	3.169	3.581	4.144	4.587
11	.260	.697	1.363	1.796	2.201	2.718	3.106	3.497	4.025	4.437
12	.259	.695	1.356	1.782	2.179	2.681	3.055	3.428	3.930	4.318
13	.259	.694	1.350	1.771	2.160	2.650	3.012	3.372	3.852	4.221
14	.258	.692	1.345	1.761	2.145	2.624	2.977	3.326	3.787	4.140
15	0.258	0.691	1.341	1.753	2.131	2.602	2.947	3.286	3.733	4.073
16	.258	.690	1.337	1.746	2.120	2.583	2.921	3.252	3.686	4.015
17	.257	.689	1.333	1.740	2.110	2.567	2.898	3.222	3.646	3.965
18	.257	.688	1.330	1.734	2.101	2.552	2.878	3.197	3.610	3.922
19	.257	.688	1.328	1.729	2.093	2.539	2.861	3.174	3.579	3.883
20	0.257	0.687	1.325	1.725	2.086	2.528	2.845	3.153	3.552	3.850
21	.257	.686	1.323	1.721	2.080	2.518	2.831	3.135	3.527	3.819
22	.256	.686	1.321	1.717	2.074	2.508	2.819	3.119	3.505	3.792
23	.256	.685	1.319	1.714	2.069	2.500	2.807	3.104	3.485	3.767
24	.256	.685	1.318	1.711	2.064	2.492	2.797	3.091	3.467	3.745
25	0.256	0.684	1.316	1.708	2.060	2.485	2.787	3.078	3.450	3.725
26	.256	.684	1.315	1.706	2.056	2.479	2.779	3.067	3.435	3.707
27	.256	.684	1.314	1.703	2.052	2.473	2.771	3.057	3.421	3.690
28	.256	.683	1.313	1.701	2.048	2.467	2.763	3.047	3.408	3.674
29	.256	.683	1.311	1.699	2.045	2.462	2.756	3.038	3.396	3.659
30	0.256	0.683	1.310	1.697	2.042	2.457	2.750	3.030	3.385	3.646
40	.255	.681	1.303	1.684	2.021	2.423	2.704	2.971	3.307	3.551
60	.254	.679	1.296	1.671	2.000	2.390	2.660	2.915	3.232	3.460
120	.254	.677	1.289	1.658	1.980	2.358	2.617	2.860	3.160	3.373
∞	.253	.674	1.282	1.645	1.960	2.326	2.576	2.807	3.090	3.291

* From *Biometrika Tables for Statisticians*, vol. 1 (2nd edition), edited
by E. S. Pearson and H. O. Hartley, Cambridge University Press,
London, 1958, table 12 with permission of the Biometrika Trustees.
Columns headed $\alpha = 0.9$, 0.99, and 0.995 are taken from table III of
Fisher and Yates: *Statistical Tables for Biological, Agricultural and
Medical Research*, published by Longman Group Ltd., London
(previously published by Oliver and Boyd, Edinburgh), and by
permission of the authors and publishers.

References

Abramson, N. *Information theory and coding*. New York: McGraw-Hill, 1963.

Blum, J. R., and Rosenblatt, J. I. *Probability and statistics*. Philadelphia: Saunders, 1972.

Burr, I. W. Average sample number under curtailed or truncated sampling. *Industrial Quality Control*, vol. 14 (1957), 5–7.

—————— *Statistical quality control methods*. New York: Marcel Dekker, 1976.

Cramer, H. *The elements of probability theory*. New York: Wiley, 1955.

Dwass, M. *Probability and statistics*. Reading, Mass.: Benjamin-Cummings, 1970.

Epstein, R. A. *The theory of gambling and statistical logic*. New York: Academic, 1967.

Feller, W. *An introduction to probability theory and its applications*. New York: Wiley, 1957.

Freeman, H. *Introduction to statistical inference*: Reading, Mass.: Addison-Wesley, 1963.

Fry, T. C. *Probability and its engineering uses*. Princeton, N.J.: van Nostrand, 1965.

Gibra, I. N. *Probability and statistical inference for scientists and engineers*. Englewood Cliffs, N.J.: Prentice-Hall, 1973.

Halmos, P. R. *Naive set theory*. Princeton, N.J.: van Nostrand, 1960.

Harris, T. *The theory of branching processes*. New York: Springer-Verlag, 1963.

Johnson, N. I., and Kotz, S. *Discrete distributions*. Boston: Houghton Mifflin, 1969.

Karlin, S. *A first course in stochastic processes*. New York: Academic, 1966.

Kreyszig, E. *Introduction to mathematical statistics*. New York: Wiley, 1970.

Larson, H. J. *Introduction to probability theory and statistical inference*: New York: Wiley, 1969.

Mann, N. R., Schafer, R. E., and Singpurwalla, N. D. *Methods for statistical analysis of reliability and life data*. New York: Wiley, 1974.

Meshalkin, L. D. *Collection of problems in probability theory*. Leyden, the Netherlands: Noordhoff, 1973.

Meyer, P. L. *Introductory probability and statistical applications*. Reading, Mass.: Addison-Wesley, 1970.

Papoulis, A. *Probability, random processes and stochastic processes*. New York: McGraw-Hill, 1965.

Parzen, E. *Modern probability theory and its applications*. New York: Wiley, 1960.

Peach, P., and Littauer, S. B. A note on sampling inspection. *Ann. Math. Stat.*, vol. 17 (1946).

Pearson, E. S., and Hartley, H. O. The probability integral of the range in samples of n observations from a normal population. *Biometrika*, vol. 32 (1941–1942), 301–310.

Reza, F. *An introduction to information theory*. New York: McGraw-Hill, 1961.

Sandler, G. H. *System reliability engineering*. Englewood Cliffs, N.J.: Prentice-Hall, 1963.

Shewhart, W. A. *Economic control of quality of manufactured product*. Princeton, N.J.: van Nostrand, 1931.

Smith, J. Maynard. *Mathematical ideas in biology*. Cambridge, England: Cambridge Univ. Press, 1971.

Sveshnikov, A. A. (ed.). *Problems in probability theory, mathematical statistics and theory of random functions*. Translated by Scripta Technica, Inc., edited by B. R. Gelbaum. Philadelphia: Saunders, 1968.

Thomasian, A. J. *The structure of probability theory with applications*. New York: McGraw-Hill, 1969.

Uspensky, J. V. *Introduction to mathematical probability*. New York: McGraw-Hill, 1937.

Wadsworth, G. P., and Bryan, J. G.: *Introduction to probability and random variables*. New York: McGraw-Hill, 1960.

Answers to Selected Odd-Numbered Problems

CHAPTER 1

1. (i) $A = B = S$ (ii) $\bar{A} \subset B$ (iii) $A = S, B = \varnothing$ (iv) $A = B = \varnothing$
   (v) $A = B$

7. (i) $A \cap C = C$ (ii) $B \cup C = \{$all numbers divisible by 5$\}$
   (iii) $A \cap B = \varnothing$

11. $A \cup B \cup C = [A \cap (\overline{B \cup C})] \cup (\bar{B} \cap C) \cup B$

13. $W = D \cap (T_1 \cup T_2)$, $\bar{W} = \bar{D} \cup (\bar{T}_1 \cap \bar{T}_2)$

15. $|A \cap B \cap C| = 9$ 17. $P(\text{no matchups}) = 9/24$

21. (i) $P(A_1) = k/n$ (ii) $P(A_2) = (n - k)/n$

    (iii) $P(A_3) = \dfrac{n-k}{n} \cdot \dfrac{n-k-1}{n} \cdots \dfrac{n-k-(i-1)}{n} \cdot \dfrac{k}{n-i}$

    (iv) $P(A_4) = \left[\left(\dfrac{n-k}{n} \right)^{i-1} \cdot \dfrac{k}{n} \right] \Big/ \left[1 - \left(\dfrac{n-k}{n} \right) \right]$

23. (i) $P(A \cup B) = .75$ (ii) $P(\bar{A} \cup \bar{B}) = 1$ (iii) $P(\bar{A} \cap B) = .45$

25. $P(A \cap \bar{B}) = a(1 - b)$, $P(\bar{A} \cap \bar{B}) = (1 - a)(1 - b)$

29. (i) $P(\text{first question matched}) = 1/n$
    (ii) $P(\text{first 2 questions matched}) = 1/n(n-1)$
    (iii) $P(\text{at least one question answered correctly}) =$

    $$1 - \frac{1}{2!} + \frac{1}{3!} - \cdots + (-1)^{n-1} \frac{1}{n!}$$

33. $P(\text{at least one card a club}) = \frac{175}{276}$

35. $\frac{1}{3}$ 37. $\frac{6}{32}$

39. $P(\text{7 with a roll of 2 dice}) = 6/36$, $P(\text{4 heads with 5 tosses}) = 5/32$, and getting a 7 is more likely.

CHAPTER 2

1. (i) $f(x) = \begin{cases} 2x, & 0 \le x < \frac{1}{2} \\ \frac{3}{2}, & \frac{1}{2} \le x < 1 \end{cases}$

   (ii) $f(x) = \begin{cases} 1, & \frac{1}{2} \le x \le 1 \text{ and } P(X = 0) = \frac{1}{2} \\ 0, & \text{elsewhere} \end{cases}$

3. (i) $A = 2/[n(n+1)]$ (ii) $A = 1$ (iii) $A = \frac{120}{53}$

5. (i) $\frac{1}{8}$ (ii) $539/(27 \times 128)$ (iii) $\frac{7}{128}$

7. (i) $F(x) = \sqrt{x}, 0 \le x \le 1$

   (ii) $F(x) = \begin{cases} \frac{1}{2}(x + 1)^3, & -1 \le x \le 0 \\ \frac{1}{2}(x - 1)^3 + 1, & 0 \le x \le 1 \end{cases}$

9. (i) $A = .05$ (ii) $\theta = .05$ (iii) $P(X > 20) = (.95)^{20}$

11. $\frac{1}{3}$

13. $F_Y(y) = \begin{cases} 0, & y < 0 \\ .2\sqrt{y}, & 0 \le y \le 25 \\ 1, & y > 25 \end{cases}$ $f_Y(y) = \begin{cases} .1/\sqrt{y}, & 0 \le y \le 25 \\ 0, & \text{elsewhere} \end{cases}$

15. $f_Y(y) = (1/\sqrt{2\pi}) \exp\left[-\left(\frac{y-7}{4}\right)^2 / 2 \right]$

19. $P(\text{profit} = k) = P\left(D = \frac{k+20}{10}\right), \quad k = -20, -10, \ldots, 80$

$= P(8D = k), \quad k = 88, 96, 104, \ldots$

23. (i) Choose $\lambda \cong 1.10$ 25. 27 27. 5670

29. (i) $C = 1/n$, $E(X) = \frac{1}{n}\left(\frac{(A+n)(A+n+1)}{2} - \frac{A(A+1)}{2}\right),$

$E(X^2) = \frac{1}{n}\left(\frac{(A+n)(A+n+1)(2(A+n)+1)}{6} - \frac{A(A+1)(2A+1)}{6}\right)$

31. $E(X) = a\sqrt{\pi/2}, \quad V(X) = (4-\pi)a^2/2, \quad P(X > a) = e^{-1/2}$

33. $E(X) = 0, \quad V(X) = 1$

35. $E(X^n) = \frac{1}{(n+1)} \cdot \frac{b^{n+1} - a^{n+1}}{b - a}, \quad E[(X - \mu)^n] = \sum_{k=0}^{n} \binom{n}{k} \mu^{n-k} E(X^k)$

37. $m_X(t) = (2/t^2)[(t-1)e^t + 1], \quad E(X) = \frac{2}{3}, \quad V(X) = \frac{1}{18}$

39. $m_X(t) = \frac{1}{(1-t)^2}, \quad |t| < 1$

CHAPTER 3

1. $9 / \binom{12}{4}$ 3. $\binom{950}{3} / \binom{1000}{4}$ 5. $\left[\binom{5}{0}\binom{95}{20} + \binom{5}{1}\binom{95}{19}\right] / \binom{100}{20}$

7. $\binom{2n}{n}$, $P(\text{all correct}) = 1 / \binom{2n}{n}$, $3 \times 2^{n-1}$

9. (i) $\frac{15!}{4!8!3!}(\frac{1}{2})^4(\frac{1}{4})^8(\frac{1}{4})^3$
   (ii) $1 - (\frac{1}{2})^n - (\frac{3}{4})^n - (\frac{3}{4})^n + (\frac{1}{4})^n + (\frac{1}{4})^n + (\frac{1}{2})^n$
   (iii) $(\frac{1}{2})^n + (\frac{3}{4})^n + (\frac{3}{4})^n - (\frac{1}{2})^n - (\frac{1}{4})^n - (\frac{1}{4})^n$

11. $P(X = j) = \binom{6}{j}\binom{2}{4-j} / \binom{8}{4}$, $j = 2, 3, 4$. On the average the same mark

will be obtained

13. $\binom{k_1}{j_1}\binom{k_2}{j_2}\cdots\binom{k_s}{j_s} / \binom{n}{m}$ 17. $\binom{n+k-1}{n-1}$

19. It is more likely the coin is biased.

21. $\sum_{j=8}^{10}\binom{10}{j}\theta^j(1-\theta)^{10-j}, \quad \theta = e^{-150\alpha}$

23. $\binom{6}{3}(.45)^3(.55)^3$ 25. Maximum occurs at the integer part of $k = n\theta - (1-\theta)$. If k is an integer, maxima occur at k and $k+1$.

27. 3   and   4   29. 29   31. $m_X(t) = \left(\dfrac{\theta e^t}{1-(1-\theta)e^t}\right)^k$, $E(X) = \dfrac{k}{\theta}$, $V(X) = \dfrac{k(1-\theta)}{\theta^2}$

33. (i) n^m (ii) $\binom{m+n-1}{n-1}$

35. $P(\text{none meet}) = \dfrac{4.3.2}{4^3}$, $P(\text{all meet}) = \dfrac{4}{4^3}$

37. $\binom{9}{7}\Big/\binom{12}{7}$ 39. $\sum_{i=0}^{4}\binom{12}{i}\left(\dfrac{1}{6}\right)^i\left(\dfrac{5}{6}\right)^{12-i}$ 41. $\frac{100}{12}$

43. $\binom{20}{10}\left(\dfrac{1}{12}\right)^{10}\left(\dfrac{11}{12}\right)^{10}$ 45. $\dfrac{(n-3)!}{n!}$ 47. $1 - \dfrac{7\cdot 6\cdot 5\cdot 4}{7^4}$

49. $\dfrac{n!}{i!j!(n-i-j)!}\left(\dfrac{1}{4}\right)^i\left(\dfrac{1}{2}\right)^j\left(\dfrac{1}{4}\right)^{n-i-j}$ 51. $\binom{15}{2}\left(\dfrac{5}{6}\right)^{13}\left(\dfrac{1}{6}\right)^3$ 53. 18

55. $f_D(d) = 2(1-d), \quad 0\le d\le 1, \quad E(D) = \frac{1}{6}, \quad V(D) = \frac{1}{18}$

57. $1 - (2r/a)^2$ 59. $\frac{1}{6}$ 61. $\frac{11}{36}$

CHAPTER 4

1. $\dfrac{4}{11}$ 3. $\dfrac{60}{85}$ 5. $\dfrac{2}{3}$ 7. $\dfrac{\binom{5}{1}\binom{7}{1}\binom{3}{1}\binom{5}{1}}{\binom{20}{4}}$

9. k/n 11. $P(X=j) = \dfrac{100^j}{j!}e^{-100}$ 13. $P(0) = .38, P(1) = .62$

17. $\binom{n-j}{k-j}\left(\dfrac{1}{2}\right)^{n-j}$ 19. $2^{n+1}/(2^{n+1}+1)$ 21. $\frac{5}{8}$

23. $P(X=j) = \dfrac{(12,500)^j}{j!}e^{-12,500}, \quad E(X) = 12,500 = V(X)$

25. $P(X=i|D) = \dfrac{i}{n\theta}\binom{n}{i}\theta^i(1-\theta)^{n-i}, \quad P(X=i) = \binom{n}{i}\theta^i(1-\theta)^{n-i}$

27. $20 + 80\theta$ 31. No   33. (i) $P(A_i \cap A_j) = \varnothing, \quad i\ne j,$
(ii) $1 - (1-\theta)^n$

CHAPTER 5

1. $E(X) = \frac{8}{15}$, median $= (1 - 1/\sqrt{2})^{1/2}$ 3. $E(X) = n + 1$, $V(X) = n + 1$
5. $e^{-1.1}$ 7. e^{-1} 9. $E(X) = 9.5$, median $= 9.5$, mode $= 9.5$
11. $C = \exp(\mu + \frac{1}{2}\sigma^2)$ 13. $\sigma = .075$ 15. 3.09
17. $\sigma = 303$ hours   19. $E[(X - \mu)^3] = 0$, $E[(X - \mu)^4] = 3\sigma^4$
21. Normal would be chosen.   23. $X = -\tau/2$
25. It is more likely to have been produced when machine is out of control.
29. $F_X(x) = 1 - e^{-\lambda x^\alpha}$, $E(X^k) = \Gamma(k/\alpha + 1)/\lambda^{k/\alpha}$

CHAPTER 6

1. (i) $E(|X|) = \sqrt{2/\pi}\,\sigma$ (ii) $E[(X^2 + Y^2)^{1/2}] = \sqrt{\pi/2}\,\sigma$
   (iii) $V(X - Y) = 2\sigma^2$
3. (i) 1327/36   (ii) 1,528,376/90
5. (i) $F(x, y) = (1 - 10^4/x)(1 - 10^4/y)$ (ii) $\frac{4}{25}$ 7. $\frac{10}{12}$
11. 2.44   13. (i) $\frac{1}{8}$ (ii) $a = 1/(2^{4/3})$
15. $\frac{1}{2}$, $P(X > 1.025 | \sqrt{X^2 + Y^2} < 1.05) = .196$ 17. $N(0, 4.4)$
19. $P(X = j | X + Y = k) = \dfrac{1}{k - 1}$, $j = 1, 2, \ldots, k - 1$
21. $\rho_{XY} = +\frac{3}{4}$, $P(\text{improvement}) = \displaystyle\int_{-\infty}^{\infty} \left[1 - \Phi\left(\frac{x - 62}{3\sqrt{7}}\right) \right] f_X(x)\, dx$
25. $\sqrt{2}\sigma/3$ 27. $\sqrt{21}/5$ 29. $-x/3 + 2$
31. (i) $f(x) = \dfrac{2}{\pi}[1 - (x - 3)^2]^{1/2}$, $2 \le x \le 4$

   (ii) $2\left(\dfrac{\sqrt{3}}{8\pi} + \dfrac{1}{6} - \dfrac{\sqrt{3}}{4\pi}\right) \Big/ \left(\dfrac{1}{2}\right)$ (iii) 0
33. (i) $9/\sqrt{616}$ (ii) $\dfrac{693}{32\sqrt{616}}\left(X - \dfrac{8}{5}\right) + \dfrac{9}{10}$
35. (i) $\binom{3}{2}\left(\dfrac{1}{4}\right)^2\left(\dfrac{3}{4}\right)$ (ii) $-\dfrac{1}{2}$ (iii) $\dfrac{5}{6}$
37. (i) Uncorrelated but not independent   (ii) $E(Y|X) = 2$

CHAPTER 7

1. $f_Y(y) = \dfrac{4y}{\sqrt{2\pi}}\exp\left(\dfrac{-y^4}{2}\right)$, $0 \le y < \infty$

3. $f_Y(y) = 1, \quad 0 \le y \le 1$

5. $f_Y(y) = \lambda e^{-\lambda y}, \quad 0 \le y \le 1 \quad$ and $\quad P(Y=1) = e^{-\lambda}$

7. $f_Y(y) = \dfrac{6y}{(1+y)^4}, \quad 0 \le y < \infty$

9. $f_C(c) = \dfrac{k}{2000 \cdot c^2}, \quad \dfrac{k}{22{,}000} \le c \le \dfrac{k}{20{,}000} \qquad$ 11. $\dfrac{306\pi}{4}$

13. $f_V(v) = \exp\left\{ -\left[\left(\dfrac{3v}{4\pi} \right)^{1/3} - \mu \right]^2 \Big/ 2\sigma^2 \right\} \Big/ \sqrt{2\pi} \cdot \sigma \cdot 4\pi \cdot \left(\dfrac{3v}{4\pi} \right)^{2/3}$

$E(V) = \dfrac{4\pi}{3} (\mu^3 + 3\mu\sigma^2)$

15. $f_Y(y) = \dfrac{1}{\pi(1+y^2)}, \quad h[g(x)] = \dfrac{h(x)}{|g'(x)|}$

19. $f_Z(z) = 2z/(1+z^2)^2, \quad z \ge 0 \qquad$ 21. $\quad U, V$ independent iff $ac + bd = 0$

23. $Z \sim N[0, 2\sigma^2(1+\rho^2)]$

25. $f_P(p) = \begin{cases} \dfrac{p-220}{80}, & 220 \le p \le 224 \\[2mm] \dfrac{1}{20}, & 224 \le p \le 240 \\[2mm] \dfrac{244-p}{80}, & 240 \le p \le 244 \end{cases}$

27. $f_Z(z) = 1/[\pi(1+z^2)] \qquad$ 29. $\quad f_Z(z) = \tfrac{1}{2}(2-z), \quad 0 \le z \le 2$

33. $f_Z(z) = 2z^3 e^{-z^2}, \quad z \ge 0 \qquad$ 37. $\quad f_Z(z) = \dfrac{\lambda_1 \lambda_2}{\lambda_1 - \lambda_2} (e^{-\lambda_2 z} - e^{-\lambda_1 z})$

39. $f_Z(z) = \dfrac{2}{\pi} \cdot \dfrac{1}{4+z^2} \qquad$ 41. $\quad \dfrac{a^2}{(1-r)(1-r^2)}$

CHAPTER 8

1. $E(Y) = n(n+2), \quad V(Y) = 2n$

5. $E(S^2) = \dfrac{n}{n-1} \mu_2, \quad V(S^2) = \dfrac{n}{(n-1)^2} (\mu_4 - \mu_2^2)$

7. 12 $\qquad$ 11. (i) .00003 (ii) .00359

13. True value $= .970$, approximation $= .80$

19. Poisson distribution with parameter 5

21. $E(Y) = \dfrac{\rho P}{1-\rho}, \quad P(Y=0) = \dfrac{1-\rho}{1-\rho q}$

23. 7500 $\qquad$ 25. .00008 $\qquad$ 27. .97778

29. .41190 (normal approximation), .384 (Poisson approximation)

CHAPTER 9

1. $\hat{\sigma}_{ML}^2(1) = .357$, $\hat{\sigma}_{ML}^2(2) = .256$ 3.   .533   5.   .000275
7. $\hat{\mu} = 88.5$, $\hat{\sigma}^2 = 10.25$ 9. $\hat{a} = 1.039$ 11.   (11.89, 28.93)
13. $\left(\dfrac{(n-1)S^2}{\chi_{\alpha;n-1}^2}, \infty\right)$ 15. (1042 ± 27.72), $n \geq 246$

19. $\hat{\lambda} = kn / \left(\sum_1^n x_i\right)$ 21. $\hat{a} = -1 - \dfrac{n}{\sum\limits_1^n \ln(x_i)}$

23. $\hat{a} = \min(Y_1, Y_2, \ldots, Y_n)$ 25. $(.604, .730)$ 27. $\hat{\lambda} = n / (\sum X_i^\alpha)$
29. $\hat{\sigma}^2 = .90$

CHAPTER 10

1. .5092   3. H_0 accepted   5. H_0 rejected
7. 22%   9.   Accept H_0 if $\sum_1^{20} (X_i - 300)^2 \leq 785.26$
11. Reject H_0 if $\sum (X_i - \mu)^2 > \sigma_0^2 \chi_{1-\alpha;n}^2$
13. Improvement not significant

15. Reject H_0 if $\bar{X} > 21.658$ 17.   Reject H_0 if $|\bar{X} - \mu_0| > \dfrac{S}{\sqrt{n}} t_{1-\alpha/2;n}$

19. $P_I = 2\Phi\left(-\dfrac{0.4}{(1/\sqrt{n})}\right)$ 21. $P_I = .56$ 23.   Hypothesis accepted

25. Both tests accept the hypothesis the die is fair   27. H_0 rejected
29. Reject H_0 if $\sum (X_i - \bar{X})^2 > \sigma_0^2 \chi_{1-\alpha/2;n-1}^2$ 31.   .447

CHAPTER 11

5. $\mu \geq 20$ 7. $9/(28 \times 59)$ 9. $\dfrac{P_0}{(s-1)!} \cdot \dfrac{\rho^s}{s-\rho}$

11. $\frac{1}{7}$ 13. $\rho/(1-\rho)$ 15. $(N-s+1)/n\mu$, $[(N-s+1)/n\mu] + s/\mu$
17. $\mu\lambda/(\mu + \lambda)$

19. (i) $\left(\dfrac{1-\rho}{1-\rho^{K+1}}\right)\rho^K$ (ii) $\dfrac{1}{\mu}\left[\left(\dfrac{1-(K+1)\rho^K + K\rho^{K+1}}{(1-\rho)(1-\rho^{K+1})}\right) + 1\right]$

     (iii) $\dfrac{\rho[1-(K+1)\rho^K + K\rho^{K+1}]}{(1-\rho)(1-\rho^{K+1})} - 1 + \left(\dfrac{1-\rho}{1-\rho^{K+1}}\right)$

21. $\mu = \lambda + \sqrt{D\lambda/C}$ 23.   Poisson distribution, $\lambda = 9$

25. Poisson distribution, $\lambda = \mu t$ 27. $P_n(t) = \dfrac{(\lambda t)^n}{n!} e^{-\lambda t}$

29. $P_0(t) = \dfrac{\mu}{\lambda + \mu}(1 - e^{-(\mu+\lambda)t}) + P_0(0)\, e^{-(\mu+\lambda)t}$

$P_1(t) = \dfrac{\mu}{\lambda + \mu}(1 - e^{-(\mu+\lambda)t}) + P_1(0)\, e^{-(\mu+\lambda)t}$

CHAPTER 12

1. $e^{-1.25}$

3. (a) $R(t) = \begin{cases} 1, & 0 \le t \le 1000 \\ -\exp[-.005(t^2/2 - 2000t + 2\times 10^6)], & t \ge 1000 \end{cases}$

   (b) 1035.44

5. $n \ge 6$ 7. $R_2(100) < R_1(100), \quad R_2(200) > R_1(200)$

9. $n = 2, \quad E(T) = \dfrac{1}{c_1} + \dfrac{1}{c_2} - \dfrac{1}{c_1 + c_2}$

   $n = 3, \quad E(T) = \dfrac{1}{c_1} + \dfrac{1}{c_2} + \dfrac{1}{c_3} - \dfrac{1}{c_1 + c_2} - \dfrac{1}{c_1 + c_3} - \dfrac{1}{c_2 + c_3} + \dfrac{1}{c_1 + c_2 + c_3}$

11. $R(t) = \dfrac{2\mu\eta}{2\mu - \eta}\left(\dfrac{1}{\eta} e^{-\eta t} + \dfrac{1}{2\mu} e^{-2\mu t}\right)$

13. $R(t) = e^{-c_1 t}[1 - (1 - e^{-c_2 t})(1 - e^{-c_3 t})]$

15. $E(T) = \dfrac{1}{\mu}\sum\limits_{1}^{n}(-1)^k \dfrac{1}{k}\binom{n}{k}$ 19. $\dfrac{1}{a}(1 - e^{-100a}) + \dfrac{1}{b} e^{-100a}$

21. $e^{-4}/3$ 25. $(\hat{1}/\lambda) = 12{,}075, \quad (7069, \infty)$

27. $(\hat{1}/\lambda) = 29{,}145 \quad (85{,}314; 15{,}036)$

CHAPTER 13

1. $P_I = .00135, \quad P_{II} = .841 \quad (8.0 \pm .082)$ 3. $(30.435, 30.732)$

5. $\bar{X}$ chart: $(9.85, 10.15)$, R chart: $(0, .4698)$

7. $\bar{X}$ chart: $(9.826, 10.204)$, $\bar{S}$ chart: $(0, .263)$

9. (i) $\bar{X}$ chart: $(\pm.173)$, R chart: $(0, .436)$

   (ii) $\bar{X}$ chart: $(\pm.122)$, R chart: $(0, .508)$

11. $\bar{X}$ chart: $(.069, .655)$, $\bar{S}$ chart: $(0, .428)$

13. $.470, .458$ 15. $n \ge 50$

21. $p_{.1} \cong .18, \quad p_{.50} \cong .083, \quad p_{.90} \cong .027$

23. $n = 700, \quad c = 20$ 27. $P(A, \theta) = (1 - \theta)^4 + 4\theta(1 - \theta)^3, \quad P_{AOQL} = \frac{1}{3}$

29. $\text{LTPD} = p_{.05} \cong .22$

33. Accept lot if $(1/N)\sum X_i > C, \ N = 49, \ C = 1000.15$

35. $n = 850, \quad c = 25$

CHAPTER 14

1. $a = 2.467, \quad b = 1.233$
3. $a_1 = 26.797, \quad a_2 = 1, \quad b_1 = -88.439, \quad b_2 = 0$
7. $a = 48.947, \quad b = .061, \quad \text{error}^2 = .00963, \quad \hat{R} = 50.289$
9. $S_y^2 = (1/n) \sum (Y_i - Bx_i)^2$
11. $a = -5.246, \quad b = .0112, \quad$ accept hypothesis $\alpha = -5.00$
13. $a = -27.038, \quad b = 2.724, \quad (-31.805, -22.271)$ 15. Accept H_0
17. $a = 20.440, \quad b = -.0618, \quad (-.0746, -.0490)$ for β, $(19.385, 19{,}641)$
    for $E(D \mid x = 15)$
19. Hypothesis that $\beta = 0$ is rejected.
21. $a = 4.255, \quad b = .456, \quad H_0: \alpha = 4.0$ is rejected and $H_0: \beta = 0.5$ is rejected
23. $(6425, 6645)$ 25. $r_{xy} = -.487, \quad (-.854, .207)$
29. $r_{xy} = .9944, \quad H_0: \rho = 0$ is rejected
31. $H_0: \rho = 0$ is accepted

CHAPTER 15

1. $H(X) = \log_2(900), \quad H(X \mid Y) = \log_2(30)$ 3. $H(X \mid Y) = 2$
9. $p_1 = p_2 = \frac{1}{2}$ 11. $H(X) = 1 - \ln(\alpha)$
13. Average word lengths are 1, 1.81, and 2.726, respectively, for both Shannon–Fano and Huffman codes.
15. Average code word length is 2.2   17. $D \geq 3$
19. $101110011011\ldots$
23. $C = 1 + p \log_2 p + (1 - p) \log_2(1 - p), \quad p = (1 - p_1)p_2 + (1 - p_2)p_1$

Index

tabulation of constants, 400
Correlation analysis, 448
Correlation coefficient, 179
  and best linear mean square estimate, 182
Covariance, 179
Craps, 87
Critical (rejection) region, 295
Cumulative distribution function, 30
  of a linear function of a random variable, 128
Cusum chart, 405

Discrete memoryless channel, 486
Discrete memoryless source, 458
Discrete probability distribution function, 34
  tabulation, 83
Discrete sample space, 5
Dodge-Romig sampling inspection tables, 416
Double sampling plan, 417

Ensemble, 458
Entropy, 459
  of joint ensemble, 460
  of n-fold Cartesian product, 470
Equally likely outcomes, 18
Equivalent events, 29
Erlang loss formula, 351
Errors of type I and II, 295
Estimate of parameter, 264
  Bayes, 286
  best linear unbiased, 276
  consistency, 276
  efficiency, 274
Estimator of parameter, 264
Events, 6
  independent, 118
Expected value of random variable, 45
  of function of random variable, 48
Experiment, 2
Exponential probability density function, 43
  failure law, 361
  mean, 48
  moment generating function, 59
  variance, 53

Failure laws, 357
  exponential, 361
  gamma, 362
  truncated normal, 363
  Weibull, 363
Failure rate function, 358
First coding theorem, 471

Gamma probability density function, 146
  failure law, 361
  mean, 147
  reproductive property, 234
  shape, 148
  variance, 147
Gaussian probability density, *see* Normal probability density function
Generating function, of discrete probability distribution, 261
Geometric distribution, 36
  mean, 47
  moment generating function, 59
  variance, 52
Geometric probability, 98
Geometric series, 35
Goodness of fit tests, 319

Hamming code, 481
Huffman encoding, 472
Hypergeometric distribution, 73
  approximation by binomial, 257
  approximation by normal, 258
  approximation by Poisson, 258
  mean, 74, 423
  variance, 74, 423
Hypothesis test, 295
  composite, 296
  simple, 296

Independence, of continuous random variables, 164
  of discrete random variables, 160
  and disjointness of sets, 119
  of random variables, 156
  of sample mean and variance of normal population, 227, 237
  of sequence of events, 121
Independent events, 118
Instantaneously decodable code, 468
Intersection of sets, 9

Jacobian, 208, 226
Joint cumulative distribution function, 156
  properties, 158
Joint ensemble, 460
Joint probability density function, 162
  of function of bivariate random variable, 206
Joint probability distribution function, 159

Kraft inequality, 470